GEOTHERMAL HEAT PUMP AND HEAT ENGINE SYSTEMS

Wiley-ASME Press Series List

Introduction to Dynamics and Control of Mechanical Engineering Systems	To	March 2016
Fundamentals of Mechanical Vibrations	Cai	May 2016
Nonlinear Regression Modeling for Engineering Applications	Rhinehart	September 2016
Stress in ASME Pressure Vessels	Jawad	November 2016
Bioprocessing Piping and Equipment Design	Huitt	November 2016
Combined Cooling, Heating, and Power Systems	Shi	January 2017

GEOTHERMAL HEAT PUMP AND HEAT ENGINE SYSTEMS

THEORY AND PRACTICE

Andrew D. Chiasson, Ph.D., P.E., P.Eng.
Department of Mechanical and Aerospace Engineering,
University of Dayton, USA

This Work is a co-publication between ASME Press and John Wiley & Sons, Ltd.

WILEY

This edition first published 2016

Registered Office
John Wiley & Sons, Ltd, The Atrium, Southern Gate, Chichester, West Sussex, PO19 8SQ, United Kingdom

For details of our global editorial offices, for customer services and for information about how to apply for permission to reuse the copyright material in this book please see our website at www.wiley.com.

Library of Congress Cataloging-in-Publication Data

Names: Chiasson, Andrew D, 1966– author.
Title: Geothermal heat pump and heat engine systems : theory and practice / Andrew D Chiasson, Ph.D., P.E., P.Eng., Department of Mechanical and Aerospace Engineering, University of Dayton, USA.
Description: Hoboken, N.J. : John Wiley & Sons Ltd, 2016. | Includes bibliographical references and index.
Identifiers: LCCN 2016015385 (print) | LCCN 2016022646 (ebook) | ISBN 9781118961940 (cloth) | ISBN 9781118961971 (pdf) | ISBN 9781118961964 (epub)
Subjects: LCSH: Ground source heat pump systems. | Heat pumps–Thermodynamics. | Heat-engines–Thermodynamics.
Classification: LCC TH7417.5 .C44 2016 (print) | LCC TH7417.5 (ebook) | DDC 697–dc23
LC record available at https://lccn.loc.gov/2016015385

A catalogue record for this book is available from the British Library.

Set in 10/12pt Times by SPi Global, Pondicherry, India
Printed and bound in Malaysia by Vivar Printing Sdn Bhd

1 2016

In Memory of Kirstin Beach Chiasson

Contents

Series Preface

The Wiley-ASME Press Series in Mechanical Engineering brings together two established leaders in mechanical engineering publishing to deliver high-quality, peer-reviewed books covering topics of current interest to engineers and researchers worldwide.

The series publishes across the breadth of mechanical engineering, comprising research, design and development, and manufacturing. It includes monographs, references and course texts.

Prospective topics include emerging and advanced technologies in Engineering Design; Computer-Aided Design; Energy Conversion & Resources; Heat Transfer; Manufacturing & Processing; Systems & Devices; Renewable Energy; Robotics; and Biotechnology.

Preface

Intent and Motivation for this Book

This book is intended as a university-level textbook for the study of Geothermal Energy Systems. The book is aimed at upper division and graduate students of the emerging field of Energy Engineering, as well as the traditional field of Mechanical Engineering. The book will also be useful to practicing engineers in these fields who are involved with low-temperature geothermal energy applications.

The emerging (or already emerged) field of Energy Engineering is highly interdisciplinary, mainly grounded in fundamentals of the thermofluid sciences, environmental science, natural sciences, and social sciences, and this book is based on this interdisciplinary viewpoint, while keeping in mind the engineering and practical constraints of the real world. Thus, readers of this book are expected to have introductory background in the natural sciences and thermofluid sciences (i.e., thermodynamics, heat transfer, and fluid mechanics).

This book is the culmination of Geothermal Energy courses I developed and delivered, beginning back in 2009 at the University of Dayton. Then, the first university-level course I developed had a focus on geothermal heat pumps, but students invariably wondered and asked about hot springs, geysers, and geothermal power plants. Conversely, courses I taught on the subjects of geothermal power plants and direct-use heating inevitably had students asking, 'what about those heat pumps I keep hearing about?'. Therefore, I felt the motivation to develop materials to unify all geothermal uses under one cover, allowing students to read about geothermal topics they were curious about, even though not necessarily the focus of their current study. Thus, this book is divided into four parts after Chapter 1, which describes general considerations of developing a geothermal energy project. The four parts are: (i) geothermal energy utilization and resource characterization, (ii) harnessing the resource, (iii) energy conversion, and (iv) energy distribution. These elements are common to all types of geothermal energy project (as well as other uses of renewable energy sources), regardless of temperature and use.

So why write a textbook? I believe that a course in Geothermal Energy in an Energy Engineering curriculum or other applied science program is very important. The field of Energy Engineering has emerged as a distinct field of engineering practice in its own right, and is a highly interdisciplinary field, with sources, applications, and uses of energy crossing all traditional fields of engineering. World population growth continues. Developing countries are urbanizing at increasing rates. How will these energy needs be met sustainably? One potential way is by meeting targets toward net-zero buildings, where buildings themselves (or groups of buildings) become providers as well as consumers of energy, both thermal and electrical. We spend much of our lives in buildings, but also take them for granted. Buildings of the future must not only meet energy challenges but also be environmentally sustainable and of high comfort for occupants. Optimal ways of reducing, supplying, and using this energy will be the key role of the next generation of Energy Engineers. Geothermal Energy has the potential to meet such needs.

The study of Geothermal Energy, in general, is somewhat of a balance of theory and practice. In other words, one can get wrapped up in the theoretical study of complex heat transfer problems, but in reality, engineers are faced with real-world uncertainties of the subsurface geology, dynamic behavior of weather, dynamic behavior of buildings, and economic and environmental constraints. Some solution methods to subsurface geothermal heat transfer problems are difficult to impossible to solve without computer aid, and developing workable solutions can take up the better part of a semester course. Thus, this book aims to present equations and theoretical approaches in a fundamentally easily understandable way, and then provides software tools on a companion website, readily adaptable for use.

Finally, this book is not just about 'geothermal energy'. Even if readers aren't particularly interested in the topic, it is hoped that this book will allow readers to gain insight about 'applied' knowledge in environmental heat transfer, applied thermodynamics, creative power generation, and energy-efficient building (or 'green building') design.

Engineering Education in the Twenty-first Century

The twenty-first century has marked a transition in engineering education. Yesterday's engineers spent the majority of their time substituting values into formulas and obtaining numerical results. Just a few decades ago, engineers used slide rules, and then graduated to electronic calculators; more complicated solutions relied on tabulated data or nomographs. Today, formulaic calculation and 'number crunching' are left to computers, thus allowing more complex problems to be solved at faster and faster rates. Computer-based solutions now enable engineers to examine numerous 'what-if?' scenarios, and readily find optimal solutions to problems, based on cost or some other metric.

Thus, the engineering curriculum has seen increasing use of computer application with engineering equation solvers and other numerical software packages. More than ever, I believe that current and future engineers need a firm grasp of basic principles so that complex problems can be formulated and optimized, and results correctly and accurately interpreted. Further, engineers need a grasp of real-world constraints such as cost, legalities, and environmental impact. This is the emphasis of this book.

Supplemental Materials

A companion website is a key component of this book, which contains software tools to allow users of this book to apply fundamental principles and solve real-world problems. The companion website includes:

- An Excel-based suite of design tools for sizing ground heat exchangers and performing pressure-drop calculations in pipes.
- Electronic files in *Engineering Equation Solver* (*EES*) for simulating the thermodynamic performance of heat pump and heat engine cycles.

Meeting Different Course Needs

Instructors who adopt this book can choose to: (1) give a light treatment of theory and focus on the practical use of the provided design software tools and economics, or (2) give a more in-depth treatment of the theory with companion practical application of the provided design tools, or (3) give a near entire treatment of theory, leaving solutions and applications for the students to develop and/or improve. In my experience, option (1) is most suitable for a college-level course at the senior, undergraduate level, or for course schedules not covering a full semester. Option (2) is suitable for a senior or graduate-level course on a semester schedule, and option (3) is suitable for more advanced graduate students completing courses in 'special topics' or completing research projects or theses.

Andrew D. Chiasson
February 2016

About the Companion Website

Don't forget to visit the companion website for this book:

www.wiley.com/go/chiasson/geoHPSTP

There you will find valuable material designed to enhance your learning, including:

- Data Files for Exercise Problems
- Example Files in EES
- Design Tool Suite (for sizing GHXs and calculating pipe pressure drop)

Scan this QR code to visit the companion website

1

Geothermal Energy Project Considerations

1.1 Overview

The main focus of this book is geothermal heat pump applications for buildings. However, this first chapter first introduces readers to general considerations for renewable/clean energy project analysis. Then, specifics of geothermal energy projects are discussed through broad considerations of geothermal energy utilization, of which geothermal heat pumps is just one type. Elements of geothermal energy systems are discussed, laying the foundation for the organization of material in this book.

The chapter describes geothermal energy from the perspective of resource temperature in the context of high-, medium-, and low-temperature applications, emphasizing that the end use of the energy, at least in theory, can be any application where thermal energy is involved.

Learning objectives and goals:

1. Be aware of favorable conditions for alternative energy projects.
2. Understand general decision analysis of renewable/clean energy systems.
3. Draw analogies between geothermal and other renewable energy system analysis.
4. Appreciate the role of resource temperature in defining a geothermal energy project.
5. Appreciate the inherent risks in undertaking a renewable/clean energy project.
6. Realize the similarities in project development in all geothermal energy projects.

1.2 Renewable/Clean Energy System Analysis

Energy project stakeholders, investors, and decision-makers are mainly interested in project viability and feasibility. RetScreen® International (2004) provides a good discussion for general decision-making. For conventional and especially for renewable/clean energy projects

Geothermal Heat Pump and Heat Engine Systems: Theory and Practice, First Edition. Andrew D. Chiasson.

Companion website: www.wiley.com/go/chiasson/geoHPSTP

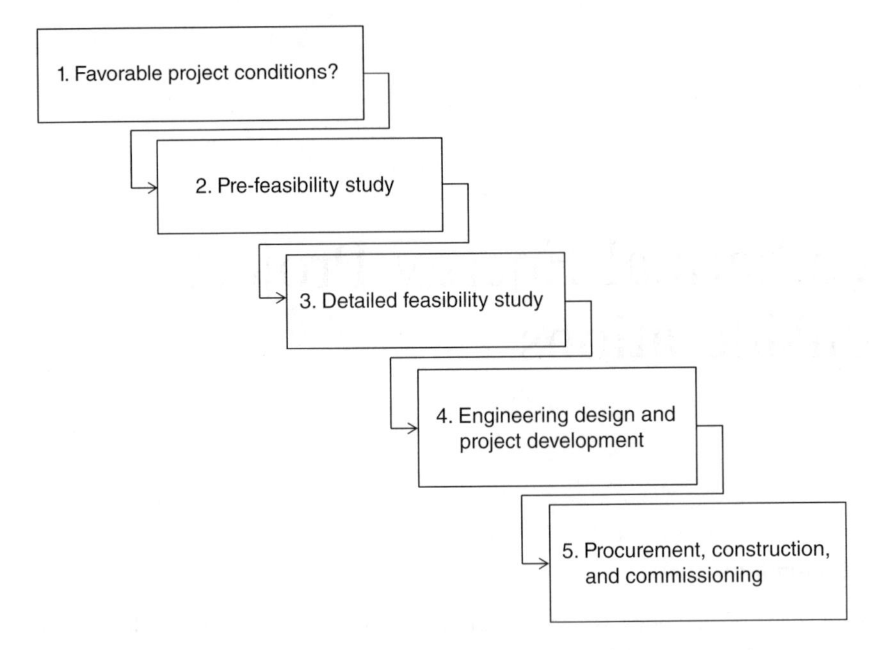

Figure 1.1 General flow of conventional and alternative energy projects

under consideration, Figure 1.1 shows general steps taken for advancing such projects to completion. At each step, a 'go/no-go' decision can be made by stakeholders as to whether or not to proceed to the next step of the development process. Thorough and accurate pre-feasibility and detailed feasibility studies are critical to helping the project owner reject projects or scenarios that do not make sense, either from a financial, regulatory, logistical, or other perspective. Accurate pre-feasibility and detailed feasibility analyses also facilitate development and engineering efforts prior to construction. The tools and techniques presented in this book are, in part, aimed toward that end.

The discussion that follows describes the boxes shown in Figure 1.1.

Favorable Project Conditions. In general, many decision-makers are not familiar with implementation of renewable/clean energy technologies and when they should be considered. The author has seen many misconceptions regarding the use of geothermal energy. One common misconception is, 'If you have a high temperature geothermal resource, you essentially have a gold mine'. The reality is: Does the capital exist to develop the resource? What is the revenue stream for the energy? Is there a business plan or a market for the energy? Who will operate and maintain any equipment? Another common misconception is, 'Geothermal heat pumps can be applied everywhere', but the reality is: How easy is your building to retrofit to a heat pump system? What's really underground at your site? How sustainable will the reservoir be? Are there qualified contractors in your area?

Renewable/clean energy systems are typically capital intensive, with low operating costs that are usually weighed against the operating cost of a conventional energy system. The following is a general set of conditions as to when a renewable/clean energy project might be considered:

1. **Need for an energy system.** An opportune time for considering a renewable/clean energy system is when an energy system is being planned or replaced. The capital cost of the

renewable/clean energy system can be offset by the avoided cost of the conventional system. In retrofit cases, for example in buildings, retrofitting the building to be compatible with a renewable/clean energy system may be prohibitive.

2. **High Conventional Energy Costs.** Obviously, when conventional energy costs are high, the relatively low energy cost of a renewable/clean energy system is attractive. Thus, interest in renewable/clean energy systems is typically proportional to conventional energy costs.
3. **Available Funding and Financing.** The relatively higher capital cost of renewable/clean energy projects is often a substantial barrier. Some jurisdictions promote clean energy projects with financial incentives, such as grants and tax rebates. Some companies offer third-party ownership of renewable/clean energy projects, where they bear the cost of the project and sell lower-cost energy to the project proponent.
4. **Qualified Contractors, Installers, and Maintenance Personnel.** Renewable/clean energy systems (particularly geothermal heat pump systems) typically involve specialized, non-traditional training and certification. If qualified personnel are not local, installation costs can become prohibitive. At early stages of a project, complexity of the system should also be considered; local availability of qualified personnel for system maintenance may eliminate a project from consideration.
5. **Persistent Project Stakeholders.** Seeing a renewable/clean energy project through to completion, especially a complex one, can be a daunting task. Diligent project management is required, involving coordination of numerous trades, monitoring budgets, navigation through complicated regulations, and, perhaps most of all, persistence.
6. **Simple Legal and Permitting Processes.** Development costs and schedule delays are minimized when laws and regulations are understood by the project team, and when these laws and regulations do not unfairly disadvantage a renewable/clean energy project.
7. **Adequate, Sustainable Resource.** An adequate resource is necessary for any renewable/clean energy project. However, special considerations are needed for geothermal resources, which are discussed in Chapter 3. In particular, a geothermal venture involves risk because the resource cannot be completely observed with depth. Further, the resource is finite, and proper resource management is necessary.

Pre-Feasibility Study. The pre-feasibility analysis determines whether the proposed project has a good chance of satisfying the owner's requirements for profitability and/or cost-effectiveness. It is characterized as a 'desktop' study, involving the use of readily available site and resource data, ±30–50% cost estimates, and simple calculations and professional judgement often involving experience with other projects. For geothermal projects, a site visit is very important to observe surface features, access, and potential site barriers. A conceptual model of the resource is generally produced.

Detailed Feasibility Study. As the name implies, this is more in-depth analysis of the project's viability. A detailed feasibility study must provide information about the physical characteristics, financial viability, and environmental, social, or other impacts of the project, such that the owner can come to a decision about whether or not to proceed. It is characterized by the collection of refined site, resource, cost, and equipment data. For geothermal projects, the resource is fully defined, which typically involves drilling and measuring of thermal exchange properties. A conceptual model of the resource is refined and completed. In some cases, detailed computer simulation is undertaken. Project costs are refined through solicitation of price information from equipment suppliers.

Engineering and Development. If, based on the feasibility study, the project owner decides to proceed with the project, then engineering and development are the next step. Engineering includes the planning and technical design of the physical aspects of the project. Development involves the planning, arrangement, and negotiation of financial, regulatory, contractual, and other non-physical or 'soft' aspects of the project. Some development activities, such as training, customer relations, and community consultations, extend through the subsequent project stages of construction and operation. Even following significant investments in engineering and development, the project may be postponed or abandoned prior to construction because financing cannot be arranged, environmental approvals cannot be obtained, the pre-feasibility and feasibility studies overlooked important cost items, qualified contractors are not available, or for other reasons.

Procurement, Construction, and Commissioning. Finally, the project is built and put into operation. Prior to turning the project over to the owner, a proper commissioning process is key. The commissioning process involves a set of procedures to verify that all components of the system are operational, and that the system functions as it was intended and designed.

1.3 Elements of Renewable/Clean Energy Systems

The study of renewable energy systems can, in general, be subdivided into the following five elements: (i) energy loads and resource characteristics, (ii) harnessing the energy, (iii) energy conversion (to useful energy), (iv) optional energy storage, and (v) energy distribution.

Figure 1.2 shows these elements specific to geothermal energy systems. It should be emphasized that these elements are not mutually exclusive, but rather they are useful for

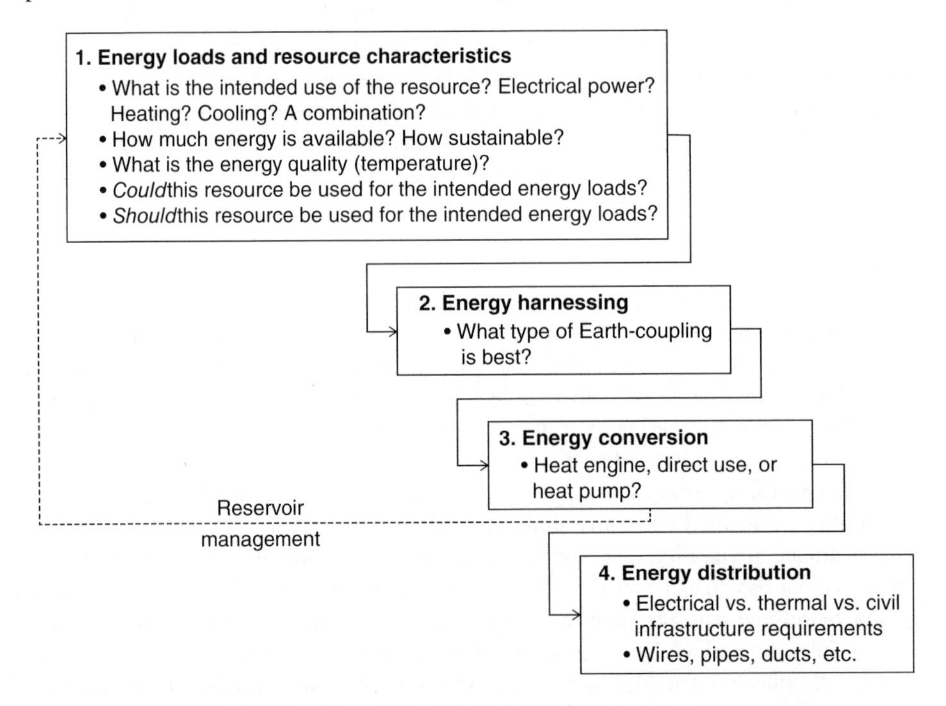

Figure 1.2 Elements of geothermal energy systems

understanding energy system components. Regarding geothermal projects, these elements can, however, and often do, represent the various specialty areas. For example, consulting scientists and engineers may be involved in resource characterization and/or resource harnessing, but not in the conversion or energy distribution stages.

Thus, the study of geothermal energy systems as presented in this book follows the aforementioned four elements of the system. Following this introductory chapter, Part I addresses energy loads and the geothermal resource. Part II covers the numerous Earth-coupling types used to harness stored thermal energy for geothermal heat pump applications. Part III discusses the various methods for converting geothermal energy to useful energy. Finally, Part IV discusses methods for distributing the energy. The focus of each part is on geothermal heat pumps, but higher-temperature geothermal applications are intermixed to give readers a broader perspective of the similarities of geothermal projects.

1.4 Geothermal Energy Utilization and Resource Temperature

We will see in Chapter 3 that there are a number of factors that dictate the end use of a geothermal resource, but the end use ultimately depends on the resource temperature. Thus, there have been a number of classification methods aimed at categorizing geothermal resources by temperature. Here, we will use the following gross temperature categories:

(a) high-temperature uses: $T_{resource} > 150\ ^\circ C$
(b) medium-temperature uses: $90\ ^\circ C < T_{resource} < 150\ ^\circ C$
(c) low-temperature uses: $30\ ^\circ C < T_{resource} < 90\ ^\circ C$
(d) ambient temperatures (heat pump uses): $\sim 0\ ^\circ C < T_{resource}$

Note that there is no distinct break between categories. The geothermal power industry typically uses only the top three temperature categories (a, b, and c), based on cut-off temperatures of economical electric power generation, which has historically not been economical for resources with temperatures below about 150 °C. However, binary organic Rankine cycle power plants, under favorable circumstances, have demonstrated that it is possible to generate electricity economically above 90 °C. A fourth category (d) is added here to distinguish the geothermal heat pump applications.

Figure 1.3 shows some of the many past and/or current uses of geothermal energy worldwide. As shown in this figure, there are many other 'high-temperature' resource use possibilities aside from electric power generation. Many of the medium-temperature uses are termed 'direct uses' because there is no energy conversion process, and the resource temperature matches or exceeds that required by the load. However, as noted in Figure 1.3, ambient groundwater can also be used for direct cooling applications.

1.5 Geothermal Energy Project History and Development

In the geothermal industry, projects are typically identified by their end use and associated resource temperature. Thus, power plant projects are associated with high-temperature reservoirs, direct-use projects are associated with low- to medium-temperature reservoirs, and geothermal heat pump projects are associated with ambient-temperature reservoirs. Note that,

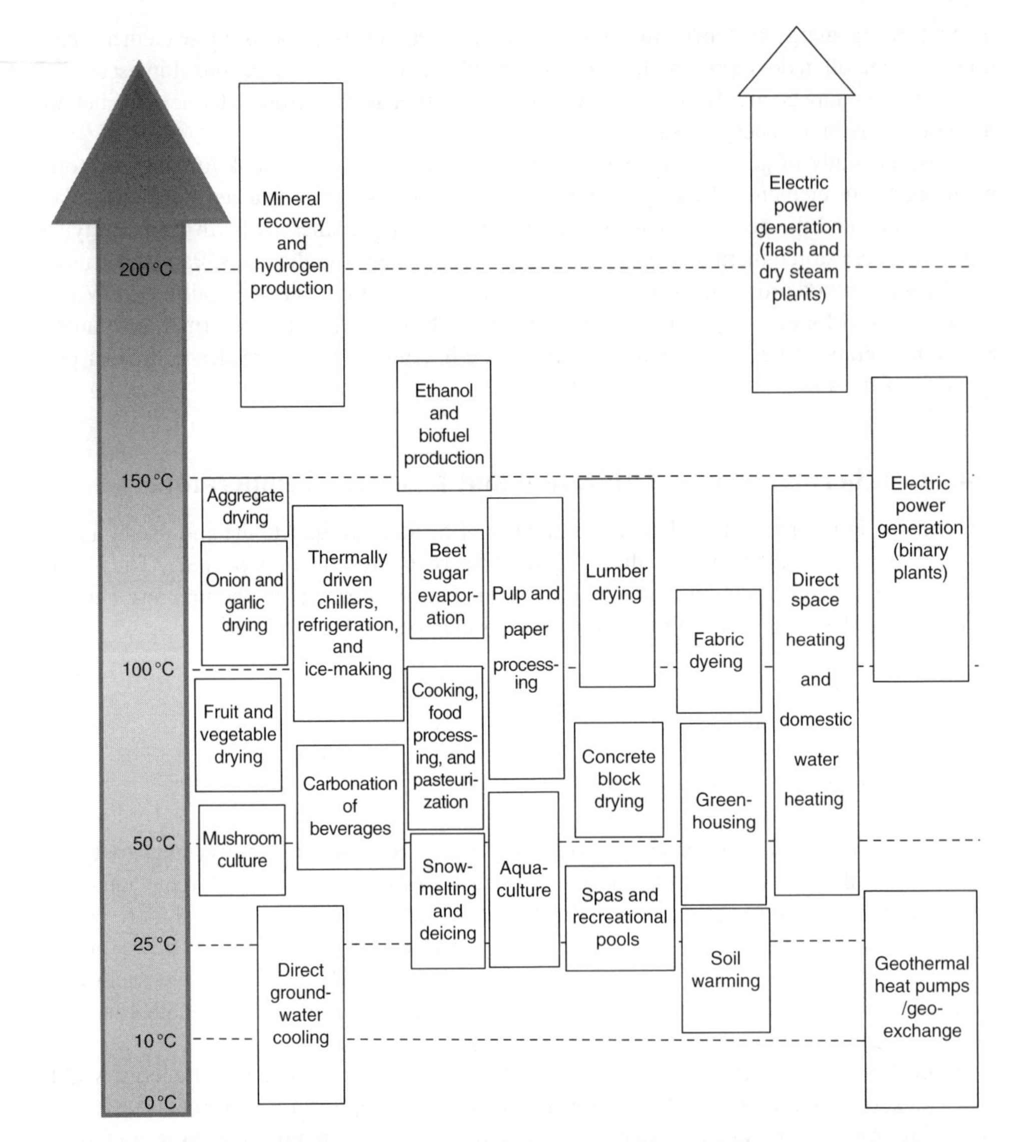

Figure 1.3 Worldwide past and present utilization of geothermal energy based on resource temperature

strictly speaking, such a classification scheme is ambiguous because the term 'direct use' means that the resource is used directly owing to its temperature match to the load, and thus a direct-use application can occur over a large temperature range. Further, thermally driven heat pump applications are utilized with moderate-temperature resources. Nevertheless, in the subsections that follow, we will use these descriptors to illustrate typical development of geothermal projects.

1.5.1 Geothermal Power Plants

1.5.1.1 Overview

The first geothermal power was generated at Larderello, Italy, in 1904. According to Lund (2007), the first commercial power plant (250 kW) was commissioned in 1913 at Larderello, Italy. Owing to the impurity of the geothermal fluids, steam was generated in a secondary loop isolated from the geothermal fluids by a heat exchanger. In the United States, an experimental 35 kW plant was installed at The Geyers geothermal field, California, in 1932, and provided power to the local resort. These developments were followed in New Zealand at Wairakei in 1958, an experimental plant at Pathe, Mexico, in 1959, and the first commercial plant at The Geysers in the United States in 1960. Japan followed with 23 MW at Matsukawa in 1966. All of these early plants used steam directly from the Earth (dry steam fields), except for New Zealand, which was the first to use flashed or separated steam.

According to Lund (2007), Iceland first produced power at Namafjall in northern Iceland, from a 3 MW non-condensing turbine. This was followed by plants in El Salvador, China, Indonesia, Kenya, Turkey, Philippines, Portugal (Azores), Greece, and Nicaragua in the 1970s and 1980s. Later plants were installed in Thailand, Argentina, Taiwan, Australia, Costa Rica, Austria, Guatemala, and Ethiopia, with the latest installations in Germany and Papua New Guinea.

The first medim-temperature geothermal binary power plant was put into operation at Paratunka near the city of Petropavlovsk on Russia's Kamchatka peninsula in 1967. It was rated at 670 kW and served a small village and some farms with both electricity and heat for use in greenhouses. It ran successfully for many years, proving the concept of binary plants as we know them today (DiPippo, 2012). Lund and Boyd (1999) list and describe the operation of 34 small geothermal power projects around the world, over 20 of which are binary. Many of these were installed in the 1980s and have (had) decades of successful operation. Notable installations since that time include Chena Hot Springs, Alaska (plant installed 2006), and the Oregon Institute of Technology, Klamath Falls, Oregon (plants installed 2009 and 2014), where ORC geothermal plants have been operating with resource temperatures of 74 and 90 °C respectively.

Bertani (2015) reports on worldwide geothermal power generation. As of 2015, approximately 25 countries have geothermal power plant installations, adding up to a total worldwide capacity of over 12 GW. The projected worldwide capacity is over 21 GW by 2020.

As we will discuss in Chapter 3, currently operating geothermal power plants make use of so-called hydrothermal resources. Other types of resource exist, but their utilization is still considered a developing and emerging use. The main types of geothermal power plant are summarized in Table 1.1: (i) binary, (ii) flash steam, and (iii) dry steam.

In binary power plants (discussed in detail in Chapter 14), the geothermal fluid remains liquid. The geothermal fluid exchanges heat with a lower-boiling-point working fluid, which is expanded through a turbine or other prime mover. The working fluid is then condensed back to liquid and operates in a closed cycle. On the other hand, in flash-steam plants (Figure 1.4), a two-phase geothermal fluid is extracted from wells and is separated near the well head into steam and water fractions. The steam fraction is expanded through a turbine, condensed, and returned to the geothermal reservoir in an open system. Dry steam plants are similar to flash plants, except that the geothermal fluid is in the superheated vapor state when extracted.

Table 1.1 Summary of Geothermal Power Plant Types

Power plant type	Geothermal resource	Working fluid	Occurrence
Binary cycle	Single-phase fluid (compressed liquid)	Engineered fluid (low-boiling-point refrigerant or hydrocarbon)	Most common type (generally considered feasible at resource temperatures up to ~175 °C)
Flash-type	Two-phase fluid	Geothermal fluid	Moderately common
Dry steam	Single-phase fluid (superheated vapor)	Geothermal fluid	Very rare (only a few plants worldwide)

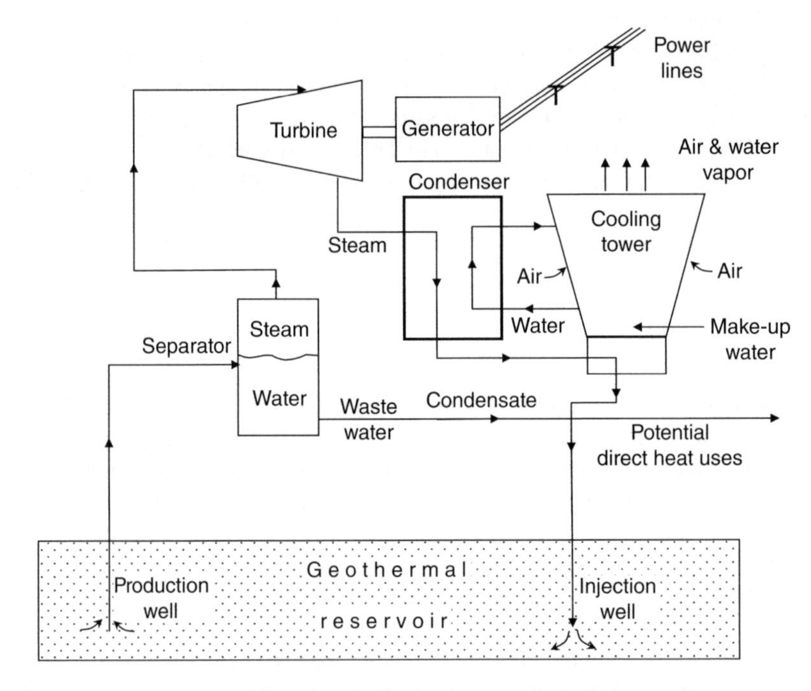

Figure 1.4 Schematic of a flash steam geothermal power plant

The development of a geothermal power plant can take up to several years, depending on what is known about the resource at the outset of a project. Resource prospectors must first determine where a viable resource might be, and then work toward acquiring access to the resource.

1.5.1.2 Geothermal Power Plant Project Development

Here, we define 'project development' as the process that takes a project from concept to construction and implementation. In reference to Figure 1.1, favorable project conditions for a high-temperature geothermal resource may begin with reconnaissance-type work. This phase is typically conducted by a geoscience consulting firm, and may take up to one year to complete, and account for about 3% of the total project costs. A regional study is conducted, typically examining thousands of square kilometers with the objective of selecting target areas for further examination. Available geoscience information is synthesized, and low-cost, rapid field studies may be conducted, such as basic geological mapping, geochemical sampling, aerial photography, and remote sensing. Construction of reservoir conceptual models is also begun.

The pre-feasibility phase consists of a more concentrated study of potential target areas, and is also typically conducted by a geoscience consulting firm. This phase may take up to two years to complete, and accounts for another 3% of the total project costs. This phase focuses on areas up to hundreds of square kilometers, with the primary objective of developing a preliminary geological model to form the basis of siting deep exploratory boreholes. Surface exploration studies to define subsurface structure, typically consisting of geophysical studies (i.e., gravity, magnetic, resistivity, seismic) and possibly additional geochemical work, are carried out. Temperature gradient holes to determine subsurface heat flow may also be drilled. A significant barrier to this phase is obtaining site access and regulatory approvals.

The detailed feasibility phase is aimed at resource confirmation. This phase may take up to two years to complete, and may account for about 5% of the total project cost. Personnel involved generally include a geoscience consulting firm, a reservoir engineering firm, and a drilling firm. This phase typically involves deep drilling and logging of exploratory wells to reach the geothermal reservoir, with the main objective of supporting and refining the geothermal model. Exploratory wells should also be used to make quantitative assessments of the available resource (one to five wells are typical). If a suitable resource is discovered, the exploratory wells are ideally used as production and injection wells for flow testing and reservoir engineering, with geochemical analysis of the geochemical fluids. A significant component of this phase also typically includes securing financing for the geothermal project. Potential barriers encountered during this phase include site access, environmental impact, and risk of not finding the resource expected.

The engineering and project development phase involves all remaining activities necessary to bring the geothermal field to full production. This phase may take up to two years to complete, and may account for about 30% of the total project cost. Personnel involved generally include a reservoir engineering firm and a drilling firm. In this phase, all final legal and regulatory permits and land acquisitions are secured. Impact studies of the project relating to the environment, ecology, and cultural issues are completed. All production and injection wells are drilled, tested, and completed, and surface piping and fluid handling systems are designed and constructed. As geothermal fields may encompass several square kilometers, the surface piping system can represent a significant portion of the total project cost (i.e., GEA (2005) reports 7%). Piping materials must be chosen that are compatible with the geothermal fluid chemistry. Potential barriers encountered during this phase include legal and regulatory permitting, cultural/community opposition to the project, and still some minor risk in not finding the resource expected.

The procurement, construction, and commissioning phase involves all remaining activities necessary to bring the geothermal power plant on line to supply electricity to the end user. This phase may take up to one year to complete, and may account for about 60% of the total project cost. Personnel involved generally include civil, mechanical, and electrical design firms. In this phase, the power plant is designed and constructed, and electrical transmission lines are designed and constructed. The final plant design (i.e., binary vs. flash steam) depends on the temperature and fluid production from the geothermal reservoir. Binary power plants are typically more cost effective at resource temperatures below about 177 °C (350 °F) (GEA, 2005). In flash plants, materials must be chosen that are compatible with the geothermal fluid chemistry. GEA (2005) reports transmission line cost estimates ranging from $168 000 to $282 000/km (in 2004 $US).

It should be noted that the project phases described above do not progress in a linear fashion. Some overlap of the phases is necessary for timely project completion. The above project development applies to large geothermal power plants, where a 'small' geothermal power plant is typically taken as less than 5 MW capacity. Geothermal project costs are highly variable and have significant economies of scale. Ormat Technologies, Inc. estimates the probable cost of a 20 MW geothermal project at $70 million (or $3500/kW). Small power plant costs can easily exceed $5000/kW. For perspective, geothermal power plants are known for providing stable, baseload power. Thus, a 1 MW geothermal power plant can supply the average electrical load to about 800–1000 homes.

1.5.2 Direct Uses of Geothermal Energy

1.5.2.1 Overview

Direct utilization of geothermal energy refers to its use for a thermal purpose, rather than to its conversion to some other form of energy such as electrical energy. In theory, then, geothermal energy can be directly used for any process or application that relies on heat, as long as the resource temperature matches the load temperatures (see Figure 1.3). The most popular direct uses of geothermal energy (excluding geothermal heat pump applications) include spas and swimming pools, space heating (including district applications), greenhouse heating, aquaculture, industrial uses, snow-melting, and agricultural drying. On a worldwide basis, of the order of 80 countries directly use geothermal energy for a thermal purpose, with spas and recreation being the most popular at approximately 50% of the total worldwide use, followed by space heating at approximately 30% of the total worldwide use.

Some authors include geothermal heat pump applications as a direct use, but that will not be the case here. The fact that a heat pump is needed for temperature amplification means that the resource temperature is too low to be used directly for heating, which contradicts the above definition of direct-use geothermal. Moderate to high geothermal fluids can be used in thermally driven (absorption or adsorption) heat pumps that produce a cooling effect for space cooling or industrial refrigeration. Thermally driven heat pumps are discussed in detail in Chapter 13.

According to Lund (2007), early humans probably used geothermal water that occurred in natural pools and hot springs for cooking, bathing, and warmth. Archeological evidence exists to suggest that the early native Americans occupied sites around these geothermal resources for over 10 000 years to recuperate from battle and to take refuge. Also according to Lund (2007), recorded history shows uses by Romans, Japanese, Turks, Icelanders, Central Europeans, and the Maori of New Zealand for bathing, cooking, and space heating. Baths in the Roman Empire and the middle kingdom of the Chinese and the Turkish baths of the Ottomans were some of the early uses of balneology. This custom has been extended to geothermal spas in Japan, Germany, Iceland, and countries of the former Austro-Hungarian Empire, the Americas and

New Zealand. Early industrial applications include chemical extraction from the natural manifestations of steam, pools, and mineral deposits in the Larderello region of Italy, with boric acid being extracted commercially from the early 1800s onwards. At Chaudes-Aigues in the heart of France, the world's first geothermal district heating system was started in the fourteenth century and is still in operation. The oldest geothermal district heating project in the United States is on Warm Springs Avenue in Boise, Idaho, going on line in 1892, which continues to provide space heating for up to 450 homes. Closed-loop, downhole heat exchangers for direct-use applications were first installed in residential wells in the late 1920s in Klamath Falls, Oregon, where some 550 direct-use geothermal wells exist today.

1.5.2.2 Direct-Use Geothermal Project Development

The development of direct-use geothermal projects is similar to that described above for power plants, with typically a much lower level of effort on resource exploration. Direct-use geothermal projects are intensely driven by economics, and thus cannot typically support an extensive exploration program. There are a number of factors that dictate direct-use project favorability, which have been summarized by Bloomquist (2006) and ASHRAE (2011). Similarly to power plant projects, these include: resource access and regulatory hurdles, level of effort needed for exploration, depth to the resource, distance between resource location and application site, well yield, allowable geothermal fluid temperature drop, resource temperature, thermal load and load factor, geothermal fluid chemical composition, and ease of geothermal fluid disposal.

Access and Regulatory Approvals. In many jurisdictions, groundwater is regulated as a resource, and thus permits and rights are required to utilize groundwater. In other jurisdictions, geothermal fluids are regulated as a mineral resource, thus requiring mineral rights. Therefore, proper water or mineral rights, in addition to land access approvals, must be obtained at the outset of a direct-use geothermal project. Local laws regarding subsurface injection requirements must also be ascertained. Competition with adjacent geothermal users and any potential royalty payments should also be understood. The cost and time to obtain such approvals may be substantial owing to the potential legal and environmental frameworks involved.

Level of Effort Needed for Exploration. Once access has been secured and all necessary regulatory approvals have been obtained, the developer may initiate a more detailed exploration program, refining whatever data were initially gathered in the reconnaissance or pre-lease phase of the development process. This phase usually consists of interdisciplinary activities of geology, geochemistry, geophysics, drilling, and reservoir engineering. These exploration activities are usually expensive, and often the economics of a direct-use activity will not support an extensive program. However, the minimum exploration and resource characteristics that are necessary include determining the depth to the resource, resource temperature, potential yield and specific capacity of wells, and chemistry of the geothermal fluid.

Depth to Resource. The cost of the wells to access the geothermal resource is typically one of the larger items in the overall cost of a direct-use geothermal system. Well cost generally increases non-linearly with resource depth. Compared with geothermal wells for electrical energy production or oil and gas wells, well depths for most direct-use geothermal projects are relatively shallow. For example, most larger direct-use geothermal systems in the United States operate with production wells at depths of less than 600 m, and many at less than 200 m (ASHRAE, 2011).

Distance Between Resource Location and Application Site. Direct use of geothermal energy must occur near the resource. This is not always the case for geothermal energy projects for electrical power generation, although these projects do have transmission costs; electrical energy can be sold to the power grid once transmitted from the resource to point of grid inter-tie. As there is no such thing as a *hot water* grid, direct-use projects must bear the transmission infrastructure costs of conveying the geothermal energy from the source to the load.

The cost of geothermal fluid transmission piping is commonly the largest cost item in the overall direct-use geothermal system (Bloomquist, 2006). Large conveyance distances generally require larger pipe diameters to offset friction losses in the pipe, in addition to increased pipe insulation to limit heat loss from the pipe. Therefore, there are economic trade-offs between the pipeline length (and cost) and the thermal load being met; long pipelines need large loads for a viable project. Geothermal fluids for direct-use projects can technically be transported relatively long distances (greater than 100 km) without great temperature loss, but such transmission is generally not economically feasible; most existing direct-use geothermal projects have transmission distances of less than about 1.5 km.

Well Yield. Thermal energy output from a geothermal production well is directly related to the fluid flow rate. However, higher fluid flow rates come at the energy expense of pumping. A typical good resource for direct-use purposes has a production rate of 25–50 L/s per production well, but some geothermal direct-use wells have been designed to produce up to 130 L/s (ASHRAE, 2011).

Allowable Temperature Drop of Geothermal Fluid. Thermal energy output from a geothermal production well is also directly related to the temperature decrease (ΔT) in the geothermal fluid as it exchanges heat with the thermal load. Therefore, the larger the allowable temperature drop of the fluid, the lower are the operating (pumping) and capital (well and production pump) costs. Cascading geothermal fluid from higher-temperature uses to lower-temperature uses can help to achieve large temperature differences in the geothermal fluid, thereby optimizing energy use of the resource. Most geothermal systems are designed for a ΔT between 15 and 30 °C, but care must be taken to avoid undesirable chemical changes or mineral precipitation in the geothermal fluid as a result of large temperature drops.

Resource Temperature. The resource temperature generally dictates the geothermal application, as previously mentioned.

Thermal Load and Load Factor. Geothermal direct-use projects can benefit significantly from economies of scale, particularly with the development cost of the resource. Therefore, it is economically desirable to match the load to the thermal output of a geothermal well.

The load factor is defined as the ratio of the average load to the design capacity of the system. Thus, the load factor effectively reflects the fraction of time for which the initial investment in the system is working. Again, as the life-cycle cost of a geothermal system is primarily attributed to its initial cost rather than its operating cost, the load factor significantly affects the viability of a geothermal system. As the load factor increases, so does the economy of using geothermal energy. There are generally two main ways to increase a geothermal direct-use load factor: (1) select applications where it is naturally high, and (2) design the system in a base-peaking arrangement, where the base load is handled by geothermal and the peak loads are met by supplemental equipment.

Geothermal Fluid Chemical Composition. The chemical quality of the geothermal fluid is site specific, and may vary from less than 1000 ppm (parts per million or mg/kg) total dissolved solids to heavily brined with total dissolved solids exceeding 100 000 ppm. Fluid chemical quality influences two aspects of the system design: (1) material selection to avoid corrosion and scaling effects, and (2) disposal or ultimate end use of the fluid. Each of these can have a significant impact on the system viability.

According to Ellis (1998), geothermal fluids commonly contain seven chemical species of concern for direct-use geothermal applications: pH, dissolved oxygen, chloride ion, sulfide species, carbon dioxide species, ammonia species, and sulfate ion. Low pH accelerates corrosion of carbon steel, while high pH indicates scaling potential. Total dissolved solids (TDS) concentrations in excess of 500 ppm indicate corrosion potential. Chloride ions (Cl^-) accelerate corrosion of steels; 304 stainless steel is acceptable to 140 ppm Cl^-, while 316 stainless steel is acceptable to 400 ppm Cl^-. Bicarbonate is linked to calcium carbonate scale above 100 ppm. Hydrogen sulfide attacks copper, and oxygen accelerates steel corrosion.

Ease of Fluid Disposal. The costs associated with disposal, particularly when injection is involved, can substantially affect geothermal direct-use development costs. Historically, most geothermal effluent was disposed of on the ground surface, including use in irrigation, or discharge to surface water bodies. This method of disposal is considerably less expensive than constructing injection wells, but can result in deleterious effects on the producing aquifer, mainly manifested as a decline in hydrostatic pressures. Discharge of geothermal fluids to the ground surface can also cause deleterious effects on and even contaminate the receiving body owing to undesirable concentrations of dissolved chemical constituents in the geothermal fluid.

Most new, large geothermal systems use injection for disposal to minimize environmental concerns and ensure long-term resource sustainability. If injection is chosen as the means of geothermal effluent disposal, the depth at which the fluid can be injected affects well cost substantially. Many jurisdictions, at least in the United States, require that the fluid be returned to the same or similar aquifers; thus, it may be necessary to complete the injection well at the same elevation as the production well.

1.5.2.3 Direct-Use Geothermal Equipment

Excluding wells, the primary equipment used in direct-use geothermal systems includes pumps, heat exchangers, and piping. Although aspects of these components are routine for many other applications, there are some special considerations with regard to direct-use geothermal applications, mainly related to the potentially aggressive nature of geothermal fluids. Thus, design and selection of equipment for direct-use geothermal applications mainly focuses on materials of construction that are capable of handling high temperatures and possible aggressive fluids. These aspects of direct-use geothermal projects are discussed in Chapter 4.

1.5.3 Geothermal Heat Pumps

1.5.3.1 Overview

Geothermal heat pump (GHP), Geoexchange®, or ground-source heat pump systems involve the coupling of low-grade thermal energy from Earth sources to a heat pump. A Swiss patent issued in 1912 to Heinrich Zoelly is the first known reference to geothermal heat pump systems (Spitler, 2005). In the United States, some ground-source and groundwater heat pump systems were installed just prior to World War II, and post-war, installations began to increase. At the same time, about a dozen research projects involving laboratory investigations and field monitoring were undertaken by US electric utilities. In addition, after some time, interest in further research seemed to wane until the 1970s after the oil crisis and initially followed much the same paths as the 1940s research, with an emphasis on experimental testing. This research did lead to solutions for several of the problems associated with the 1940s installations, particularly leakage problems, which were substantially resolved with the use of heat fusion of polybutylene and high-density polyethylene pipe. The 1980s saw the formation of the International Ground Source Heat Pump Association (IGSHPA), which worked to develop protocols for sizing closed-loop ground heat exchangers (GHX). Simultaneous research at Lund University in Sweden (e.g., Eskilson, 1987 and Hellström, 1991) made significant contributions, still in use to this day, in new GHX sizing algorithms and computer software tools. The 1990s saw the emergence and rapid growth of the GHP market in the United States and in Europe, combined with the growth and formation of various engineering trade organizations (e.g., ASHRAE Technical Committee 6.8 expanded in scope from direct use geothermal to GHPs).

Today, the term 'geothermal heat pump' system has become an all-inclusive term to describe a heat pump system that uses the Earth, groundwater, surface water, or other Earth-based heat exchange, such as sewer heat, as a heat source and/or sink. Other names exist, such as 'Geoexchange®' and 'ground-source heat pump'. Still others define the name based on the Earth coupling: groundwater heat pump (GWHP) systems, ground-coupled heat pump (GCHP) systems, surface water heat pump (SWHP) systems, and standing column well (SCW) systems. The types of Earth coupling used to harness shallow Earth energy are the focus of Part II of this book, Chapters 4 to 10. Common types of ground heat exchanger (GHX) couplings are shown in Figure 1.5.

In **groundwater heat exchange systems** (Figure 1.5a), conventional water wells and well pumps are used to supply groundwater to a heat pump or directly to some application. Corrosion protection of the heat pump may be necessary if groundwater chemical quality is poor. The 'used' groundwater is typically discharged to a suitable receptor, such as back to an aquifer, to the unsaturated zone (as shown in Figure 1.5a), to a surface water body, or to a sewer. Design considerations for groundwater heat exchange systems are considered in Chapter 4, and include groundwater availability, groundwater chemical quality, groundwater disposal method, well-drilling technologies, and well-testing methods. The main advantage of groundwater heat exchange systems is their potentially lower cost, simplicity, and small amount of ground area required relative to other Earth couplings and conventional systems. Disadvantages include limited availability, regulations, and poor chemical quality of groundwater in some regions. With growing environmental concerns over recent decades, many legal issues have arisen over groundwater withdrawal and injection in some localities.

A special type of Earth heat exchange is the so-called **standing column well system** as shown in Figure 1.5b. Sometimes referred to as an 'open–closed' system or a 'semi-open-loop' system, a standing column well has features of both open- and closed-loop systems. This type of system draws water to a heat pump from a standing column of water in a deep well bore, and returns the water to the same well. These systems, primarily installed in hard rock areas (e.g., granite), use uncased boreholes with typical diameters of about 15 cm and depths up to about 500 m. The uncased borehole allows the heat exchange fluid to be in direct contact with the Earth (unlike grouted closed-loop heat exchangers), and allows groundwater infiltration over the entire length of the borehole. Properly sited and designed, standing column systems have

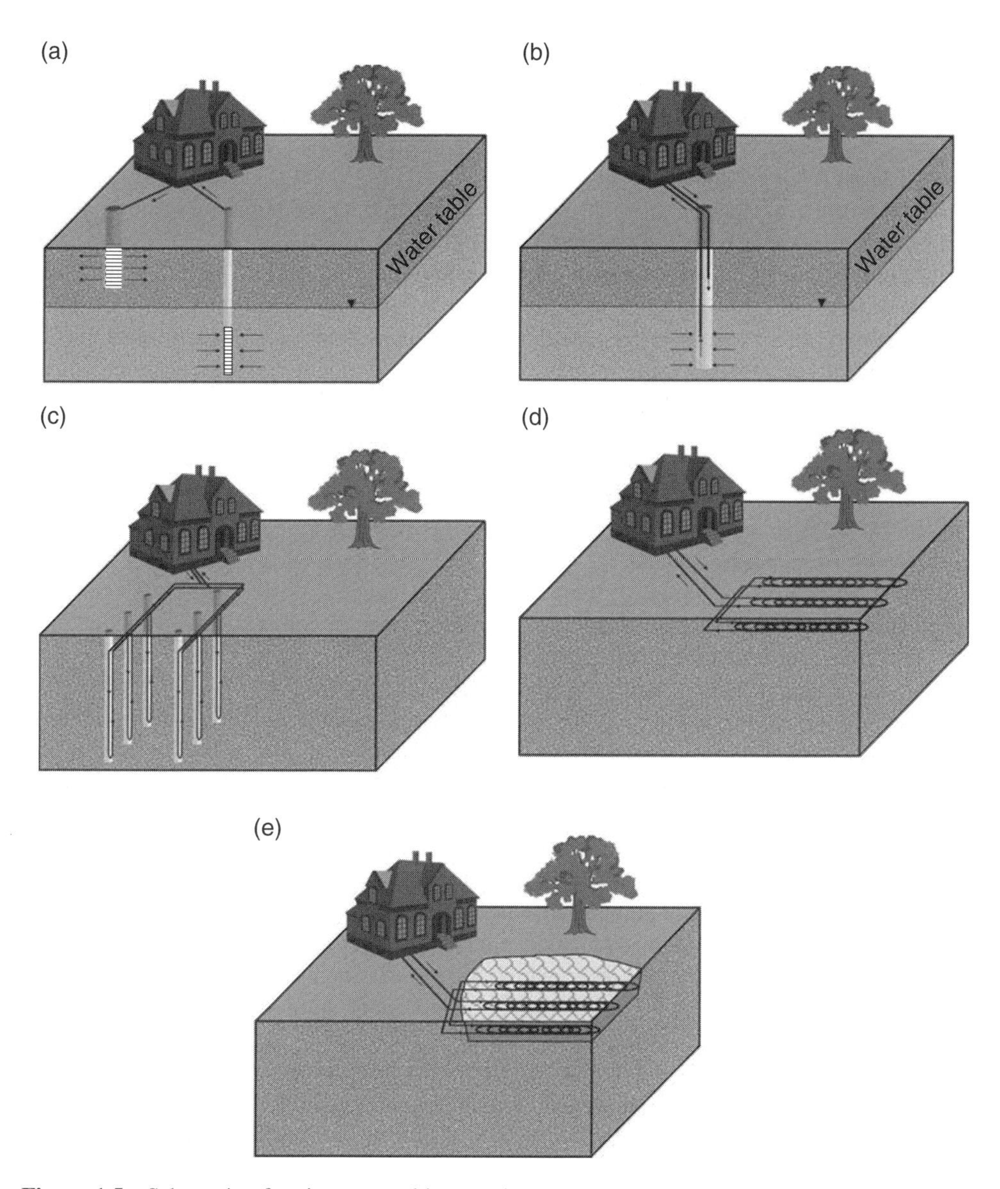

Figure 1.5 Schematic of various ground heat exchangers: (a) groundwater well, (b) standing column well, (c) vertical closed-loop borehole, (d) horizontal Slinky, and (e) surface water closed loop

been shown to have significant installation cost savings over closed-loop systems. Integrated into a drinking water system, the system has a natural 'bleed', as water is removed from the well, thereby allowing new groundwater into the well and moderating the well water temperature.

In **closed-loop, vertical bore heat exchange systems** (Figure 1.5c), heat rejection/extraction is accomplished by circulating a heat exchange fluid through a plastic piping system installed in an array of vertical boreholes, typically drilled to depths ranging from 50 to 100 m. Figure 1.5c shows a commonly-used u-pipe heat exchanger grouted in place. However, in some jurisdictions, groundwater-filled boreholes are acceptable. The key design aspects of closed-loop vertical bore heat exchange systems involve proper subsurface characterization, design of the borehole heat exchanger, and sizing and dimensioning of the ground heat exchanger (i.e., number of boreholes, depth, and spacing). In addition, optimal piping and pumping schemes need to be determined. These aspects are covered in Chapters 5, 6, and 10.

Horizontal closed-loop ground heat exchange systems (Figure 1.5d) are similar in principle to vertical ones, except, obviously, for their configuration. Horizontal GHX configurations typically consist of a series of parallel pipe arrangements laid out in dug trenches, excavations, or horizontal boreholes about 1–2 m deep. A number of piping arrangements are possible. 'Slinky' configurations (as shown in Figure 1.5d) are popular and simple to install in trenches and shallow excavations. In horizontal boreholes, straight pipe configurations are installed. Typical pipes have a diameter ranging from ¾ in (19 mm) to 1½ in (38 mm). Because of their proximity to the ground surface, horizontal GHXs are more affected by weather and air temperature fluctuations. Design aspects of horizontal GHXs are covered in Chapters 7 and 10.

Surface water heat exchange systems (Figure 1.5e) can be a closed-loop or an open-loop type. Typical closed-loop heat exchanger configurations are the loose bundle coil type, plate type, or Slinky coil type (as shown in Figure 1.5e). In closed-loop systems, heat rejection/extraction is accomplished by circulating a heat exchange fluid through a heat exchanger positioned at an adequate depth within a lake, pond, reservoir, or other suitable open channel. In open-loop systems, water is extracted from the surface water body through a screened intake area at an adequate depth and is discharged to a suitable receptor. Open-loop systems can be used for direct cooling (e.g., Cornell University). Heat transfer mechanisms and the thermal characteristics of surface water bodies are quite different from those of soils and rocks. Design aspects of surface water GHXs are covered in Chapters 8 and 10.

A photo of a geothermal heat pump in a residential building is shown in Figure 1.6. Note the fluid connections and associated circulating pumps, and the ductwork. This particular heat pump is equipped with a *desuperheater*, which is used to generate hot water in the adjacent storage tank, also coupled to a solar thermal system. Vapor compression heat pumps used in geothermal applications are discussed in detail in Chapter 12.

Worldwide, there are well over 1 million geothermal heat pump installations. Geothermal heat pumps are relatively well established as a means of significantly reducing energy consumption in space conditioning of buildings. This improvement in efficiency, however, generally comes at a higher first cost, as with most renewable/clean energy systems, which must be offset by lower operating and maintenance costs within an acceptable period of time to the building owner. As with most alternative energy systems, high capital cost is a significant barrier to market penetration.

Why are geothermal heat pumps labeled as an energy-efficient technology? First, any heat pump is more thermodynamically efficient than fossil fuel combustion because heat pumps 'move' heat from a lower-temperature source to a higher-temperature sink, and do not generate heat. You will hear arguments that heat pumps in a space heating application are less efficient than fossil fuel combustion owing to inefficiencies at the power plant. In most cases this is untrue, and will become progressively less true in the future as supply-side electricity generation becomes more efficient and as more renewable energy sources are used. Second,

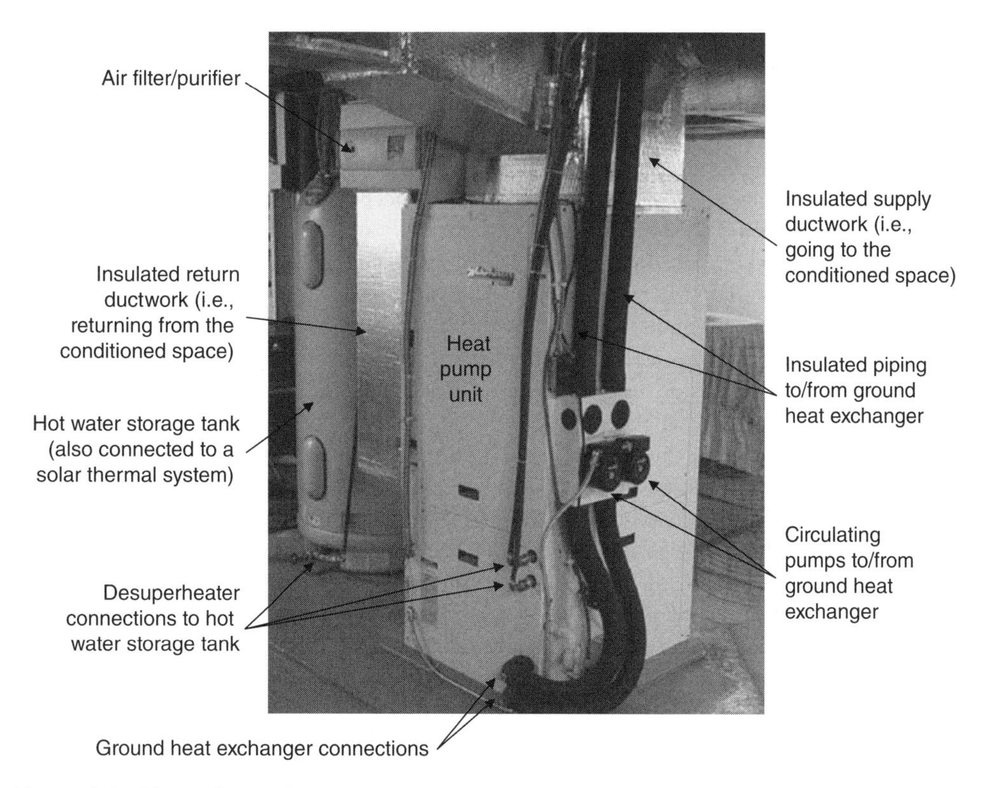

Figure 1.6 Photo of a geothermal heat pump in a residential building (photo by J. Bohrer; permission obtained)

geothermal heat pumps are more energy efficient than other heat pumps because the heat source/sink (the ground) is seasonally closer to room temperature than outdoor air. Additionally, heat is absorbed and rejected through water or an aqueous antifreeze solution, which is a more desirable heat transfer medium than air because of its relatively high heat capacity.

1.5.3.2 Geothermal Heat Pump Project Development

Again, we define 'project development' as the process that takes a project from concept to construction and implementation. In reference to Figure 1.1, favorable project conditions for a geothermal heat pump project may begin with the need for an HVAC system. Other favorable factors include: (i) owner preference for a sustainable building (life-cycle cost approach); (ii) applications where the total annual heating loads are balanced with total annual cooling loads; (iii) buildings with diverse floor plans and loads; (iv) contractor/designer experience and availability; (v) high alternative fuel costs.

The pre-feasibility phase should be conducted by qualified personnel, and is generally aimed at ultimately selecting the best GHX for the site conditions. Schillereff *et al.* (2008) describe a site suitability assessment method that we will further discuss in Chapter 3. In the pre-feasibility stage, all plausible options are first considered with an open mind, and then non-viable options are screened out. Thus, the energy intent of the project should be clear, and estimates of the thermal loads are needed in order to determine probable system costs. Detailed knowledge of every site is not always required if there are numerous other systems in the area and the local geology is uniform. The level of effort in site characterization is typically proportional to the size of the project; the economics of small residential systems cannot support a detailed study. This phase may take months to complete in complex projects. At the conclusion of this phase, GHX options are rejected if they: (i) are non-compliant with laws and regulations, (ii) have an unacceptable development schedule, (iii) require an unacceptable land area, (iv) have unacceptable thermal or hydraulic effects on neighbors or on the environment, or (v) do not meet owner's financial or sustainability criteria.

The detailed feasibility phase is aimed at resource confirmation and determining the most suitable GHX to work with the strengths of the site. Based on the recommendations of the pre-feasibility study, intrusive investigation for site-specific information is undertaken as needed (exploratory drilling, thermal response testing, aquifer testing, water chemistry sampling, etc.). Again, the economics of small projects generally cannot support a detailed feasibility study; favorable projects based on professional judgement typically proceed directly to the design and construction phase. A significant component of this phase may also include refining the load calculations of the application, further HVAC design work, refining costs, and securing financing for the geothermal project. Potential barriers encountered during this phase include site access, environmental impact, and risk of not finding the resource expected. This phase may take months to complete, depending on the project complexity.

The engineering and project development phase involves all activities necessary to complete the geothermal field and associated HVAC system design. This phase may take months to complete, depending on the project complexity. Personnel involved generally include certified designers and installers and licensed professional engineers. Also in this phase, all final legal and regulatory permits are secured. Impact studies of the project relating to the environment, ecology, and cultural issues are completed. Potential barriers encountered during this phase include legal and regulatory permitting, cultural/community opposition to the project, and still some minor risk in not finding the resource expected.

The procurement, construction, and commissioning phase involves all activities necessary to construct the geothermal field and associated HVAC systems. Personnel involved generally include civil, mechanical, and electrical design firms, in addition to certified installers and construction managers.

1.6 Chapter Summary

This chapter presented an overview of the utilization of geothermal energy based on resource temperature. Information flow of energy projects was discussed in some detail, and parallels were drawn with the development of the following types of geothermal energy project: (i) power plants, (ii) direct uses, and (iii) geothermal heat pumps. Site and resource characteristics of geothermal projects were emphasized because all other pieces of the project flow from them.

This chapter also introduced elements of renewable energy systems that are helpful in studying components of the overall system. With regard to geothermal energy systems, these elements represent the four subsections of this book: (i) energy loads and resource characteristics, (ii) harnessing the energy, (iii) energy conversion (to useful energy), and (iv) energy distribution.

Discussion Questions and Exercise Problems

1.1 Is geothermal energy renewable? Give reasons why or why not and cite some examples.
1.2 Identify locations of some low, medium, and high geothermal resource locations nearest to you, and in your country. Find an example utilization of each (if it exists) near you, and in your country.

Part I

Geothermal Energy – Utilization and Resource Characterization

In Part I, we discuss the general development of the geothermal resource, a key element in geothermal energy systems. The intended use of the resource must be determined prior to project development, and *resource characterization* efforts seek to answer the questions, '*Could* the resource be used for the intended energy needs? *Should* the resource be used for the intended energy needs?'. Thus, intended energy use may need modification as the resource characterization stage unfolds.

In **Chapter** 2, an overview of loads and their calculation methods are presented, where the term 'load' refers to the time-dependent energy needs to be met by the geothermal energy system. In **Chapter** 3, we discuss the Earth and elements of the geothermal resource in the context of a finite reservoir, the intended loads, and owner needs and desires.

Geothermal Heat Pump and Heat Engine Systems: Theory and Practice, First Edition. Andrew D. Chiasson.

Companion website: www.wiley.com/go/chiasson/geoHPSTP

2

Geothermal Process Loads

2.1 Overview

To characterize a geothermal resource properly and ultimately design an Earth heat exchange system, the process load dynamics must be known. Peak-hour loads are necessary for sizing heating and cooling equipment in all systems, but the design of geothermal energy systems also requires time-dependent loads throughout the year in order to predict thermal storage effects and long-term temperatures in the subsurface.

This chapter describes the types of load typical of geothermal energy systems. Prior to undertaking a geothermal project, a clear objective should be established as to the intent of the geothermal energy system. Often, building owners 'bite off more than they can chew' and become ambitious that a geothermal energy system can meet all loads of a building – heating, cooling, hot water, snow melting, and swimming pool heating, with energy to spare. As we shall discuss in Chapter 3, geothermal reservoirs are finite, and the larger the intended loads, the larger the reservoir volume needs to be.

Learning objectives and goals:

1. Appreciate the role of weather in load calculations, along with the various forms of weather data and where to find weather data.
2. Be familiar with methods and tools for determining peak heating and cooling loads for buildings.
3. Be able to estimate time-dependent energy loads in buildings.

Geothermal Heat Pump and Heat Engine Systems: Theory and Practice, First Edition. Andrew D. Chiasson.

Companion website: www.wiley.com/go/chiasson/geoHPSTP

4. Be familiar with methods to determine greenhouse heating loads.
5. Be able to calculate swimming pool loads.
6. Be able to calculate snow-melting loads.

2.2 Weather Data

The net heat gain or loss to buildings and other applications depends mainly on weather. The three most important weather variables are: (i) outdoor air dry-bulb temperature, (ii) outdoor air humidity, and (iii) solar radiation (wind velocity is also important).

Three types of weather data are commonly used for heating and cooling design and analysis:

- typical weather (for estimating typical energy use or savings)
  - hourly
  - monthly
- actual weather (for calibrating energy use calculated by a theoretical model to measured energy use)
- design weather (for calculating maximum expected building loads to size heating and air-conditioning equipment)

Typical Weather Data (Hourly). To determine typical weather conditions, meteorologists created *typical meteorological year* (or *TMY*) *files*, mostly for US locations. The third generation of these files are called TMY3 or TM3 files, and are derived from data from 1991–2005. TMY3 sites are noted as class I, class II, or class III, with the most robust data from class I sites. TMY3 files for 1020 US locations, and a manual which describes the data, can be downloaded from the National Renewable Energy Laboratory at: http://rredc.nrel.gov/solar/old_data/nsrdb/1991-2005/tmy3/.

Many building energy simulation programs use these files or similar files as input, as these files represent the typical meteorological conditions for a given site and contain data on a short enough time interval to quantify many transient effects. Because they represent typical rather than extreme conditions, they are not suited for designing systems to meet the worst-case conditions occurring at a location.

The US Department of Energy has developed a building energy simulation program called Energy Plus, and the weather files for Energy Plus are called *Energy Plus Weather* (or *EPW*) *files*. EPW files contain the same data as TMY2 files, with the advantage of being available for many international locations. EPW files, and a manual which describes the data, can be downloaded from: http://www.eere.energy.gov/buildings/energyplus/.

Typical Weather Data (Monthly). In some situations with large enough thermal mass, hourly transients are small, and average weather data on a monthly basis are suitable. An excellent source of monthly weather data can be found at: https://eosweb.larc.nasa.gov/sse/RETScreen/. Users enter the latitude and longitude of the location of interest, and output as shown on Figure 2.1 is available. We shall use monthly weather data in several instances in this book.

Actual Weather Data. To improve the accuracy of theoretical models of building energy use, it is useful to compare the predicted energy use of the theoretical model with measured

NASA Surface meteorology and Solar Energy: RETScreen Data

Latitude 40.7 / Longitude -74 was chosen.

	Unit	Climate data location
Latitude	°N	40.7
Longitude	°E	-74
Elevation	m	64
Heating design temperature	°C	-4.35
Cooling design temperature	°C	26.70
Earth temperature amplitude	°C	13.52
Frost days at site	day	60

Month	Air temperature	Relative humidity	Daily solar radiation - horizontal	Atmospheric pressure	Wind speed	Earth temperature	Heating degree-days	Cooling degree-days
	°C	%	$kWh/m^2/d$	kPa	m/s	°C	°C-d	°C-d
January	0.6	66.7%	1.97	101.0	5.8	2.5	543	2
February	1.6	64.6%	2.86	101.0	5.9	2.6	466	1
March	4.8	65.2%	3.93	100.9	5.8	4.8	409	10
April	10.0	65.4%	4.83	100.8	5.3	9.2	240	43
May	15.4	65.0%	5.55	100.8	4.7	14.2	96	167
June	20.6	65.8%	6.03	100.7	4.4	19.5	9	318
July	23.6	67.5%	5.82	100.8	4.0	22.9	0	422
August	23.2	68.7%	5.27	100.9	3.8	22.8	0	411
September	20.0	65.9%	4.37	101.0	4.2	19.9	16	300
October	14.2	64.1%	3.26	101.1	4.7	14.7	128	140
November	8.9	65.5%	2.14	101.1	5.5	9.8	276	41
December	3.5	66.0%	1.73	101.1	5.9	5.3	453	7
Annual	12.2	65.9%	3.98	100.9	5.0	12.3	2636	1862
Measured at (m)					10.0	0.0		

Figure 2.1 Monthly weather data summary from NASA Atmospheric Science Data Center

energy use of the building. The process of comparing the predicted energy use from theoretical models with measured energy use and adjusting the theoretical model to improve the fit is called 'calibrating the model'.

The best weather data for calibrating models are measured hourly temperature, humidity, and solar radiation data from the period during which the building energy use was measured. Unfortunately, it is often difficult to obtain measured hourly temperature, humidity, and solar radiation data over specific periods. However, average daily temperature data are widely available, and outdoor air temperature is typically the single most important weather variable influencing building energy use. Thus, hourly weather data files synthesized from average daily temperature data are often the best weather data available for calibrating models to measured energy use. Useful tools for this purpose can be found at: http://academic.udayton.edu/kissock/http/RESEARCH/EnergySoftware.htm.

Design Weather Data. Design weather datasets include the hottest and coldest expected weather conditions for a location, within statistical reason. This information is used to size heating and cooling systems so that they can handle the largest expected loads. Design weather data are available in the ASHRAE Fundamentals Handbook for over 4400 sites worldwide. Design heating and cooling temperatures are also available from the NASA website, as shown in Figure 2.1.

For heating, ASHRAE tabulates the 99.6th percentile and 99.0th percentile design conditions, meaning that the actual hourly temperatures were greater (warmer) than the design temperature for 99.6 or 99.0% of all annual hours. To ensure that the heating system is large enough to handle the coldest expected temperatures, the 99.6% design temperature is typical. However, this is often a designers' choice. Systems could be greatly oversized if sized for the coldest temperature on record at a particular location.

For cooling, ASHRAE tabulates 0.4th percentile and 1st percentile design conditions for dry-bulb temperature and humidity, such that the actual hourly temperatures were greater (warmer) than the design temperatures for 0.4 or 1.0% of all annual hours. In addition, ASHRAE determined the mean coincident wet-bulb (MCWB) temperature for each design condition, which is the mean wet-bulb temperature at the specified dry-bulb temperature. The MCWB temperature is used for calculating latent cooling loads. To ensure that the cooling system is large enough to handle the warmest expected conditions, the 0.4% design temperatures are typically used.

ASHRAE also presents guidance for determining solar radiation incident on an arbitrary surface using the so-called ASHRAE clear sky model. Following this method, the total radiation on a surface is the sum of the direct radiation, the diffuse radiation, and the ground-reflected radiation. The method uses local standard time, the local standard meridian, the local longitude, the angle of the surface from south, and the tilt angle of the surface from horizontal as inputs.

2.3 Space Heating and Cooling Loads

2.3.1 Peak Design Loads

Calculation of peak design space heating and cooling loads is well beyond the scope of this book. Here, we will discuss the basic principles only because of their importance in designing geothermal energy systems. For a detailed treatment of load calculations, readers are referred to Spitler (2014) or Kavanaugh (2006), for example.

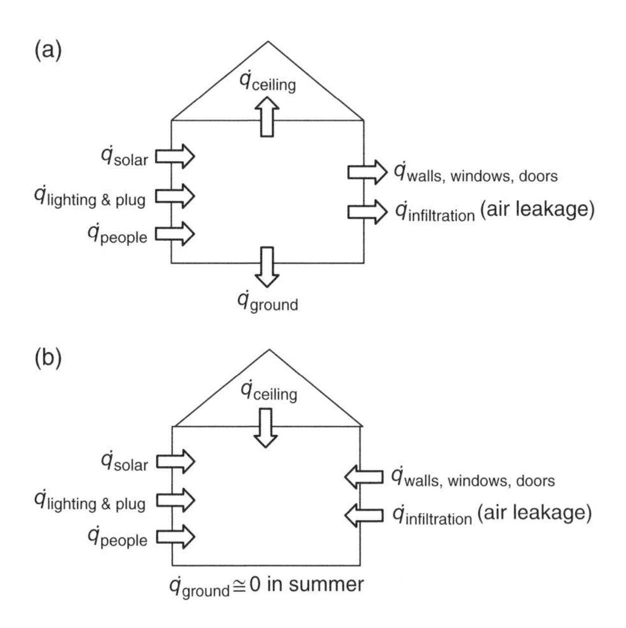

Figure 2.2 Heat flows in buildings in (a) winter and (b) summer

Heat flows into and out of a space are typically called 'loads'. To determine the net load on a space, the components of the loads are equated in an energy balance. Heat flows in buildings in winter and summer are shown in Figure 2.2.

The net loads resulting from an energy balance on the loads shown in Figure 2.2 are referred to as *space loads*. It is important to note that the loads do not occur instantaneously owing to the thermal mass of the building shell and interior components. In some cases, particularly in heating load calculations, transients may be ignored. This is not always true for the cooling load in situations with large solar gains and internal gains. In these cases, the space cooling load is not coincident with instantaneous heat gains; the cooling load seen by the space is delayed and damped owing to thermal storage by the building shell and internal components.

Quantification of the envelope loads shown in Figure 2.2 involves application of Fourier's law and standard methods for estimating the building *UA* value, where *U* is the overall heat transfer coefficient and *A* is the surface area. Internal gains due to lighting and electrical equipment are typically tabulated or estimated for various appliances. Internal heat gains from people are typically tabulated based on the activity level of the occupants.

Heating and cooling equipment is sized to meet envelope loads if there are no ventilation air loads. Ventilation air is distinguished from infiltration air by the fact that infiltration is unintended air leakage, while ventilation air is intentionally introduced into a space for indoor air quality and occupant comfort purposes. Infiltration is typically reported in air changes per hour (ACH), which represents the number of times that the entire volume of air in a building leaks out each hour. Older buildings or poorly constructed buildings have 1–3 ACH, while modern buildings have 0.3–0.6 ACH. Very tight construction is typically characterized by ~0.1 ACH. Local codes and regulations typically dictate the ventilation air requirements for a given occupancy. This outdoor air represents a significant load and results in the need for larger-sized heating and cooling equipment relative to that needed for the building loads only. The ventilation air plus the space loads are sometimes referred to as *equipment loads*. The role of ventilation air will be discussed in Chapter 15.

2.3.2 *Monthly and Annual Loads*

Calculation of monthly and annual loads is typically uncommon in conventional HVAC engineering; peak design loads are used to size equipment, and time-varying loads, or 'off-design' conditions, are only considered in energy-use calculations. However, monthly and annual load estimates are essential in the design of GHXs owing to the thermal storage effects and large time constants of the Earth.

2.3.2.1 Use of Building Simulation Software

Building simulation programs as described here refer to whole building energy software tools that calculate energy use in buildings on an hourly basis. Some examples of freeware include *EnergyPlus* (by the US Department of Energy), *eQuest* (also by the US Department of Energy), and *ESim* (by K. Kissock). A number of other building simulation programs are also commercially available.

The general approach to the use of these tools is similar: users must input details of the building envelope, occupancy, HVAC systems and controls, and all other factors that impact the thermal behavior of the building. These building simulation programs are mainly driven by weather files and other time-dependent functions. The output of interest to GHX design include: hourly heating loads, hourly cooling loads, and hourly domestic hot water loads. Figure 2.3 is a plot of hourly building loads for a year for a residential building in the US Midwest.

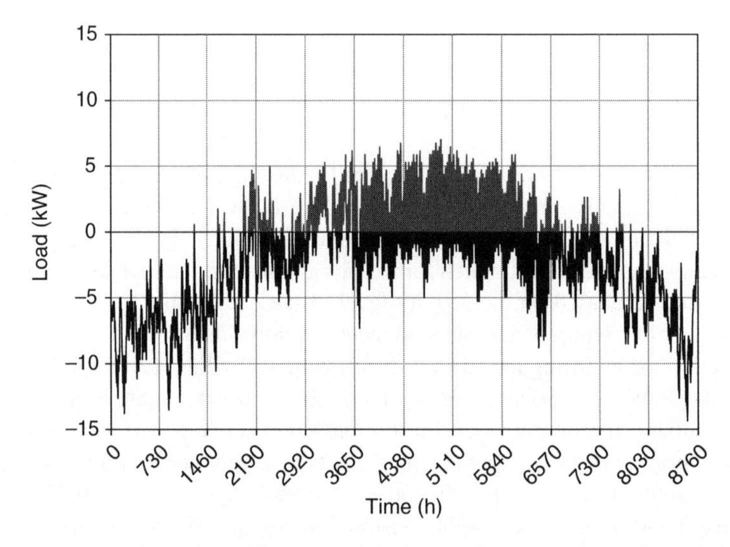

Figure 2.3 Example hourly loads profile for a residential building. Heating (cooling) loads are negative (positive)

The use of building simulation programs is a definite example of 'garbage in, garbage out'. All simulation tools must strike a balance between complexity and ease of use. Complicated simulation tools may require large amounts of input data, much of which must be estimated, and can quickly become unwieldy and difficult to calibrate. Simplistic tools may have difficulty capturing essential energy flows and can produce unreliable results. Thus, designers should gain a feel for when simulation programs are suitable and when they are not.

2.3.2.2 Degree-Day Procedure and Balance Temperature

The degree-day procedure was conceived to estimate the amount of heating or cooling required over a given period of time, usually a month. Monthly degree-days are summed to determine the annual degree-days.

The procedure involves measuring the average daily temperature and quantifying the amount of heating or cooling relative to a reference or base temperature. The historical reference temperature was taken as 65 °F (or 18.3 °C), which means that heating was required at outdoor air temperatures below 65 °F, and cooling was required at outdoor air temperatures below 65 °F. Mathematically, the number of degree-days per month (DD_{month}) is given by

$$DD_{month} = \sum_{day=1}^{N} \left| T_{base} - \bar{T}_{day} \right| \tag{2.1}$$

where DD_{month} is in units of °F-day or °C-day, N is the number of days per month, T_{base} is the base temperature at which heating is assumed to begin (typically 65 °F or 18.3 °C), and $\bar{T}_{day}$ is the average daily temperature. Thus, for example, if the average daily temperature on a particular day was measured at 60 °F, then there would be 5 heating degree-days for that day (in units of °F-day).

The degree-day procedure may be used to estimate the monthly heating (or cooling) load of a building, and therefore the fuel requirement. **An inherent assumption in use of this method for energy calculations is that the space load is dominantly driven by weather**. The degree-day procedure is not recommended for buildings with large internal gains or solar heat gains.

The monthly heating or cooling load (Q_m) is given by

$$Q_m = DD_{month} \times \frac{24\,\text{h}}{day} \times \frac{\dot{q}_{peak}}{\Delta T} \tag{2.2}$$

where ΔT is the difference between the indoor and outdoor design temperatures used in the calculation of the peak hour design load, $\dot{q}_{\text{peak}}$. The average hourly load ($\bar{\dot{q}}$) for a given month is given by

$$\bar{q} = \frac{DD_{month}}{N} \times \frac{\dot{q}_{peak}}{\Delta T} \tag{2.3}$$

where N is the number of days per month.

Example 2.1 Estimation of Monthly Total and Average Loads Using Degree-Days
The peak load of a residential building has been calculated to be 20 kW for indoor and outdoor design temperatures of 21 and −4.35 °C respectively. Using the heating degree-day data shown in Figure 2.1, estimate:

(a) the total monthly heating requirement for January, and
(b) the average hourly heating load for January.

Solution

(a) the total monthly heating requirement for January given by Equation (2.2):

$$Q_m = DD_{month} \times \frac{24\,\text{h}}{day} \times \frac{\dot{q}_{peak}}{\Delta T}$$

$$Q_m = 543\ ^\circ\text{C}\cdot\text{day} \times \frac{24\,\text{h}}{day} \times \frac{20\,\text{kW}}{(21 - -4.35)^\circ\text{C}}$$

$$Q_m \cong 10\,282\,\text{kWh}$$

(b) the average hourly heating load for January is given by Equation (2.3):

$$\bar{q} = \frac{DD_{month}}{N} \times \frac{\dot{q}_{peak}}{\Delta T}$$

$$\bar{q} = \frac{543\,^\circ\text{C}\cdot\text{day}}{31\ \text{days}} \times \frac{20\,\text{kW}}{(21 - -4.35)\,^\circ\text{C}}$$

$$\bar{q} = 13.4\,\text{kW}$$

The choice of 65 °F or 18.3 °C as an outdoor air reference temperature is highly unrealistic with today's energy-efficient buildings and controls. Therefore, variable degree-day methods have been developed to allow calculation of degree-days referenced to other temperatures. Duffie and Beckman (2013) provide equations to calculate monthly degree-days for any reference temperature:

$$DD_{month} = \sigma_m N^{3/2}\left[\frac{Z}{2} + \frac{\ln[\cosh(1.698Z)]}{3.396} + 0.2041\right] \tag{2.4}$$

where N is the number of days in the month, σ_m is the standard deviation of the monthly average ambient temperature, and Z is further defined as

$$Z = \frac{T_{base} - \bar{T}_{month}}{\sigma_m\sqrt{N}} \tag{2.5}$$

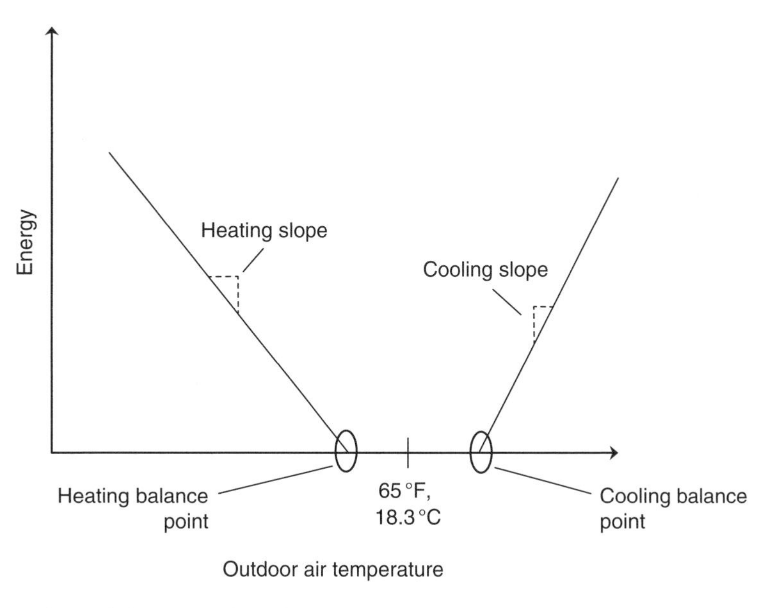

Figure 2.4 Illustration of heating and cooling balance point temperatures

where T_{base} is the base (or reference) temperature, and $\bar{T}_{month}$ is the monthly average outdoor air temperature. σ_m is not readily available from monthly data, but may be approximated by

$$\sigma_m = 1.45 - 0.0290\bar{T}_{month} + 0.0664\sigma_{yr} \tag{2.6}$$

where σ_{yr} is the standard deviation of $\bar{T}_{\text{month}}$ from the average annual air temperature.

The choice of the base temperature in degree-day energy calculations is based on the concept of the balance point temperature, which is the outdoor air temperature at which heating or cooling is required, as illustrated in Figure 2.4. A review of Figure 2.4 shows that the default choice of 65 °F (18.3 °C) generally lies in the middle of actual heating and cooling balance point temperatures. Some buildings may have heating balance point temperatures as low as 10 °C (50 °F) and cooling balance point temperatures of the order of 25 °C (77 °F).

The best means of determining the balance point temperature of an existing building is to plot monthly energy consumption from utility bills versus average outdoor air temperature and fit a line through the points. Two or more years of utility billing data are recommended. The result will be two lines as shown in Figure 2.4, where the *x*-intercepts represent the balance point temperatures. It is also interesting to note that the slopes of the lines are proportional to the *UA* value of the building.

Example 2.2 Calculation of Variable-Base Heating Degree-Days

Using Equation (2.4), calculate the monthly heating degree-days for the data shown in Figure 2.1 for:

(a) the usual base temperature of 18.3 °C,
(b) a base temperature of 13 °C,
(c) compare the results with the measured values shown in Figure 2.1.

Solution

First, we calculate $\sigma_{yr} = 8.5$ °C, and solve using a spreadsheet as shown below.

Note that Z is set to zero when $\bar{T}_{month} > T_{base}$.

A review of the solution table (Table E.2.2) shows that the monthly heating degree-days as calculated from Equation (2.4) is quite accurate, with the annual total being within 2% of the measured total for a base temperature of 18.3 °C.

When the base temperature is lowered to 13 °C, the annual heating degree-days decreases by 58%. Thus, it is important to use representative base temperatures when using the degree-day procedure.

Table E.2.2 Variable-Base Heating Degree-Day Calculations for Example 2.2

Month	$\bar{T}_{month}$ (°C)	HDD (°C-day)	N (days)	σ_m (°C)	T_{base}			
					18.3 °C		13 °C	
					Z (–)	$HDD_{18.3\,°C}$ (°C-day)	Z (–)	$HDD_{13.0\,°C}$ (°C-day)
January	0.6	543	31	2.0	1.6	549	1.1	387
February	1.6	466	28	2.0	1.6	468	1.1	321
March	4.8	409	31	1.9	1.3	420	0.8	261
April	10	240	30	1.7	0.9	253	0.3	114
May	15.4	96	31	1.6	0.3	112	0.0	0
June	20.6	9	30	1.4	0.0	0	0.0	0
July	23.6	0	31	1.3	0.0	0	0.0	0
August	23.2	0	31	1.3	0.0	0	0.0	0
September	20	16	30	1.4	0.0	0	0.0	0
October	14.2	128	31	1.6	0.5	143	0.0	0
November	8.9	276	30	1.8	1.0	285	0.4	141
December	3.5	453	31	1.9	1.4	460	0.9	299
Annual	12.2	2636	365			2689		1523
Std deviation	8.5							

2.3.2.3 Load Factor Method

The load factor method is a commonly used, simple means of expressing monthly and annual loads, and is defined as the ratio of the average load over a time period of interest to the peak load:

$$LF = \frac{\bar{q}}{\dot{q}} \tag{2.7}$$

where LF is the load factor (—), $\bar{q}$ is the average load over the time period of interest (kW or Btu/h), and $\dot{q}$ is the peak hour load (kW or Btu/h).

Example 2.3 Calculation of Load Factors

The hourly loads from a building energy simulation program are summarized in Table E.2.3.

Determine the following:

(a) the peak hourly loads for both heating and cooling,

(b) the monthly load factors for both heating and cooling, and

(c) the annual load factors for both heating and cooling.

Table E.2.3 Monthly Load Summaries for Example 2.3

A	B h/month	C Total heating (1000 Btu)	D Total cooling (1000 Btu)	E Peak heating (1000 Btu/h)	F Peak cooling (1000 Btu/h)
Jan	744	18 963	0	47	0
Feb	672	15 782	5	46	2
Mar	744	9640	596	36	16
Apr	720	4352	772	19	17
May	744	0	3730	0	21
Jun	720	0	3987	0	23
Jul	744	0	5545	0	24
Aug	744	67	4748	6	21
Sep	720	4453	2211	30	20
Oct	744	4460	411	22	9
Nov	720	9711	89	30	11
Dec	744	15 745	0	47	0

Solution

By inspection of the above table, we see that the peak heating loads occur in January, and the peak cooling loads occur in July.

(a) The peak hourly loads from column E are:

$$\text{For heating: } \dot{q}_{h,htg} = 47\,000\,\text{Btu/h} \quad \text{For cooling: } \dot{q}_{h,clg} = 24\,000\,\text{Btu/h}$$

(b) The monthly load factors for both heating and cooling are given by Equation (2.7):

$$\text{For heating: } LF_{m,htg} = \frac{18\,963\,\text{kBtu/month}}{47\,\text{kBtu/h} \times 744\,\text{h/month}} = 0.54$$

$$\text{For cooling: } LF_{m,clg} = \frac{5545\,\text{kBtu/month}}{24\,\text{kBtu/h} \times 744\,\text{h/month}} = 0.31$$

(c) The annual load factor for both heating and cooling are also given by Equation (2.7):

$$\text{For heating: } LF_{a,clg} = \frac{83\,173\,\text{kBtu/year}}{47\,\text{kBtu/h} \times 8760\,\text{h/year}} = 0.20$$

$$\text{For cooling: } LF_{a,clg} = \frac{22\,094\,\text{kBtu/year}}{24\,\text{kBtu/h} \times 8760\,\text{h/year}} = 0.11$$

Monthly and annual load factors may be estimated from degree-days, but again, that method is suitable where the heating and cooling loads are predominantly driven by outdoor air temperature. For other cases, monthly and annual load factors that have been determined using a building simulation program are summarized in Table 3.1 for residential, office, and school buildings. Load factors are tabulated as a function of US Department of Energy climate zone, and extrapolation to other worldwide climate zones is possible, based on the following zone descriptions:

Hot-Humid (portions of zones 1, 2, and 3 that are in moist category A): a region that receives more than 20 in (50 cm) of annual precipitation and where one or both of the following occur:
- a 67 °F (19.5 °C) or higher wet-bulb temperature for 3000 or more hours during the warmest six consecutive months of the year; or
- a 73 °F (23 °C) or higher wet-bulb temperature for 1500 or more hours during the warmest six consecutive months of the year.

Mixed-Humid (zones 4 and 3 in category A): a region that receives more than 20 in (50 cm) of annual precipitation, has approximately 5400 heating degree-days (65 °F basis) or fewer, and where the average monthly outdoor temperature drops below 45 °F (7 °C) during the winter months.

Hot-Dry (zones 2 and 3 in the dry category): a region that receives less than 20 in (50 cm) of annual precipitation and where the monthly average outdoor temperature remains above 45 °F (7 °C) throughout the year.

Mixed-Dry (zone 4 in dry category B): a region that receives less than 20 in (50 cm) of annual precipitation, has approximately 5400 heating degree-days (50 °F basis) or fewer, and where the average monthly outdoor temperature drops below 45 °F (7 °C) during the winter months.

Cold (zones 5 and 6): a region with between 5400 and 9000 heating degree-days (65 °F basis).

Very Cold (zone 7): a region with between 9000 and 12 600 heating degree-days (65 °F basis).

Subarctic (zone 8): a region with 12 600 heating degree-days (65 °F basis) or more. The only subarctic regions in the United States are found in Alaska, which is not shown on the map.

Marine (zones 3 and 4 located in moisture category C): a region that meets all of the following criteria:
- a coldest month mean temperature between 27 °F (−3 °C) and 65 °F (18 °C);
- a warmest month mean of less than 72 °F (22 °C);
- at least 4 months with mean temperatures higher than 50 °F (10 °C);
- a dry season in summer;
- the month with the heaviest precipitation in the cold season has at least 3 times as much precipitation as the month with the least precipitation in the rest of the year; and
- the cold season is October through March in the Northern Hemisphere and April through September in the Southern Hemisphere.

2.3.2.4 Base Load–Peak Load Designs for Heating

The peak heating load occurs only for a limited time of the year. Thus, for the majority of the time, depending upon climatic conditions, the heating load of the system is only a fraction of the peak heating load. The variation in heating loads over a year is typically described by a *load–duration curve*, which shows the cumulative duration of heating loads over a full year.

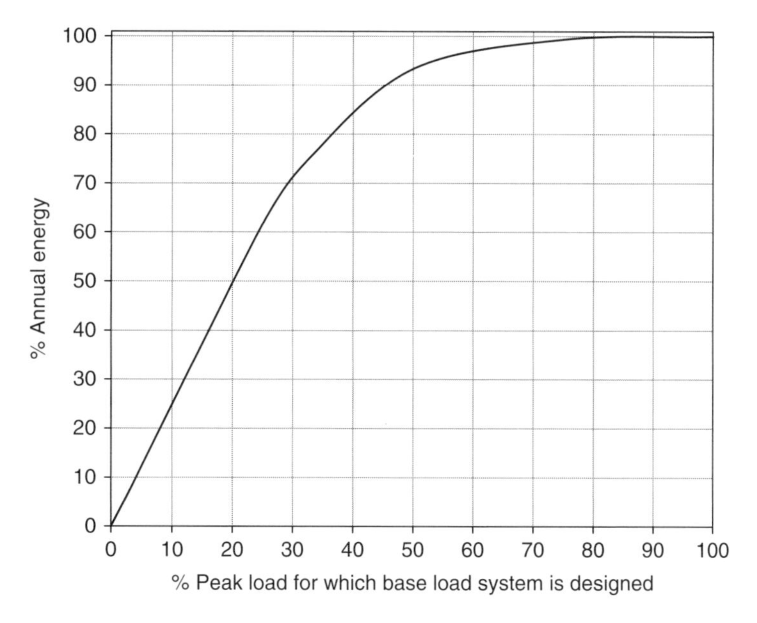

Figure 2.5 Example heating load–duration curve

Figure 2.5 presents a plot of the annual energy requirements that could be met by a base load system designed for various percentages of the peak load. This plot assumes that the base load system continues to operate (at its maximum capacity) in parallel with the peak load system below the balance point. A review of the graph shown in Figure 2.5 reveals that a heating system designed for 50% of peak load would capture over 90% of the annual heating requirements of the structure. This observation has significance with regard to capital-intensive heating systems such as geothermal heat pump systems; a base load peaking arrangement can be designed economically, such that the base load system can meet nearly 100% of the annual loads.

The process of sizing and selecting a heat pump for both heating and cooling duty is different from that of sizing an air-conditioner or furnace. The size of conventional air-conditioners and furnaces is typically selected so that the equipment can deliver the required cooling or heating at the most extreme conditions expected in a given climate. However, heat pumps supply both heating and cooling, and peak heating and cooling loads are rarely identical. Thus, a heat pump could be sized to meet the peak cooling load or the peak heating load, but rarely can it be sized to meet both peak loads. If sizing for the peak cooling load results in insufficient heating capacity, the additional heat is typically supplied by electric resistance heating elements. Electric resistance heating is a factory-supplied option with geothermal heat pumps.

The foregoing monthly and annual load factors shown in Table 2.1 are applicable to situations where the heat pump load is the same as the building load. In cases with supplemental heating, the load factors may be adjusted by multipliers determined by the following empirical equations based on numerous building simulations:

$$\text{Monthly load factor multiplier} = 0.561x^2 + 1.1366x + 0.9954, \textbf{ valid for } \mathbf{0 \le x \le 0.5} \quad (2.8)$$

$$\text{Annual load factor multiplier} = 1.3281x^2 + 0.941x + 1.0007, \textbf{ valid for } \mathbf{0 \le x \le 0.5} \quad (2.9)$$

Table 2.1 Monthly and Annual Load Factors for Various Building Types in Various Climates

Climate zone	Residential				Office (low–mid rise)				Schools			
	Peak monthly load fraction		Annual load fraction		Peak monthly load fraction		Annual load fraction		Peak monthly load fraction		Annual load fraction	
	Htg	Clg	Htg	Clg	Htg	Clg	Htg	Clg	Htg	Clg	Htg	Clg
1	0.04	0.48	0.01	0.29	0.04	0.46	0.01	0.39	0.05	0.52	0.01	0.24
2A	0.18	0.40	0.07	0.16	0.18	0.44	0.05	0.31	0.20	0.33	0.06	0.25
2B	0.26	0.62	0.08	0.24	0.21	0.50	0.05	0.35	0.24	0.53	0.07	0.22
3A	0.43	0.46	0.17	0.13	0.33	0.44	0.09	0.27	0.35	0.31	0.11	0.21
3AWH	0.36	0.38	0.11	0.13	0.21	0.43	0.06	0.30	0.23	0.30	0.08	0.23
3B	0.36	0.36	0.13	0.14	0.26	0.44	0.08	0.32	0.28	0.34	0.10	0.25
3C	0.45	0.00	0.20	0.01	0.27	0.34	0.13	0.26	0.31	0.21	0.16	0.20
4A	0.50	0.36	0.17	0.09	0.42	0.42	0.13	0.25	0.42	0.31	0.15	0.18
4B	0.39	0.38	0.14	0.12	0.38	0.42	0.11	0.28	0.41	0.31	0.13	0.21
4C	0.52	0.06	0.25	0.02	0.46	0.35	0.20	0.19	0.46	0.22	0.22	0.15
5A	0.53	0.27	0.21	0.06	0.42	0.38	0.16	0.22	0.41	0.27	0.17	0.16
5B	0.53	0.34	0.23	0.08	0.37	0.44	0.14	0.25	0.39	0.32	0.16	0.18
6A	0.58	0.14	0.23	0.03	0.48	0.36	0.17	0.21	0.47	0.25	0.18	0.14
6B	0.49	0.21	0.22	0.05	0.41	0.37	0.18	0.21	0.40	0.25	0.20	0.14
7A	0.58	0.12	0.26	0.03	0.50	0.33	0.21	0.19	0.48	0.21	0.22	0.12
7B	0.54	0.14	0.29	0.03	0.37	0.36	0.17	0.25	0.38	0.24	0.19	0.16
8	0.54	0.06	0.30	0.01	0.64	0.34	0.28	0.18	0.60	0.21	0.28	0.10

where x is the percentage of the peak load attributed to supplemental heating. For example, say the heating monthly and annual load factors are 0.5 and 0.2 respectively, with no supplemental heating. If a 50% supplemental heating element is added, then the adjusted monthly and annual load factor multipliers are calculated as 1.70 and 1.24 respectively. Thus, the adjusted monthly and annual load factors with a 50% peak heating element are 0.85 and 0.25 respectively.

2.3.2.5 Greenhouse Heating

Greenhouse heating is a popular use of geothermal energy because greenhouses have a large necessity for heat, and they can be heated with relatively low-temperature water. Greenhouses are quite varied in construction, heating system, and indoor design temperature, which is dictated by the crop produced. Further details of greenhouse heating load calculations and greenhouse construction can be found in Rafferty (1998a). Greenhouse cooling is commonly accomplished with evaporative methods.

Peak heating design loads typically only consider heat transmission plus infiltration through the structure. On a unit footprint area basis, peak greenhouse heating loads may be up to 5 times that of buildings. These relatively high peak heating loads are due to the lightweight construction of a greenhouse, which is designed to maximize sunlight exposure. However, during sunlight hours, solar heat gains often provide adequate space heating, hence the term 'greenhouse effect'; short-wave radiation enters the space through the transparent envelope, but long-wave thermal radiation does not transfer as readily back through the envelope. Typical construction materials for greenhouses include: glass panes, fiberglass panels, rigid polycarbonate panels, or polyethylene plastic sheets.

Greenhouse structures include peaked or arched type. In commercial operations, individual buildings are of the order of 0.5 ha, and are rectangular, typically with a 3:1 aspect ratio. Major commercial operations may consist of dozens of greenhouse buildings. A photo of a large commercial greenhouse is shown in Figure 2.6.

Figure 2.6 Arched-type (Venlo-type) greenhouse constructed of glass panels in southern Ontario, Canada (photo taken by the author)

Figure 2.7 Radiant floor pipes for heating a greenhouse (photo taken by the author)

Greenhouse heating systems are quite varied, but generally consist of radiant and/or forced-air systems. A base load peaking system arrangement is common in cold climates, where a radiant method is employed as the base load system. Radiant heating systems may be installed in the floor, under potting benches, or on the side walls of the greenhouse. Forced-air systems typically consist of flexible ducts installed overhead, or under benches. Figure 2.7 is a photo showing radiant floor pipes heating a greenhouse where hydroponic tomatoes are grown. In this figure, the pipes on the floor used for heating are also used as rails for handcarts that are wheeled down the aisles to harvest the tomatoes.

2.4 Hot Water Process Loads

Hot water process loads refer to energy required to heat water to a storage or delivery temperature for domestic purposes (cooking, washing, bathing) or other commercial or industrial purposes. These loads are quite significant in residential buildings, and in some climates may exceed the energy required for space heating.

Hot water process loads over a given time interval ($\dot{Q}_{HW}$) are calculated by an energy balance on the water:

$$\dot{Q}_{HW} = \dot{V}\rho c_p(T_{hot} - T_{cold}) \tag{2.10}$$

where $\dot{V}$ is the flow rate of water over the time period of load calculation (L/day, L/min, L/s), ρ is the density of water (1 kg/L), c_p is the specific heat of water, T_{cold} is the water temperature entering the water heating device (also referred to as the temperature of the water main), and T_{hot} is the required hot water temperature. The ASHRAE Applications Handbook lists typical use requirements for hot water processes. A commonly assumed value for domestic purposes is up to 60 L/day/person.

Example 2.4 Calculation of Monthly Domestic Hot Water Heating Load
Calculate the hot water heating load for a family of four in the month of January, assuming the average cold water temperature entering the house is 10 °C and heated to 50 °C.

Solution
Assuming the family uses 60 L/day/person, and $c_p = 4.2\,\text{kJ}\cdot\text{kg}^{-1}\ °\text{C}^{-1}$ for water, by Equation (2.10) we have

$$\dot{Q}_{HW} = 60\,\text{L/day/person} \times 4\,\text{persons} \times 31\,\text{days} \times 1\,\text{kg/L} \times 4.2\,\text{kJ}\,\text{kg}^{-1}\,°\text{C}^{-1} \times (50-10)\,°\text{C}$$

$$\dot{Q}_{HW} = 1.25\,\text{GJ} \cong 347\,\text{kWh}$$

For computer simulations, an equation for calculating T_{cold} is handy. If the source of cold water is from a water main buried in soil, an approximation of the water temperature is given by the shallow Earth temperature. Soil temperature as a function of depth and day of the year has been given by Kasuda and Achenbach (1965):

$$T(d,t) = T_M - A_S \cdot e^{\left[-d\cdot\left(\frac{\pi}{365\alpha}\right)^{1/2}\right]} \cdot \cos\left[\frac{2\pi}{365}\left(t - t_o - \frac{d}{2}\cdot\left(\frac{365}{\pi\alpha}\right)^{1/2}\right)\right] \tag{2.11}$$

where $T(d,t)$ is the Earth temperature (in °F or °C) at soil depth (d) after t days from 1 January; T_M is the mean Earth temperature (°F or °C); A_S is the Earth surface temperature amplitude above/below T_M (°F or °C) and is a measured weather parameter, typically 18 °F or 10 °C; t is the number of days after 1 January (days); t_o is the number of days after 1 January where the minimum Earth surface temperature occurs (days), typically 30–35 days; d is the depth (ft or m); α is the soil thermal diffusivity (ft^2/day or $\text{m}^2\cdot\text{s}^{-1}$). Recall from Section 2.2 that soil temperature data are available at: http://eosweb.larc.nasa.gov/sse/RETScreen/. Be aware that that website gives the surface amplitude as the *difference of the maximum and minimum surface temperature*, not the difference about the mean. Therefore, when using the amplitude from this data source, it must be divided by 2 to get the appropriate value of (A_s) for Equation (2.11).

Hourly performance modeling of hot water systems with storage is given by

$$\frac{\mathrm{d}T}{\mathrm{d}t} = \frac{\Sigma\dot{q}}{\rho c_p V} \tag{2.12}$$

where $\mathrm{d}T/\mathrm{d}t$ is the time rate of change of water temperature in the storage tank, ρ is the density of the stored water, c_p is the specific heat capacity of the stored water, V is the volume of stored water, and $\Sigma\dot{q}$ represents the net heat exchanges to/from the tank, which include rate of energy removal due to utilization, heat losses from the tank to the surroundings, rate of heat addition to the tank to heat the water, and recirculation losses (if the hot water tank is part of a building hot water recirculation system). Hourly draw profiles for various uses have been developed by many researchers. For residential applications, the highest hot water demands are in the morning (06:00–09:00) and in the evening (17:00–21:00).

2.5 Swimming Pool and Small Pond Heating Loads

The heating of swimming and recreational pools and small ponds is a commonly considered geothermal application. A photo of a geothermally heated outdoor pool is shown in Figure 2.8, and the heat transfer processes occurring in swimming pools and small outdoor ponds are shown in Figure 2.9. Similar processes also occur in indoor pools. Thermal stratification of the pool water is ignored.

Figure 2.8 Outdoor swimming pool heated year round with direct-use geothermal energy in Southern Oregon, USA. Note this photo was taken when the outdoor air temperature was 0 °C (photo taken by the author)

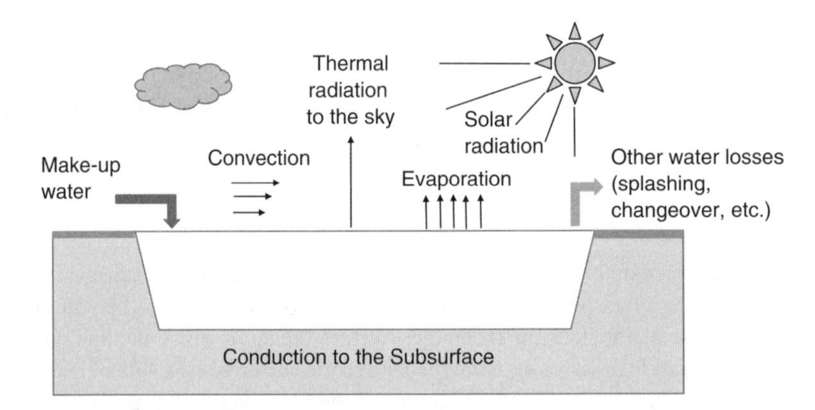

Figure 2.9 Heat transfer processes acting on an outdoor swimming pool or small pond

Recreational pools operate in a narrow temperature range, typically between 25 and 32 ° C. Hot tubs and soaking pools operate at temperatures up to 40 °C. The main heating mechanism of pools, particularly outdoor applications, is by solar radiation. Evaporation and convection are the main cooling mechanisms. Evaporation can be eliminated by blankets or covers when the pool is not in use. The heating load required to maintain a pool at a required temperature is determined by a heat balance on the processes shown in Figure 2.9. Pools have a large thermal mass, and therefore their temperature does not change quickly. Therefore, daily or monthly average energy balances are usually sufficient.

Solar Radiation. The solar radiation incident on the surface of a pool is either reflected or absorbed. The amount of radiation reflected from the surface of the water body varies with the reflectivity coefficient (ρ_w). The reflectivity of the water surface is a function of many factors, but it is largely dependent on solar altitude angle. The approach used to calculate reflectivity averaged over a day is adopted from Hamilton and Schladow (1997):

$$\rho_w = 0.08 + 0.02 \times \sin\left(\frac{2\pi n}{365} \pm \frac{\pi}{2}\right) \tag{2.13}$$

where n is the day of the year ($1 \le n \le 365$), and the term $\pi/2$ is added for the northern hemisphere and subtracted for the southern hemisphere.

The solar heat flux on the pool ($\dot{q}''_{solar}$) in $W \cdot m^{-2}$ is given by

$$\dot{q}''_{solar} = \dot{q}''_{solar,incident}(1 - \rho_w) \tag{2.14}$$

where $\dot{q}''_{solar,incident}$ is the incident solar radiation on the horizontal ($W \cdot m^{-2}$).

Thermal (or Long-Wave) Radiation. The net thermal or long-wave radiation flux (in $W \cdot m^{-2}$) is calculated between the pool surface temperature and the fictitious sky temperature:

$$\dot{q}''_{thermal\ radiation} = \varepsilon\sigma\left(T^4_{sky} - T^4_{pool}\right) \tag{2.15}$$

where ε is the emissivity coefficient of water (~0.95), and σ is the Stefan–Boltzmann constant (5.678×10^{-8} ($W \cdot m^{-2} \cdot K^{-4}$). T_{sky} is the sky temperature determined from the ASHRAE (2011) correlation of T_{sky} (K) as a function of ambient air temperature (T_{air}, in K), and the relative humidity (ϕ , in decimal form):

$$T_{sky} = T_{air} - \left(1105.8 - 7.562T_{air} + 0.01333T^2_{air} - 31.292\varphi + 14.58\varphi^2\right) \tag{2.16}$$

Convection. Convective heat transfer at the pool surface may be driven by wind (forced convection) or by buoyancy forces due to air–surface water temperature differences (free convection). There are a number of empirical convection correlations available, and here we adopt that by Molineaux *et al.* (1994) who correlated the convection coefficient (h_c) with the wind speed ($\dot{V}_{wind}$):

$$h_c = 3.1 + 2.1\dot{V}_{wind} \tag{2.17}$$

where h_c is the convection coefficient in $W \cdot m^{-2} \cdot K^{-1}$, and $\dot{V}_{wind}$ is the wind speed in $m \cdot s^{-1}$. The convective heat flux (in $W \cdot m^{-2}$) is given by

$$\dot{q}''_{convection} = h_c\left(T_{air} - T_{pool}\right) \tag{2.18}$$

where T_{air} is the ambient air temperature.

Evaporation. Heat transfer due to evaporation is the dominant lake cooling mechanism and is approximated with the Chilton–Colburn analogy. The evaporative heat flux is given by

$$\dot{q}''_{evaporation} = h_{fg}\,\dot{m}''_w \tag{2.19}$$

where h_{fg} is the heat of vaporization of water ($J{\cdot}kg^{-1}$), and $\dot{m}''_w$ is the mass flux of evaporating water ($kg{\cdot}s^{-1}{\cdot}m^{-2}$), where

$$\dot{m}''_w = h_d\left(w_{air} - w_{surf}\right) \tag{2.20}$$

where w_{air} and w_{surf} represent the humidity ratio (in kg H_2O/kg dry air) of the ambient air and pool surface, respectively, and h_d is the mass transfer coefficient ($kg{\cdot}s^{-1}{\cdot}m^{-2}$). For moist air, the humidity ratio is a function of the air temperature and relative humidity. w_{surf} represents the humidity ratio of saturated air at the pool surface and is calculated by

$$w_{surf} = 0.62198\left(\frac{p_{ws}}{p - p_{ws}}\right) \tag{2.21}$$

where p_{ws} is the saturation pressure of water (in bars) evaluated at the pond–air film temperature, and p is the atmospheric pressure (in bars).

The mass transfer coefficient (h_d) is analogous to and related to the convection heat transfer coefficient (h_c) using the so-called Chilton–Colburn analogy:

$$h_d = \frac{h_c}{c_p\,Le^{2/3}} \tag{2.22}$$

where c_p is the specific heat ($J{\cdot}kg^{-1}{\cdot}K^{-1}$) of air evaluated at the pool–air film temperature, and *Le* is the dimensionless Lewis number given by

$$Le = \frac{\alpha}{D} \tag{2.23}$$

where α is the thermal diffusivity ($m^2 \cdot s^{-1}$) of air evaluated at the pool-air film temperature, and D is a binary diffusion coefficient ($m^2 \cdot s^{-1}$) given by

$$D = \frac{1.87 \times 10^{-10}\, T^{2.072}}{p} \tag{2.24}$$

where T refers to the pool–air film temperature (in K) and p is atmospheric pressure (in bars).

Conduction to the Ground. Heat transfer from pools and ponds to the ground is generally a weak process, and is often ignored. Hull *et al.* (1984) express ground heat losses from any pond as a function of the pond area, pond perimeter, the ground thermal conductivity (k_{ground}), and the distance from the pond bottom to a constant temperature sink. For practical purposes, Kishore and Joshi (1984) define a ground heat transfer coefficient (U_{ground}) as approximately equal to $1.5\ W \cdot m^{-2} \cdot K^{-1}$. For a rectangular pool or pond with vertical side walls, the conductive heat transfer to the ground ($\dot{q}''_{ground}$) in $W \cdot m^{-2}$ is given by

$$\dot{q}''_{ground} = U_{ground}\left(T_{ground} - T_{pool}\right) \tag{2.25}$$

Make-Up Water. Fresh water is added to a pool to compensate for evaporative losses, water lost because of swimmers' activity, and voluntary or mandatory water changes. The heat load imposed on the pool by make-up water ($\dot{q}_{make-up}$) in W is given by an energy balance on the water:

$$\dot{q}_{make-up} = \dot{V}\rho c_p\left(T_{make\text{-}up} - T_{pool}\right) \tag{2.26}$$

Where $\dot{V}$ is the flow rate of water (L/s), ρ is the density of water (1 kg/L), c_p is the specific heat of water, and $T_{make\text{-}up}$ is the make-up water temperature entering the pool.

Energy Balance. The net load resulting from an energy balance on the loads described above is the pool heating load:

$$\dot{q}_{net} = A\left(\dot{q}''_{solar} + \dot{q}''_{thermal\ radiation} + \dot{q}''_{convection} + \dot{q}''_{evaporation} + \dot{q}''_{ground}\right) + \dot{q}_{make-up} \tag{2.27}$$

where A is the pool surface area (m^2).

Example 2.5 Calculation of the Heating Load of an Uncovered Outdoor Pool
Using the monthly weather summaries shown in Figure 2.1, estimate the monthly energy required to maintain an uncovered $1250\ m^2$ pool at a temperature of 27 °C. The average water make-rate is $0.1\ kg \cdot s^{-1}$ at an average temperature of 20 °C.

Solution

Given:

From Figure 2.1, we have the following average weather conditions for July:

$$\dot{q}''_{solar,incident} = 5.82\,\text{kWh}\cdot\text{m}^{-2}\cdot\text{day}^{-1}$$

$$\dot{V}_{wind} = 4.0\,\text{m}\cdot\text{s}^{-1}$$

$$\phi = 0.675$$

$$T_a = 23.6\,°\text{C}$$

$$\text{T}_{ground} = 22.9\,°C$$

$$\text{P} = 100.8\,\text{kPa} = 1.0\,\text{bar}$$

Assume 15 July 15 as the day of the year for the solar reflectivity calculation.

Analysis:

As we need thermodynamic and psychrometric properties, we shall solve this problem using Engineering Equation Solver (EES) software. The input data are as follows:

```
"Design Parameters:"
T_pool = 27 [C]                               "the design temperature of the pool"
Solar_Radiation = 5.82 [kWh/m^2-day]          "the average daily solar radiation on horizontal"
Day = 196                                     "the day of the year"
Area = 1250 [m^2]                             "the surface area of the pool"
T_air = 23.6[C]                               "the ambient air temperature"
T_ground = 22.9 [C]                           "the average ground temperature"
P = 100.8 [kPa]                               "the atmospheric pressure"
RH = 0.675                                    "the relative humidity"
WindSpeed = 4.0 [m/s]                         "the wind speed"
T_makeup = 20 [C]                             "the make-up water temperature"
m_dot_makeup = 0.1 [kg/s]                     "the make-up water mass flow rate"

"CALCULATIONS:
T_film = (T_air + T_pool)/2 "Typical to evaluate thermo properties at this temperature."

"Calculate the Solar Radiation Heat Transfer Rate"
rho_surface = 0.08 + 0.02 * sin(2*pi*day/     "the reflectivity; add (subtract) pi/2 for northern
365[day] + pi/2)                              (southern) hemisphere"
q_solar = (1 - rho_surface) * Solar_Radiation * 1000 [W/kW] / 24 [h/day] * Area
q_solar_flux = q_solar / Area

"Calculate the Convection Heat Transfer Rate"
h = 3.1 + 2.1 * WindSpeed                     "correlation for the convection coefficient"
q_convection = h * Area * (T_air - T_pool)
q_convection_flux = q_convection/Area

"Calculate the Evaporation Heat Transfer Rate"
k_air = Conductivity(Air,T = T_film); rho_air = Density(Air,T = T_film,P = P); cp_air = Cp(Air,-
T = T_film)
alpha_air = k_air / (rho_air * cp_air) "thermal diffusivity"
```

```
D_AB = 0.000000000187 * ConvertTemp('C', 'K', T_film)      "binary diffusion coefficient
^ 2.072 / (P * convert(kPa, bar))                           of A - > B"

Le = alpha_air / D_AB                  "the Lewis number"
hd = h / (cp_air* Le ^ (2 / 3))
MassFlux = hd * (HumRat(AirH2O,T = T_air,r = rh,P = P) - HumRat(AirH2O,T = T_film,r = 1,P
= P))
q_evap = Area * MassFlux * Enthalpy_vaporization(Water, T = T_film)
q_evap_flux = q_evap / Area

"Calculate the Radiation Heat Transfer Rate"
T_sky = ConvertTemp('C', 'K', T_air)-(1105.8 - 7.562*ConvertTemp('C', 'K', T_air) +
0.01333*ConvertTemp('C', 'K', T_air)^2 - 31.292*RH + 14.58*RH^2)
q_rad = 0.95 * sigma# * Area * (T_sky^4 - ConvertTemp('C', 'K', T_pool)^4)
q_rad_flux = q_rad / Area

"Calculate the Heat Transfer Rate to the Ground"
U_ground = 1.5 [W/m^2-C] "U-value for ground conduction based on the work of Kishore and
Joshi, 1984"
q_ground = U_ground * Area * (T_ground - T_pool)
q_ground_flux = q_ground / Area

"Make-up Water Energy Balance"
q_makeup = m_dot_makeup * Cp(water,T = T_makeup, P = P) * (T_makeup - T_pool)

"The Total Heat Loss"
q_total = q_solar + q_convection + q_evap + q_rad + q_ground + q_makeup
q_total_flux = q_total / Area
Q = -q_total * 31[day] * convert('day','s') * convert('J','GJ') "the total energy per month"
```

The following results are obtained:

The average heat fluxes are:	
Solar:	227.8 [W/m^2]
Thermal (long-wave) radiation:	−126.4 [W/m^2]
Convection:	−39.1 [W/m^2]
Evaporation:	−250.9 [W/m^2]
Ground:	−6.15 [W/m^2]
The make-up water heating load is:	−2928 [W]
The average heating requirement is:	−246427 [W]
The total July monthly heating requirement is:	660 [GJ]

Discussion: An inspection of the relative loads shows that evaporation is by far the greatest source of heat loss at approximately double the thermal radiation heat loss, the next highest heat loss. This demonstrates the benefit of using a pool cover.

Example 2.6 Estimation of the Heating Load of a Covered Outdoor Pool

Estimate the reduction in the monthly heating load of the pool in Example 2.5 if the pool is covered for 12 h each night.

Solution

Here, we may assume that evaporation losses are eliminated for half the month. Thus, the average evaporation heat flux decreases to $-125.45\ \mathrm{W \cdot m^{-2}}$, reducing the total heating requirement for July to 240 GJ, or about 63% relative to the uncovered case. Convection and thermal radiation losses would also be slightly reduced.

2.6 Snow-Melting Loads

The heating of and melting of snow on travelled surfaces has many applications related to: sidewalks, stairways, access ramps, roadways, helicopter pads, runways, and turf/sod. There are numerous practical advantages to these types of system, such as safety, elimination of de-icing chemicals, and labor-cost savings. Turf-heating systems installed under sports fields reduce player injuries and extend the life of the turf.

Snow-melting is a commonly considered geothermal application, particularly with direct-use geothermal. In geothermal applications, pavement heating and snow-melting systems are *hydronic systems*, where a heated fluid is circulated through a network of pipes embedded in a concrete slab, sand bedding, or turf layer. A photo of an under-construction, geothermally heated driveway approach is shown in Figure 2.10. The heat transfer processes occurring in these heated slabs are shown in Figure 2.11. Similar processes also occur in heated turf systems. Note that other types of engineered snow-melting system exist, such as overhead radiant methods and embedded electric resistance methods.

In this subsection, we will focus on hydronic systems. The heating load required to melt snow is determined by a heat balance on the processes shown in Figure 2.11. Ignoring solar

Figure 2.10 Geothermal snow-melting system under construction in a driveway (photo taken by author)

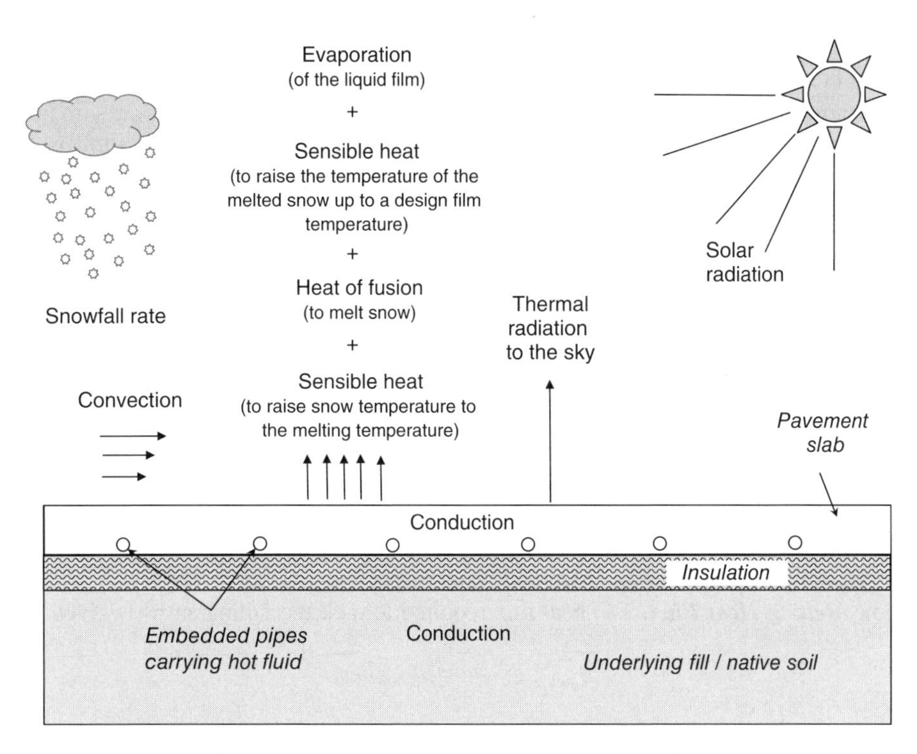

Figure 2.11 Heat transfer processes acting on a hydronically heated slab

radiation, ASHRAE defines the net heat flux required to melt snow at a given snowfall rate and weather condition as

$$\dot{q}''_{net} = \dot{q}''_{sensible} + \dot{q}''_{melting} + A_{ratio}\left(\dot{q}''_{convection} + \dot{q}''_{evaporation} + \dot{q}''_{thermal\ radiation}\right) \tag{2.28}$$

where $\dot{q}''_{sensible}$ is the heat flux required to warm snow up to the melting temperature plus that required to raise the temperature of the melted snow up to a design liquid film temperature, $\dot{q}''_{melting}$ is the latent heat flux of fusion required to melt the snow, $\dot{q}''_{thermal}$ radiation is the long-wave radiation heat flux to the sky, $\dot{q}''_{convection}$ is the convective heat flux, and $\dot{q}''_{evaporation}$ is the evaporative heat flux. A_{ratio} is the area ratio, which is a fraction (0–1) of the total area desired to be snow free. If A_{ratio} is equal to 1, then the design has zero tolerance for snow accumulation (e.g., runways, hospitals, etc.). If A_{ratio} is zero, then the design allows snow accumulation on the entire surface, but the system will catch up later, such as may be the case for a residential driveway. In this case, snow acts as an insulator, so convection, evaporation, and thermal radiation are taken as zero.

The base and edges of the slab should be insulated to minimize ground conduction losses. Spitler *et al.* (2002a and 2002b) conducted numerous, detailed transient analyses of snow-melting systems, and concluded that heat fluxes can be 10–30% greater in uninsulated snow-melting systems relative to those with insulation.

Snow-melting loads are unusual, in the sense that there is not a unique set of circumstances resulting in a peak heating load. A number of factors may coincide to give similar peak loads, such as snowfall rate, air temperature, and wind speed. The greatest snowfall rates do not typically occur during the coldest periods at a particular location. Further, warmer snow is denser than colder snow, thus typically resulting in higher snow-melting loads coinciding with warmer air temperatures.

Each of these heat transfer mechanisms is discussed in what follows, assuming that the design surface temperature of the slab ($T_{surface}$) is 1 °C. The liquid film temperature is assumed to be 0.5 °C.

Sensible Heat Flux. Sensible heat is required to raise the temperature of snow falling on the slab to the melting temperature plus to raise the temperature of the liquid, after the snow has melted, to the design temperature of the liquid film (T_{film}). The snow is assumed to fall at air temperature (T_{air}). The sensible heat flux is given by

$$\dot{q}''_{sensible} = (sn\dot{o}w)\rho_{snow}\left(c_{p,\,ice}\left(T_{air} - T_{surface}\right) + c_{p,\,film}\left(T_{film} - T_{surface}\right)\right) \tag{2.29}$$

where $sn\dot{o}w$ is the snowfall rate ($m \cdot s^{-1}$), ρ_{snow} is the density of the snow ($kg \cdot m^{-3}$), and $c_{p,ice}$ and $c_{p,film}$ are the specific heats of ice and water respectively ($J \cdot kg^{-1} \cdot {}^{\circ}C^{-1}$).

Snow Melting Heat Flux. The heat flux required to melt the falling snow is given by

$$\dot{q}''_{melting} = (sn\dot{o}w)\rho_{snow}h_{f,ice} \tag{2.30}$$

where $h_{f,ice}$ is the enthalpy of fusion of ice.

Convective Heat Flux. The convective heat flux represents heat loss from the slab owing to wind:

$$\dot{q}''_{convection} = h_c\left(T_{air} - T_{surface}\right) \tag{2.31}$$

The convection coefficient (h_c) is determined from Nusselt number (Nu) correlations for flow over a flat plate (Bergman *et al.*, 2011):

$$Nu = Re^{\frac{4}{5}}Pr^{\frac{1}{3}} \tag{2.32}$$

where Re and Pr are the Reynolds and Prandtl numbers, respectively, evaluated at the average air-surface temperature:

$$Re = \frac{\rho\dot{V}_{wind}L}{\mu} = \frac{\dot{V}_{wind}L}{\upsilon} \tag{2.33}$$

and

$$Pr = \frac{c_p\mu}{k} \tag{2.34}$$

where ρ, μ, υ, k, and c_p are the density, dynamic viscosity, kinematic viscosity, thermal conductivity, and specific heat of air evaluated at the average air-surface temperature, $\dot{V}_{wind}$ is the wind speed ($m \cdot s^{-1}$), and L is the length of the slab (m).

Finally, the convection coefficient (h_c) is determined from

$$h_c = \frac{Nu \cdot k}{L} \tag{2.35}$$

Evaporative Heat Flux. The evaporative heat flux represents heat loss from the slab owing to evaporation of the water film. These calculations are the same as those described in Section 2.5, Equations (2.19) to (2.24).

Thermal or Long-Wave Radiation Heat Flux. Under snowfall conditions, the sky is assumed to be completely or nearly completely cloud covered, such that the surrounding temperature may be taken as T_{air}. Thus, the net thermal or long-wave radiation flux (in $W \cdot m^{-2}$) is calculated between the slab surface temperature and the air temperature:

$$\dot{q}''_{thermal\ radiation} = \varepsilon\sigma\left(T^4_{air} - T^4_{surface}\right) \tag{2.36}$$

where ε is the emissivity coefficient of the slab surface (~0.9), and σ is the Stefan–Boltzmann constant (5.678×10^{-8} ($W \cdot m^{-2} \cdot K^{-4}$).

Pavement Installations and Piping Considerations. Piping materials are either metal or plastic. Steel, iron, and copper pipes have been used extensively in the past, and are still used in some cases, but steel and iron corrode rapidly if they are not protected by coatings and/or cathodic protection. The use of salts and other chemicals for pavement de-icing accelerates the corrosion of these materials.

Present practice in most jurisdictions is to use plastic pipe, typically crosslinked polyethylene (PEX), which is a durable plastic capable of conveying fluids near the boiling point of water at reasonable pressures. This type of pipe is lightweight, easy to handle, flexible, and non-corrodible. Generally, an antifreeze solution (ethylene or propylene glycol) is used in the pipes, circulated in a closed system.

According to Lund (2000), Portland cement concrete (PCC) or asphalt concrete (AC) may be used for snow-melting systems. The thermal conductivity of AC is less than that of PCC, thus pipe spacing and temperatures are different. However, the main reason for not using AC pavements with pipes embedded in them is that the hot asphalt may damage the pipes, as AC is usually placed at above 150 °C in order to get adequate compaction. Also, the compaction process may deform and even break pipes and their connections. With PCC pavements, the pipes can be attached to the reinforcing/expansion steel within the pavement or wire mesh, but should have at least 50 mm or 2 in of concrete above and below the pipes. This then requires a pavement of at least 120 mm or 5 in thickness. In the case of sidewalks, the piping is usually placed below the slab in a base or sub-base, as these pavements are usually only 100 mm (or 3–4 in) thick. In this latter case, the advantage of not placing the pipes in the concrete is that future utility cuts or repairs can be made without damaging the pipes.

In the above equations, heat losses from the slab are negative. Thus, the average fluid temperature required to maintain a known surface heat flux is given by

$$\dot{q}''_{net} = \frac{1}{R''}\left(T_{film} - T_{avg,fluid}\right) \tag{2.37}$$

where $\dot{q}''_{net}$ is the net snow-melting load determined by Equation (2.28), R'' is the thermal resistance per unit area (of the pavement surface) between the heat transfer fluid and the water film on the pavement surface, $T_{avg,fluid}$ is the average temperature of the heat transfer fluid circulating in the pipes, and T_{film} is the temperature of water film on the pavement surface. For a typical design of 25 mm diameter pipe on 300 mm centers, 50 mm below the concrete surface, Chapman (1952) derived a value of R'' of $0.089\ \mathrm{m^2 \cdot K \cdot W^{-1}}$.

The average fluid temperature ($T_{avg,fluid}$) is defined simply as $(T_{in} + T_{out})/2$, where T_{in} and T_{out} are the inlet and outlet fluid temperatures, respectively, entering and leaving the pavement piping network. With design specification of the allowable temperature change between T_{in} and T_{out} (5 °C is a typical value through hydronically heated slabs), the required mass flow rate ($\dot{m}$) is given by an energy balance on the fluid:

$$\dot{q}_{net} = \dot{m}c_p\left(T_{out} - T_{in}\right) \tag{2.38}$$

where c_p is the specific heat of the antifreeze solution.

Example 2.7 Calculation of Snow-Melting Loads, Required Fluid Temperature, and Required Flow Rate

A snow-melting system is being planned for a 75 m^2 driveway. A snow-free area ratio of 0.5 has been deemed acceptable for this location. The design weather conditions are as follows:

- $T_{air} = -10$ °C
- Relative humidity = 0.90
- Wind speed = $5\ \mathrm{m \cdot s^{-1}}$
- Snowfall rate = 2.5 cm/h
- Pressure = 99 kPa

Determine the steady-state heating requirement, the supply fluid temperature, and the fluid mass flow rate. Use a propylene glycol antifreeze solution, and determine an acceptable propylene glycol concentration to prevent freezing of the system in case of a power outage.

Solution

As we need thermodynamic and psychrometric properties, we shall again solve this problem using Engineering Equation Solver (EES) software. We shall also make use of EES thermophysical property functions for propylene glycol, found in the EES Brines Library, and the thermophysical property functions for snow and ice. The input data are as follows:

```
"DESIGN PARAMETERS:"
"Weather Parameters:"
T_air = -10 [C]                          "outdoor air temperature"
P = 99 [kPa]                             "atmospheric pressure"
RH = 0.90                                "relative humidity"
WindSpeed = 5 [m/s]                      "the wind speed"
SnowFallRate = 2.5 [cm/h]                "the rate of snowfall in cm/h"
"Slab and Brine Design Parameters"
Area = 75 [m^2]                          "area to be snow-melted"
Area_Ratio = 0.5 [-]                     "the snow-free area ratio"
Brine$ = 'PG'                            "string variable signifying Propylene
                                         Glycol"
Conc = 30[%]                             "the brine concentration (mass basis)"
DT = 5 [C]                               "the allowable temperature change in
                                         the brine fluid"

"CALCULATIONS:"
T_surf = 1.0 [C] "The design slab surface temperature"
T_film = 0.5 [C] "Design for a liquid film temperature just above freezing"
T_film_air = (T_film + T_air) / 2   "This is the film temperature for calculating thermodynamic
                                    properties of air for evaporation calcs"

"Calculate the Sensible Heat Transfer Rate to Bring the Snow up to the Melting Temperature"
rho_snow = Density(Snow_rho_166, T = T_air)          "density of snow"
cp_ice = cp(Ice, T = T_air, P = P); cp_film = cp(Water,     "specific heat of ice"
T = T_film, P = P)
q_sens = SnowFallRate * convert(cm/h, m/s) * Area * rho_snow * (cp_ice * (T_air -T_surf )
        + cp_film * (T_film - T_surf ))
q_sens_flux = q_sens/Area

"Calculate the Heat Transfer Rate to Melting the Snow"
q_melt = -SnowFallRate * convert(cm/h, m/s) * Area * rho_snow * Enthalpy_fusion(Ice)
q_melt_flux = q_melt / Area

"Calculate the Convection Heat Transfer Rate"
Call External_Flow_Plate('Air', T_air, T_surf, P,        "function for the convection
WindSpeed, Area^0.5: tau, h, C_f, Nusselt, Re)           coefficient"
q_convection = h * Area * (T_air - T_surf ) * Area_Ratio
q_convection_flux = q_convection/Area

"Calculate the Evaporation Heat Transfer Rate"
k_air = Conductivity(Air,T = T_film_air); rho_air = Density(Air,T = T_film_air,P = P); cp_air =
Cp(Air,T = T_film_air)
alpha_air = k_air / (rho_air * cp_air)                   "thermal diffusivity"
D_AB = 0.000000000187 * ConvertTemp('C', 'K',            "binary diffusion coefficient of
T_film_air) ^ 2.072 / (P * convert(kPa, bar))            A->B"
Le = alpha_air / D_AB
hd = h / (cp_air * Le ^ (2 / 3))
```

```
MassFlux = hd * (HumRat(AirH2O,T = T_air,r = rh,P = P) - HumRat(AirH2O,T = T_film,r =
1,P = P))
q_evap = Area * MassFlux * Enthalpy_vaporization(Water, T = T_film_air) * Area_Ratio
q_evap_flux = q_evap / Area

"Calculate the Radiation Heat Transfer Rate"
q_rad = 0.9 * sigma# * Area * (ConvertTemp('C', 'K',       "assuming full cloud cover"
T_air)^4 - ConvertTemp('C', 'K', T_surf )^4) * Area_-
Ratio
q_rad_flux = q_rad / Area

"The Total Heat Rate Needed"
q_total = q_sens + q_melt + q_convection + q_evap + q_rad
q_total_flux = q_total / Area

"Fluid Energy Balance"
q_total_flux = 1/0.089 [m^2-K/W]* (T_film - T_avg_-      "0.089 is a calculated thermal resist-
fluid)                                                    ance from Chapman (1952)"
"The above resistance equation applies to a typical design: 25 mm diameter pipe on 300 mm
centers, 50 mm below concrete surface"
T_avg_fluid = (T_fluid_in + T_fluid_out)/2
q_total = mdot_fluid * Cp(Brine$,T = T_avg_fluid,C =     "fluid energy balance for 20% pro-
Conc) * (T_fluid_out - T_fluid_in)                        pylene glycol"
T_fluid_in- T_fluid_out = DT                              "the allowable fluid temperature
                                                          drop"
T_freeze = FreezingPt(Brine$,C = Conc)                    "fluid freeze point"
```

The following results are obtained:

The total heating load:	–41359 [W]
The heat fluxes are:	
Sensible heat flux:	-28.51 [W/m^2]
Melting heat flux:	-384.6 [W/m^2]
Convective heat flux:	-68.42 [W/m^2]
Thermal radiation heat flux:	-21.78 [W/m^2]
Evaporative heat flux:	-48.18 [W/m^2]
The brine fluid details are:	
Freeze point:	-12.79 [C]
Supply temperature:	52.08 [C]
Mass flow rate:	2.102 [kg/s]

Discussion: A propylene glycol concentration of 30% by mass has a freeze point of –12.79 °C, which is sufficient for the conditions given. The design scenario results in a total heating load of 41.4 kW, or 551 $W \cdot m^{-2}$, which is a relatively large load. An inspection of the heat fluxes shows that about 70% of this load is attributed to the enthalpy of fusion to melt the snow. Thus, it is very important to consider the snow-free area ratio in the design.

Transient Considerations. The thermal mass of pavement slabs is large, and systems must be started several hours before a storm in order for the pavement to be warm enough to melt snow at the onset of a storm. Otherwise, a storm could come and go without any snow being melted. Thus, a significant energy load on snow-melting systems is during the so-called idling time, where the system is used to keep the slab surface temperature at or close to 0 °C.

Example 2.8 Calculation of Idling Energy for a Snow-Melting System, Required Fluid Temperature, and Required Flow Rate
For a 75 m^2 driveway, calculate the energy required to keep the pavement surface temperature at 1 °C over a 7 day period for the following average weather conditions:

- $T_{air} = -2$ °C
- Relative humidity = 0.90
- Wind speed = 2.5 $m \cdot s^{-1}$
- Pressure = 100 kPa

Determine the energy required over the 7 day period, the steady-state average heating load, the supply fluid temperature, and the fluid mass flow rate. Use a 30% (by mass) propylene glycol antifreeze solution.

Solution
The EES file used in the previous example can be adjusted to solve this problem. For a conservative estimate, we will ignore any benefits of solar radiation, and assume thermal radiation heat transfer to the sky. The only heat fluxes to consider are those due to convection and thermal radiation.

The following results are obtained:

The 7 day energy requirement:	4.57 GJ
The average heating load:	−7559 [W]
The heat fluxes are:	
Convective heat flux:	−18.62 [W/m^2]
Thermal radiation heat flux:	−82.18 [W/m^2]
The brine fluid details are:	
Supply temperature:	12.47 [C]
Mass flow rate:	0.4 [kg/s]

Discussion: The average heating load is approximately one-fifth that of the peak design heating load when snow is falling. The mass flow rate is significantly lower than during the peak time, and the supply temperature is only 12.47 °C. The energy required under the given scenario is 4.57 GJ. However, this is likely a conservatively high estimate, as solar fluxes were not considered.

2.7 Chapter Summary

This chapter discussed several loads that are commonly applied to geothermal energy systems: space heating and cooling, hot water heating, swimming pool and pond heating, and snow-melting. The time-dependent nature of these loads is important in determining the geothermal reservoir size, and its ability to meet the intended loads.

The role of weather in driving thermal loads was discussed. Three types of weather data are commonly used, depending on the task. Design weather data are used for peak load calculations. Typical hourly or monthly average weather data are used to estimate time-dependent loads. Actual weather data are used for model calibration or for determining building envelope parameters in conjunction with utility bill data.

Limited theory behind calculation of peak heating and cooling loads of buildings was presented, along with some sources for further details. Three methods were described for estimating the time-dependent energy loads of buildings: building simulation software programs, the degree-day procedure, and load factors. Benefits of base load–peak load arrangements for heating systems were discussed. A unique case of space heating – heating of greenhouses – was also briefly described.

Hot water loads were described, along with their method of calculation. Heat transfer processes and their calculation were addressed in some detail for determination of swimming pool/small pond heating and snow-melting loads. Useful EES models were provided for calculating swimming pool/small pond heating and snow-melting loads.

Discussion Questions and Exercise Problems

2.1 For the location where you live, calculate the monthly heating degree-days for:

   (a) the usual base temperature of 18.3 °C, and
   (b) a base temperature of 13 °C.

2.2 Using monthly average weather data from https://eosweb.larc.nasa.gov/sse/RETScreen/, use the EES file from Example 2.5 to calculate the monthly heating load of an uncovered, outdoor pool for where you live, for July if you live in the Northern Hemisphere or for January if you live in the Southern Hemisphere.

2.3 Redo Example 2.7 with a snow-free area ratio of 1 and 0. Comment on the results.

2.4 Redo Example 2.8 with consideration of solar energy on the horizontal available at 1.5 kWh/m^2/day.

3

Characterizing the Resource

3.1 Overview

This chapter describes the geothermal resource: the Earth. Proper resource characterization is perhaps the most important part of a geothermal energy project because all other aspects of the project stem from the resource availability and sustainability. Readers will learn where geothermal resources of varying temperature can be found in nature, and why they occur. The concept of reservoir engineering, normally used in the petroleum and high-temperature geothermal industry, is discussed in the context of all types of geothermal application. Resource characterization for geothermal heat pump systems is the main focus.

Learning objectives and goals:

1. Appreciate the effect of the thermal energy generation within the Earth on plate tectonics, the rock cycle, the water cycle, and on the Earth as we know it today.
2. Apply the concept of reservoir engineering to all geothermal energy projects.
3. Understand the thermal regime of the Earth, and the role of the atmosphere and deeper thermal energy generation.
4. Understand the nature and occurrence of groundwater resources, aquifer systems, and hydrothermal resources, and how geoscientists explore for these resources.
5. Obtain a general understanding of the various types of drilling method, and when they should or should not be used.
6. Become familiar with site factors that affect the best choice of Earth coupling for geoexchange projects, as well as the evaluation steps in pre-feasibility and detailed feasibility analyses.

Geothermal Heat Pump and Heat Engine Systems: Theory and Practice, First Edition. Andrew D. Chiasson.

Companion website: www.wiley.com/go/chiasson/geoHPSTP

3.2 Origin and Structure of the Earth

The Earth is thought to have formed about 4.6 billion years ago from accretion of cosmic material that also formed the solar system. During the Earth's early development, gravity differentiation of materials occurred, with denser material concentrating toward the center of the Earth, and lighter materials tending toward the surface. Over time, the planet began to cool, forming a solid crust, thus allowing liquid water and a layer of gases (the atmosphere) to exist on the surface. Remnants of the primordial continents are still visible today (i.e., the Canadian Shield, the Australian Shield, and parts of Greenland) where rocks are age-dated using radioisotope methods to be of the order of 2–3 billion years old.

Much of our knowledge of the Earth's history is restricted to changes in its surface – changes that are the result of internal and external forces. The results we observe are due to the interaction of surface processes, with uplift, subsidence, and metamorphism controlled by forces arising deep within the Earth's interior. Earth Scientists can directly investigate surficial processes such as weathering, erosion, and sedimentation, but processes occurring in the interior must be inferred from studies of seismic waves, magnetic fields, variations in gravity, and the products of volcanoes.

Although some of the physical properties of the Earth's interior are established, Earth Scientists generally do not fully agree on the processes at work beneath the crust; reconstruction of the nature and rates of subcrustal processes in the geological past is speculative. Earth Scientists do agree, however, that the Earth beneath the crust is not static; it generates heat, ejects material through volcanoes, moves in earthquakes, controls the movement of continents, controls the birth and destruction of continents and oceans, and generates a magnetic field that surrounds the planet. Since the 1960s, studies of the ocean floors and advancements in the field of paleomagnetism have entrenched the theory of *plate tectonics*, and greatly contributed to the development of a synthetic theory of how the Earth works.

A simplified cross-section of the Earth, showing its interior zones, is provided in Figure 3.1. Figure 3.2 is a diagrammatic cross-section of the upper zones of the Earth. These layers are typically depicted as concentric spheres, but in reality, the interfaces are likely irregular and gradational.

The **core** of the Earth is about 7000 km in diameter. Earthquake waves travelling through the Earth reveal a solid inner core about 2500 km in diameter, and a liquid outer core. The entire core is believed to be comprised of an alloy of nickel and iron at a temperature of about 4000 °C. As the Earth rotates, the motion of the liquid outer core is thought to be responsible for the magnetic field of the Earth.

The **mantle** forms a 3000 km thick shell that surrounds the core and constitutes about 85% of the Earth's volume. The mantle consists of denser minerals than those found in the crust, and is comprised of two distinct regions called the *upper mantle* and *lower mantle*. Two main zones are distinguished in the *upper mantle*: the inner asthenosphere composed of plastic flowing rock, and the lowermost part of the lithosphere composed of rigid rock. The *lower mantle*, inferred from transmission of earthquake waves and by vertical pressure and temperature gradients, exhibits properties unlike any surface rocks. Here, the main constituents are probably high-density silicates and oxides of magnesium and iron.

The **crust** of the Earth is an irregular layer 12–75 m thick, completely enclosing the mantle. The crust is thickest under large mountain chains like the Himalayas, and thinnest under ocean basins. Thus, crustal thicknesses of over 50 km are quite rare (<10% of Earth); a more typical average continental crust thickness is of the order of 30 km. The lower boundary of the crust is

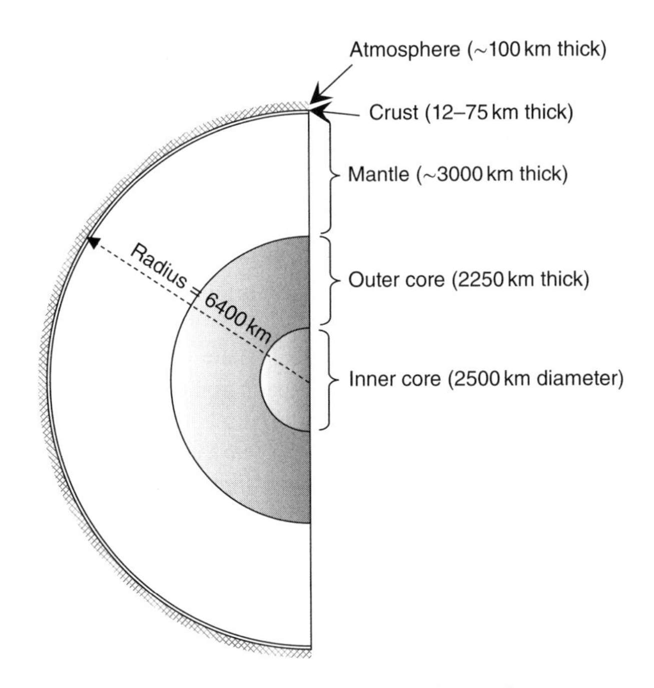

Figure 3.1 Cross-section of the Earth

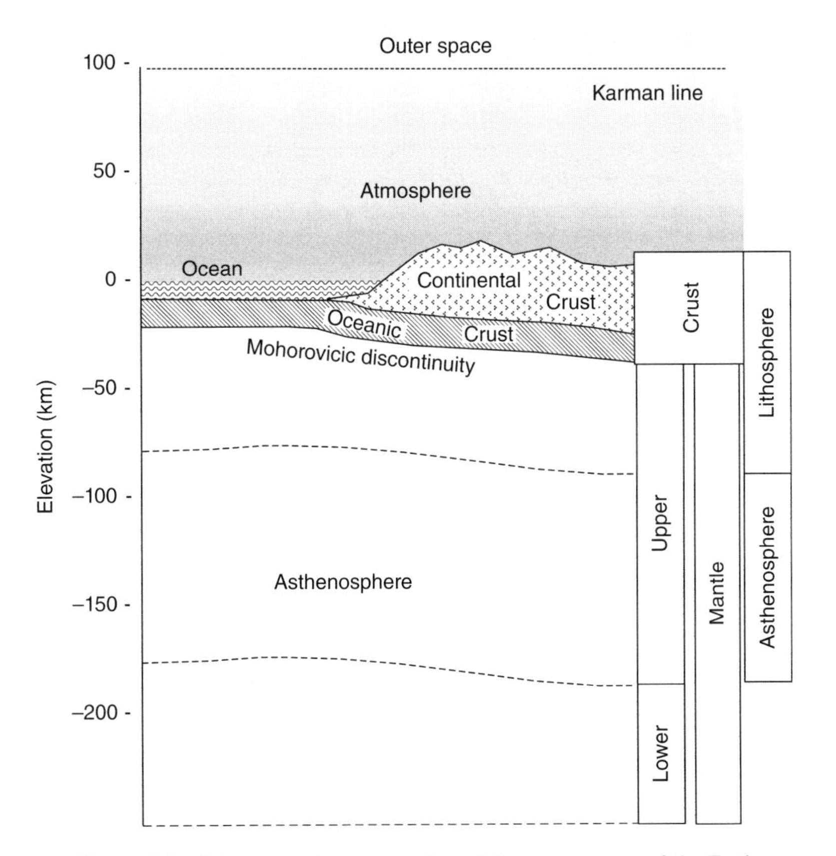

Figure 3.2 Diagrammatic cross-section of the upper zones of the Earth

marked by a change in velocity of earthquake waves and is called the Mohorovicic discontinuity (or Moho) after the seismologist who first recognized its presence. Owing to density stratification, the crust is composed of a lower part, consisting mainly of rocks rich in magnesium silicate minerals (i.e., basaltic rocks), and an upper part, consisting mainly of rocks rich in aluminum silicate minerals (i.e., granitic rocks). These are more commonly referred to as oceanic crust and continental crust, respectively, and it is this presence of denser crust that gives rise to the occurrence of Earth's oceans.

Models of the Earth's crust have been proposed, where continental and oceanic crust are like blocks with slightly different specific gravity floating on a liquid of higher specific gravity (i.e., the mantle). Mantle convection has resulted in the crust being broken into a number of plates, which move about the Earth's surface. This theory of *plate tectonics* results in continual destruction and rebirth of the Earth's crust, as mentioned above. Figure 3.3 is a map of the Earth, showing plate boundaries as they appear today. Note the occurrence of volcanoes near plate boundaries. It is near these regions where most of the world's geothermal power plants exist today.

Readers are likely familiar with the supercontinent *Pangaea*, which existed some 300 million years ago, when all Earth's continents were combined into continental mass. That 300 million years only represents ~7% of Earth's history gives rise to the concept that the continents continually drift above the mantle, undergoing successive periods of consolidation and separation.

Thus, our knowledge of the Earth beyond a depth of a few kilometers is based on indirect evidence. What we accept as the model for the Earth's inner structure is filled with uncertainty, particularly the temperature as a function of depth. Sometimes the analogy is drawn between

Figure 3.3 Plate tectonic map of the Earth's surface (NASA public domain)

the Earth and a chicken's egg, where the crust is analogous to the shell of the egg. According to DiPippo (2012), relating the thickness of the Earth's crust, 35 km for continental regions, to its diameter, roughly 12 700 km, we get a ratio of 35/12 700 or 0.00276. If we were to apply the same ratio to an egg with a diameter of say, 50 mm, we would find a shell thickness of 0.138 mm or 0.0054 in. In fact, the shell of an egg is about 1/64 in or about 0.016 in. Thus an egg's shell is about 3 times thicker proportionally than the crust of the Earth. In other words, if the Earth's crust were in proportion to the shell of an egg, it would be about 100 km thick instead of 35 km thick.

3.3 Geology and Drilling Basics for Energy Engineers

Successful installation and operation of geothermal energy systems requires some theoretical considerations but is always limited by the practical. Regarding geothermal heat pump systems, a designer can specify and draw anything in the comfort of the office, but in reality, the system *constructability* is dictated by field conditions, which are never fully known; when dealing with the subsurface environment, we can't see it beforehand, so there is always some uncertainty of actual conditions. The more effort a designer puts in upfront, the less will be the risk of dealing with unknown conditions.

Designers should, at least, have a working knowledge of the soil and rock types that will be encountered during geothermal installations. Unexpected groundwater and unstable soils can pose significant problems during drilling if the correct drilling method is not specified. This subsection, therefore, describes basic geological considerations and potential drilling methods for each.

3.3.1 'Geology 101' for Energy Engineers

A fundamental concept in *Geology* that describes the formation of rocks is known as the rock cycle (Figure 3.4). The rock cycle is the natural process in which rocks transform from one rock type into another rock type over time; a type of natural recycling. This occurs over geologic time: thousands to billions of years.

Igneous rocks, which are formed from magma cooling underground, or from lava above ground, are formed from previous igneous, metamorphic, or sedimentary rocks that have become melted, usually as a result of tectonic plate collision and subduction. When these igneous rocks are exposed to weathering and erosion, they break down into smaller particles that are transported by wind and water to a place of deposition, where they can form into sedimentary rock strata, through a process of lithification, where excess water is squeezed out by overburden pressures and the particles are cemented together by various minerals precipitating out of solution. Igneous and metamorphic rocks can both be turned into sedimentary rocks in this way.

Igneous and sedimentary rocks can also be changed by heat and/or pressure into metamorphic rocks, by transforming their existing mineral structures into new minerals or realigning the existing minerals. There are different degrees of metamorphism, so even an existing metamorphic rock can become a different metamorphic rock. If these metamorphic rocks are melted, then solidify, they become igneous rocks, and the cycle starts all over again.

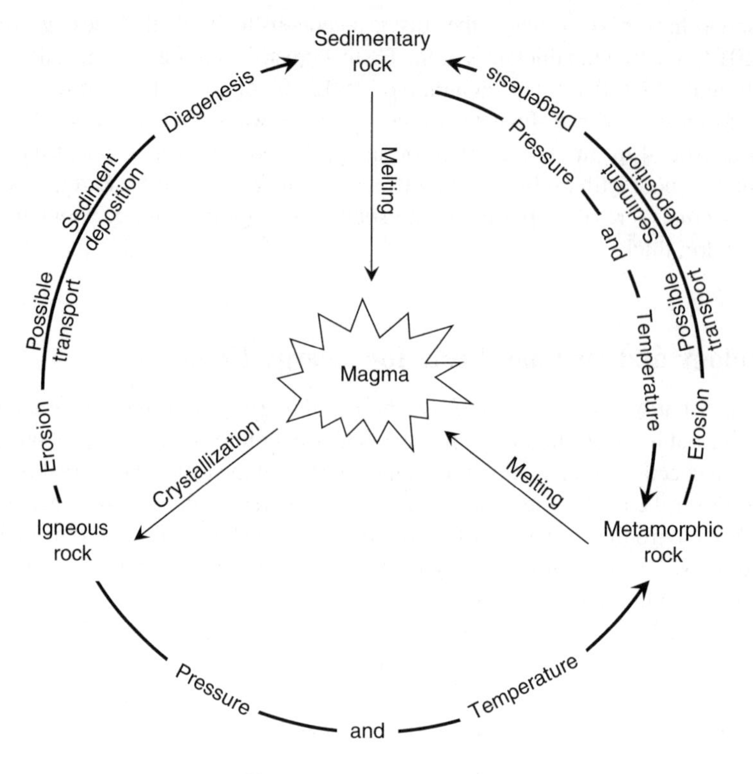

Figure 3.4 The rock cycle

Forces that drive the rock cycle are largely lithospheric plate movements, which cause subduction and uplift, and also climatic conditions and the associated erosional elements. Lithospheric plate movements are described by the theory of *plate tectonics*. Water, wind, and glaciers are the main erosional elements of rocks. Figure 3.5 depicts how the theory of plate tectonics contributes to the recycling of Earth's materials.

In summary, **bedrock** is the first consolidated and competent material that would remain if all loose material were stripped from the Earth's surface. **Igneous rocks** are formed by crystallization of molten rock. Molten rock within the Earth is called magma; molten rock that has been extruded onto the Earth's surface is called lava. Magmatic rocks are coarse grained (e.g., granite) and volcanic rocks are usually fine grained (e.g., basalt). The lay term is 'hard rock'.

Regarding **soil**, different professions have widely differing definitions, depending on their specialty (agriculture, engineering, geology, soil scientists). We'll define it here as unconsolidated material produced by the mechanical (i.e., wind, glacial) and/or chemical weathering of rocks. **Sedimentary rocks** are formed from weathering, and erosion of rocks and other soils forms sediments that undergo burial and crystallization by heat and pressure. Examples are sandstone, shale, and limestone. The lay term is 'soft rock'.

Finally, **metamorphic rocks** are rocks that undergo very deep burial and recrystallization by intense heat and pressure. Examples are gneiss and marble. The lay term is 'hard rock'.

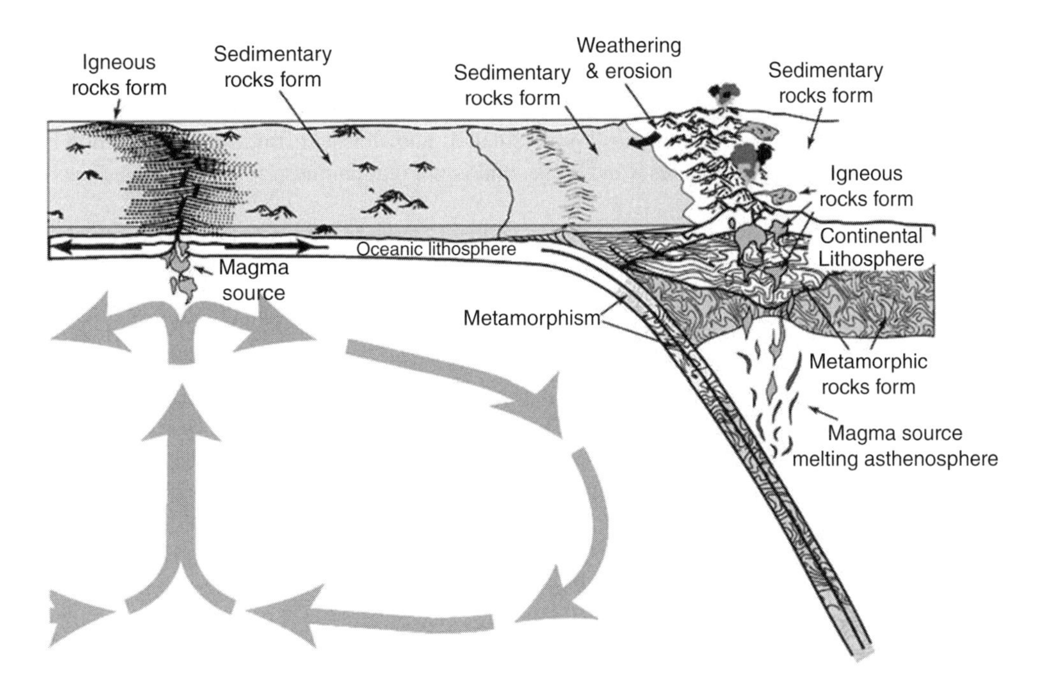

Figure 3.5 Schematic of tectonic plate interactions (public domain figure from US Geological Survey)

3.3.2 Overview of Drilling Methods

Knowledge of the rock type at a particular location has implications regarding the ground heat exchanger installation method. Drilling methods generally consist of two broad groups: (1) rotary methods, and (2) percussion methods. These are discussed in the paragraphs that follow.

3.3.2.1 Rotary Drilling Methods

Rotary drilling methods consist of the following: (i) auger, (ii) mud rotary, (iii) reverse circulation, (iv) air rotary, and (v) sonic methods. Each is described below.

The **auger method** is a not very common drilling method in the geothermal industry. The method is generally limited to soils that are not too wet, and is very slow beyond depths of 15–30 m. Auger flights (i.e., the cutting edges) result in undesirable, large-diameter boreholes. Smaller-diameter boreholes are generally preferred with geoexchange applications because they limit the amount of grout to be emplaced in the annular space. Larger-diameter boreholes also increase the amount of material to be disposed of (as the soil/rock cuttings cannot be put back into the borehole reliably).

Mud rotary methods are very popular in the geothermal industry. A bentonite-based engineered fluid known as 'drilling mud' is used as the cutting/lubricating fluid, and is very effective and most common in soils and soft rocks. Wet soils are usually not a problem. Mud fluid is pumped down the drill rods through the center of the bit, and the mud flows back up the annular borehole space, removing the drill cuttings and transporting them to a settling pit. At the settling pit, solids settle out, and the fluid is recycled for use.

The **reverse circulation method** is a variant of mud rotary drilling, where the difference is with the fluid handling. In reverse circulation, mud is pumped in the reverse direction, through the annular space and then up through the drill rods. This results in better drill cutting removal owing to higher mud velocity in the drill pipe (smaller diameter, less friction). This method is preferred in drilling large-diameter water wells, and is not that common in closed-loop geoexchange installation.

Air rotary methods are common, effective, and fast methods in areas of soft, consolidated materials (e.g., shale). Air with a foaming agent is used as the cutting/lubricating fluid. Like mud rotary methods, the fluid (air) is pumped down the drill rods (an air compressor is needed) through the center of the bit, and the air flows back up the annular borehole space, removing the drill cuttings. Wet soils/rocks *are* problematic because required air pressures and velocities to remove cuttings below about a 15 m water column become prohibitively excessive.

Sonic methods work by sending high-frequency resonant vibrations from a sonic head down the drill string to the drill bit, while the operator controls these frequencies to suit the specific conditions of the soil/rock geology. The frequency is generally between 50 and 120 Hz and can be varied by the operator. Resonance magnifies the amplitude of the drill bit, which fluidizes the soil particles at the bit face, allowing for fast and easy penetration through most geological formations. An internal spring system isolates these vibrational forces from the rest of the drill rig. This method does not work well in wet materials. One advantage is that is produces a continuous soil/rock core sample.

3.3.2.2 Percussion Drilling Methods

Percussion methods consist of: (i) air hammer, and (ii) cable tool methods.

Air hammer methods are commonly used in areas with 'hard rock' where the rock hardness precludes the use of a rotating bit. As with air rotary methods, air with a foaming agent is used as the cutting/lubricating fluid, and wet rocks are a problem. The drilling tool is a pneumatic hammer coupled to a diamond carbide bit that pulverizes the rock into fine dust. The bit is cooled by the air and any moisture encountered in the rock.

Cable tool methods (Figure 3.6) are 'brute force' methods where a hammer is repeatedly raised and lowered, breaking up the rock. At regular intervals, a 'bailer' is sent down the hole to remove cuttings. The cable tool method is the oldest drilling method, dating back thousands of years. It is most commonly used in 'hard rock' areas with brittle or highly fractured rock that precludes drill tool rotation (i.e., volcanic rocks).

3.4 Earth Temperature Regime and Global Heat Flows: Why is the Center of the Earth Hot?

The temperature of the core of the Earth is estimated to be as high as 6650 °C (12 000 °F). If we could drill a very deep well from the surface into the Earth and measure the temperature as a function of depth, we would find a fairly constant increase of about 3 °C for each 100 m of depth (1.6 °F per 100 ft). This rate is typical in normal areas, but there are anomalous regions associated with volcanic or tectonic activity where the temperature gradients far exceed these normal values. For example, at Larderello, Italy, where the first engine driven by geothermal

Figure 3.6 Photograph of a cable tool rig set up for drilling a water well (photo by the author)

steam was built, the gradient is 10–30 times higher than normal (DiPippo, 2012). This means that temperatures over 300 °C (575 °F) can be found at a depth of 1 km (3300 ft), easily reachable with today's drilling technology.

If geothermal energy is the residual thermal energy in the Earth left over from the planet's origins, Lord Kelvin calculated that the Earth would have cooled in about 100 million years. The Earth's crust is composed of various types of rock that contain some radioactive isotopes, in particular, uranium (U-235, U-238), thorium (Th-232), and potassium (K-40), and the heat released by these nuclear reactions is thought to be responsible for the natural heat that reaches the surface. Currently, physicists are attempting to isolate *geoneutrinos* – particles created from the decay of radioactive uranium and thorium. Thus, the Earth's thermal regime today is the complex result of primordial thermal energy, ongoing radioactive decay of elements, and the formation of a solid crust.

The normal conductive geothermal gradient is usually taken to be about 30 °C/km or 3 °C/100 m. Good geothermal prospects for power plant projects occur where the thermal gradient is several times greater than normal. The rate of natural heat flow per unit area is called the normal heat flux; it is roughly 50 $mW{\cdot}m^{-2}$ in non-thermal areas of the Earth. Assuming an International Standard Atmosphere (ISA) model, where the sea level temperature is 15 °C, Figure 3.7 shows the Earth's temperature from the Karman line (100 km above the Earth's surface) to the Moho (about 30 km below the Earth's surface).

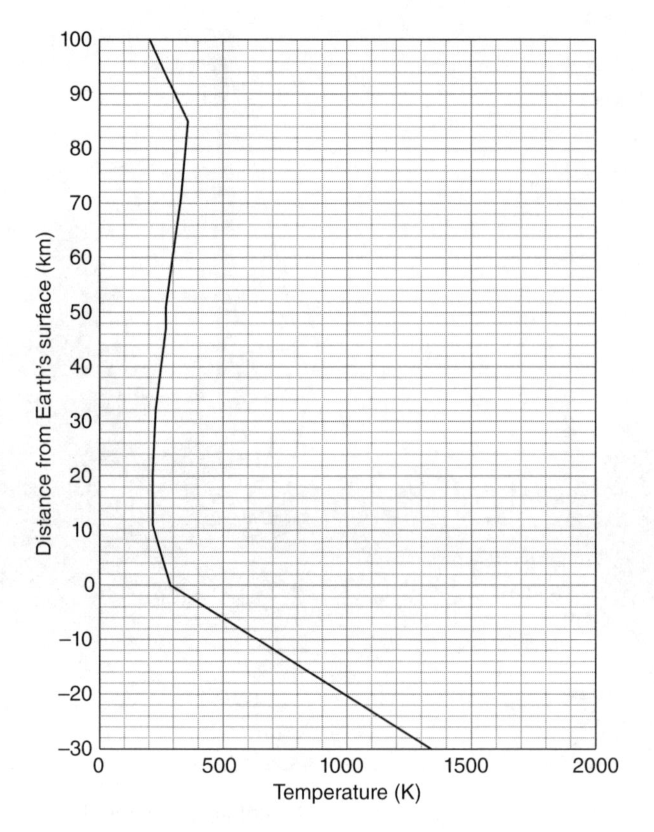

Figure 3.7 Temperature as a function of distance from the Earth's surface

3.5 Shallow Earth Temperatures

Many say that the Earth's temperature at depths of typical geoexchange applications is stored solar energy. This is only partly true. The solar heat flux reaching the Earth is of the order of 1360 W·m^{-2} but is attenuated and scattered by the Earth's atmosphere, such that about 1000 W·m^{-2} reaches the surface on a clear day. However, solar energy absorbed by the ground surface is quickly reradiated and convected back to the atmosphere. Comparing the solar heat flux of 1000 W·m^{-2} with the milliwatts of geothermal heat flux, we may ask, why is the Earth's surface not at a temperature of thousands of degrees like the mantle? We quickly realize that solar energy is recycled and stored in the atmosphere. Thus, shallow Earth temperatures at a particular location are correlated with average annual air temperatures.

The Earth's temperature variation with depth is a function of surface process (i.e., atmospheric and solar indirectly) and deep geothermal processes. Under circumstances where the geothermal gradient is considered *normal*, such as at locations away from tectonic plate boundaries, the subsurface temperature profile over the depth interval of interest for geoexchange can roughly be divided into three thermal regimes:

- Shallow thermal regime. Here, the Earth's temperature is strongly influenced by atmospheric processes. Surface and near-surface temperatures fluctuate seasonally by magnitudes of the order of ±10 °C. These temperature fluctuations dampen with depth and completely diminish at about 5 m. Equation (2.9) is commonly used to calculate soil temperature as a function of depth and day of the year.

- Transition zone. This zone generally occurs from depth intervals of 5 m to about 10 m, and is characterized by thermal influences from above and below. In this zone, the Earth's temperature is approximately equivalent to the average annual air temperature.
- Deeper Earth regime. In this zone, the Earth's temperatures are dominated by geothermal processes and begin at a depth of about 10 m. With increasing depth, the Earth's temperature rises at an average rate of about 30 °C per kilometer, or about 3 °C per 100 m. This value is lower in areas with a thick continental crust, and greater in geothermic areas.

Signorelli (2004) suggests that the average underground temperature equivalent over the typical geoexchange borehole depth (~100 m) is approximated by the average annual air temperature plus about 1.5 °C. This estimate captures the subsurface temperature due to heat fluxes from above and below, and is applicable only in areas of normal geothermal gradient.

The Earth's temperature versus depth (0–100 m) is plotted in Figure 3.8 using a mathematical model described in Chapter 7. The average surface temperature is assumed to be 15 °C, and the temperature increase due to a geothermal gradient of 30 °C/km is assumed to begin at a depth of 7 m. A review of Figure 3.8 shows that the shallow Earth temperature is quite variable and dynamic. Thus, it is important to consider the Earth's temperatures carefully in geoexchange applications.

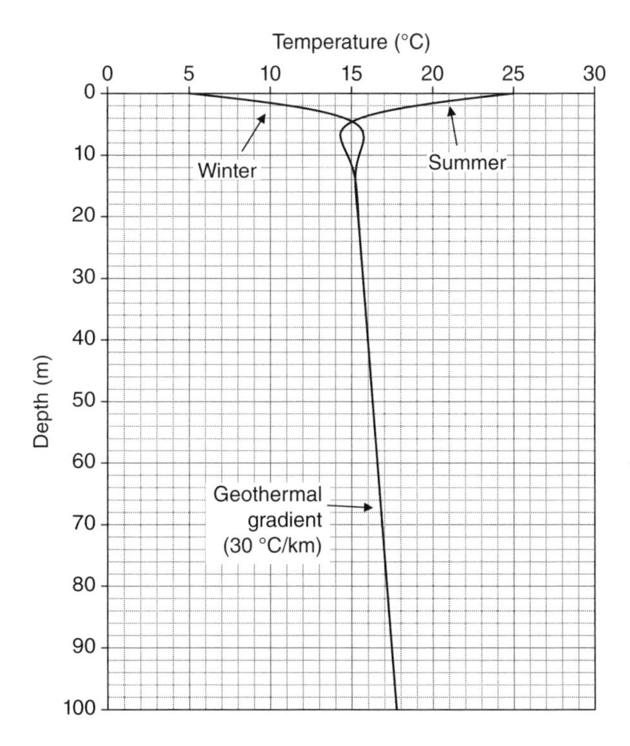

Figure 3.8 Temperature as a function of depth (0–100 m) below the Earth's surface

3.6 The Geothermal Reservoir Concept

What is commonly referred to as 'geothermal energy' is actually stored energy in the Earth. Thus, a unique attribute of geothermal energy projects, relative to other types of renewable energy project, is that the reservoir is finite and requires proper management. This 'feedback loop' is shown in Figure 1.2. Therefore, the energy flows into and out of the reservoir must be engineered to manage the change in this stored energy as the energy is exploited for a useful purpose over the life of the reservoir.

A classic example of misuse of a high-temperature reservoir occurred at the Geysers geothermal field in northern California. Dry steam was harnessed from the reservoir at excessive rates, and condensate was not returned to the reservoir, and the result was a decline in reservoir pressures and available fluids. A crisis was mitigated by injecting treated effluent from the City of Santa Rosa back into the reservoir to restore fluid levels.

Similarly, with regard to geothermal heat pump projects, the 'reservoir' can quickly experience excessive thermal energy build-up or thermal energy depletion if the Earth heat exchanger is not coupled to a large enough Earth volume. Calculating adequate Earth heat exchanger size is one of the main focuses of this book and is treated in Part II – Harnessing the Resource.

One documented case of gradual thermal energy build-up in a reservoir acting as a heat source and sink for geothermal heat pumps occurred at The Richard Stockton College of New Jersey, USA. Sowers *et al.* (2006) and Epstein and Sowers (2006) describe this thermal build-up and associated environmental impact in the vertical borehole field, which was the largest geothermal heat pump installation at the time of its construction in 1993–1994. The borehole field spans 2.3 ha under a campus parking lot, and serves 40 900 m^2 of floor space. Owing to the imbalanced annual cooling loads relative to the annual heating loads, the average temperature of the borehole field increased by 11 °C over the years 1994 to 2005, as determined by groundwater temperature measurements in eight monitoring wells installed within the borehole field. Sowers *et al.* (2006) found increased numbers of and changes in types of microbe in groundwater samples as compared with the control samples outside the borehole field, and this observation prompted the addition of a cooling tower to the system.

A basic model of a geothermal reservoir can be thought of as a simple control volume, as shown in Figure 3.9. Modeled as a simple cube, energy exchanges ($\dot{E}$) can take place through all six faces. Such energy exchanges include natural diffusion and advection of heat through all faces, the Earth's surface processes (water infiltration, solar energy fluxes, atmospheric interactions) at the top face, and engineered heat extraction and heat rejection through the top face. In addition, energy can be generated internally ($\dot{E}_{generated}$) in the control volume (e.g., radioactive decay of elements, biological decay, and *in situ* coal combustion). All of these energy flows will act to change the amount of stored energy in the control volume. Thus, the goal of any geothermal project should be to manage the energy transfers into and out of the reservoir such that the quality of energy (i.e., the temperature) being extracted from the reservoir is used sustainably over the life of the geothermal project.

In mathematical terms, the geothermal reservoir control volume concept can be written as

$$\dot{E}_{in} - \dot{E}_{out} + \dot{E}_{generated} = \Delta\dot{E}_{stored} \tag{3.1}$$

where $\dot{E}$ is the energy transfer rate. Thus, geothermal energy projects should seek to minimize $\Delta\dot{E}_{stored}$.

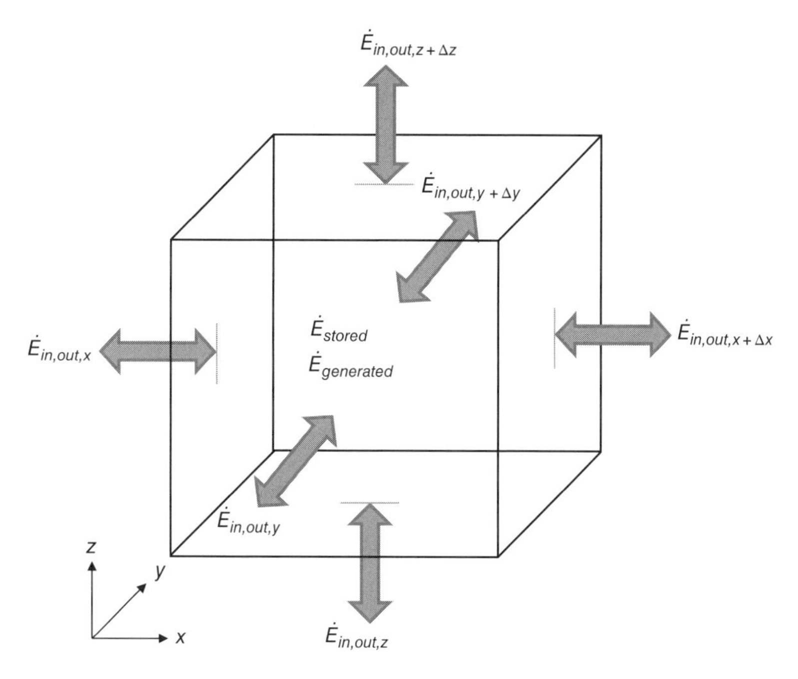

Figure 3.9 Control volume concept of a geothermal reservoir

As discussed in Chapter 1, the level of effort and the methods used in geothermal resource characterization entirely depend on the energy need, and the potential of a geothermal resource to meet this intended need depends entirely on its temperature. While temperature is a convenient metric for categorizing a geothermal resource, something also needs to be said about the heat transfer processes occurring in and around the reservoir. Thus, simply stated, geothermal reservoirs can be further described as (i) conductive dominated or (ii) advective/convective dominated, based on the main heat transfer mechanism governing the energy transport processes in the reservoir. Figure 3.10 shows general types of geothermal resource occurrence based on temperature and available fluids.

Some further explanation of the general geothermal reservoir occurrence as described in Figure 3.10 is necessary. First, the level of effort in characterizing the resource generally increases with reservoir depth, which implies that the higher-temperature uses require a greater level of effort in exploration and characterization.

Second, the term 'relative depth' is used, because the geothermal gradient varies by location, and thus reservoir types can occur over a range of depths. At some locations, geysers and hot springs can occur at or near the Earth's surface. In volcanic regions, magma bodies can be quite shallow.

Third, magmas range in temperature from about 650 to 1300 °C, depending on chemical composition. For comparison purposes, common steel melts at about 1500 °C. With current drilling technology, drilling depths are limited by temperatures of about 260 °C.

Fourth, a dashed line generally separates conductive-dominated reservoirs from advective/ convective reservoirs. Note that this line is not at the dividing line of aquifers. Geologic materials may contain water, but this water may not be in adequate quantities to result in an advective/ convective-dominated reservoir.

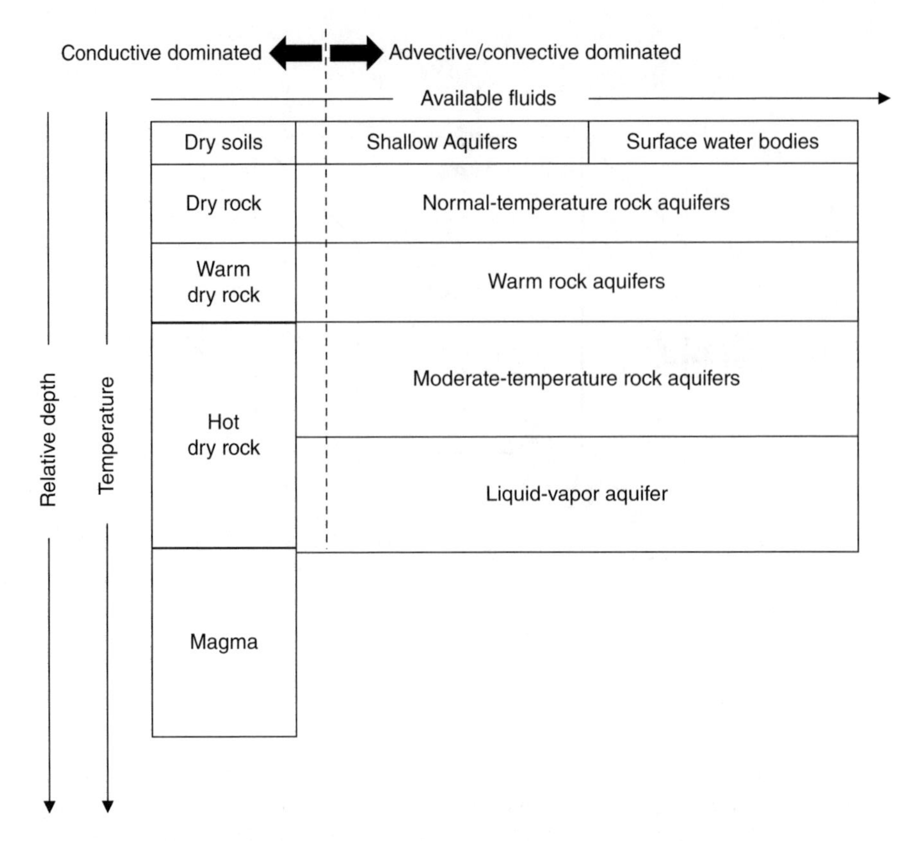

Figure 3.10 General geothermal reservoir occurrence in nature

Fifth, at the time of writing, all medium- to high-temperature geothermal utilization is with *hydrothermal resources* (i.e., reservoirs containing appreciable available fluids). In these cases, hot groundwater is used as the heat-carrier fluid to transport geothermal energy to the Earth's surface. Some exceptions occur where downhole heat exchangers are used to transport heat, but these are still installed in hydrothermal aquifers. Research and development continues in earnest on exploiting conductive-dominated (i.e., hot dry rock) resources.

Sixth, ambient-temperature or geothermal heat pump uses take advantage of shallow stored Earth energy, and therefore generally don't require extensive exploration methods because these resources are ubiquitous. Further, these resources could be either the conductive-dominated or the advective/convective-dominated type.

3.7 Geothermal Site Suitability Analysis

The key goals and outcomes of this stage of geothermal project development are answers to the following questions:

- *Could* a resource be utilized for the intended energy needs?
- *Should* a resource be utilized for the intended energy needs?

The answer to the first question in most cases can be answered by qualified scientists and engineers. However, the answer to the second question has more to do with sociological factors, human factors, environmental factors, business models for use of the resource, and the capability and desire of the owner to operate and maintain the resource and any associated equipment.

3.7.1 Groundwater Resources

Regardless of the geothermal energy use, groundwater is either essential or beneficial. Thus, it is important for Energy Engineers to be aware of the nature and occurrence of groundwater, and methods for groundwater exploration. For details on basic groundwater hydrology, readers are referred to Heath (1983), which is a public domain document published by the US Geological Survey: http://pubs.usgs.gov/wsp/2220/report.pdf

3.7.1.1 The Nature and Occurrence of Groundwater

Groundwater resources are not ubiquitous in nature, and therefore some exploration effort is needed to find usable groundwater. In many jurisdictions, groundwater is treated as a natural resource, and thus its use is regulated. Of all water on Earth, most of it (over 96%) is contained in the oceans, and only about 2.5% is fresh water. Of all Earth's fresh water, about 69% is contained in glaciers and ice caps, about 1% is surface water, and the remaining 30% is groundwater. Thus, groundwater represents a significant amount of usable water on Earth.

All water beneath the land surface is referred to as underground water, or subsurface water, as shown in Figure 3.11. The equivalent term for water on the land surface is *surface water*. Underground water occurs in two different zones: (i) the unsaturated zone, and (ii) the saturated zone.

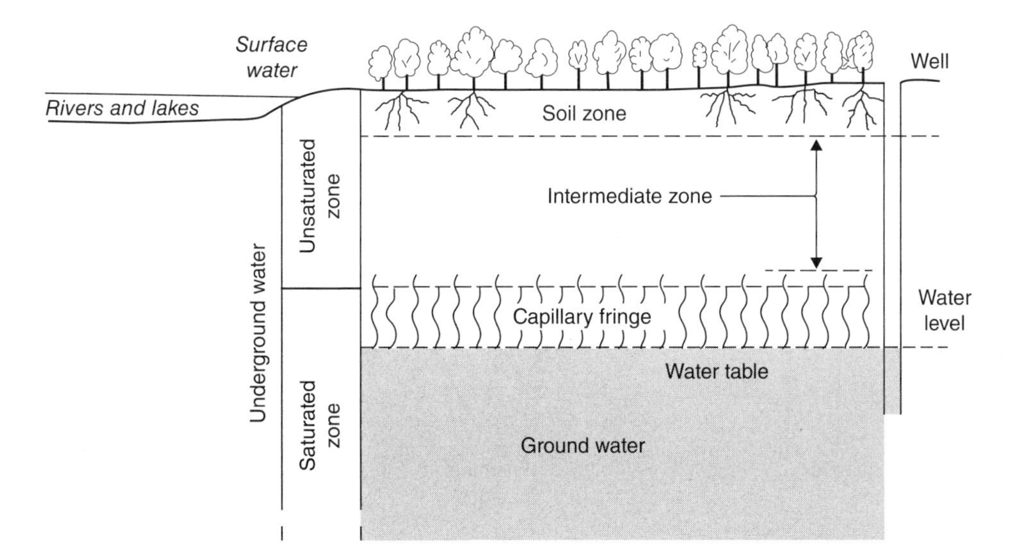

Figure 3.11 Groundwater in the context of all underground water (Heath, 1983)

The unsaturated zone occurs immediately below the land surface in most areas, and contains both water and air. Underlying the unsaturated zone is the saturated zone, which is a zone in which all interconnected openings are full of water. Materials in the capillary fringe are saturated, but water is held under tension, and is thus not available to wells. Thus, groundwater is that water in the saturated zone that is available to wells. The water table is the level in the saturated zone at which the hydraulic pressure is equal to atmospheric pressure, and is represented by the static water level in wells with an intake zone at the water table. Below the water table, the hydraulic pressure increases with increasing depth.

Groundwater occurs in soils and rocks in what are referred to as primary or secondary openings (Figure 3.12). Primary openings are essentially pore spaces resulting from formation of the material, as discussed above in the rock cycle. Secondary openings include fractures due to tectonic forces, and solution caverns due to dissolution of the rock by water.

Groundwater movement through the Earth is part of the larger hydrologic cycle or water cycle, which you probably learned as a kid. The water cycle describes the continuous path of water through the Earth's surface, subsurface, and atmosphere. Figure 3.13 depicts a groundwater system showing recharge and discharge areas, which is part of the larger water cycle. Precipitation enters the subsurface and then begins to infiltrate into the ground. Infiltration rates vary widely, depending on land use, the character and moisture content of the soil, and the intensity and duration of precipitation. Water in the zone of saturation moves downward

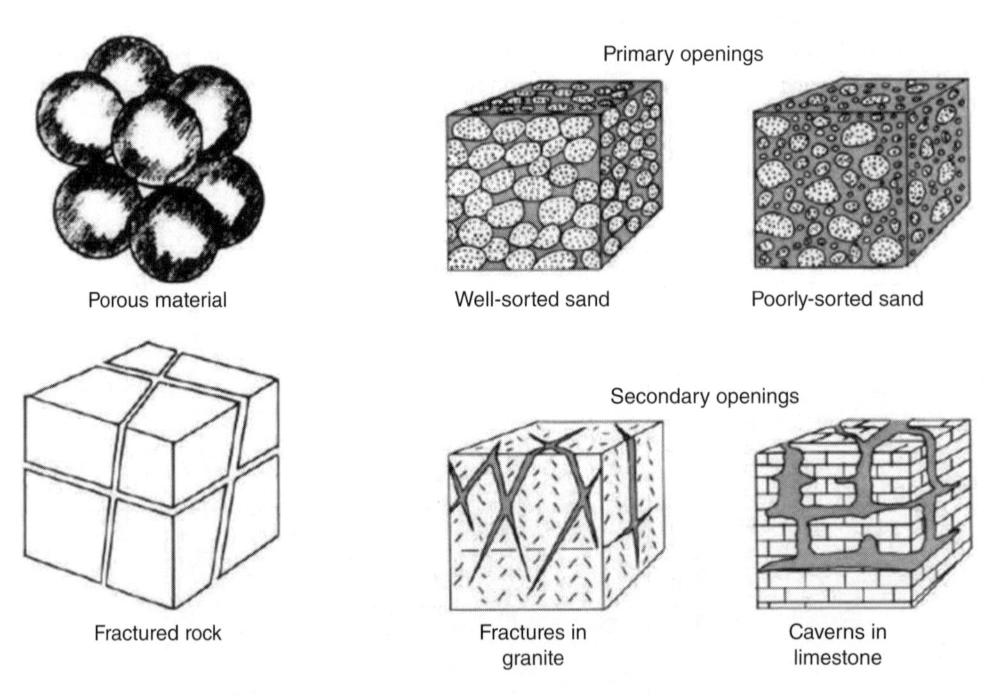

Figure 3.12 Groundwater occurrence in porous and fractured Earth materials (Heath, 1983)

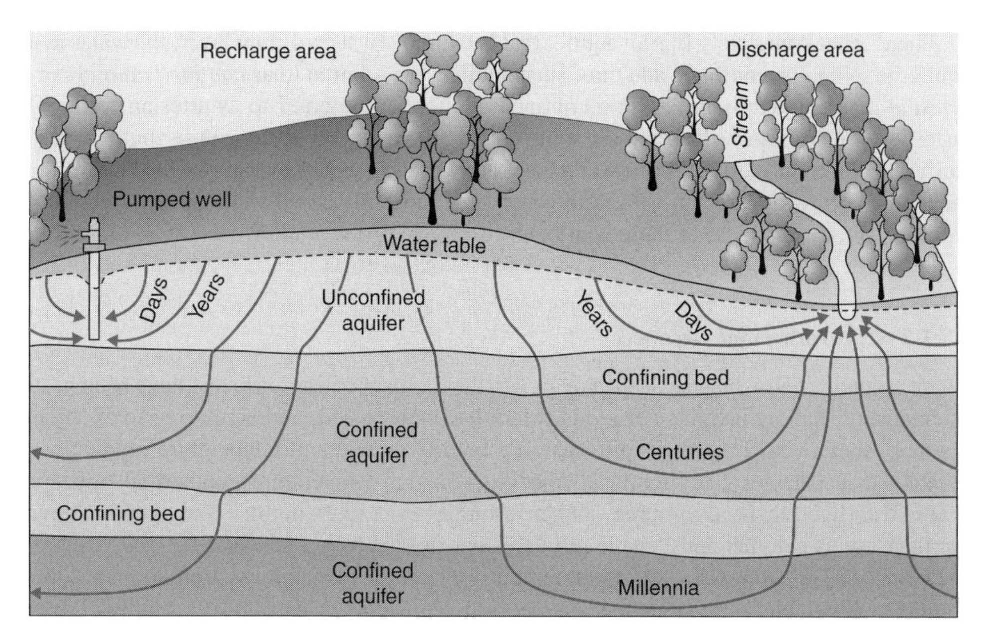

Figure 3.13 Schematic representation of a groundwater system of aquifers and confining beds (US Geological Survey, public domain)

and laterally to sites of groundwater discharge such as springs on hillsides or seeps in the bottoms of streams and lakes or beneath the ocean. Water reaching streams, both by overland flow and from groundwater discharge, eventually moves to oceans, where it is again evaporated to perpetuate the cycle.

Groundwater movement is, of course, a key element in the concept of the hydrologic cycle, as depicted in Figure 3.13. The physics of groundwater flow is governed by Darcy's law, which we will discuss in Chapter 4. Typical rates of groundwater movement are slow relative to that found in rivers and streams. Groundwater velocities are of the order of centimeters to meters per year, whereas river and stream velocities are of the order of tens to hundreds of kilometers per day. As depicted in Figure 3.13, groundwater residence times in some flow systems can be of the order of hundreds to thousands of years.

At the macroscopic level, groundwater, when present, moves through either aquifers or confining beds as shown in Figure 3.13. An aquifer is a geologic unit that is capable of yielding water in a usable quantity to a well or spring. A confining bed is a geologic unit having very low hydraulic conductivity that restricts the movement of groundwater either into or out of adjacent aquifers.

Groundwater occurs in aquifers under two different conditions: (i) unconfined, or (ii) confined. In an unconfined aquifer, as the name implies, the top surface is the water table (at atmospheric pressure) and is free to rise and decline. Unconfined aquifers are also widely referred to as *water-table aquifers*. Wells open to unconfined aquifers are referred to as *water-table wells*, and the water level in these wells indicates the position of the water table in the surrounding aquifer.

Where water completely fills an aquifer that is overlain by a confining layer, the water in the aquifer is said to be confined, and thus, such aquifers are referred to as *confined aquifers* or as *artesian aquifers*. Wells drilled into confined aquifers are referred to as artesian wells. The water level in artesian wells stands at some height above the top of the aquifer but not necessarily above the land surface. If the water level in an artesian well stands above the land surface, the well is a flowing artesian well. The water level in tightly cased wells open to a confined aquifer stands at the level of the potentiometric surface of the aquifer.

3.7.1.2 Hydrothermal Resources

When groundwater percolates deep enough into the Earth, or enters rocks that have been heated by relatively shallow magma, the groundwater becomes heated, and is referred to as 'hydrothermal'. Its movement is then influenced by both hydraulic and temperature gradients. As mentioned in Chapter 1, currently all medium- to high-temperature geothermal utilization occurs with hydrothermal resources. Other resources exist (e.g., hot-dry rock, geopressured), but their use as commercially viable resources has not yet been realized.

DiPippo (2012) summarizes five features that are essential to making hydrothermal resources commercially viable: (a) a large heat source, (b) a permeable reservoir, (c) a supply of water, (d) an overlying layer of impervious rock, and (e) a reliable recharge mechanism. If any one of these five features is absent or deficient, the field generally will not be worth exploiting.

A conceptual model of a hydrothermal reservoir was originally described by White (1973). As shown in Figure 3.14, cold meteoric recharge water enters the surface as rain, and percolates through faults and fractures deep into the formation where it comes in contact with heated rocks. Tectonic forces are the cause of the presence of relatively shallow magma and these

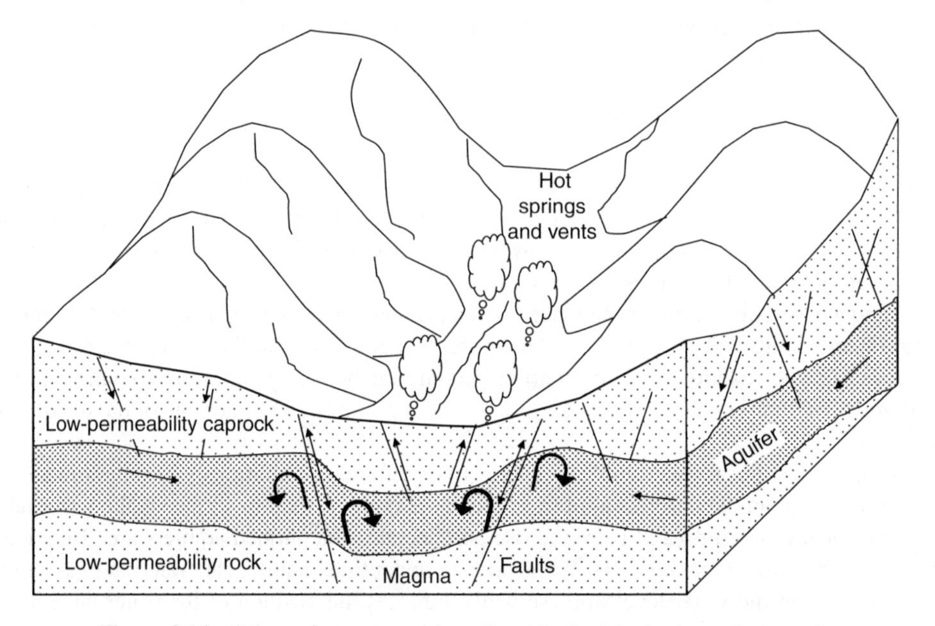

Figure 3.14 Schematic representation of an idealized hydrothermal reservoir

faults and fractures. The permeable layer offers a path of lower hydraulic resistance, and as the liquid heats, it becomes less dense and tends to rise within the formation. If it encounters a major fault, it will ascend toward the surface, losing pressure as it rises. In some cases, the fluid pressure drops and it may boil in the formation. In other cases, the fluid may reach the ground surface and flash into steam, being observed as a fumarole, a hot spring, a mud pot, or a steam-heated pool.

Such geologic structures as idealized in Figure 3.14 are sometimes referred to as *horst-and-graben* structures or *basin-and-range* structures. Thus, exploration methods for these types of hydrothermal resource focus on finding geologic structures (i.e., faults and fractures) that are conducive to the occurrence of geothermal fluids. Hydrothermal resources (like oil and gas resources) are not easy to find, as there is no method for direct observation of the temperature of the subsurface or the presence of fluids; all exploration methods rely on indirect observation, and the resource is not confirmed until it is drilled into and tested. Thus, the process of hydrothermal reservoir exploration is sometimes viewed as risk management.

The main types of scientific method used to explore for hydrothermal resources are either geophysical or geochemical.

Geophysical methods use 'physics of the Earth' to define/refine a conceptual model of a hydrothermal reservoir. These methods look at interpreting some physical, measured response of the Earth, and in the geothermal industry they typically include remote sensing, heat flow studies, seismic studies, gravity studies, and magnetics studies. It is uncommon to rely on one method only; redundancy and reassurance is necessary.

Remote sensing involves thermal imaging of the Earth's surface over a prospective resource to determine thermal anomalies. This can be done by examining existing satellite images, or flying over an area and recording thermal images.

Heat flow studies are conducted either by examining existing well data or by drilling shallow boreholes (if site access is available) and measuring the temperature profile with depth. The heat flux ($\dot{q}''$) is readily determined by Fourier's law of heat conduction:

$$\dot{q}'' = k\frac{\Delta T}{\Delta z} \tag{3.2}$$

where k is the average thermal conductivity of the rock ($W{\cdot}m^{-1}{\cdot}K^{-1}$), ΔT is the temperature difference between the surface and bottom hole temperatures (°C or K), and Δz is the bottom hole depth (or distance over which the temperature measurements are taken).

Seismic studies involve measuring the travel times of elastic waves travelling through the Earth as the result of a force or vibration. Much of our knowledge of the internal structure of the Earth is the result of seismic recordings during earthquakes. Seismic studies are perhaps the most popular type of geophysical study, because propagation of seismic waves is significantly changed when induced waves (by a vibrating truck or shot blast) encounter a fault or other physical change in the subsurface. Thus, careful analysis of seismic data can produce a fairly reliable image of the subsurface.

Resistivity surveys measure electrical properties of the subsurface by imposing an electrical current through an array of electrodes penetrating the ground surface. Voltage drops are measured, allowing the mapping of subsurface resistivity, and giving clues to various geological parameters such as the mineral and fluid content, porosity, and degree of water saturation in the rock. Abrupt subsurface changes such as faults are sometimes readily discerned.

Gravity studies use the principles of inertia and gravitational attraction. An inertial mass on a damper, installed in a simple gravimeter, can undergo measurable deflections when passed over rock bodies of significantly varying density. Thus, gravity studies are capable of detecting fault structures and other subsurface anomalies. It is interesting to note that a gravity anomaly in the Yucatan Peninsula, Mexico, was, in part, responsible for locating the remnant impact crater made by the meteorite believed to be responsible for causing extinction of the dinosaurs. A significant gravity low has been detected under Hudson Bay, Canada, which is cause for some debate.

Magnetics studies are sometimes used as a secondary exploration tool in geothermal studies. Paleomagnetism is a key branch of geophysics in understanding the Earth's history and plate motions. When rocks form, magnetic minerals align with the current Earth's magnetic field. As these rocks subsequently move or become deformed, changes in magnetic fields can be detected with a magnetometer. As with the previously described geophysical studies, magnetic studies can give clues about subsurface structure.

Geochemical methods use the 'chemistry of the Earth' to define/refine a conceptual model of a hydrothermal reservoir. These methods look at interpreting some chemical signatures of geothermal fluids, and in the geothermal industry they typically include: general groundwater chemistry analysis, isotope analysis, geothermometer analysis, and geobarometer analysis. It is uncommon to rely on one method only; redundancy and reassurance are necessary. In the early exploration phase, geoscientists may look for obvious surface manifestations of a hydrothermal resource (unusual mineral deposits, pools and springs, stressed vegetation).

General groundwater chemistry analysis involves sampling typically shallow waters to determine their origin. Major ions in groundwater can be grouped to determine reservoir fluid types. Piper (1944) has developed a method to diagram water chemistry based on a trilinear plot (also called a Piper trilinear plot). This method is based on relative amounts of sodium and potassium (Na + K), magnesium (Mg), calcium (Ca), chloride (Cl), sulfate (SO_4), bicarbonate (HCO_3), and carbonate (CO_3) in a fluid. A water sample plots on the diagram as a single point, based on its overall major ion chemistry. It can be subjectively inferred that samples falling within the same region of the Piper trilinear plot have similar origins.

Geothermal fluids tend to have uncommon dissolved constituents and dissolved gases, and thus a trace element analysis of fluids is a potentially useful tool in exploration. Trace metals include arsenic (As), boron (B), lithium (Li), and mercury (Hg). Dissolved gases include carbon dioxide (CO_2) and hydrogen sulfide (H_2S).

Isotopic analyses, similar to major ion analyses, can be used to infer origin of groundwater and give some insight into potential groundwater flow paths. In geothermal system analyses, the isotopes most often useful are the stable isotopes (i.e., those that are not radioactive) of hydrogen and oxygen: hydrogen-2 or deuterium (D) and oxygen-18 (O-18). These isotopes are part of the water molecule, and are present in differing proportions in meteoric waters (i.e., rain and snow), depending upon the temperature of precipitation. As meteoric water enters the subsurface and becomes groundwater, the isotopic signature is generally preserved unless the groundwater experiences a significant temperature change or mineral exchange with surrounding rock.

Chemical geothermometry and geobarometry are techniques that can be used to estimate subsurface geothermal reservoir temperatures using concentrations of certain elements in groundwater. Chemical geothermometers are developed on the basis of temperature-dependent chemical equilibrium between the water and the minerals at the deep reservoir conditions. The two main types of chemical geothermometer are based on silica (SiO_2) equiibrium and alkali

(Na–K–Ca) equilibrium. Chemical geothermeters are determined by empirical equations that relate temperature to concentrations of elements in groundwater. A review of use of chemical geothermometers is given by Fournier (1981).

3.7.2 Geoexchange Applications

Given the numerous types of Earth coupling for geothermal heat pump systems (see Section 1.5.3), and that the ground is obviously ubiquitous, there is no universal reservoir type that is targeted for geoexchange applications. However, we are still dealing with the subsurface, and therefore there is always some uncertainty and inherent risk when undertaking a geoexchange project. Site characterization is essential in determining the GHX that best works with the site and application to minimize capital cost, maximize operational savings, and maximize long-term sustainability. In the subsections that follow, we will discuss a few subsurface features that affect the best choice of GHX, and present some methods for assessment of geoexchange site suitability.

3.7.2.1 Factors Affecting Choice of GHX

There are several factors that affect the choice of GHX to be installed, some of which are described by Shillereff *et al.* (2008). These factors are summarized below.

Suitable Building or Application. One of the first considerations in applying geoexchange is to assess the type of application. Energy-efficient buildings, where thermal loads are minimized, result in smaller heat pump equipment and smaller associated GHX, and thus lower first costs. Applications with balanced annual heating and cooling loads make most practical sense. If total annual heat rejection to the GHX significantly exceeds total annual heat extraction from the GHX (or vice versa), a larger volume of Earth coupling is required, as described above in Section 3.9. Geoexchange in these cases may still be a viable option if a hybrid system is considered.

Geology. Geological conditions at a site are important because they impact the thermal exchange and the 'constructability' of the GHX. Properties affecting thermal conductivity of soils and rocks include density, mineral content, moisture content, and porosity. These properties are quite variable in the Earth's materials, and are also dynamic. Complex and highly variable geology can also make constructability difficult. This is especially true if a hard rock layer underlies a soil layer, requiring more than one drilling method to drill through soft soils and hard rock.

Urban Geology. Here, we identify 'urban geology' as aspects of the surface and subsurface that are not natural. These include, but are not limited to: underground utilities (sewers, water pipelines, electrical power lines, gas lines, irrigation piping, storage tanks, etc.), landfills, subsurface contamination, landscaping, and pavements and concretes. Damage to and restoration of these structures can be a significant cost item affecting the choice of GHX.

Groundwater. The presence of groundwater affects the choice of the GHX in many ways. Groundwater in pore spaces of soils and rocks significantly improves the thermal properties; if groundwater is not present, pore spaces are filled with air, which has a very low thermal conductivity. The presence of stored groundwater in soils and rocks may also pose significant problems with drilling boreholes and excavating trenches and pits. When groundwater flows in the subsurface at significant rates, heat advection occurs in the subsurface. If wells can yield an acceptable amount of groundwater, then open-loop systems might be considered. Scaling and corrosion potential of groundwater in open-loop systems needs consideration.

Subsurface Property Uncertainty and Heterogeneity. Geology is not uniform, and can vary over small scales. Contingencies are necessary in planning an underground project.

Land Area Availability. The available land area is a very important consideration in the choice of GHX. The GHX tools provided in this book allow calculation of GHX size, and therefore available land area. Available land area also dictates working and staging logistics.

Site Access and Limitations. Simply put, the question here is, who owns the resource? Adequate permits and water rights can easily impact the choice of GHX. The presence of overhead utilities, buildings, and other site logistics may also impact the choice of GHX.

Owner Concept and Vision (or Lack Thereof). Geoexchange concepts should not ignore input from the owner of the project. Owners may have a strong vision of what they want to achieve, both with regard to aesthetics, energy performance, and capital cost. On the other hand, owners may need a significant amount of education regarding geoexchange technology. Educating the owner is important to steer a geoexchange project in the proper direction, and to prevent such a project from progressing for the wrong reasons.

3.7.2.2 A Method for Assessing Geoexchange Site Suitability

Previous work on this topic includes Sachs (2002) and Shillereff *et al.* (2008). A staged approach with up to three phases is presented by Shillereff *et al.* (2008). Here, we present a tiered approach based on the flow of alternative energy projects as described in Chapter 1.

Pre-Feasibility. The first stage in assessing geoexchange site suitability is suggested here in the form of a general decision matrix (Table 3.1), which would ideally be part of a pre-feasibility study. A concept-type scoring approach is used, such as:

+1 = favorable impact,
0 = neutral impact, and
−1 = negative impact or unacceptable.

Weighting factors can also be applied on a project-specific basis.

Each Earth coupling option is evaluated against the following criteria:

- **Favorable Site Attributes.** This decision parameter captures key aspects of each Earth coupling. For example, a surface water heat exchanger would not be considered if there is no existing surface water or ability to excavate one. Favorable site features for a standing column well system include hard-rock geology with low well yields. Locations of utilities, expensive landscaping, site contamination, and other 'urban geology' aspects described above are important.
- **Capital Cost within Budget.** This decision parameter involves determining whether or not the project fits with the owner's budget.
- **Favorable Economic Indicators.** This decision parameter involves estimating economic indicators, such as simple payback period, net present value, and return on investment.

Table 3.1 Pre-Feasibility General Decision Matrix for Assessment of Earth Coupling Options for Geothermal Heat Pump Systems

	Decision Parameter										
Earth Coupling Option	Favorable Site Attributes?	Capital Cost within Budget	Favorable Economic Indicators	Ability to Meet Intended Loads	Effects on Neighbors and Environment	System Complexity and Constructability	Effect on Property Equity	Local Contractor Capability	Required Time for Construction	Regulatory and Legal Issues	**Total Score**
Opportunistic heat sources/sinks (e.g., sewer heat recovery, already planned excavation or pit)											
Groundwater Open Loop											
Standing Column Well											
Surface Water Open Loop											
Surface Water Closed Loop											
Horizontal Closed Loop											
Vertical Closed Loop											
Hybrid (with boiler, dry fluid cooler, solar thermal, etc.)											

NOTES: Base scoring criteria: +1 = favorable impact; 0 = neutral impact; −1 = negative impact or unacceptable. Weighting factors can be applied as necessary.

- **Ability to Meet Intended Loads.** Each Earth coupling option should be preliminarily sized to determine whether the intended loads can be met. If not, a hybrid option may be considered, or another HVAC system may be needed.
- **Effects on Neighbors and the Environment.** This decision parameter involves deleterious effects of the Earth coupling option on surrounding neighbors and on the environment. Such impact could occur during construction, or over the life of the system. For example, unacceptable noise or mud run-off during drilling could pose difficulty. An example of an environmental impact might include long-term temperature changes in surface water bodies.
- **System Complexity and Constructability.** A key aspect of a successful project is simplicity. If the system is overly complex, it becomes difficult and costly to construct and operate.
- **Effect on Property Equity.** Alternative energy systems can often increase property equity. However, this is sometimes associated with increased property taxes. This decision parameter should weigh these costs and benefits.

Table 3.2 Field Tests and Activities for Determining Site Suitability of Various Earth Coupling Options

Earth Coupling Option	Field Tests and Engineering Studies for Site Characterization
Opportunistic heat sources/ sinks	• Determine thermal characteristics of the heat source/sink and its availability. • Determine logistics of use.
Groundwater Open Loop	• Select suitable type of drilling method and plan and design at least one test well. • Three additional observation wells (or piezometers) may also be needed to determine aquifer parameters. • During drilling, soil and rock types and groundwater presence should be logged. • Note drilling difficulties. • Conduct a step-drawdown test to determine well yield and stable temperatures. These will be discussed in Chapter 4. • Collect and analyze groundwater samples for major ion analyses to determine scaling and corrosion potential. Biological testing of groundwater may also be needed.
Standing Column Well	• Similar to groundwater open loop. • A thermal response test is performed rather than a step-drawdown test.
Surface Water Open Loop	• Determine bathymetry of water body, including an assessment of bottom material for locating intake and outlet points. • Determine/measure temperature profiles on a seasonal basis. • Assess ice cover. • Collect and analyze water samples for major ion analyses to determine scaling and corrosion potential. Biological testing of water may also be needed.
Surface Water Closed Loop	• Determine bathymetry of water body, including an assessment of bottom material for purposes of anchoring the GHX. • Determine/measure temperature profiles on a seasonal basis. • Assess ice cover. • Biological testing of water may also be needed to assess fouling potential. • Assess location of the GHX to determine potential interference with other recreational activities, etc.
Horizontal Closed Loop	• Dig test pits spaced out across intended GHX area. These should be dug about ~2 m deep to determine soil types, groundwater seepage, and potential bedrock presence. • Assess thermal properties and level of difficulty in excavating.
Vertical Closed Loop	• Select suitable type of drilling method and plan at least one test borehole to be drilled. Larger projects may require multiple test holes if non-uniform geology is expected. • During drilling, soil and rock types and groundwater presence should be logged. • Note drilling difficulties.

Table 3.2 *(continued)*

Earth Coupling Option	Field Tests and Engineering Studies for Site Characterization
	• Outfit one borehole with a borehole heat exchanger and conduct a thermal response test. These will be discussed in Chapter 5.
Hybrid (with boiler, dry fluid cooler, solar, etc.)	• For the geothermal component, field testing as described above for the Earth coupling under consideration. • For the hybrid component, an engineering analysis of how the supplemental loads will be handled (i.e., will they offset building loads or ground loads?), and associated thermal storage. Solar hybrid options are discussed in Chapter 6 with closed-loop vertical systems.

- **Local Contractor Capability.** Lack of local contractors can result in cost-prohibitive construction and maintenance.
- **Required Time for Construction.** A key aspect in considering a GHX option is whether or not it fits into construction schedules. Long lead times for equipment and/or permits can result in an unacceptable option.
- **Regulatory and Legal Issues.** Laws and permits should be understood by the project team, and can dictate whether or not some options are viable or not.

During this pre-feasibility stage, the following background information should be reviewed where available: heating and cooling load calculations or estimates, topographic maps, bedrock and surficial geology maps, online Earth images, underground utility drawings, well records, previous relevant consultant reports, zoning and legal property maps and descriptions, contaminated site registry, and discussions with local drillers and other relevant contractors.

Earth coupling options receiving favorable scores (i.e., >0) are identified for further site characterization as warranted, if the project should proceed to a detailed feasibility study.

Detailed Feasibility. In detailed geoexchange feasibility studies, the HVAC system design is refined, and intrusive field studies are conducted toward further site characterization. Small projects (e.g., residential) typically cannot bear the economics of intrusive studies, and would typically move on to the design and construction phase. Intrusive studies and activities can be scaled by the size of the project, and are summarized in Table 3.2.

At the conclusion of additional field testing, the decision matrix shown in Table 3.1 is refined with more detail of the final option(s) with the purpose of identifying the best GHX option. There could be more than one suitable GHX, and owner consultation should be used to decide on the best choice. At this point, the decision could be made to move toward design and construction phases of the project, or to abandon the geothermal concept altogether. In some cases, additional work may be needed to simulate GHX performance in complex projects, assess environmental impact, and satisfy government agencies who require various studies for obtaining permits.

3.8 Chapter Summary

This chapter presented an overview of site characterization methods for geothermal energy projects. Earth origin and history were discussed, along with theories as to why the Earth is still hot today. Internal thermal energy generation is responsible for many (if not all) of the processes occurring on Earth, and this heat makes its way to the surface, where it interacts with atmospheric processes. A primer on geology and drilling methods for Energy Engineers was also presented.

Conceptual models of geothermal reservoirs were discussed, starting with a simple cubic control volume. Next, groundwater resources were discussed in the context of the nature and occurrence of groundwater. Conceptual models of aquifer systems and hydrothermal reservoirs were presented, along with an overview of geophysical and geochemical exploration methods. The chapter concluded with discussion of a useful decision model for assessment of sites under consideration for geothermal heat pump applications.

Discussion Questions and Exercise Problems

3.1 Make a detailed, scaled, cross-sectional sketch of the Earth. Show the different layers and note the thickness of each.

3.2 If the average ground surface temperature is 10 °C, what temperature would you expect at 2000 m at a location with a *normal* geothermal gradient? At a location with a geothermal gradient 2× the normal?

3.3 What is the terrestrial heat flow (in $W \cdot m^{-2}$) in the previous question? Compare your terrestrial heat flux values to an average solar heat flux reaching the ground surface and explain why geothermal energy is still useful or not.

3.4 Describe the concept of plate tectonics. Why does it occur? Describe its role in creating active, high-temperature geothermal regions.

3.5 Conduct some research on the branch of geophysics known as *paleomagnetism*. Briefly explain what it is and how it was/is used to confirm the theory of plate tectonics. Explain how paleomagnetism is used to reconstruct motion of the Earth's plates.

3.6 How do we know how old rocks are? How do we know how old the Earth is?

3.7 Locate and describe the nearest groundwater resource to where you live. How is it used?

3.8 Define the following in short concise sentences:

(a) geothermometer,
(b) seismic survey,
(c) resistivity survey,
(d) gravity survey,
(e) P waves.

3.9 List three surface features that are important indicators of moderate- to high-temperature near-surface geothermal resources and discuss what they indicate.

3.10 Why might a warm spring be at the ground surface but not have a subsurface geothermal reservoir directly below it?

3.11 Develop a plot of soil temperature vs. depth (from 0 to 5 m) for where you live on 21 March and 21 September. Assume a value of $\alpha = 6.0$ E-7 $m^2 \cdot s^{-1}$.

3.12 List and describe five factors that would make a site unsuitable for a geothermal heat pump project.

Part II

Harnessing the Resource

In Part II, we discuss harnessing the geothermal resource, which involves methods of transferring the geothermal energy from the Earth to the point of conversion to useful energy. One of the key tasks in this element of a geothermal project is the proper sizing of a ground heat exchanger (GHX) to transfer heat to/from the Earth at an acceptable rate over the life cycle of the reservoir. A suite of design tools is provided on this book's companion website to facilitate this GHX sizing process.

Broadly grouped, methods of harnessing a geothermal resource are classified as open- and closed-loop systems. However, we shall examine the GHX in the chapters that follow according to the heat exchange configuration, and shall examine the theoretical and practical considerations of each.

In **Chapter** 4, groundwater heat exchange systems are examined. In **Chapter** 5, we discuss the theoretical and practical aspects of single, vertical, closed-loop borehole heat exchangers, leading into **Chapter** 6, which examines the design of GHXs with multiple vertical borehole heat exchangers. **Chapter** 7 borrows upon many of the same principles of Chapters 5 and 6 to examine shallow, horizontal GHX systems. **Chapter** 8 deals with surface water heat exchange systems, examining both open- and closed-loop options. **Chapter** 9 describes a number of *opportunistic heat sources* that may be incorporated into geothermal energy systems. Finally, **Chapter** 10, the last chapter of this section, discusses hydraulic considerations and piping systems to convey heat transfer fluids from the GHX to the point of use.

Geothermal Heat Pump and Heat Engine Systems: Theory and Practice, First Edition. Andrew D. Chiasson.

Companion website: www.wiley.com/go/chiasson/geoHPSTP

4

Groundwater Heat Exchange Systems

4.1 Overview

Use of groundwater as a heat exchange medium is the oldest use of geothermal energy. Groundwater-based geothermal systems are often referred to simply as 'open-loop systems'.

Ancient civilizations used geothermal waters from springs for bathing, cooking, and heating, and for therapeutic and medicinal purposes. Hot geothermal water from wells has been used for centuries. Groundwater from wells has been used for decades for either free cooling or condenser cooling. The oldest documented use of a groundwater heat pump dates back to the 1940s.

It is often not the responsibility of the Energy Engineer to design a groundwater well – this is often left to other specialties, like Civil Engineers or Hydrogeologists. However, Energy Engineers do need to be familiar with water well technology and terms, as it is often within the realm of the Energy Engineer to specify required groundwater flow rates and temperatures for adequate system operation.

Ironically, even though the use of groundwater in an open-loop configuration is the oldest means of conditioning buildings in heat pump applications, it is perhaps now the least understood. Use of groundwater for geothermal systems requires an understanding of its occurrence in aquifers, hydraulics of water wells, its chemistry (in the context of scaling and corrosion potential), flow requirements for the building loads, and local regulations regarding its use.

In this chapter, we will first review some hydrogeologic definitions. We will then examine the practical and theoretical considerations related to open-loop, groundwater heat pump systems, and use a spreadsheet tool to calculate their optimized groundwater flow rates.

Geothermal Heat Pump and Heat Engine Systems: Theory and Practice, First Edition. Andrew D. Chiasson.

Companion website: www.wiley.com/go/chiasson/geoHPSTP

Learning objectives and goals:

1. Define basic terms used in Darcy's law and be able to calculate groundwater velocity.
2. Draw analogies between Fourier's law of heat conduction to Darcy's law.
3. Define basic terms used in water well hydraulics.
4. Identify general materials to avoid in open-loop systems based on groundwater chemistry.
5. Calculate groundwater flow requirements for geothermal heat pump systems.

4.2 Why Groundwater?

In open-loop groundwater systems, groundwater is used as the heat-carrier fluid. Recall from Chapter 3 that usable groundwater occurs in aquifers, which were defined as geologic formations that yield groundwater in usable quantities to wells. Also, recall that groundwater does not occur everywhere in usable quantity and/or quality in all locations, and therefore aquifers are geographically limited. In many jurisdictions, groundwater is treated and legislated as a mineral resource, and therefore many laws and regulations may dictate its use.

So why consider groundwater in heat exchange applications? One main reason is energy density. If a large quantity of good-quality groundwater can be accessed through one or a few wells, it can yield a significant quantity of energy; much more energy than by heat conduction through closed-loop heat exchangers. This is especially true in large applications, where the available space to drill multi-borehole fields may be limited. On the other end of the load spectrum, groundwater systems are a prudent choice for residential buildings that have a water well for drinking water purposes. When groundwater is found at a temperature that matches the load requirements, it can be used directly for heating or cooling without the need for conversion.

Thus, the advantages of groundwater systems are:

- higher potential energy density per land area (relative to closed-loop systems),
- low cost, especially for large loads and residential applications that need a drinking water well,
- water well drilling technology is well established,
- stable source temperature (with proper well field design), and
- standing column well option in certain circumstances.

There are, however, some significant disadvantages:

- groundwater use is water quality dependent. Poor chemical quality can result in scaling, corrosion, and well fouling due to bacteriological growth in wells (i.e., iron bacteria),
- water disposal can be limited (water pumped from the ground has to go somewhere),
- laws and regulations, and
- permits and water rights may limit or prohibit groundwater use.

4.3 Theoretical Considerations

4.3.1 Equations of Groundwater Flow

Darcy's Law. The origin of groundwater hydrology as a science can be traced back to the 1850s when Henri Darcy published a report on experiments he carried out to examine the flow of water through porous media. The results of those experiments have been generalized into an empirical law that now bears his name.

In Darcy's experiments, a glass column tube of cross-section A was packed with a sandy soil, and the column was outfitted with a manometer on each end, at a known distance apart (ΔL). Darcy then conducted a number of experiments with the column tilted at various angles from the vertical, by introducing water into the column, which was allowed to flow through the column until the inflow rate ($\dot{Q}$) equilibrated with the outflow rate. Darcy noted that $\dot{Q}$ was directly proportional to the difference in equilibrated water levels in the manometers (i.e., the change in hydraulic head, or Δh) and the cross-sectional area (A) of the column tube, and that $\dot{Q}$ was indirectly proportional to the length of the column (or ΔL). Thus,

$$\dot{Q} \propto A \frac{\Delta h}{\Delta L} \tag{4.1}$$

Darcy converted the proportionality to an equality with the introduction of a proportionality constant (K), known as the hydraulic conductivity, and thus the law became

$$\dot{Q} = KA \frac{\Delta h}{\Delta L} \tag{4.2}$$

Darcy concluded that this proportionality (K) must be a function of the soil. Subsequent researchers concluded that the hydraulic conductivity must be a function not only of the soil or rock but also of the fluid. Thus, hydraulic conductivity is further defined as

$$K = \frac{k \rho g}{\mu} \tag{4.3}$$

where k is the permeability of the medium (with dimensions of [L^2]) and is a function of the grain size and interconnected pore spaces, ρ and μ are the density and viscosity, respectively, of the fluid, and g is the acceleration due to gravity. Typical values of hydraulic conductivity for various geologic materials are included in Appendix B.

It should be emphasized that Darcy's law is an empirical law; it rests only on experimental evidence. Attempts have been made, however, to derive Darcy's law more fundamentally, from Navier–Stokes equations, which are widely known in fluid mechanics. Regardless, Darcy's law is a powerful, basic law that describes flow through porous media. It is such a fundamental law that we may observe several analogies in other forms of energy transfers due to a driving gradient. Some of these analogies are summarized in Table 4.1.

Darcy's law has lower and upper limits of applicability. It is a macroscopic law, and thus has a lower validity limit of scale, where a minimum representative elementary volume must statistically capture all elements of the porous medium. Further, in materials of very low hydraulic

Table 4.1 Summary of Analogous One-Dimensional Energy Rate Equations

Energy Form	Energy Flux (common SI units)	Driving Gradient	Proportionality Constant (common units)	Governing Law
Hydraulic (groundwater)	Groundwater flux, $\dot{Q}/m^2$ ($m{\cdot}s^{-1}{\cdot}m^{-2}$)	Hydraulic gradient ($\Delta H/\Delta L$)	Hydraulic Conductivity, K (m/s)	Darcy's law $\dot{q}''=K\frac{\Delta H}{\Delta L}$
Thermal	Conduction heat flux, q/m^2 ($W{\cdot}m^{-2}$)	Temperature gradient ($\Delta T/\Delta x$)	Thermal conductivity, k ($W{\cdot}m^{-1}{\cdot}K^{-1}$)	Fourier's law of heat conduction $\dot{q}''=k\frac{\Delta T}{\Delta x}$
Thermal	Convection heat flux, q/m^2 ($W{\cdot}m^{-2}$)	Temperature gradient (ΔT)	Convection heat transfer coefficient, h_c ($W{\cdot}m^{-2}{\cdot}K^{-1}$)	Newton's law of cooling $\dot{q}''=h_c\Delta T$
Thermal	Radiation heat flux, q/m^2 ($W{\cdot}m^{-2}$)	Temperature gradient (ΔT)	Radiation heat transfer coefficient (linearized), h_r ($W{\cdot}m^{-2}{\cdot}K^{-1}$)	Stefan–Boltzmann's law $\dot{q}''=h_r\Delta T$
Electrical	Current, I (A)	Voltage drop (ΔV)	Electrical resistance, R (ohm)	Ohm's law $I=\frac{1}{R}V$
Chemical/ mass	Chemical flux, J ($mg{\cdot}s^{-1}{\cdot}m^{-2}$)	Concentration gradient ($\Delta C/\Delta x$)	Diffusion coefficient, D ($m^2{\cdot}s^{-1}$)	Fick's law $J=D\frac{\Delta C}{\Delta x}$
Elastic	Force, F (N)	Distance (Δx)	Spring constant, k ($N{\cdot}m^{-1}$)	Hooke's law $F=k\Delta x$

conductivity, there is a threshold hydraulic gradient below which flow does not take place. On the other end of the scale, Darcy's law has upper limits with respect to groundwater velocity; Darcy's law breaks down in turbulent flow regimes.

Hydraulic Head, Hydraulic Gradient, and Groundwater Velocity. Darcy's law is often expressed in terms of the Darcy velocity or the Darcy flux, or more precisely, the *specific discharge*:

$$\dot{q}''=K\frac{\Delta h}{\Delta L} \tag{4.4}$$

Note that even though $\dot{q}''$ has dimensions of $[L{\cdot}T^{-1}]$, it is not a true velocity. The groundwater velocity (v), also known as the average linear groundwater velocity, is given by

$$v=\frac{Ki}{\phi} \tag{4.5}$$

where ϕ is the porosity of the geologic medium, and i is the hydraulic gradient ($\Delta h/\Delta L$). Typical values of porosity for various geologic materials are given in Appendix B.

Groundwater flows from areas of higher hydraulic head to areas of lower hydraulic head. The total groundwater hydraulic head (h) is given by a simplified form of the Bernoulli equation and is the sum of the elevation head (z) and the pressure head (ψ):

$$h=z+\psi \tag{4.6}$$

Thus, in the case of constant density groundwater flow, Δh is mainly a function of z.

Groundwater head is determined by installing observation wells and measuring the elevation of the static water level in the well. Three observation wells are needed to determine the slope of the potentiometric surface of the aquifer, as shown in Figure 4.1.

Thus, to determine the direction of groundwater flow and the hydraulic gradient, the following data are needed, as shown in Figure 4.1:

1. The relative geographic position of the wells, typically done by a land surveyor.
2. The distance between the wells.
3. The total head at each well, measured relative to a common datum.

Steps in the solution are outlined below and illustrated in Figure 4.1:

(a) On a scaled map, identify the well that has the intermediate water level (i.e., neither the highest head nor the lowest head).
(b) Calculate the position (x) between the well having the highest head and the well having the lowest head at which the head is the same as that in the intermediate well.
(c) Draw a straight line between the intermediate well and the point identified in step (b) as being between the well having the highest head and that having the lowest head. This line represents an equipotential line (or a water-level contour) along which the total head is the same as that in the intermediate well.
(d) Draw a line perpendicular to the water-level contour and through either the well with the highest head or the well with the lowest head. This line represents the direction of groundwater flow.
(e) Divide the difference between the head of the well and that of the contour by the distance between the well and the contour. The result is the hydraulic gradient (i).

The Diffusion Equation. A major objective in groundwater flow analysis is determining the hydraulic head distribution in a flow system. By applying the law of conservation of mass to a control volume and by making use of Darcy's law [Equation (4.2)], an equation defining the hydraulic head distribution can be derived. Transient groundwater flow with constant density can then be expressed in Cartesian coordinates as

$$\alpha\left[\frac{\partial^2 h}{\partial x^2}+\frac{\partial^2 h}{\partial y^2}+\frac{\partial^2 h}{\partial z^2}\right]+G=\frac{\partial h}{\partial t} \tag{4.7}$$

where α is the hydraulic diffusivity = T/S, and G is a general source/sink term that accounts for inflows and outflows due to wells, surface recharge, etc. T is the aquifer transmissivity, which is equivalent to $K \times b$, where b is the saturated aquifer thickness. S is the aquifer storativity (or the storage coefficient), which is defined as the volume of water released from storage per unit decline in hydraulic head in the aquifer per unit area of the aquifer. In confined aquifers, storativity is related to expansion of the water and compression of the aquifer, and is thus a relatively small value ranging between 10^{-5} and 10^{-3}. In unconfined aquifers, storativity is known as the *specific yield*, as water released from storage represents actual dewatering of pores. Thus,

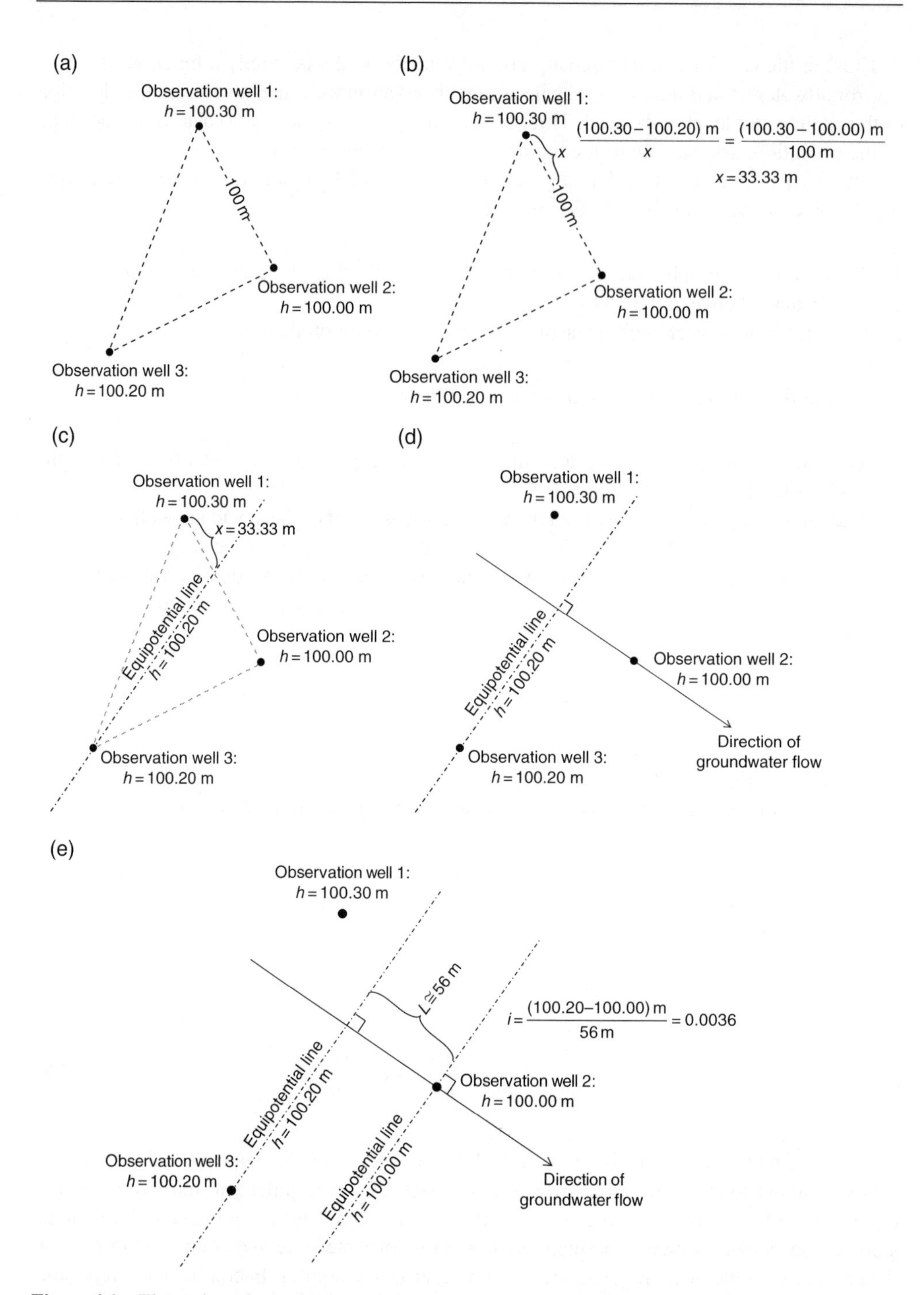

Figure 4.1 Illustration of steps in determining the hydraulic gradient and direction of groundwater flow from three observation wells

values of S for unconfined aquifers are much greater than those for confined aquifers, and are of the order of 0.01–0.30.

Numerical Methods and Simple Spreadsheet Modeling. Under simplified conditions and aquifer geometry, Equation (4.7) can be solved analytically. However, except for applications in well hydraulics (which we shall discuss in the next subsection), analytical solutions to Equation (4.7) are not widely used in practical, multidimensional applications. Numerical solutions are much more versatile, and can be readily solved with computer application.

Numerical methods involve three general steps. First, the partial differential equation is approximated by finite difference, finite element, or finite volume techniques for the appropriate boundary and initial conditions. Second, the domain of interest is discretized into finite-sized units to which the finite equation is applied. Third, the resulting set of algebraic equations cast as a matrix equation is solved to arrive at the hydraulic head distribution in space and time.

Here, we will focus on a simple case of determining the steady-state head distribution in a two-dimensional, confined, homogeneous, isotropic aquifer. The governing equation describing the head distribution is the well-known Laplace equation:

$$\frac{\partial^2 h}{\partial x^2} + \frac{\partial^2 h}{\partial y^2} = 0 \tag{4.8}$$

The typical geometry and notation of the finite difference cells in the x–y Cartesian coordinate plane are shown in Figure 4.2.

As depicted in Figure 4.2, the nodal equations are formulated using a node-centered, mass/energy balance approach on a control volume where $\Delta x = \Delta y$. Thus, for a general node at $(x, y) = (m, n)$, the mass/energy balance is given by

$$\sum_{i=1}^{4} \dot{q}_i = 0 \tag{4.9}$$

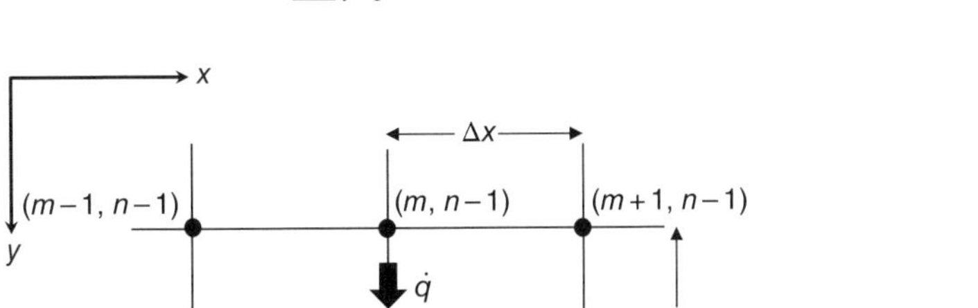
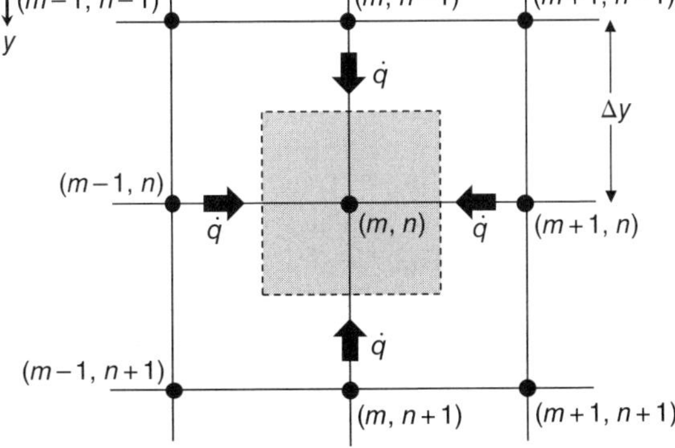

Figure 4.2 General finite difference cell geometry and notation

For two-dimensional conditions, mass/energy exchange is influenced by Darcy flow between (m, n) and its four adjoining nodes, as well as by point sources, such as pumping or injection wells. At boundary surface nodes, mass/energy exchange is further influenced by external fluxes. Table 4.2 summarizes steady-state, finite difference equations for common situations.

Table 4.2 Summary of steady-state, nodal finite difference equations (for $\Delta x = \Delta y$)

Nodal Geometry	Finite Difference Equation for Groundwater Head (h)	
Interior Node (i.e., nodes surrounded by four other nodes)	$h_{m,n} = \dfrac{h_{m+1,n} + h_{m-1,n} + h_{m,n+1} + h_{m,n-1}}{4}$	(4.10)
Top Boundary Node with Uniform Flux	$h_{m,n} = \dfrac{2h_{m,n+1} + h_{m+1,n} + h_{m-1,n} + 2\dfrac{\dot{q}''\Delta x}{K}}{4}$	(4.11)
Bottom Boundary Node with Uniform Flux	$h_{m,n} = \dfrac{2h_{m,n-1} + h_{m+1,n} + h_{m-1,n} + 2\dfrac{\dot{q}''\Delta x}{K}}{4}$	(4.12)
Left-Side Boundary Node with Uniform Flux	$h_{m,n} = \dfrac{2h_{m+1,n} + h_{m,n+1} + h_{m,n-1} + 2\dfrac{\dot{q}''\Delta x}{K}}{4}$	(4.13)
Right-Side Boundary Node with Uniform Flux	$h_{m,n} = \dfrac{2h_{m-1,n} + h_{m,n+1} + h_{m,n-1} + 2\dfrac{\dot{q}''\Delta x}{K}}{4}$	(4.14)
Point Source in Interior Node (e.g., well)	$h_{m,n} = \dfrac{h_{m+1,n} + h_{m-1,n} + h_{m,n+1} + h_{m,n-1} + \dfrac{\dot{Q}}{HK}}{4}$	(4.15)

where K is the hydraulic conductivity, $\dot{Q}$ is the well pumping rate (negative for groundwater extraction, positive for water injection), and H is the well intake length (in the z direction).

Many commercially available software programs exist for groundwater flow modeling. However, we can employ spreadsheets in a simple but powerful way to perform iterative calculations. In spreadsheet modeling, each spreadsheet cell represents a nodal equation. Let's illustrate with the following example.

Example 4.1 Spreadsheet Modeling of Groundwater Flow Equations

Consider a simple flow system shown in plan view in Figure E.4.1a. A supply well (well 1) is extracting groundwater and using it in a cooling application. The groundwater is returned to the aquifer via two injection wells (wells 2 and 3). The aquifer system has been characterized as 30 m wide by 40 m long, with a relatively constant thickness of 10 m. The east and west sides of the aquifer are no-flow boundaries (i.e., no groundwater flow crosses these boundaries). The north side of the aquifer is a recharge area, with a constant head of 50.0 m. The south boundary is defined by a creek, and a constant head of 49.5 m can be assumed. The hydraulic conductivity of the aquifer is estimated to be 5×10^{-4} m·s^{-1}. The locations and pumping rates of the three wells are:

Well Number	(x, y) Position (m from origin)	Pumping Rate ($m^3 \cdot s^{-1}$)
1	(15, 10)	−1.0 E-3
2	(10, 30)	+5.0 E-4
3	(20, 30)	+5.0 E-4

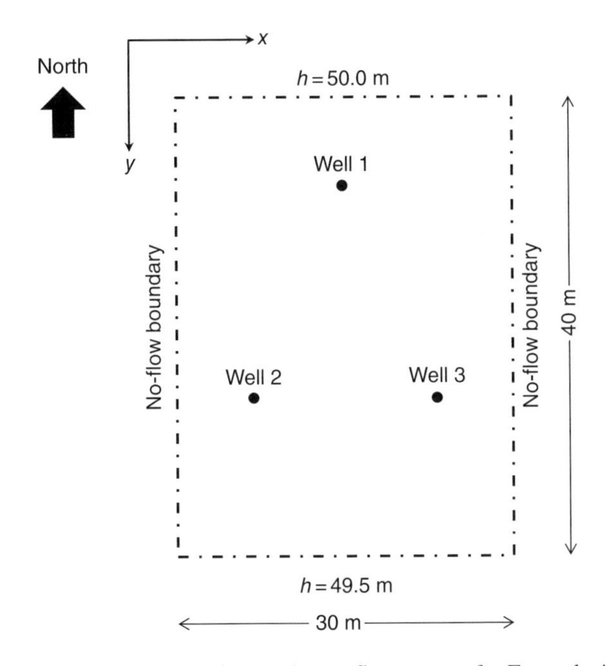

Figure E.4.1a Map of groundwater flow system for Example 4.1

Determine the steady-state head distribution in the aquifer.

Solution

Here, we will implement the finite difference equations shown in Table 4.2 into an Excel Spreadsheet as follows:

- enable iterative calculation in the Options, Formulas window,
- outline a boundary of 30 × 40 cells, thus setting $\Delta x = \Delta y = 1$ m,
- enter Equation (4.10) in all interior cells,
- enter a value of 50.0 in all north boundary cells,
- enter a value of 49.5 in all south boundary cells,
- enter Equation (4.13) in all west boundary cells with $K = 5 \times 10^{-4}$ m·s^{-1} and $\Delta x = 1$ m,
- enter Equation (4.14) in all east boundary cells with $K = 5 \times 10^{-4}$ m·s^{-1} and $\Delta x = 1$ m,
- enter Equation (4.15) in appropriate well cells, where $\dot{Q}$ is the pumping rate and $H = 10$ m.

The following results are obtained:

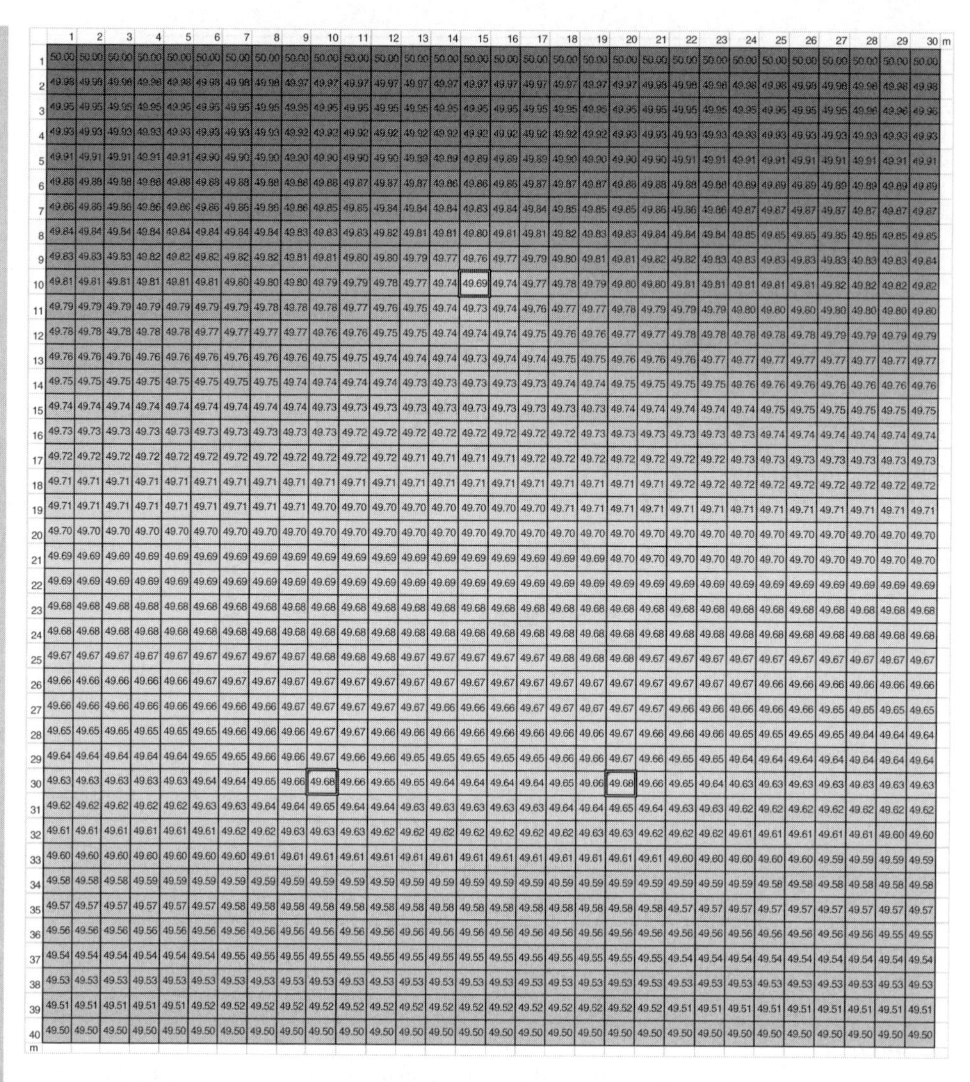

	1	2	3	4	5	6	7	8	9	10	11	12	13	14	15	16	17	18	19	20	21	22	23	24	25	26	27	28	29	30 m
1	50.00	50.00	50.00	50.00	50.00	50.00	50.00	50.00	50.00	50.00	50.00	50.00	50.00	50.00	50.00	50.00	50.00	50.00	50.00	50.00	50.00	50.00	50.00	50.00	50.00	50.00	50.00	50.00	50.00	50.00
2	49.98	49.98	49.98	49.98	49.98	49.98	49.98	49.98	49.97	49.97	49.97	49.97	49.97	49.97	49.97	49.97	49.97	49.97	49.97	49.97	49.98	49.98	49.98	49.98	49.98	49.98	49.98	49.98	49.98	49.98
3	49.95	49.95	49.95	49.95	49.95	49.95	49.95	49.95	49.95	49.95	49.95	49.95	49.95	49.95	49.95	49.95	49.95	49.95	49.95	49.95	49.95	49.95	49.95	49.95	49.95	49.95	49.95	49.96	49.96	49.96
4	49.93	49.93	49.93	49.93	49.93	49.93	49.93	49.93	49.92	49.92	49.92	49.92	49.92	49.92	49.92	49.92	49.92	49.92	49.92	49.93	49.93	49.93	49.93	49.93	49.93	49.93	49.93	49.93	49.93	49.93
5	49.91	49.91	49.91	49.91	49.91	49.90	49.90	49.90	49.90	49.90	49.90	49.90	49.89	49.89	49.89	49.89	49.89	49.90	49.90	49.90	49.90	49.91	49.91	49.91	49.91	49.91	49.91	49.91	49.91	49.91
6	49.88	49.88	49.88	49.88	49.88	49.88	49.88	49.88	49.88	49.88	49.87	49.87	49.87	49.86	49.86	49.86	49.87	49.87	49.87	49.88	49.88	49.88	49.88	49.89	49.89	49.89	49.89	49.89	49.89	49.89
7	49.86	49.86	49.86	49.86	49.86	49.86	49.86	49.86	49.86	49.85	49.85	49.84	49.84	49.84	49.83	49.84	49.84	49.85	49.85	49.85	49.86	49.86	49.86	49.87	49.87	49.87	49.87	49.87	49.87	49.87
8	49.84	49.84	49.84	49.84	49.84	49.84	49.84	49.84	49.83	49.83	49.83	49.82	49.81	49.81	49.80	49.81	49.81	49.82	49.83	49.83	49.84	49.84	49.84	49.85	49.85	49.85	49.85	49.85	49.85	49.85
9	49.83	49.83	49.83	49.82	49.82	49.82	49.82	49.82	49.81	49.81	49.80	49.80	49.79	49.77	49.76	49.77	49.79	49.80	49.81	49.81	49.82	49.82	49.83	49.83	49.83	49.83	49.83	49.83	49.83	49.84
10	49.81	49.81	49.81	49.81	49.81	49.81	49.80	49.80	49.80	49.79	49.79	49.78	49.77	49.74	49.69	49.74	49.77	49.78	49.79	49.80	49.80	49.81	49.81	49.81	49.81	49.81	49.82	49.82	49.82	49.82
11	49.79	49.79	49.79	49.79	49.79	49.79	49.79	49.78	49.78	49.78	49.77	49.76	49.75	49.74	49.73	49.74	49.76	49.77	49.77	49.78	49.79	49.79	49.79	49.80	49.80	49.80	49.80	49.80	49.80	49.80
12	49.78	49.78	49.78	49.78	49.78	49.77	49.77	49.77	49.77	49.76	49.76	49.75	49.75	49.74	49.74	49.74	49.75	49.76	49.76	49.77	49.77	49.78	49.78	49.78	49.78	49.78	49.79	49.79	49.79	49.79
13	49.76	49.76	49.76	49.76	49.76	49.76	49.76	49.76	49.76	49.75	49.75	49.74	49.74	49.74	49.73	49.74	49.74	49.75	49.75	49.76	49.76	49.76	49.77	49.77	49.77	49.77	49.77	49.77	49.77	49.77
14	49.75	49.75	49.75	49.75	49.75	49.75	49.75	49.75	49.74	49.74	49.74	49.74	49.73	49.73	49.73	49.73	49.73	49.74	49.74	49.75	49.75	49.75	49.75	49.76	49.76	49.76	49.76	49.76	49.76	49.76
15	49.74	49.74	49.74	49.74	49.74	49.74	49.74	49.74	49.74	49.73	49.73	49.73	49.73	49.73	49.73	49.73	49.73	49.73	49.73	49.74	49.74	49.74	49.74	49.74	49.75	49.75	49.75	49.75	49.75	49.75
16	49.73	49.73	49.73	49.73	49.73	49.73	49.73	49.73	49.73	49.73	49.72	49.72	49.72	49.72	49.72	49.72	49.72	49.72	49.73	49.73	49.73	49.73	49.73	49.73	49.74	49.74	49.74	49.74	49.74	49.74
17	49.72	49.72	49.72	49.72	49.72	49.72	49.72	49.72	49.72	49.72	49.72	49.72	49.71	49.71	49.71	49.71	49.72	49.72	49.72	49.72	49.72	49.72	49.72	49.73	49.73	49.73	49.73	49.73	49.73	49.73
18	49.71	49.71	49.71	49.71	49.71	49.71	49.71	49.71	49.71	49.71	49.71	49.71	49.71	49.71	49.71	49.71	49.71	49.71	49.71	49.71	49.71	49.72	49.72	49.72	49.72	49.72	49.72	49.72	49.72	49.72
19	49.71	49.71	49.71	49.71	49.71	49.71	49.71	49.71	49.71	49.70	49.70	49.70	49.70	49.70	49.70	49.70	49.70	49.71	49.71	49.71	49.71	49.71	49.71	49.71	49.71	49.71	49.71	49.71	49.71	49.71
20	49.70	49.70	49.70	49.70	49.70	49.70	49.70	49.70	49.70	49.70	49.70	49.70	49.70	49.70	49.70	49.70	49.70	49.70	49.70	49.70	49.70	49.70	49.70	49.70	49.70	49.70	49.70	49.70	49.70	49.70
21	49.69	49.69	49.69	49.69	49.69	49.69	49.69	49.69	49.69	49.69	49.69	49.69	49.69	49.69	49.69	49.69	49.69	49.69	49.69	49.70	49.70	49.70	49.70	49.70	49.70	49.70	49.70	49.70	49.70	49.70
22	49.69	49.69	49.69	49.69	49.69	49.69	49.69	49.69	49.69	49.69	49.69	49.69	49.69	49.69	49.69	49.69	49.69	49.69	49.69	49.69	49.69	49.69	49.69	49.69	49.69	49.69	49.69	49.69	49.69	49.69
23	49.68	49.68	49.68	49.68	49.68	49.68	49.68	49.68	49.68	49.68	49.68	49.68	49.68	49.68	49.68	49.68	49.68	49.68	49.68	49.68	49.68	49.68	49.68	49.68	49.68	49.68	49.68	49.68	49.68	49.68
24	49.68	49.68	49.68	49.68	49.68	49.68	49.68	49.68	49.68	49.68	49.68	49.68	49.68	49.68	49.68	49.68	49.68	49.68	49.68	49.68	49.68	49.68	49.68	49.68	49.68	49.68	49.68	49.68	49.68	49.68
25	49.67	49.67	49.67	49.67	49.67	49.67	49.67	49.67	49.67	49.68	49.68	49.68	49.67	49.67	49.67	49.67	49.67	49.68	49.68	49.68	49.67	49.67	49.67	49.67	49.67	49.67	49.67	49.67	49.67	49.67
26	49.66	49.66	49.66	49.66	49.66	49.67	49.67	49.67	49.67	49.67	49.67	49.67	49.67	49.67	49.67	49.67	49.67	49.67	49.67	49.67	49.67	49.67	49.67	49.67	49.66	49.66	49.66	49.66	49.66	49.66
27	49.66	49.66	49.66	49.66	49.66	49.66	49.66	49.66	49.67	49.67	49.67	49.67	49.67	49.66	49.66	49.66	49.67	49.67	49.67	49.67	49.67	49.66	49.66	49.66	49.66	49.66	49.65	49.65	49.65	49.65
28	49.65	49.65	49.65	49.65	49.65	49.65	49.66	49.66	49.66	49.67	49.67	49.66	49.66	49.66	49.66	49.66	49.66	49.66	49.66	49.67	49.66	49.66	49.66	49.65	49.65	49.65	49.65	49.64	49.64	49.64
29	49.64	49.64	49.64	49.64	49.64	49.65	49.65	49.66	49.66	49.67	49.66	49.66	49.65	49.65	49.65	49.65	49.65	49.66	49.66	49.67	49.66	49.65	49.65	49.64	49.64	49.64	49.64	49.64	49.64	49.64
30	49.63	49.63	49.63	49.63	49.63	49.64	49.64	49.65	49.66	49.68	49.66	49.65	49.65	49.64	49.64	49.64	49.64	49.65	49.66	49.68	49.66	49.65	49.64	49.63	49.63	49.63	49.63	49.63	49.63	49.63
31	49.62	49.62	49.62	49.62	49.62	49.63	49.63	49.64	49.64	49.65	49.64	49.64	49.63	49.63	49.63	49.63	49.63	49.64	49.64	49.65	49.64	49.63	49.63	49.62	49.62	49.62	49.62	49.62	49.62	49.62
32	49.61	49.61	49.61	49.61	49.61	49.61	49.62	49.62	49.63	49.63	49.63	49.62	49.62	49.62	49.62	49.62	49.62	49.62	49.63	49.63	49.62	49.62	49.62	49.61	49.61	49.61	49.61	49.61	49.60	49.60
33	49.60	49.60	49.60	49.60	49.60	49.60	49.60	49.61	49.61	49.61	49.61	49.61	49.61	49.61	49.61	49.61	49.61	49.61	49.61	49.61	49.61	49.60	49.60	49.60	49.60	49.60	49.59	49.59	49.59	49.59
34	49.58	49.58	49.58	49.59	49.59	49.59	49.59	49.59	49.59	49.59	49.59	49.59	49.59	49.59	49.59	49.59	49.59	49.59	49.59	49.59	49.59	49.59	49.59	49.59	49.58	49.58	49.58	49.58	49.58	49.58
35	49.57	49.57	49.57	49.57	49.57	49.57	49.58	49.58	49.58	49.58	49.58	49.58	49.58	49.58	49.58	49.58	49.58	49.58	49.58	49.58	49.58	49.57	49.57	49.57	49.57	49.57	49.57	49.57	49.57	49.57
36	49.56	49.56	49.56	49.56	49.56	49.56	49.56	49.56	49.56	49.56	49.56	49.56	49.56	49.56	49.56	49.56	49.56	49.56	49.56	49.56	49.56	49.56	49.56	49.56	49.56	49.56	49.56	49.56	49.55	49.55
37	49.54	49.54	49.54	49.54	49.54	49.54	49.55	49.55	49.55	49.55	49.55	49.55	49.55	49.55	49.55	49.55	49.55	49.55	49.55	49.55	49.55	49.54	49.54	49.54	49.54	49.54	49.54	49.54	49.54	49.54
38	49.53	49.53	49.53	49.53	49.53	49.53	49.53	49.53	49.53	49.53	49.53	49.53	49.53	49.53	49.53	49.53	49.53	49.53	49.53	49.53	49.53	49.53	49.53	49.53	49.53	49.53	49.53	49.53	49.53	49.53
39	49.51	49.51	49.51	49.51	49.51	49.52	49.52	49.52	49.52	49.52	49.52	49.52	49.52	49.52	49.52	49.52	49.52	49.52	49.52	49.52	49.52	49.51	49.51	49.51	49.51	49.51	49.51	49.51	49.51	49.51
40	49.50	49.50	49.50	49.50	49.50	49.50	49.50	49.50	49.50	49.50	49.50	49.50	49.50	49.50	49.50	49.50	49.50	49.50	49.50	49.50	49.50	49.50	49.50	49.50	49.50	49.50	49.50	49.50	49.50	49.50
m																														

Figure E.4.1b Finite difference spreadsheet model results for Example 4.1

***Discussion*:** The cells containing the wells are denoted in Figure E.4.1b with double-lined borders. The head results presented in Figure E.4.1b are shown to two decimal places for clarity, but in practice the groundwater elevations should be measured to the nearest millimeter. Note how the pumping scheme changes the natural groundwater flow pattern slightly; groundwater flow gradients develop in a radial pattern around each well, such that we see a groundwater depression around the pumping well, and groundwater mounds around the injection wells. However, pumping rates are not high enough, and the wells are spaced far enough apart, such that the natural groundwater flow direction remains north to south. A further visual inspection of the hydraulic heads reveals that the pumping well does not reverse the flow gradient enough to draw injected water back to the pumping well.

Thus, this pumping and injection scheme is acceptable in avoiding 'shortcircuiting' or thermal breakthrough of the pumped and injected water. In a geothermal application, some thermal breakthrough may be acceptable, especially in situations that will switch between heating and cooling. We will subsequently discuss a well doublet with seasonal changeover between pumping and injection wells. A geothermal application where thermal breakthrough would be unacceptable is a direct-use application that is used for 100% heating or 100% cooling.

For better visualization, we could develop contour lines of equal head to visualize the aquifer potentiometric surface. Further, streamlines could be drawn, which are lines that are perpendicular to the hydraulic head lines. Streamlines show the direction that fluid elements travel.

As we shall discuss in Section 4.3.3, heat is not transported in aquifers at the same rate as groundwater flows. Thus, this example serves as a powerful first approximation of the potential for thermal breakthrough in a scheme of pumping and injection wells. A more detailed analysis must include the coupled equations of heat transport in groundwater in order to determine the actual temperature distribution in the aquifer.

4.3.2 Well Hydraulics

The foregoing analysis is useful for examining groundwater flow patterns over relatively large scales, with multiple sources and sinks. In groundwater hydrology, there is a large class of boundary-value problems describing radial flow to a well, and the analysis of such problems constitutes the field of *well hydraulics*. There are numerous variants of these types of problem, depending on the boundary condition, where the mathematical boundary condition often represents a physical boundary in an aquifer. Thus, there are numerous books and research papers on the topic of well hydraulics, and readers are referred to these for further details. For example, Driscoll (2008), Dawson and Istok (1991), and Freeze and Cherry (1979).

The response of aquifers to withdrawals from wells is an important topic in groundwater hydrology, and is important to groundwater heat exchange systems. Here, we will consider two cases, an ideal aquifer and a leaky aquifer, and then consider the hydraulic efficiency of wells.

Response of an Ideal Aquifer to Pumping. When withdrawals start, the water level in the well begins to decline as water is removed from storage in the well. The head in the well falls below the level in the surrounding aquifer. As a result, water begins to move from the aquifer into the well. As pumping continues, the water level in the well continues to decline, and the rate of flow into the well from the aquifer continues to increase until the rate of inflow equals the rate of withdrawal. The movement of water from an aquifer into a well results in the formation of a cone of depression, as shown in Figure 4.3, in a confined aquifer. Because water must converge on the well from all directions, and because the area through which the flow occurs decreases toward the well, the hydraulic gradient gets steeper toward the well.

As groundwater flow toward a well exhibits radial symmetry, it is convenient to convert Equation (4.7) to cylindrical coordinates:

$$\alpha\left[\frac{\partial^2 h}{\partial r^2}+\frac{1}{r}\frac{\partial h}{\partial r}+\frac{1}{r^2}\frac{\partial^2 h}{\partial \varphi^2}+\frac{\partial^2 h}{\partial z^2}\right]+G=\frac{\partial h}{\partial t} \tag{4.16}$$

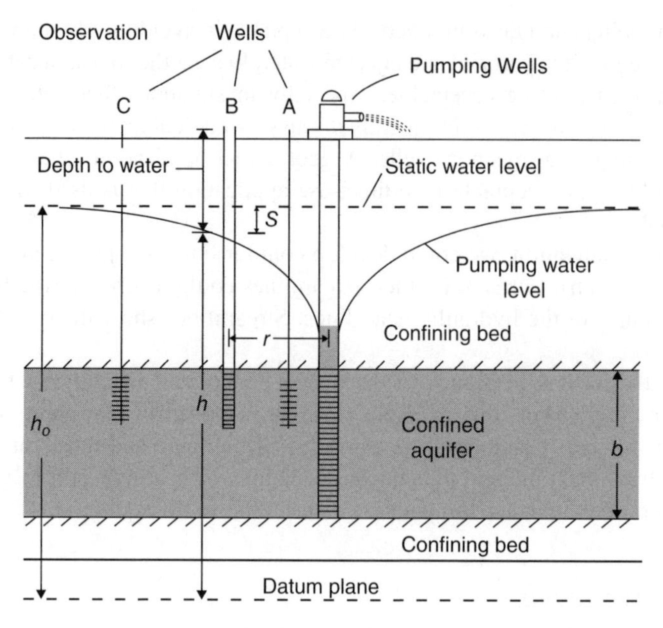

Figure 4.3 Drawdown cone developed during pumping of a well (Heath, 1983)

Considering a simple aquifer configuration, that is, one that is horizontal, confined, infinite in horizontal extent, constant thickness, homogeneous, and isotropic, Equation (4.16) simplifies to

$$\frac{\partial^2 h}{\partial r^2} + \frac{1}{r}\frac{\partial h}{\partial r} = \frac{1}{\alpha}\frac{\partial h}{\partial t} \tag{4.17}$$

The initial conditions are prescribed as

$$h(r,t) = h(r,0) = h_o \tag{4.18}$$

and the boundary conditions, assuming no drawdown at infinite radius, are given by

$$h(r,t) = h(\infty,t) = h_o \tag{4.19}$$

$$\lim_{r \to 0}\left(r\frac{\partial h}{\partial r}\right) = \frac{\dot{Q}}{2\pi T} \tag{4.20}$$

where h_o is the undisturbed hydraulic head.

Theis (1935), in what is considered to be one of the fundamental breakthroughs in groundwater hydrology, adapted Kelvin's line source model for practical-use well hydraulics. The result is an analytical solution to Equation (4.17), subject to the initial and boundary conditions described by Equations (4.18) to (4.20):

$$\Delta h_r = s_r = \frac{\dot{Q}}{4\pi T}\int_u^\infty \frac{e^{-u}}{u}\mathrm{d}u \tag{4.21}$$

where Δh_r is the change in total head at radius r, s_r is the drawdown at radius r, and u is defined as

$$u = \frac{r^2}{4\alpha t} \tag{4.22}$$

where r is the radius from the well, α is the hydraulic diffusivity of the medium, and t is the time duration of pumping at a rate of $\dot{Q}$.

The infinite integral appearing in Equation (4.21) is well known in mathematics as the *exponential integral*. For the definition of u given in Equation (4.22), the integral is known as the well function $W(u)$. Thus, Equation (4.21) may be written as

$$s_r = \frac{\dot{Q}}{4\pi T} W(u) \tag{4.23}$$

For a wide range of u values, $W(u)$ may be approximated (Srivastava and Guzan-Guzman, 1998) as

$$W(u) = \ln\left(\frac{\exp(-\gamma)}{u}\right) + 0.9653u - 0.1690u^2 \text{ for } u \leq 1 \tag{4.24}$$

where γ is Euler's constant = 0.5772, and

$$W(u) = \frac{1}{ue^u}\frac{u + 0.3575}{u + 1.280} \text{ for } u > 1 \tag{4.25}$$

In Equation (4.24), the higher-order terms are negligible for small values of u.

Response of a Leaky Aquifer to Pumping. The inherent assumption in the Theis solution that geologic formations overlying and underlying a confined aquifer are impermeable is seldom truly satisfied. More typically, aquifers undergoing pumping receive water from adjacent beds, and the aquifer is termed 'leaky'. In some cases the leakage rate may be ignored, but in other cases it may not.

Physically, in leaky aquifers, flow to the pumping well is no longer purely radial, and some vertical hydraulic gradients develop in the adjacent geologic units. Mathematically, the $\partial^2 h/\partial z^2$ term in Equation (4.16) cannot be ignored as with the Theis solution. The following solution has been developed by Hantush and Jacob (1956):

$$s_r = \frac{\dot{Q}}{4\pi T} W(u, B) \tag{4.26}$$

where $W(u, B)$ is known as the *leaky well function* or the *Hantush–Jacob well function*, defined mathematically as

$$W(u, B) = \int_u^{\infty} \frac{1}{y} \exp\left[-y - \frac{B^2}{4y}\right] \mathrm{d}y \tag{4.27}$$

$W(u, B)$ is also known as an *incomplete Bessel function*. In the well hydraulics of leaky aquifers, B is defined as:

$$B = \left(\frac{K'}{Kbb'}\right)^{-0.5} \tag{4.28}$$

where K and b are the hydraulic conductivity and thickness of the aquifer, and K' and b' are the hydraulic conductivity and thickness of the confining (leaky) beds.

The function $W(u, B)$ has some important features. First, $W(u,0) = W(u)$. Second, for large values of u, $W(u, B)$ approaches $W(u, 0)$. Third, $W(0, B) = 2K_0(B)$, where K_0 is the modified Bessel function of the second kind of order 0.

Srivastava and Guzan-Guzman (1998) give the following practical approximations for $W(u, B)$:

$$W(u,B) = W(0,B) - W\left(\frac{B^2}{4u}, 0\right) \quad \text{for } u < u_{min} \tag{4.29}$$

$$W(u,B) = W(u,0) \quad \text{for } u > u_{max} \tag{4.30}$$

$$0.5 \cdot W(0,B) \cdot \text{erfc}\left(\alpha X + \beta X^3\right) \quad \text{for } u_{min} \le u \le u_{max} \tag{4.31}$$

where

$$X = \ln\left(\frac{2u}{B}\right) \tag{4.32a}$$

$$\alpha = 0.7708 + 0.3457\ln(B) + 0.09128\{\ln(B)\}^2 + 0.09937\{\ln(B)\}^3 \tag{4.32b}$$

$$\beta = 0.02796 + 0.01023\ln(B) \tag{4.32c}$$

The values of u_{min} and u_{max} are given by

$$u_{min} = \max\left(0.06541B^{0.2763}, 0.02985\right) \quad \text{for } B \le 0.5 \tag{4.33a}$$

$$u_{min} = 0.1192B^{1.142} \quad \text{for } B > 0.5 \tag{4.33b}$$

$$u_{max} = \max\left(39.93B^{2.391}, 0.02985\right) \tag{4.33c}$$

Hantush (1956) tabulated the values of $W(u, B)$ (actually, $W(u, r/B)$). Figure 4.4 is a plot of this function versus $1/u$, which has been developed using the foregoing equations. If K' is impermeable, then $K' = 0$, and from Equation (4.28), $B = 0$. For the case of $W(u, B)$ with $B = 0$, the solution reduces to the Theis solution (i.e., the well function $W(u)$).

Note the shape of the curves shown in Figure 4.4. The shape of an actual response curve of drawdown in a well versus time when plotted on a log–log graph has the same shape as one of the curves in Figure 4.4 if the aquifer behavior conforms to the assumptions inherent in the mathematical model. That is, if the aquifer conforms to the assumptions of an ideal aquifer, the drawdown vs. time response will be similar to that shown in Figure 4.4 for $W(u, B)$ with

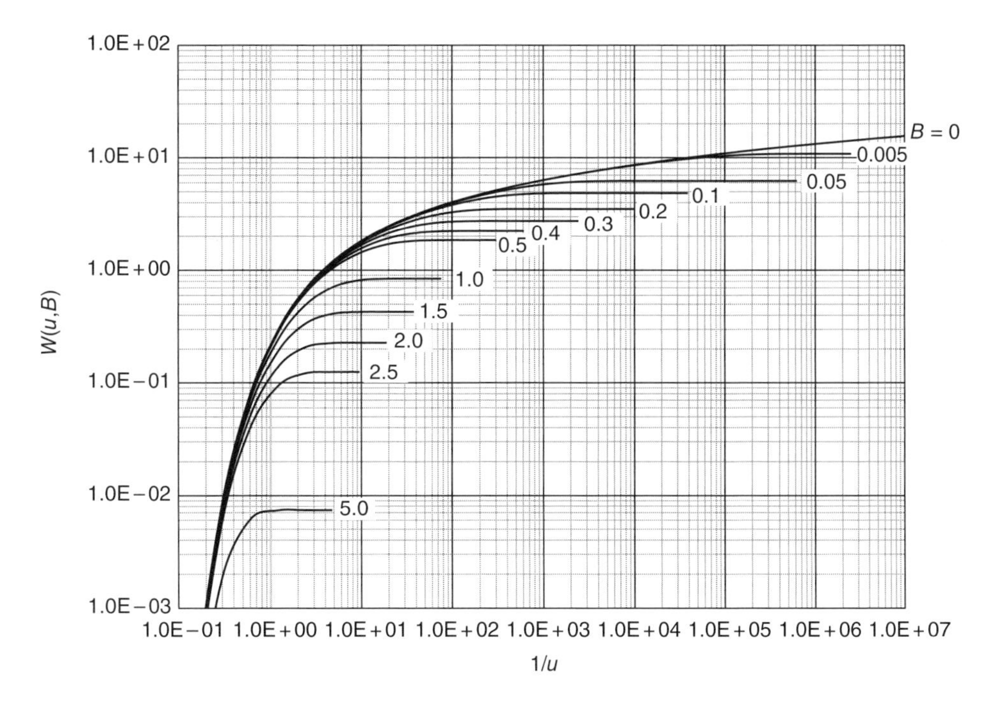

Figure 4.4 Theoretical curves for the leaky well function for various values of B. Note that the well function $W(u)$ is the specific case of $W(u, B)$ with $B = 0$

$B = 0$. If the aquifer is leaky, the response curve will be similar to a curve with $B > 0$. This gives rise to a number of classical *curve-matching* techniques for analyzing aquifer test data.

Determining Aquifer Hydraulic Properties. Groundwater hydrologists and engineers are mainly interested in the aquifer hydraulic properties of transmissivity (T) and storativity (S). With these parameters known, predictions can be made as to aquifer performance (i.e., long-term response of the aquifer to pumping, contaminant migration, etc.). These properties are also of interest to the energy engineer if numerical modeling of the aquifer system is needed; however, they are typically not necessary in designing a groundwater heat exchange system. In small systems in particular, the cost of aquifer testing is often not justified. Energy engineers are more interested in the performance of the actual well, which we shall discuss in the next subsection. Thus, we will only briefly discuss the methods of determining aquifer properties.

The most useful means of determining aquifer properties is with so-called *multiple-well* tests. These tests involve the drilling and installation of the main pumping well, in addition to observation wells (typically three to determine the aquifer potentiometric surface, hydraulic gradient, and groundwater flow direction). Static water levels are accurately determined prior to the test, and drawdown is recorded with time in all wells, ideally with pressure transducers and computerized data loggers. Aquifer tests are typically conducted for at least 72 h. An additional 24 h period of water level 'recovery' may be useful, which involves recording water levels as they recover back to static after termination of pumping.

The aquifer test data are analyzed using mathematical models of drawdown vs. time or distance. We discussed two possible solutions above (i.e., the ideal Theis solution and the leaky aquifer solution) that cover a wide range of aquifer cases, but there are numerous other specialized cases, which are ultimately variants of the Theis and leaky aquifer solutions. There are two broad types of data analysis approach: (i) log–log plot analysis (the Theis method), and (ii) semi-log plot analysis (the Jacob method).

In the log–log plot analysis method (the Theis method), drawdown data vs. time are plotted on a log–log plot and fitted to the appropriate mathematical model to determine aquifer T and S.

In the semi-log plot analysis method (the Jacob method), drawdown data are plotted against the logarithm of time (in a time–drawdown analysis). Alternatively, the Jacob method can be used to plot the stabilized drawdown in multiple wells versus the radial distance from the pumping well (in a distance–drawdown analysis). In the Jacob method, the plot results in a straight line, the slope of which is proportional to the pumping rate and the transmissivity. This method was developed by Cooper and Jacob (1946) with the realization that terms in Equation (4.24) beyond the logarithmic term may be neglected for small values of u. In the time–drawdown analysis, T and S are given by

$$T = \frac{2.3\dot{Q}}{4\pi s} \tag{4.34a}$$

$$S = \frac{2.25Tt_0}{r^2} \tag{4.34b}$$

where s is the drawdown across one log cycle of time, t_0 is the value of time at the intercept of the straight line with the axis at drawdown = 0, and r is the radial distance of the pumping well to the observation well. All parameters must be in consistent units. In the distance–drawdown analysis, T and S are given by

$$T = \frac{2.3\dot{Q}}{2\pi s} \tag{4.35a}$$

$$S = \frac{2.25Tt}{r_0^2} \tag{4.35b}$$

where s is the drawdown across one log cycle of time, r_0 is the value of distance from the pumping well at the intercept of the straight line with the axis at drawdown = 0, and t is the time at which the drawdowns were measured. All parameters must be in consistent units.

Aquifer tests have advantages and disadvantages. The obvious advantage is that they provide measurements of aquifer transmissivity and storativity. Their main disadvantage is cost, and thus they would not likely be undertaken except in large heat exchange systems. Another disadvantage is the non-uniqueness of interpretation of the data. Without clear geologic evidence, different situations may give the same drawdown response, and thus an erroneous mathematical model could be applied.

The Hydraulic Efficiency of Wells and Single-Well Tests. Of perhaps most interest to energy engineers and groundwater heat exchange system designers is the performance of the supply well delivering the energy (groundwater) to the point of use. The study of thermodynamics tells us that systems are not 100% efficient, and groundwater wells are no exception. Analogous to thermodynamic irreversibilities, energy in flowing groundwater is lost owing to turbulence through the well screen, etc.

The relative relation of drawdown in a pumping well to drawdown in an aquifer is shown schematically in Figures 4.5 and 4.6. The total drawdown (s_t) in most, if not all, pumping wells consists of two components:

$$s_t = s_a + s_w \tag{4.36}$$

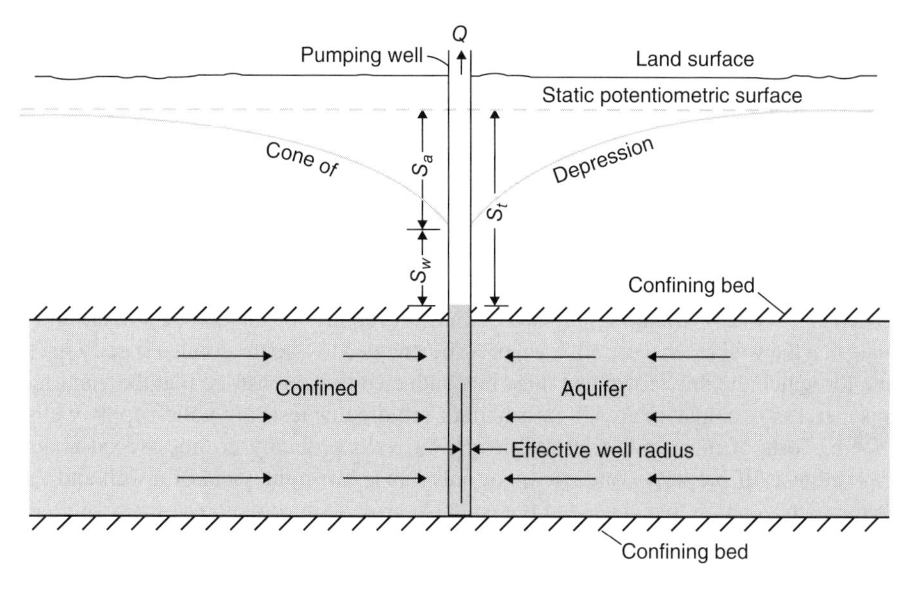

Figure 4.5 Schematic of drawdown in a pumping well relative to drawdown in the aquifer (Heath, 1983)

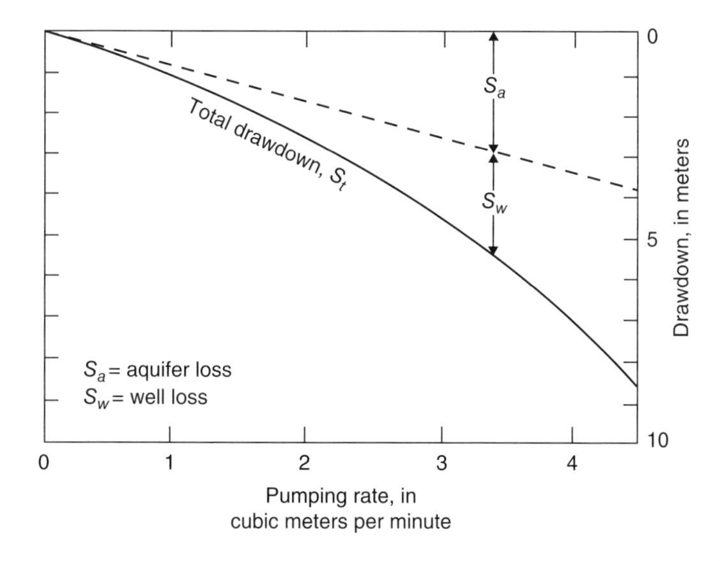

Figure 4.6 Relation of pumping rate to drawdown in an aquifer (Heath, 1983)

where s_a is the drawdown in the aquifer, and s_w is the drawdown that occurs as water moves from the aquifer into the well and up the well bore to the pump intake. Thus, the drawdown in most pumping wells is greater than the drawdown in the aquifer at the radius of the pumping well. The well efficiency (η_{well}) is described as

$$\eta_{well} = \frac{s_a}{s_t} \tag{4.37}$$

The total drawdown in a pumping well as a function of pumping rate ($\dot{Q}$) is given by

$$s_t = \beta\dot{Q} + C\dot{Q}^2 \tag{4.38}$$

where β is a factor related to the hydraulic characteristics of the aquifer and the length of the pumping period, and C is a factor related to the characteristics of the well. In order to determine C, it is necessary to pump the well at multiple rates.

The single-well test or step-drawdown test is an essential tool for design of a groundwater heat exchange system. Many supply-well contracts require a 'guaranteed' yield, and some stipulate that the well reach a certain level of 'efficiency'. Most contracts also specify the length of time for which the drawdown test must be conducted to demonstrate that the yield requirement is met. For example, many US jurisdictions require that tests of public-supply wells be at least 24 h. Tests of most industrial and irrigation wells probably do not exceed about 8 h. Single-well tests, if properly conducted, not only can confirm the yield of a well and the size of the production pump that is needed but can also provide information of great value in well operation and maintenance (Heath, 1983).

Single-well tests are typically conducted by qualified personnel with a portable well pump and generator. Discharge from the well is measured with an orifice plate or other type of flow meter. In these tests, the pumping rate is either held constant during the entire test or is increased in steps of equal length. The latter is known as a step-drawdown test. The pumping rate during each step should be held constant, and the length of each step should be dictated by the time to stabilize the drawdown. Results of a single-well, constant-rate test and a step-drawdown test are shown in Figure 4.7.

Determining the long-term yield of a well from data collected during a short-period test is one of the most important, practical problems in groundwater heat exchange designs. Two of the most important factors that must be considered are the extent to which the yield will decrease if the well is pumped continuously for periods longer than the test period, and the effect on the yield of changes in the static (regional) water level from that existing at the time of the test. A useful parameter describing the well yield is the *specific capacity* (S_c), defined as the well yield per unit of drawdown, which is determined by dividing the pumping rate at any time during the test by the drawdown at the same time:

$$S_c = \frac{\dot{Q}}{s_t} \tag{4.39}$$

Note that values of well specific capacity are shown in Figure 4.7 for that particular well test. The well specific capacity is essential to determine the level of placement of the well pump in the final well design.

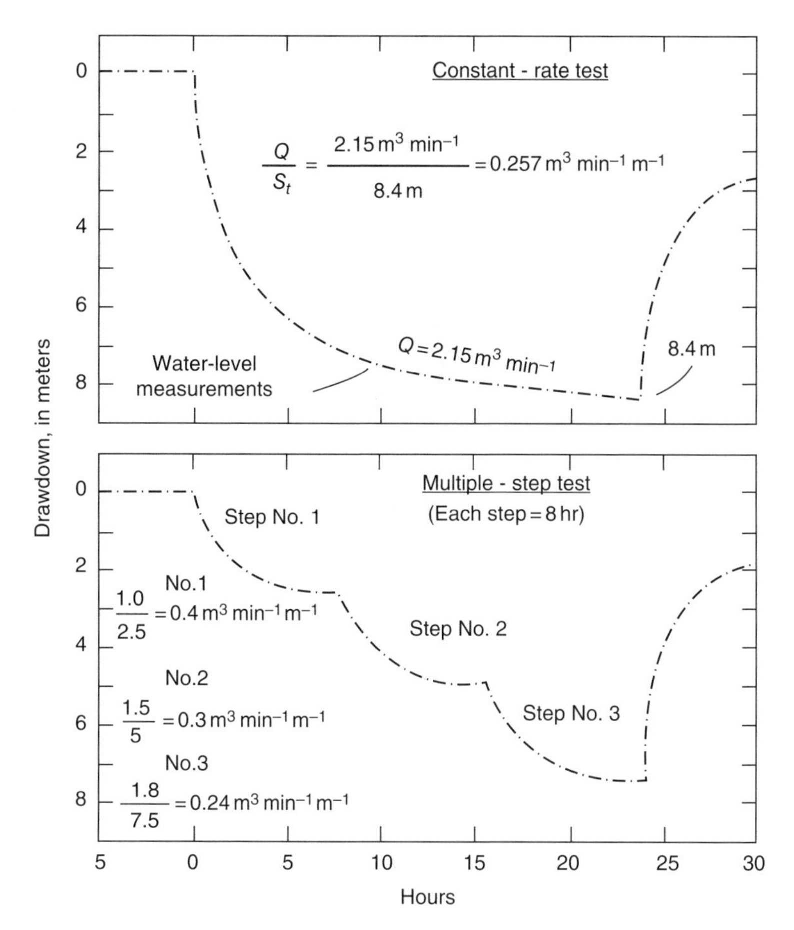

Figure 4.7 Drawdown versus time in a constant-rate single-well test and a multiple-step single-well test (Heath, 1983)

Bierschenk (1964) developed a graphical method to determine loss coefficients β and C by rearranging Equation (4.38) and dividing both sides by $\dot{Q}$ to obtain a linear equation:

$$\frac{s_t}{\dot{Q}} = \beta + C\dot{Q} \tag{4.40}$$

where $s_t/\dot{Q}$ is the inverse of specific capacity, also known as the *specific drawdown*. Fitting a straight line through the observed data, the slope of the best-fit line will be C (well losses) and the intercept of this line with $\dot{Q}=0$ will be β (aquifer losses).

Example 4.2 Specific Capacity and Specific Drawdown from a Step-Drawdown Test
With the test data shown in Figure 4.7, develop a plot of stabilized drawdown vs. pumping rate and determine the aquifer and well loss constants β and C. Also, develop a plot of the specific drawdown.

Solution

From the test data in Figure 4.7, we have:

Step	Pumping Rate ($m^3 \cdot min^{-1}$)	Drawdown (m)	Specific Drawdown ($m \cdot min \cdot m^{-3}$)
—	0	0	—
1	1.0	2.5	2.500
2	1.5	5.0	3.333
3	1.8	7.5	4.167

Plotting these values and determining the best-fit line, we have

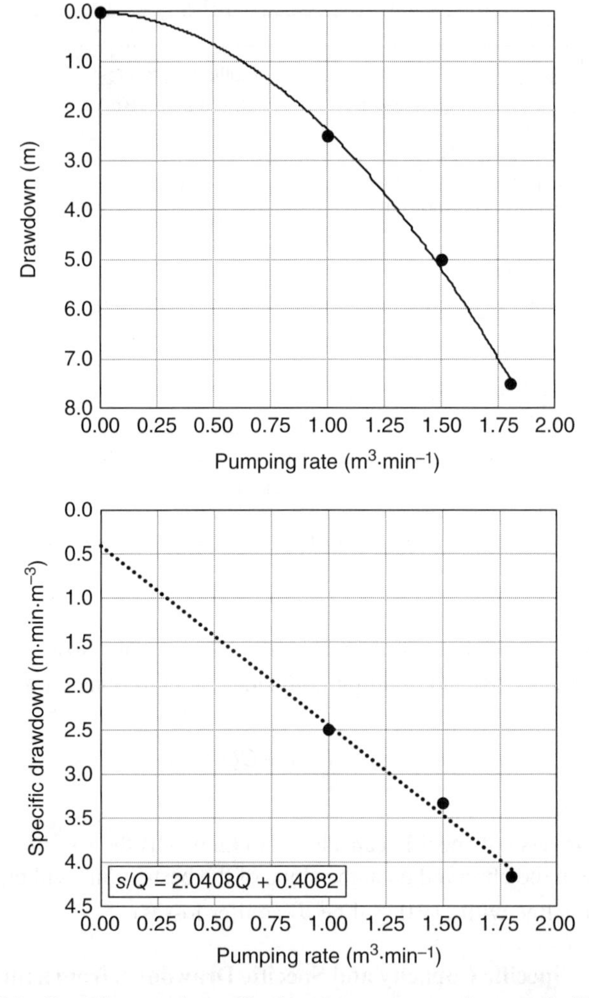

Figure E.4.2 Drawdown and specific drawdown vs. pumping rate for Example 4.2

From the plot of the specific drawdown vs. pumping rate (Figure E.4.2), we see that $\beta = 0.41$ min.$\cdot$m$^{-3}\cdot$m and $C = 2.04$ min.$^2\cdot$m$^{-6}\cdot$m.

4.3.3 Heat Transport in Groundwater

Governing Equations. Heat can be transported through a saturated porous medium by the following three processes: (1) heat transfer through the solid phase by conduction, (2) heat transfer through the liquid phase by conduction, and (3) heat transfer through the liquid phase by advection.

The transport of heat in groundwater can be described by the so-called advection-dispersion equation, as described by Freeze and Cherry (1979) and Bear (1972) for mass transport in flowing groundwater. Advection is the transport mechanism of mass or energy by a fluid owing to the fluid's bulk motion. Dispersion is a phenomenon occurring in porous media flow, which causes mixing along tortuous groundwater flow paths; the groundwater flowing in a given porous medium does not all travel at the same velocity, nor does it travel in straight paths as shown schematically in Figure 4.8. Thermal dispersion causes dissipation of energy in porous media.

By applying the law of conservation of energy to a control volume, and using an analogy to mass-heat transport, an equation for heat transport can be found and expressed in two-dimensional Cartesian coordinates (Chiasson *et al.*, 2000) as

$$\frac{\partial}{\partial x}\left(D_{xx}\frac{\partial T}{\partial x}\right)+\frac{\partial}{\partial y}\left(D_{yy}\frac{\partial T}{\partial y}\right)-\left(v_x\frac{\partial T}{\partial x}+v_y\frac{\partial T}{\partial y}\right)=R\frac{\partial T}{\partial t} \tag{4.41}$$

Geologic materials can exhibit heterogeneous properties owing to bedding planes and other horizontal layering effects. Assuming a homogeneous medium with a uniform velocity, Equation (4.41) for two-dimensional flow with the direction of flow parallel to the x-axis reduces to

$$D_L\frac{\partial^2 T}{\partial x^2}+D_T\frac{\partial^2 T}{\partial y^2}-v_x\frac{\partial T}{\partial x}=R\frac{\partial T}{\partial t} \tag{4.42}$$

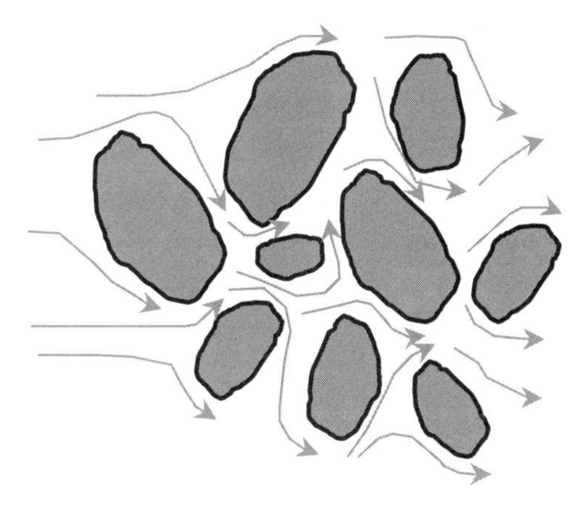

Figure 4.8 Concept of dispersion in porous media flow

where

$$D_L = a_L v_x + D^* \text{ and } D_T = a_T v_x + D^* \tag{4.43a,b}$$

It is the presence of these terms that distinguish the advection-dispersion equation from what is referred to as the advection-diffusion equation or the convection-diffusion equation found in the literature. The $a_L v_x$ and $a_T v_x$ terms are referred to as mechanical dispersion in the longitudinal and transverse directions. The actual values of dispersivity are difficult to determine, and are functions of space and timescales. Xu and Eckstein (1995) developed an empirical relationship of the longitudinal dispersivity (α_L) as a function of the flow path length (L):

$$a_L = 0.83(\log_{10}(L))^{2.414} \tag{4.44}$$

The transverse dispersivity (α_T) is typically taken as one order of magnitude lower than α_L.

In the mass-heat transport analogy, the diffusion coefficient (D^*) is modeled as an effective thermal diffusivity given by

$$D^* = \frac{k_{eff}}{\phi \rho_l c_l} \tag{4.45}$$

where k_{eff} is defined as $\phi k_l + (1-\phi)k_s$. A retardation coefficient (R) is necessary to adjust the advection and diffusion terms to account for the fact that thermal energy is stored and conducted through both the water and rock, but heat is only advected by the water (Bear, 1972). This is given by

$$R = 1 + \frac{(1-\phi)\rho_s c_s}{\phi \rho_l c_l} = \frac{(\rho c)_{eff}}{\phi \rho_l c_l} \tag{4.46}$$

where $(\rho c)_{eff}$ is defined as $\phi(\rho c)_l + (1-\phi)(\rho c)_s$.

Analytical Solution for Continuous Injection of Heat into a Uniform Flow Field. Analytical solutions have been developed for Equation (4.42) by a number of researchers for a number of various boundary and initial conditions. Fetter (1999) describes a solution to Equation (4.42) for a continuous injection of mass (located at the origin, $x = 0$, $y = 0$) into a two-dimensional flow field with uniform groundwater flow velocity (v_x) parallel to the x-axis. This solution would describe a situation in nature where heat or mass would travel in groundwater from a point source such as an injection well. A schematic of a thermal plume emanating from an injection point is shown in Figure 4.9.

Initial and boundary conditions for injection of heat in water at a temperature T above the undisturbed groundwater temperature T_o are described as

$$\lim_{r \to 0}\left(-r\frac{\partial T}{\partial r}\right) = \frac{\dot{Q}'(T-T_o)}{2\pi T} \text{ and } T_{r=\infty} = T_o \tag{4.47}$$

where T in the denominator of Equation (4.47) is the aquifer transmissivity.

The solution adapted here for groundwater temperature at time t and distance x and y from the origin, and adjusting for thermal retardation, is given by

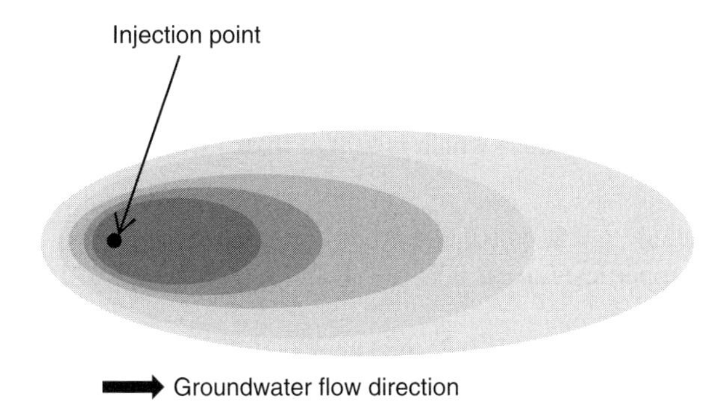

Figure 4.9 Schematic of a thermal plume emanating from an injection point in a groundwater aquifer

$$\Delta T(x,y,t) = \frac{(T-T_0)\dot{Q}'}{4\pi\left(\frac{D_L D_T}{R^2}\right)^{1/2}} e^{\frac{v_x x}{2D_L}} \left(\int_{t_D=0}^{t_D=\infty} \frac{e^{-u-B^2/(4u)}}{u} \mathrm{d}u \right) \tag{4.48}$$

Using the leaky well function notation described in Section 4.3.2, Equation (4.48) can be written as

$$\Delta T(x,y,t) = \frac{(T-T_0)\dot{Q}'}{4\pi\left(\frac{D_L D_T}{R^2}\right)^{1/2}} e^{\frac{v_x x}{2D_L}} [W(0,B) - W(t_D,B)] \tag{4.49}$$

where t_D is a dimensionless form of time given by

$$t_D = \frac{\left(\frac{v_x}{R}\right)^2 t}{4\frac{D_L}{R}} \tag{4.50}$$

W(0, B) = $2K_0(B)$, where K_0 is the modified Bessel function of the second kind of order 0, and B is defined as

$$B = \left[\left(v_x^2 x^2\right)/\left(4D_L^2\right) + \left(v_x^2 y^2\right)/\left(4D_L D_T\right) \right]^{0.5} \tag{4.51}$$

Example 4.3 Temperature in a Groundwater Thermal Plume
Water exiting a groundwater heat pump system is returned to the aquifer via an injection well. Over the cooling season (4 months), the average temperature of the groundwater being injected is 30 °C at an average flow rate of 5.0×10^{-4} m^3·s^{-1}. The porosity of the aquifer material is estimated to be 0.3, and the effective thermal conductivity (k_{eff}) and volumetric heat capacity $(\rho c)_{eff}$ of the aquifer materials are 2.0 W·m^{-1}·K^{-1} and 2.7×10^6 J·m^{-3}·K^{-1} respectively.

If the undisturbed aquifer temperature is 15 °C, estimate the temperature of groundwater in the aquifer at a distance of 2, 5, 10, and 20 m down-gradient of the injection well at the end of the cooling season. The groundwater flow rate is estimated to be 50 m/year. The injection well fully penetrates the aquifer, which is 100 m thick.

Given:
Geometry parameters: x = 2, 5, 10, and 20 m; y = 0 m
Aquifer thermal properties: $k_{eff} = 2.0\ \text{W·m}^{-1}\text{·K}^{-1}$; $(\rho c)_{eff} = 2.7 \times 10^6\ \text{J·m}^{-3}\text{·K}^{-1}$; $T = 30$ °C; $T_o = 15$ °C
Aquifer hydraulic properties: $\phi = 0.3$; thickness = 100 m; $v_x = 50\ \text{m·year}^{-1}$

Assumptions:
For water: $\rho_l c_l = 1000\,\text{kg}\cdot\text{m}^3 \times 4190\,\text{J}\cdot\text{kg}^1\cdot\text{K}^1 = 4.19\times 10^6\,\text{J}\cdot\text{m}^3\cdot\text{K}^1$
The time $t = 4\,\text{months} \times 30\frac{\text{days}}{\text{month}} \times 24\frac{\text{h}}{\text{day}} \times 3600\frac{\text{s}}{\text{h}} = 1.0368\times 10^7 s$

Solution
Here, we will show the calculations for $x = 10$ m, and only the results for the other values of x.

$$v_x = 50\frac{\text{m}}{\text{year}} \times \frac{1\,\text{year}}{8766\,\text{h}} \times \frac{1\,\text{h}}{3600\,\text{s}} = 1.584\times 10^{-6}\text{m}\cdot\text{s}^{-1}$$

$$a_L = 0.83(\log_{10}(L))^{2.414}\,\text{for}\,L = x = 10\,\text{m}, \alpha_L = 0.83\,\text{m}$$

$$a_T = \frac{1}{10}\alpha_L = 0.083\,\text{m}$$

$$D^* = \frac{k_{eff}}{\text{Ø}\rho_l c_l} = \frac{2.0\,\text{W}\cdot\text{m}^{-1}\cdot\text{K}^{-1}}{0.3\times\left(4.19\times 10^6\,\text{J}\cdot\text{m}^{-3}\cdot\text{K}^{-1}\right)} = 1.59\times 10^{-6}\,\text{m}^2\cdot\text{s}^{-1}$$

$$D_L = a_L v_x + D^* = 0.83\,\text{m}\times 1.584\times 10^{-6}\,\text{m}\cdot\text{s}^{-1} + 1.59\times 10^{-6}\,\text{m}^2\cdot\text{s}^{-1} = 2.91\times 10^{-6}\,\text{m}^2\cdot\text{s}^{-1}$$

$$D_T = a_T v_x + D^* = 0.083\,\text{m}\times 1.584\times 10^{-6}\,\text{m}\cdot\text{s}^{-1} + 1.59\times 10^{-6}\,\text{m}^2\cdot\text{s}^{-1} = 1.72\times 10^{-6}\,\text{m}^2\cdot\text{s}^{-1}$$

$$R = \frac{(\rho c)_{eff}}{\text{Ø}\rho_l c_l} = \frac{2.7\ 10^6\,\text{J}\cdot\text{m}^{-3}\cdot\text{K}^{-1}}{0.3\times\left(4.19\times 10^6\text{J}\cdot\text{m}^{-3}\cdot\text{K}^{-1}\right)} = 2.15$$

$$R = \frac{(\rho c)_{eff}}{\phi\rho_l c_l} = \frac{2.7\times 10^6\,\text{J}\cdot\text{m}^{-3}\cdot\text{K}^{-1}}{0.3\times\left(4.19\times 10^6\,\text{J}\cdot\text{m}^{-3}\cdot\text{K}^{-1}\right)} = 2.15$$

$$\dot{Q}' = \frac{\dot{Q}}{\text{thickness}} = \frac{5.0\times 10^{-4}\,\text{m}^3\cdot\text{s}^{-1}}{100\,\text{m}} = 5.0\times 10^{-6}\,\text{m}^3\cdot\text{s}^{-1}\cdot\text{m}^{-1}$$

$$t_D = \frac{\left(\frac{v_x}{R}\right)^2 t}{4\frac{D_L}{R}} = \frac{\left(\frac{1.584\times 10^{-6}\,\text{m}\cdot\text{s}^{-1}}{2.15}\right)^2\cdot 1.0368\times 10^7\,\text{s}}{4\cdot\frac{2.91\times 10^{-6}\,\text{m}^2\cdot\text{s}^{-1}}{2.15}} = 1.042$$

$$B = \left(\frac{v_x^2 x^2}{4D_L^2} + \frac{v_x^2 y^2}{4D_L D_T}\right)^{0.5} = \left(\frac{\left(1.584\times 10^{-6}\,\text{m}\cdot\text{s}^{-1}\right)^2\cdot(10\,\text{m})^2}{4\cdot\left(2.91\times 10^{-6}\,\text{m}^2\cdot\text{s}^{-1}\right)^2} + 0\right)^{0.5} = 2.726$$

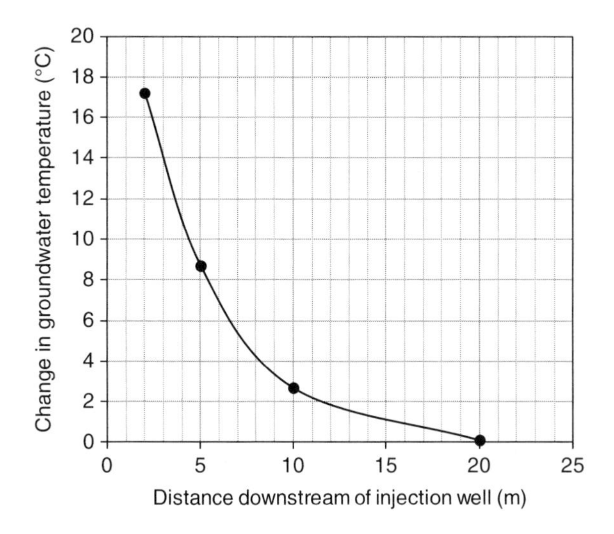

Figure E.4.3 Groundwater temperature vs. downstream distance from injection well

$W[0, B] = 9.556 \times 10^{-2}$

$W[t_D, B] = 6.478 \times 10^{-2}$

Finally, for $\Delta T(x,y,t) = \Delta T\left(10\,\text{m}, 0\,\text{m}, 1.0368 \times 10^{7}\,\text{s}\right)$

$$\Delta T(x,y,t) = \frac{(T - T_0)\dot{Q}'}{4\pi\left(\dfrac{D_L D_T}{R^2}\right)^{1/2}} e^{\frac{v_x x}{2D_L}}\left[W(0,B) - W(t_D,B)\right]$$

$$\Delta T(x,y,t) = \frac{(30\,°C - 15\,°C)\left(5.0 \times 10^{-6}\,\text{m}^3\cdot\text{s}^{-1}\cdot\text{m}^{-1}\right)}{4\pi\left(\dfrac{\left(2.91 \times 10^{-6}\,\text{m}^2\cdot\text{s}^{-1}\right)\left(1.72 \times 10^{-6}\,\text{m}^2\cdot\text{s}^{-1}\right)}{2.15^2}\right)^{1/2}}$$

$$\times \exp\left(\frac{\left(1.584 \times 10^{-6}\,\text{m}\cdot\text{s}^{-1}\right)(10\,\text{m})}{2\left(2.91 \times 10^{-6}\,\text{m}^2\cdot\text{s}^{-1}\right)}\right)\left[9.556 \times 10^{-2} - 6.478 \times 10^{-2}\right]$$

$$\Delta T(x,y,t) = 2.69\,°C$$

A plot of the temperatures vs. downstream distance from the pumping well is shown in Figure E.4.3.

Numerical Methods. As mentioned above for groundwater flow problems, analytical solutions are limited to problems with simple boundary conditions. Numerous commercially available software packages exist for modeling of heat transport in groundwater. Finite element methods in solving the advection-dispersion equation tend to have less numerical instability than finite difference methods. The typical solution approach first involves solving the flow problem for hydraulic head distribution, which then allows calculation of groundwater velocity vectors. With the velocity field known, numerical equations are applied to determine the temperature distribution. Some example software packages include: Comsol Multiphysics®, FEFLOW, Aqua3D, and HST3D.

4.4 Practical Considerations

The primary equipment used in groundwater heat exchange systems includes wells, pumps, heat exchangers, and piping. Although aspects of these components are routine for many other applications, there are some special considerations that will be discussed here with regard to geothermal applications, mainly related to the potentially aggressive nature of groundwater.

4.4.1 Equipment Needed

4.4.1.1 Wells

Production (Groundwater Supply) Wells. Water wells are not just blindly drilled holes. Rather, their construction requires careful design and engineering to maximize their efficiency. Elements of water wells that require some design consideration are shown in Figure 4.10. To design a production well, it is necessary to know the expected yield from the well, the depth to aquifers underlying the area, the composition and hydraulic characteristics of those aquifers, and the quality of water in the aquifer. This information should be determined from the site characterization phase(s) of a geothermal project. The completed design should specify the diameter, the total depth of the well and the position of the screen or open-hole sections, the method of construction, and the materials to be used in the construction. Composition and thickness of a gravel pack, if necessary, should also be specified.

The *well diameter* is mainly determined by the size of the pump that needs to go down the well, which is determined by the desired yield of the well. Although well diameter is proportional to well yield, the relationship is far from direct; doubling the well diameter results in only a 10% increase in yield. This can be seen from an inspection of the equations presented in Section 4.3.2, and that well yield is proportional to its length. Heath (1983) and Rafferty (2009) provide suggested well diameters to accommodate pumps that will provide various yields (Table 4.3). The depth to the source aquifer also affects the well diameter to the extent that wells expected to reach deep aquifers must be large enough to accept the larger-diameter drilling tools required to reach these depths.

The *well screen*, when present, is another engineered element of a well. The purpose of the well screen is to filter the water, and to control the entrance velocity of groundwater entering the well. There are several types of screen available, but the most common are wire wound, louvered, slotted, and perforated (with holes). The wire-wound type tends to maximize screen intake area. However, the screen should be designed such that the entrance velocity maintains laminar flow. Turbulent flow decreases well efficiency and promotes aeration of the water, which in turn accelerates scaling and corrosion potential. The screen entrance velocity should not normally exceed about 6 $ft \cdot min^{-1}$ or 1.8 $m^3 \cdot min^{-1}$ (Heath, 1983).

Not all wells are screened; wells installed in competent rock may have no screens, and are sometimes referred to as *open-hole* wells.

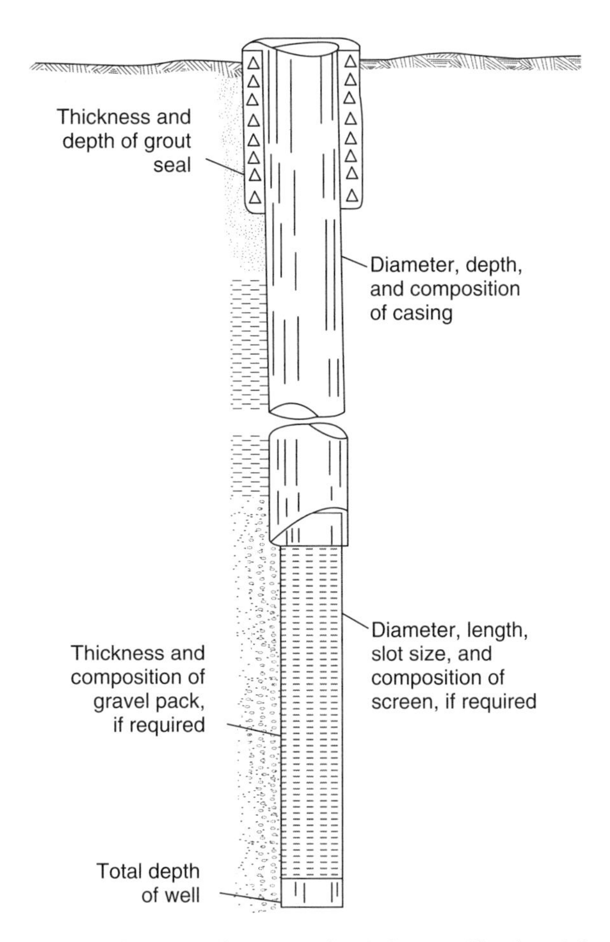

Figure 4.10 Elements of water wells that require design specifications (after Heath, 1983)

Table 4.3 Suggested Well Casing Diameters for Various Desired Flow Rates

Pump Bowl Diameter, in (mm)	Optimum Well Casing Diameter, in (mm)	Submersible Flow Range (3450 rpm), gpm (L/s)	Lineshaft Flow Range (1750 rpm), gpm (L/s)
4 (100)	6 (150)	<80 (<5)	<50 (<3)
6 (150)	10 (250)	80–350 (5–22)	50–175 (3–11)
7 (180)	12 (300)	250–600 (16–38)	150–275 (9–17)
8 (200)	12 (300)	360–800 (22–50)	250–500 (16–30)
9 (230)	14 (360)	475–850 (30–53)	275–550 (17–34)
10 (250)	14 (360)		500–1000 (30–63)
12 (300)	16 (400)		900–1300 (57–82)

The position of the well screen or open-hole section depends on the thickness and composition of the source aquifer and whether the well is being designed to obtain the maximum possible yield. As withdrawals from unconfined aquifers result in dewatering of the aquifers, the intake area of wells in these aquifers is normally in the lower part in order to obtain the maximum available drawdown. In confined aquifers, intake areas are set either in the most permeable part of the aquifer or, where vertical differences in hydraulic conductivity are not significant, in the middle part of the aquifer.

The length of the well screen or open-hole section specified in the well design depends on the thickness of the aquifer, the desired yield, whether the aquifer is unconfined or confined, and economic considerations. When an attempt is being made to obtain the maximum available yield, intake areas are normally installed in the lower 30–40% of unconfined aquifers and in the middle 70–80% of confined aquifers (Heath, 1983).

The thickness and depth of the grout seal is usually dictated by local codes and regulations. Its purpose is to provide a sanitary seal, and to prevent surface run-off from leaking into groundwater. Typical grout seal depths are of the order of 10 m.

Casing composition is typically carbon steel or PVC plastic in most geothermal applications. The choice depends on structural integrity and groundwater chemical compatibility. Either material is acceptable at resource temperatures below about 30 °C. PVC is not rated for pressure service at temperatures above about 60 °C or 140 °F. In aggressive environments, corrosion protection of the well casing may be necessary. In some extreme cases where fluids are extracted for geothermal power plants, casings are constructed of titanium.

Well Development. Well development is the process where fines and any remaining drilling fluid are removed from the well and the geologic formations surrounding the well. This is accomplished by periodic surging and removal of water from the well. Surging is especially important in wells with filter packs, where the objective is to move particles back and forth to promote settling of the filter pack, with action similar to plunging of a toilet. The development activity is continued until the water reaches a stabilized, acceptable level of clarity.

Injection Wells. Best engineering practice dictates that we return groundwater used for thermal purposes back to its original aquifer. This eliminates depletion of the groundwater resource. In some jurisdictions, return of groundwater to a surface water receptor is allowed if the surface water body is directly connected to the groundwater system. In other cases, injection wells are needed. As energy engineers, we need to be sure that the injection well is properly designed, and that it is far enough away from the supply well(s) such that is does not hydraulically communicate.

Construction of injection wells differs from that of production wells primarily in the recommended screen velocity (which is generally half that of production wells) and well sealing design. Some designers automatically double the design screen length for injection wells relative to production wells. Injection wells, particularly those likely to be subject to positive injection pressure, should be fully cased and sealed from the top of the injection zone to the surface. The most common factors in reduced water well (production and injection) performance are: incrustation and biofouling of screens, formation plugging with fines, sand pumping, casing/screen collapse, and pump problems (Driscoll, 2008). Incrustation and biofouling can be largely reduced by minimizing drawdown through careful well design and the avoidance of excessive groundwater flows. Supply wells will unavoidably contain some fine soil/rock particles; inadvertent pumping of fine sand should be limited by screen, gravel pack and development practices or removed by strainers prior to injection.

4.4.1.2 Well Pumps

Well or downhole pumps are needed in open-loop systems to convey the geothermal fluids to the ground surface and beyond, unless the well is artesian with enough flow to serve the thermal load. Pumps are further discussed in Chapter 10, and thus our focus here is on the aspects of well pumps. Vertical turbine (vertical lineshaft) and submersible pumps (Figure 4.11) are the most common types of well pump in geothermal applications. The choice mainly depends on the resource temperature and on the required flow rate.

Vertical lineshaft pumps are preferred in applications with (i) large flow requirements (see Table 4.3) and/or (ii) resource temperatures above 30 °C. These preferences are both related to the pump motor. In large flow applications, the pump motor is prohibitively large to fit down the well, and thus vertical lineshaft pumps are preferred. In applications at resource temperatures progressively above 30 °C, vertical lineshaft pumps are preferred because submersible pump motors become too difficult to cool.

The lineshaft pump system consists of a multistage downhole centrifugal pump, a surface-mounted motor, and a long driveshaft assembly extending from the motor to the pump bowls. In higher-temperature applications, the lineshaft is usually the enclosed type to allow for oil lubrication, as opposed to lineshaft pumps used for irrigation and cold water applications, which are generally open to allow for the well water itself to act as the lubricant. Lineshaft pumps for higher-temperature, direct-use geothermal application are typically constructed of carbon or stainless steel shafts and bronze bearings in the lineshaft assembly, and stainless steel shafts and leaded red bronze bearings in the bowl assembly. Figure 4.12 presents photos of a vertical lineshaft pump removed from a geothermal well for service.

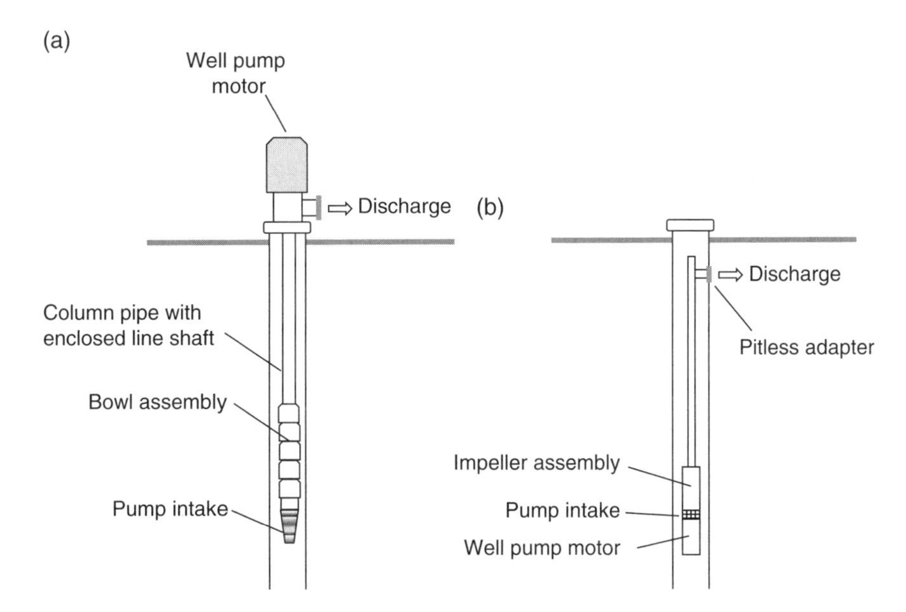

Figure 4.11 Schematic of well pump types in geothermal applications: (a) vertical turbine; (b) electric submersible

Figure 4.12 Photos of a vertical lineshaft pump, showing (a) the 10 in (250 mm) diameter column pipe with internal shaft (enclosed and oil lubricated when in service), and (b) the bowl assembly with screened intake (photos taken by the author)

The submersible pump is an electrically driven pump that consists of three primary components located downhole: the pump, the drive motor, and the motor protector (Figure 4.13). The pump is a vertical multistage centrifugal type, and the motor is usually a three-phase induction type that is filled with oil for cooling and lubrication. The motor is cooled by the groundwater itself, and, typically above 30 °C, special precautions are required (i.e., increased water flow past the motor). There are a couple of advantages of submersible pumps over vertical lineshaft pumps. First, submersible pumps require little to no surface equipment; the pump is connected to the discharge line via a *pitless adapter*. Second, submersible pumps are more adaptable to different depths, given the basic piping connections downhole. These simpler piping connections also give rise to economic advantages with increasing depth settings.

The size of the pump motor is a function of the required flow rate and the well pump lift, in addition to the pump efficiency and the motor efficiency. The well pump lift or total dynamic head (TDH) is given by:

$$\begin{aligned}\text{TDH} &= \text{depth to static water level} + \text{total drawdown} \\ &\quad + \text{surface piping and heat exchanger losses}\end{aligned} \tag{4.52}$$

The pump shaft power in IP units is given by

$$\dot{W}_{pump} = \frac{\dot{Q} \times \text{TDH}}{3960\eta_{pump}} \tag{4.53a}$$

Figure 4.13 Photo of a 6 in (150 mm) diameter electric submersible well pump. Note the motor at the bottom (photo taken by the author)

where $\dot{W}_{pump}$ is in units of horsepower, $\dot{Q}$ is units of gallons per minute (gpm), TDH is in units of feet, and η_{pump} is the hydraulic efficiency of the pump, typically of the order of 70% at best.

The pump shaft power in SI units is given by

$$\dot{W}_{pump} = \frac{\dot{Q} \times \text{TDH} \times \rho g}{\eta_{pump}} \tag{4.53b}$$

where $\dot{W}_{pump}$ is in units of kW, $\dot{Q}$ is units of $m^3 \cdot s^{-1}$, TDH is in units of m, and η_{pump} is the hydraulic efficiency of the pump, typically of the order of 70% at best.

The pump motor size is then determined by multiplying $\dot{W}_{pump}$ by the motor efficiency. High-efficiency motors are over 90% efficient. The combined pump and motor efficiency is sometimes referred to as the *wire-to-water efficiency*.

Further details of pumps and piping systems are discussed in Chapter 10.

4.4.1.3 Heat Exchangers

Heat exchangers are common in groundwater applications where it is necessary to isolate the geothermal fluid from the system equipment in order to limit corrosion and/or scaling potential. As we shall discuss in a subsequent section of this chapter, the additional cost of a heat exchanger is typically not warranted in small, residential, or single-zone systems, but provisions on the system are needed to allow for flushing and descaling. As we shall also discuss in a subsequent section of this chapter, use of heat exchangers creates a primary–secondary pumping loop arrangement, which allows some advantages.

The principal types of heat exchanger used in geothermal applications are the plate, shell-and-tube, and downhole types. In groundwater heat exchange applications for heating and cooling, the plate type is most common. Plate-type heat exchangers can be built in a very compact arrangement with close approach temperatures. Figure 4.14 is a simple schematic of a heat exchanger in a counterflow arrangement.

Plate-type heat exchangers in groundwater applications are typically of two main constructions: (i) brazed plate and (ii) gasketed plate. The brazed-plate heat exchanger consists of a series of plates that are brazed together, resulting in the heat exchanger being one unit. The plate-and-frame type contains plates held together with gaskets in a frame with clamping rods. These are common in large applications, as plates can be added as necessary, or removed for maintenance and cleaning. The counterflow design through the plates and the high turbulence achieved in plate heat exchangers provide for compact, efficient heat exchange in a small footprint when compared with shell-and-tube heat exchangers. The plates are usually made of stainless steel, although titanium is used when the fluids are especially corrosive. Heat exchangers in groundwater applications should be designed for pressure drops as low as practical, of the order of 70 kPa (~10 psi).

A photograph of an intact and cut-away view of a brazed-plate heat exchanger is shown in Figure 4.15. Figure 4.16 is a photo of gasketed plates that would be installed adjacent to one another. Note that the gasketing allows fluid to pass through the appropriate inlet and outlet ports of the heat exchanger. Figure 4.17 is a gasketed-type plate heat exchanger in a geothermal application, serving a building of approximately 1000 m^2.

A unique type of heat exchanger used in direct-use geothermal heating applications is known as the *downhole heat exchanger* (Figure 4.18). Although these are closed-loop systems similar

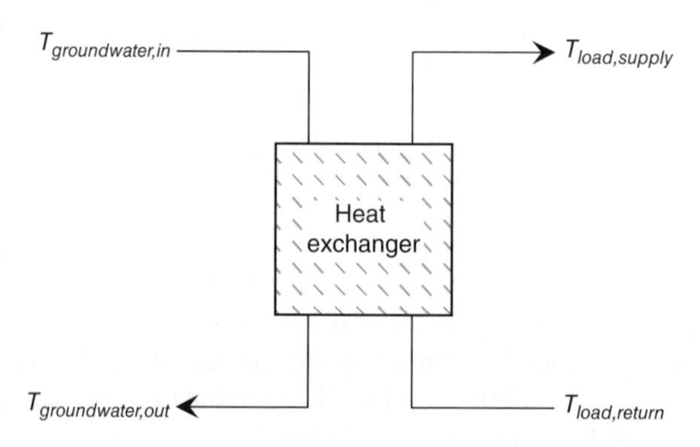

Figure 4.14 Schematic of a counterflow heat exchanger

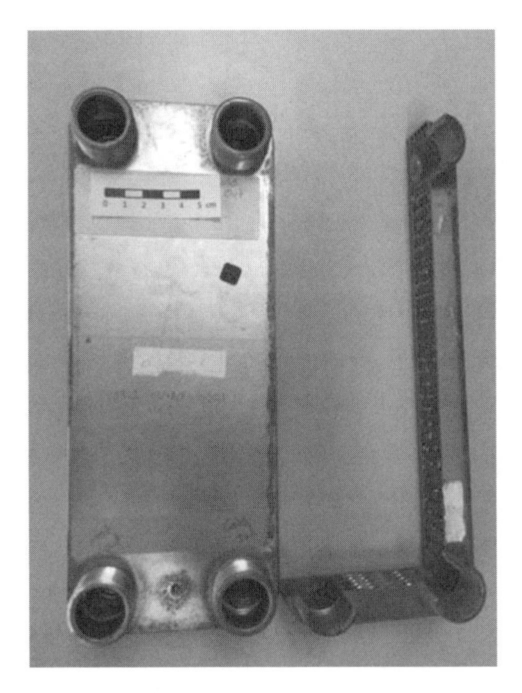

Figure 4.15 Photo of an intact and cut-away view of a brazed-plate heat exchanger (photo taken by the author)

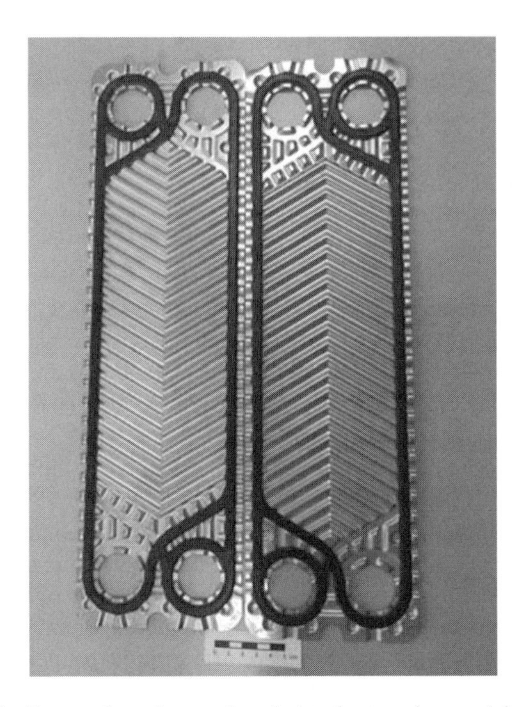

Figure 4.16 Photo of adjacent plates in a gasketed plate heat exchanger (photo taken by the author)

Figure 4.17 Photo of a gasketed, plate-type heat exchanger in the mechanical room of a building (photo taken by the author)

to the vertical borehole heat exchangers that are discussed in Chapter 5, we mention them here because they rely on the presence of groundwater, and thus may be considered as groundwater heat exchange systems. As these systems are closed loop, only heat is extracted from the geothermal well, thereby eliminating the issue of disposal of geothermal fluid. Their use, however, is typically limited to smaller loads, such as the heating of individual or small groups of homes, small apartment buildings, or schools.

The downhole exchanger (DHE) consists of one or more u-pipes suspended in a well, through which secondary water is pumped or allowed to circulate by natural convection. In order to obtain maximum thermal output of the system per unit length of well bore, the well is typically designed to have an open annulus between the well bore and the casing, with perforations in the casing near the top and the bottom of the water column in the well. Natural convection currents form in the well, thus allowing full mixing of the geothermal water within the well bore. Culver (2005) presents a summary of downhole heat exchanger materials, with the first one being installed in 1929 in Klamath Falls, Oregon. Chiasson *et al.* (2005) reports on

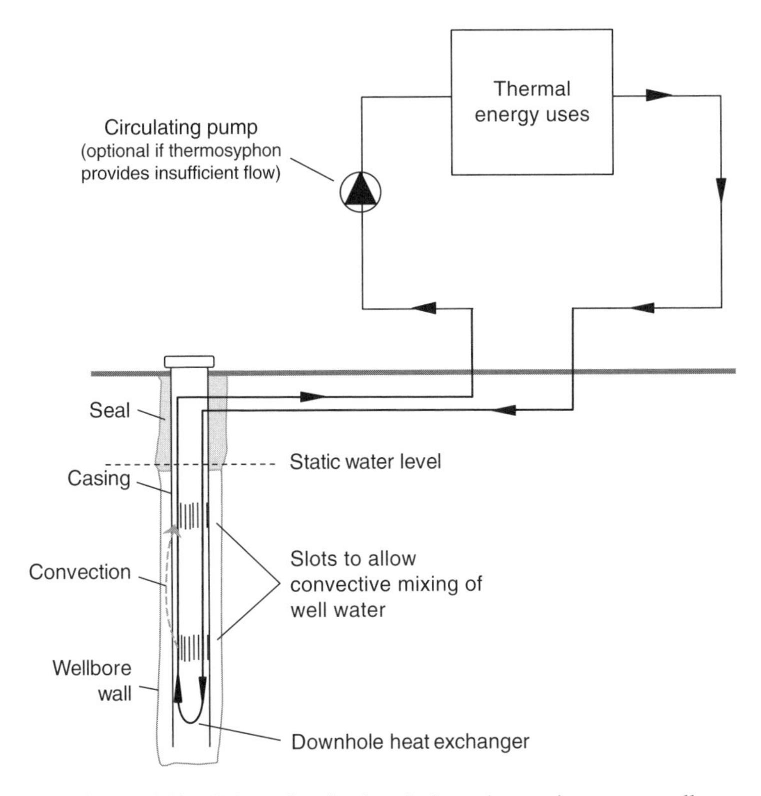

Figure 4.18 Schematic of a downhole exchanger in a water well

the first known use of crosslinked polyethylene (PEX) plastic as a downhole heat exchange material. These systems rely on relatively high groundwater velocities.

In groundwater heat exchange applications, there are a three methods for heat exchanger analysis: (1) approach temperature, (2) log-mean temperature difference (LMTD), and (3) effectiveness-NTU method. The choice depends on preference and what parameters are known. The *approach method* is simple and describes how closely the inlet temperature of one side of the heat exchanger approaches the leaving temperature of the other. The *LMTD method* is easier to apply when inlet and outlet temperature and flow rate are known on both sides of the heat exchanger, which are typically known in the design stage. Thus, this method is typically used in heat exchanger design. The *effectiveness method* is easier to apply if the size of the heat exchanger can be estimated and the outlet fluid temperatures are unknown. Thus, this method is typically used in heat exchanger performance calculations. Each of these heat exchanger analysis methods will be discussed in the paragraphs that follow.

The *approach temperature* is a very simple but powerful concept used to describe heat exchanger performance. The word 'approach' is derived from how closely one fluid approaches the other in a heat exchanger. The closer the approach temperature, the more effective is the heat exchanger. Relatively small approach temperatures and relatively high heat effectiveness values are associated with larger and more costly heat exchangers. Different fields of engineering may have different definitions of the approach temperature, and there is no universal definition. The use of the approach temperature can be ambiguous and confusing because there are at least two inlets and two outlets in a heat exchanger, so what temperature differences do we take? In general, this depends on the purpose of the heat exchanger. In a groundwater application, as shown in Figure 4.14, groundwater is used either to heat or to cool a secondary fluid. Thus, we will define the approach temperature for these applications as the difference between the entering groundwater temperature and the leaving temperature of the process fluid.

In the *LMTD method*, the total heat transfer rate ($\dot{q}$) is related to the heat exchanger inlet and outlet temperatures, the overall heat transfer coefficient (U), and the total surface area (A) of the heat exchanger as

$$\dot{q} = UA\Delta T_{lm} \tag{4.54}$$

where ΔT_{lm} is the log-mean temperature difference defined as

$$\Delta T_{lm} = \frac{\Delta T_2 - \Delta T_1}{\ln(\Delta T_2/\Delta T_1)} = \frac{\Delta T_1 - \Delta T_2}{\ln(\Delta T_1/\Delta T_2)} \tag{4.55}$$

In counterflow arrangements, ΔT_1 and ΔT_2 are defined as

$$\begin{bmatrix} \Delta T_1 = T_{h,i} - T_{c,o} \\ \Delta T_2 = T_{h,o} - T_{c,i} \end{bmatrix} \tag{4.56}$$

where $T_{h,i}$, $T_{h,o}$, $T_{c,i}$, and $T_{c,o}$ are the hot inlet, hot outlet, cold inlet, and cold outlet temperatures respectively. Note that ΔT_1 and ΔT_2 are described differently for parallel-flow heat exchangers. The LMTD for counterflow arrangements is greater than that for parallel-flow arrangements, thus resulting in a smaller heat transfer area to transfer the same amount of heat, assuming the same value of U. Also, in counterflow heat exchangers, $T_{c,o}$ can exceed $T_{h,o}$, but not in parallel-flow heat exchangers. Note that the LMTD method requires a heat exchanger correction factor, which is a fraction of 1, to account for departures in true counterflow arrangements. This is true for shell-and-tube heat exchangers with multiple passes. In the plate heat exchangers discussed here, we will assume that the heat exchanger correction factor is 1.

In the *effectiveness-NTU* method, the effectiveness of heat transfer is governed by the *overall heat transfer coefficient*, which is a function of the total thermal resistance (conductive plus convective) to heat transfer between the two fluids. Here, it is useful to employ the *heat exchanger effectiveness* (ε) concept, where

$$\varepsilon = \frac{\dot{q}}{\dot{q}_{max}} \tag{4.57}$$

where $\dot{q}$ is the actual heat transfer rate, $\dot{q}_{max}$ is the maximum possible heat transfer rate, and ε can range from 0 to 1. In the study of heat exchangers in classic heat transfer theory, you may recall that ε is a function of NTU (the number of transfer units) and C, the heat capacity rates ($\dot{m}c_p$) of the fluids on both sides of the heat exchanger. Furthermore, $\dot{q}_{max}$ is actually experienced by the fluid with the minimum heat capacity rate (C_{min}) such that

$$\dot{q}_{max} = C_{min}(T_{h,i} - T_{c,i}) \tag{4.58}$$

where $T_{h,i}$ and $T_{c,i}$ are the hot inlet and cold inlet fluid temperatures respectively. NTU is further defined as

$$\text{NTU} = \frac{UA}{C_{min}} \tag{4.59}$$

In counterflow heat exchangers where the heat capacity rates are unequal, NTU may be calculated by

$$\text{NTU} = \frac{1}{C_r - 1} \ln\left(\frac{\varepsilon - 1}{\varepsilon C_r - 1}\right) \tag{4.60}$$

where C_r is the heat capacity ratio of the two fluids ($C_r < 1$).

The heat transfer rate in the heat exchanger is then

$$\dot{q} = \varepsilon C_{min}(T_{h,i} - T_{c,i}) \tag{4.61}$$

Alternatively,

$$\varepsilon = \frac{\dot{q}}{\dot{q}_{max}} = \frac{(\dot{m}c_p(T_i - T_o))_{hot}}{C_{min}(T_{h,i} - T_{c,i})} = \frac{(\dot{m}c_p(T_o - T_i))_{cold}}{C_{min}(T_{h,i} - T_{c,i})} \tag{4.62}$$

Example 4.4 Heat Exchanger Analysis

Consider the heat exchanger shown in Figure E.4.4 in a groundwater cooling application. The process fluid is water. If the heat transfer rate is 100 kW, calculate:

(a) the heat exchanger approach temperature,
(b) the heat exchanger *UA* value, and
(c) the heat exchanger effectiveness.

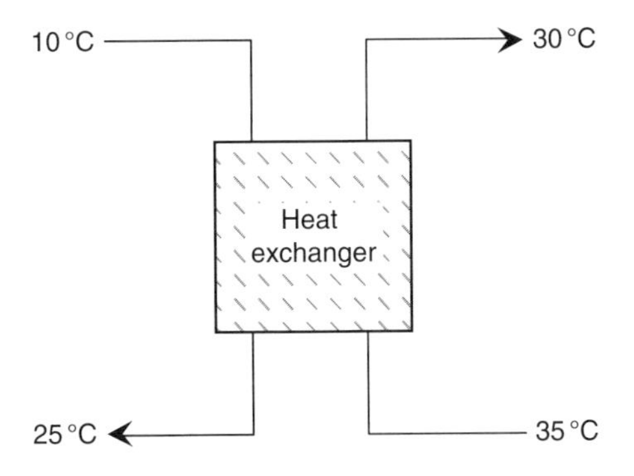

Figure E.4.4 Heat exchanger configuration for Example 4.4

Given:
$\dot{q}$ = 100 kW
$T_{h,i}$ = 35 °C, $T_{h,o}$ = 30 °C, $T_{c,i}$ = 10 °C, and $T_{c,o}$ = 25 °C

Assumptions:
For water: $c_p = 4.190\,\text{kJ}\cdot\text{kg}^{-1}\cdot\text{K}^{-1}$

Solution

(a) $T_{approach} = \left|T_{groundwater,in} - T_{process,out}\right| = |10\,°\text{C} - 30\,°\text{C}| = 20\,°\text{C}$
(b) The heat exchanger *UA* may be determined by the LMTD method:

$$\begin{bmatrix} \Delta T_1 = T_{h,i} - T_{c,o} \\ \Delta T_2 = T_{h,o} - T_{c,i} \end{bmatrix}, \begin{bmatrix} \Delta T_1 = 35\,°\text{C} - 25\,°\text{C} \\ \Delta T_2 = 30\,°\text{C} - 10\,°\text{C} \end{bmatrix}, \begin{bmatrix} \Delta T_1 = 10\,°\text{C} \\ \Delta T_2 = 20\,°\text{C} \end{bmatrix}$$

$$\Delta T_{lm} = \frac{\Delta T_2 - \Delta T_1}{\ln(\Delta T_2/\Delta T_1)} = \frac{20\,°\text{C} - 10\,°\text{C}}{\ln(20\,°\text{C}/10\,°\text{C})} = 14.43\,°\text{C}$$

Finally,

$$\dot{q} = UA\Delta T_{lm}$$

$$100\,\text{kW} = UA\,(14.43\,°\text{C})$$

$$\therefore UA = 6.93\,\text{kW}\cdot°\text{C}^{-1}$$

(c) The heat exchanger effectiveness may be determined from Equation (4.61).

To apply Equation (4.61), we need C_{min}. First, we need the mass flow rates. From an energy balance of the fluids, we have

$$\dot{q} = \dot{m}c_p\Delta T$$

For the groundwater, $100\,\text{kW} = \dot{m}_{gw} \times 4.190\,\text{kJ}\cdot\text{kg}^{-1}\cdot\text{K}^{-1} \times 15\,°\text{C}$, $\therefore \dot{m}_{gw}$ = 1.59 kg·s^{-1}
For the process water, $100\,\text{kW} = \dot{m}_{process} \times 4.190\,\text{kJ}\cdot\text{kg}^{-1}\cdot\text{K}^{-1} \times 5\,°\text{C}$, $\therefore$ $\dot{m}_{gw}$ = 4.77 kg·s^{-1}
Clearly, the mass flow rate of the groundwater is lower, and thus

$$C_{min} = \dot{m}_{gw} \times c_{p,\,gw} = 1.59\,\text{kg}\cdot\text{s}^{-1} \times 4.190\,\text{kJ}\cdot\text{kg}^{-1}\cdot\text{K}^{-1} = 6.66\,\text{kW}\cdot\text{K}^{-1}$$

Thus,

$$\dot{q} = \varepsilon C_{min}(T_{h,i} - T_{c,i})$$

$$100\,\text{kW} = \varepsilon \times 6.66\,\text{kW}\cdot\text{K}^{-1} \times (35\,°\text{C} - 10\,°\text{C})$$

$$\therefore \varepsilon = 0.6$$

We can also calculate the heat exchanger *UA* from the effectiveness-NTU method. First, the heat capacity ratio (C_r) is

$$C_r = \frac{1.59\,\text{kg}\cdot\text{s}^{-1}}{4.77\,\text{kg}\cdot\text{s}^{-1}} = 0.33$$

With Equation (4.60)

$$\text{NTU} = \frac{1}{C_r - 1}\ \ln\left(\frac{\varepsilon - 1}{\varepsilon C_r - 1}\right) = \frac{1}{0.33 - 1}\ \ln\left(\frac{0.6 - 1}{0.6 \times 0.33 - 1}\right) = 1.04$$

With Equation (4.59)

$$\text{NTU} = \frac{UA}{C_{min}}$$

$$UA = \text{NTU} \times C_{min} = 1.04 \times 6.66\,\text{kW}\cdot\text{K}^{-1}$$

$$\therefore UA = 6.93\,\text{kW}\cdot{}^{\circ}\text{C}^{-1}$$

Discussion: Plate-type heat exchangers are capable of achieving very low approach temperatures, of the order of 2–3 °C. However, in this case, the groundwater is considerably colder than the process water, and a large heat exchanger with a close approach temperature is not necessary. This results in a relatively low effectiveness value of 0.6. Typical *U* values of plate-type heat exchangers are in the range 6.5–7.0 $\text{kW}\cdot\text{m}^{-2}$, and thus a 1 m^2 heat exchanger could be adequate for this example.

4.4.1.4 Transmission Piping

Transmission piping for direct-use geothermal projects is discussed in detail by Rafferty (1998b), much of which is applicable to all geothermal projects. The fluid state in these transmission lines can be liquid water, steam, or a two-phase mixture. These pipelines carry fluids from the wellhead to either a site of application, a steam-water separator, or a heat exchanger. Thermal expansion of metallic pipelines heated rapidly from ambient to geothermal fluid temperatures (which could vary from 50 to 200 °C) causes stresses that must be accommodated by careful engineering design.

The choice of pipe material is primarily based on the temperature of the geothermal fluid and its chemical composition. Metallic pipe is capable of withstanding high temperatures and pressures, but is subject to corrosion. On the other hand, thermoplastic pipe is corrosion resistant, but has lower pressure and temperature limits relative to metallic pipe, and has various joining methods; thermoplastic joints can be made mechanically, chemically, or thermally, depending on the material.

Carbon steel is the most widely used material for geothermal transmission lines and distribution networks, especially if the fluid temperature is over 100 °C (Lund, 2010). Conventional steel piping requires expansion provisions, either bellowed arrangements or by loops. A typical piping installation would have fixed points and expansion points about every 100 m. In addition, the piping would have to be placed on rollers or slip plates between points. When hot water metallic pipelines are buried, they can be subjected to external corrosion from groundwater and/or electrolysis, and therefore they must be protected by coatings and wrappings, or by insulation. Concrete tunnels or trenches have been used to protect steel pipes in many geothermal district heating systems. Insulated pipe with protective jackets also inhibits the pipe from corrosion.

Common types of thermoplastic pipe material include fiberglass reinforced plastic (FRP), polyvinyl chloride (PVC), and crosslinked polyethylene plastic (PEX). PVC piping is often used for the distribution network and for uninsulated waste disposal lines where temperatures are below 60 °C. CPVC plastic typically has higher temperature and pressure ratings than PVC. PEX has become popular in recent years as it can tolerate temperatures up to 100 °C at pressures up to 550 kPa. However, PEX pipe is currently only available in sizes less than 50 mm in diameter.

Supply and distribution systems can consist of either a single-pipe or a two-pipe system. A two-pipe system is most common, and consists of a supply and return pipeline.

The quantity of thermal insulation on transmission lines and distribution networks depends on many factors. In addition to minimizing the heat loss of the fluid, the insulation must be waterproof and watertight, as moisture can destroy the value of any thermal insulation and cause rapid external corrosion. Above-ground and overhead pipeline installations can be considered in special cases, but considerable insulation is achieved by burying hot water pipelines. For example, burying bare steel pipe results in a reduction in heat loss of about one-third as compared with above ground in air with minimal convection (Ryan, 1981). If the soil around the buried pipe can be kept dry, then the heat loss can be minimized owing to the relatively low thermal conductivity of dry soil. Carbon steel piping can be insulated with polyurethane foam, rock wool, or fiberglass and encased in a protective polyvinyl chloride (PVC) jacket. Above-ground pipelines can be insulated with a protective, aluminium radiant barrier. In general, 25–100 mm of insulation is adequate.

At flowing conditions, the temperature decrease in insulated pipelines is in the range 0.1–1.0 °C/km, and in uninsulated lines the decrease is 2–5 °C/km (in the approximate range of 5–15 L/s flow for 15 cm diameter pipe) (Ryan 1981). It is smaller for larger-diameter pipes. The cost of uninsulated pipe is about half the cost of insulated pipe, and thus it is used where temperature drop is not critical. Pipe material does not have a significant effect on heat loss, but the flow rate does, owing to the convection terms dominating the overall thermal resistance of the pipe. At lower flow rates (off peak), the heat loss can be higher than with greater flows owing to more contact time of the fluid with the Earth.

4.4.2 *Groundwater Quality*

The chemical quality of the groundwater is site specific, and may vary from less than 1000 ppm (parts per million or mg/L) of total dissolved solids to heavily brined with total dissolved solids exceeding 100 000 ppm. Water with a total dissolved solids (TDS) concentration of less than 3000 mg/L can be considered to be fresh water. Water with 3000–10 000 mg/L of TDS is considered to be brackish, and water with in excess of 10 000 mg/L of TDS is considered to be saline. Groundwater with salinity greater than seawater (about 35 000 mg/L) is typically referred to as brine.

Fluid chemical quality influences two aspects of the system design: (1) material selection to avoid corrosion and scaling effects, and (2) disposal or ultimate end use of the fluid. Each of these can have a significant impact on the system viability. According to Ellis (1998), geothermal fluids commonly contain seven chemical species of concern for groundwater heat exchanger applications: pH, dissolved oxygen, chloride ion, sulfide species, carbon dioxide species, ammonia species, and sulfate ion. Low pH accelerates corrosion of carbon steel, while high pH indicates scaling potential. TDS concentrations in excess of 500 ppm indicate corrosion potential. Chloride ions (Cl^-) accelerate corrosion of steels; 304 stainless steel is acceptable to 140 ppm Cl^-, while 316 stainless steel is acceptable to 400 ppm Cl^-. Bicarbonate is linked to calcium carbonate scale above 100 ppm. Hydrogen sulfide attacks copper, and oxygen accelerates steel corrosion. Thus, it is very important to eliminate air and other gases from the system.

Scaling in geothermal heat pump systems is described by Rafferty (2000). Scale can be formed from a variety of dissolved chemical species, but two of the most common, reliable indicators are hardness and alkalinity. Calcium carbonate scale is the most common form of scale found in groundwater heat exchange systems. Carbonate hardness, depending upon the nature of the water, is composed of calcium or magnesium carbonates and bicarbonates. It is this form of hardness that contributes most to scale formation. Non-carbonate hardness is normally a small component of the total hardness and is characterized by much higher solubility. As a result, its role in scale formation is generally negligible. Scaling and corrosion potential in the HVAC industry are estimated from (i) the Ryznar stability index (RSI) and (ii) the Langelier saturation index (LSI). Each of these indices is calculated from concentrations of calcium carbonate ($CaCO_3$) dissolved in water and the pH.

4.5 Groundwater Heat Pump Systems

4.5.1 Small Residential Systems

Groundwater heat pumps make practical sense in residential buildings on well water if the groundwater is of good quality. In these applications, the heat pump can be integrated into the household water system and considered as just another water-using appliance. In either new construction or retrofit application, design care must be taken to ensure that the well has adequate yield, and that the pressure tank has adequate capacity to handle the additional flow demand of the heat pump.

Residential groundwater heat pump systems are distinguished from larger commercial systems in that: (1) they are typically integrated with a household domestic water system, (2) they are usually single-zone systems, and (3) they typically do not isolate the groundwater from the heat pump unit(s) (a practice that is recommended for commercial systems, as we shall see). A schematic of a residential groundwater heat pump system is shown in Figure 4.19.

As we shall discuss in Chapter 12, geothermal heat pumps have certain flow requirements, but in open-loop groundwater systems the flow rate can be selected on the basis of the groundwater temperature such that the system COP is maximized. Systems should strive to handle as minimal an amount of groundwater as possible, as this minimizes well pump energy and minimizes the volume of groundwater to get rid of. Flow control valves are recommended in the discharge line to ensure that the well is not overpumped. Placement of a slow-closing motorized

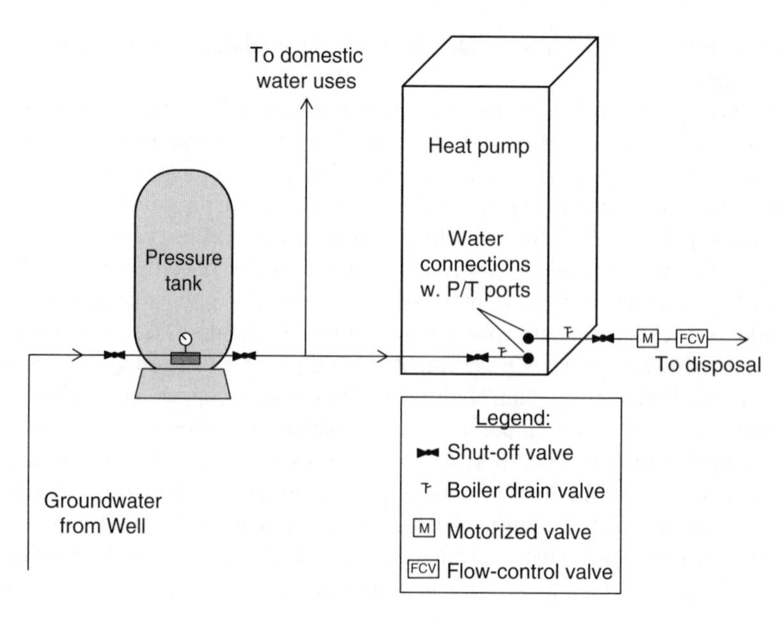

Figure 4.19 Schematic of a residential groundwater heat pump system

valve also on the discharge line ensures positive pressure on the heat pump water coil, and will stop the flow of water when the heat pump is not operating (Figure 4.19).

The purpose of the pressure tank is to provide water at pressure on demand without short-cycling the well pump. The pressure tank should be a pre-pressurized bladder-type tank, and should be large enough to take at least 1 min to be filled with the well pump.

Residential groundwater heat pump systems are small enough for the additional cost of an isolation heat exchanger typically not to be economically justified. As untreated groundwater is generally used as the heat transfer fluid, provisions must be made to allow for flushing and descaling of the heat pump water-refrigerant coil if necessary. Such provisions, as shown in Figure 4.19, include shut-off valves to isolate the heat pump water connections. Access ports such as hose bibs or boiler drains are installed so that a flushing apparatus can be connected.

Groundwater leaving the heat pump should be returned to a point of discharge in accordance with best practices and/or local codes. Surface discharge into a pond or wetland or infiltration into a dry well may be more of an option in these smaller systems than with larger commercial systems, owing to the correspondingly much lower flow rates.

4.5.2 Large Commercial Distributed Heat Pump Systems

The design configuration of large commercial systems, or any system with distributed heat pumps, is quite different from that of small residential systems. As shown in Figure 4.20, best practice involves isolating the groundwater from the heat pump equipment with a heat exchanger. The main reason for this, as discussed previously, is to avoid potential scaling

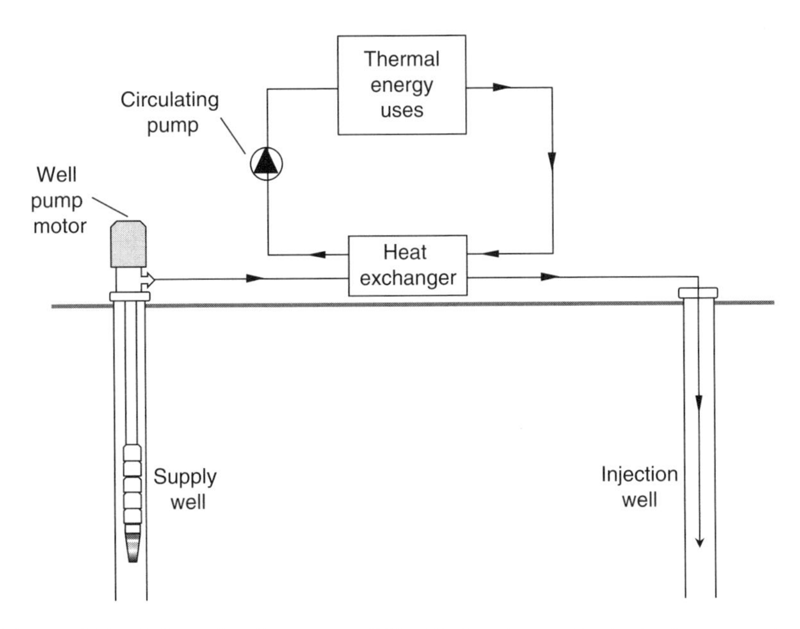

Figure 4.20 Schematic of a commercial/institutional groundwater heat pump system

and corrosion problems in heat pump equipment owing to aggressive groundwater; plate-type heat exchangers are constructed of stainless steel and are relatively corrosion resistant.

The practice of isolating groundwater from heat pump equipment creates a primary–secondary pumping loop, and gives rise to some novel control strategies. Note that now the heat pumps are in a closed-loop arrangement, and thus can be controlled separately. That is, building loop temperature can be allowed to 'float' between heating and cooling set-points within heat pump desirable operating limits, and then the groundwater is used to moderate building loop temperature, minimizing groundwater pumping requirements. We shall discuss heat pump details further in Chapter 12, but geothermal heat pumps of the water-source type are easily capable of providing adequate heat at entering fluid temperatures below 0 °C (with an antifreeze solution), and adequate cooling at entering fluid temperatures above 30 °C, for example. In this primary–secondary indirect design, required groundwater flow rates can be reduced by up to one-third of the flow rates required in a direct design (i.e., no heat exchanger), but the actual flow rate is a function of the groundwater temperature. This reduction in required pumping rate is a significant point in the design of groundwater heat pump systems.

4.5.3 System Energy Analysis and the Required Groundwater Flow Rate

4.5.3.1 Design Approaches

The design groundwater flow rate is perhaps the most important parameter in a groundwater heat pump project. It dictates the well design, and whether or not you have a potential project; if wells cannot meet the flow requirement, then another heating/cooling solution may be needed.

There are a couple of design approaches based on (i) minimizing groundwater flow requirements or (ii) minimizing system energy consumption. The latter approach may seem like the obvious way to go, but the groundwater flow rate that minimizes energy consumption may not be available from wells; the total system energy consumption is a function of the heat pump and well pump energy consumption.

Determining the Minimum Groundwater Flow Requirement. To calculate the minimum groundwater flow requirement for a particular peak load (either heating or cooling), we have the following approach, assuming that the groundwater flow rate is smaller than that of the building loop:

1. Choose the maximum heat pump entering fluid temperature for cooling and the minimum heat pump entering fluid temperature for heating. Based on the above discussion, 30 °C (90 °F) in cooling and 0 °C (32 °F) in heating are reasonable.
2. Calculate the 'ground load' q_{ground} from

$$\dot{q}_{ground} = \dot{q}_{building} \times (\mathrm{COP}+1)/\mathrm{COP}, \text{ for cooling} \quad (4.63a)$$

$$\dot{q}_{ground} = -\dot{q}_{building} \times (\mathrm{COP}\text{-}1)/\mathrm{COP}, \text{ for heating} \quad (4.63b)$$

   where COP is the heat pump coefficient of performance. Note that ground load is negative for heating applications to indicate heat extraction from the Earth. We shall develop these equations later in Chapter 12 from an energy balance on the heat pump, but for now, suffice it to say that the load on the geothermal heat exchanger is not the same as the building load in a heat pump application.
3. Calculate the heat pump exiting fluid temperature by an energy balance on the fluid [i.e., $T_{heat\ pump\ out} = T_{heat\ pump\ in} + q_{ground}/(\dot{m}c_p)$]. Note that $T_{heat\ pump\ out}$ is one of the entering fluid temperatures to the heat exchanger.
4. Select a heat exchanger effectiveness, ε (0.80 is a good first approximation for plate heat exchangers in groundwater heat pump applications).
5. Apply the heat exchanger effectiveness equation [Equation (4.61)] to calculate the groundwater flow rate.

Example 4.5 Calculating the Required Groundwater Flow Rate for a Given Cooling Load

A commercial building is planning a groundwater heat pump system similar to the arrangement shown in Figure 4.20. Determine the required groundwater flow rate if the peak cooling load is 100 kW. The groundwater temperature is 15 °C, and the HVAC engineer has designed the heat pump loop to float up to 30 °C before energizing the well pump. The building loop fluid is pure water, and the peak design flow requirement for the heat pumps is 300 L·min^{-1} (5.0 kg·s^{-1}). From the heat pump manufacturer's catalog data, the COP of the heat pumps at 30 °C is 4.0. Assume a heat exchanger effectiveness of 0.80.

Assumptions:

The heat exchanger effectiveness $\varepsilon = 0.80$.

The specific heat of groundwater is 4.2 kJ·kg^{-1}·K^{-1}.

C_{min} is on the groundwater side.

Solution

The load on the heat exchanger is the ground load. As this a cooling application, we have

$$\dot{q}_{ground} = \dot{q}_{building} \times \frac{(\text{COP}+1)}{\text{COP}} = 100\,\text{kW} \times \frac{(4+1)}{4} = 125\,\text{kW}$$

The fluid temperature exiting the heat pumps and entering the heat exchanger on the building side is

$$T_{heat\,pump\,out} = T_{heat\,pump\,in} + \frac{\dot{q}_{ground}}{\dot{m}c_p} = 30\,°\text{C} + \frac{125\,\text{kW}}{(5\,\text{kg}\cdot\text{s}^{-1})(4.2\,\text{kJ}\cdot\text{kg}^{-1}\cdot\text{K}^{-1})} = 35.95\,°\text{C}$$

From the heat exchanger effectiveness equation:

$$\dot{q} = \varepsilon C_{min}(T_{h,i} - T_{c,i}), \text{ or}$$

$$\dot{q}_{ground} = 0.80 \times C_{min}\left(T_{heat\,pump\,out} - T_{groundwater}\right)$$

$$\therefore C_{min} = \frac{125\,\text{kW}}{0.80 \times (35.95\,°C - 15\,°C)} = 7.46\,\text{kW}\cdot°C^{-1}$$

$$\therefore \dot{m}_{groundwater} = \frac{C_{min}}{c_{p,\,groundwater}} = \frac{7.46\,\text{kW}\cdot°C^{-1}}{4.2\,\text{kJ}\cdot\text{kg}^{-1}\cdot\text{K}^{-1}} = 1.77\,\text{kg}\cdot\text{s}^{-1}$$

Thus, the required groundwater mass flow rate is 1.77 $\text{kg}\cdot\text{s}^{-1}$, or 106 $\text{L}\cdot\text{min}^{-1}$. This flow rate is about one-third that of the building loop flow rate.

Determining the Groundwater Flow Requirement that Minimizes Energy Consumption. The groundwater flow rate for minimum energy consumption is that which maximizes system COP. Thus, this is inherently an optimization problem such that the following objective function is maximized:

$$\text{COP}_{system} = \frac{\text{Useful cooling or heating}}{\dot{E}_{heat\,pump} + \dot{E}_{well\,pump} + \dot{E}_{building\,loop\,pump}} \tag{4.64}$$

where $\dot{E}$ is the electrical power usage.

Note that the 'optimum' groundwater flow rate may not coincide with the minimum groundwater flow rate. The maximum system COP must balance well pumping power with heat pump performance. As groundwater flow increases through a system, more favorable average temperatures are produced for the heat pumps. Higher groundwater flow rates, to a point, increase system COP, as increased well pump power is outweighed by decreased heat pump power requirements (because of the more favorable temperatures). At some point, additional increases in groundwater flow result in a greater increase in well pump power than the resulting decrease in heat pump power. The key strategy in groundwater heat pump system design is identifying the point of maximum system performance with respect to heat pump and well pump power requirements.

Once this optimum relationship has been established for the design condition, the method of controlling the well pump determines the extent to which the relationship is preserved at off-peak conditions. This optimization process involves evaluating the performance of the heat pumps and well pump(s) over a range of groundwater flows. Key data necessary to make this calculation include: well performance (i.e., the well and aquifer loss coefficients) and heat pump performance vs. entering water temperatures at different flow rates. Well information is derived from well yield test results as described above. Heat pump performance data are available from the manufacturer.

4.5.3.2 GHX Design Tool for Groundwater Heat Exchange Systems

The above equations and design approaches have been implemented into the suite of GHX design tools provided on the companion website. Figure 4.21 is a screen capture of the worksheet for groundwater heat exchange project analysis, showing the input data and output calculations for Example 4.5, with additional input data to determine the system COP. As seen in Figure 4.21, the spreadsheet is configured as a system schematic. Although the heat pump COP is 4, the system COP is 3.62 when other energy uses due to pumps are considered. A summary of input and output data is provided in Table 4.4.

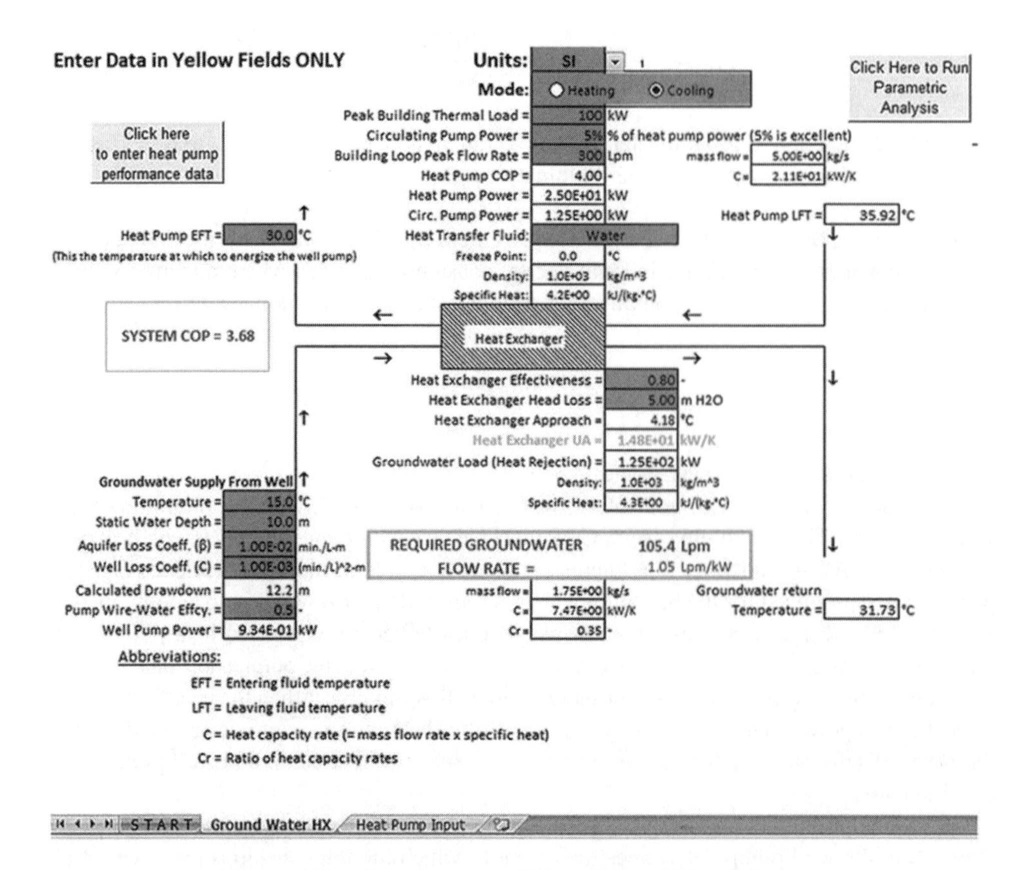

Figure 4.21 Screen capture of Groundwater HX worksheet

Table 4.4 Summary of Input and Output Data for Groundwater HX Design Tool

Item	Input Data	Calculated Output
General	• Units (IP or SI) • Mode (heating or cooling)	—
Building Loop	• Peak load on building • Heat pump COP as a function of entering fluid temperature (entered on separate worksheet shown in Figure 4.22; these values can be left as the default, generic values if the actual heat pump is unknown) • Heat transfer fluid (water or aqueous glycol solution) • Flow rate (typicaly 2–3 Lpm/kW or 2–3 gpm/ton)	• Heat pump power • Fluid temperature exiting heat pumps and entering heat exchanger
Heat Exchanger	• Effectiveness (ε) • Heat exchanger head loss, plus any other significant head loss through pipes, fittings, etc.	• Heat exchanger *UA* value • Heat exchanger approach temperature
Groundwater Loop	• Groundwater supply temperature • Depth to static water level • Aquifer loss coefficient (β) • Well loss coefficient (C) • Pump wire-to-water efficiency	• **Required groundwater flow rate** • Groundwater load • Groundwater injection temperature • Drawdown • Well pump power
Energy Performance	—	• **System COP**

A powerful parametric analysis feature is also included in the groundwater HX tool. Clicking the *Parametric Analysis* button calls a VBA code that automatically varies the design heat pump entering fluid temperature, and calculates and plots the system COP, the required groundwater flow rate, and the heat exchanger *UA*. This analysis allows determination of the groundwater flow rate that results in maximum system COP. A screen capture of the results of a parametric analysis of the data for Example 4.5 is shown in Figure 4.23. A review of these results shows that the system COP can be increased to about 4.27 with the choice of a lower design entering fluid temperature of 22 °C. The higher system COP, however, comes at the expense of pumping about 60% more groundwater than when the entering heat pump fluid temperature is 30 °C. But as the heat pump power requirement is an order of magnitude greater than the well pump power, the overall system COP increases with the lower heat pump entering fluid temperature.

Note that the parametric analysis results are system specific. If the well efficiency were lower, or the static water level were deeper, more pump energy would be required to deliver the same flow, thus lowering the system COP. Thus, it is very important to consider the system COP when choosing a design heat pump entering fluid temperature and the associated groundwater flow rate.

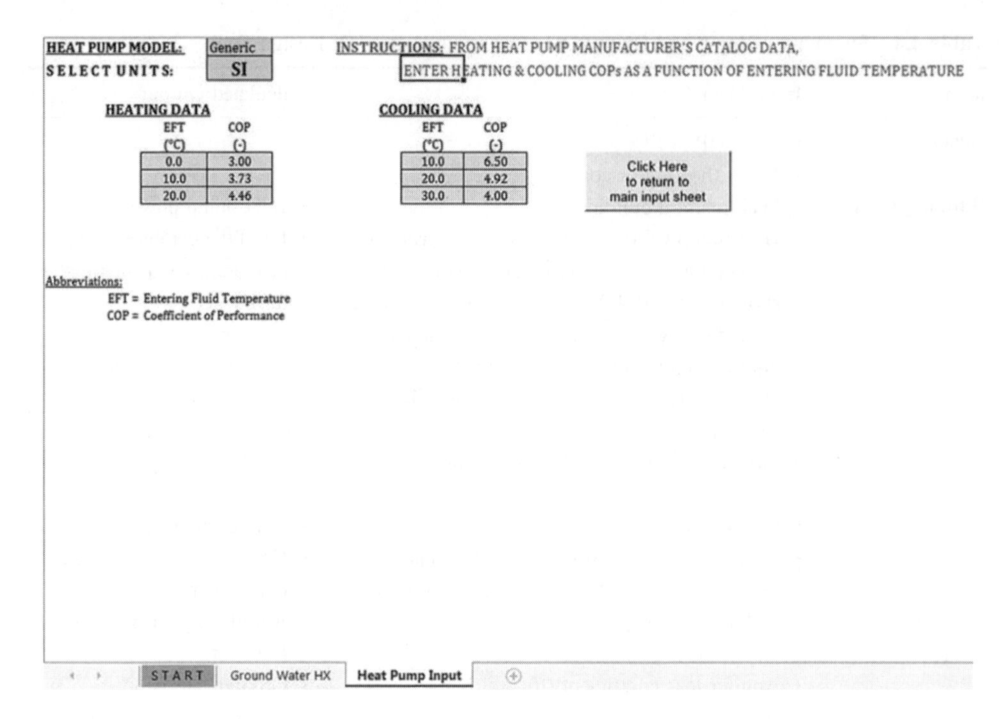

Figure 4.22 Screen capture of Heat Pump Input worksheet

4.5.4 *Well Pump Control*

In groundwater heat pump system design, the method of control of the well pump should ideally preserve the optimum relationship between well pump power and heat pump power at off-peak conditions. There are several ways the pump can be controlled.

Some older groundwater heat exchange designs used multiple pumps staged to meet system loads, either with multiple wells or with multiple pumps installed in a single well. A dual set-point control, similar to that used in boiler/cooling tower systems, has also been implemented in many geothermal systems. As described above, the well pump is energized above a given temperature in the cooling mode and below a given temperature in the heating mode. Between those temperatures the building loop 'floats' without the addition of groundwater.

Newer groundwater heat exchange systems use variable-speed well pumps to minimize well pump cycling, and hard, rapid starts of the well pump. This facilitates the establishment of a temperature range (difference between pump-on and pump-off temperatures) over which the pump operates in both the heating and cooling modes. The size of this range is primarily a function of the building loop water volume relative to the peak system load. For example, reconsider Example 4.5, where the heat pump entering fluid temperature was chosen to be 30 °C. The well pump could be sped up at 28 °C and modulated up to full speed at 32 °C, resulting in a temperature range of 4 °C. Thus, full pump speed would only be needed during peak load hours, potentially significantly reducing annual pump energy expense. This type of variable-speed pump control is shown schematically in Figure 4.24. Note that the minimum pump speed is typically 30–50% of peak flow. The actual minimum flow requirement depends on the TDH and the motor specifications.

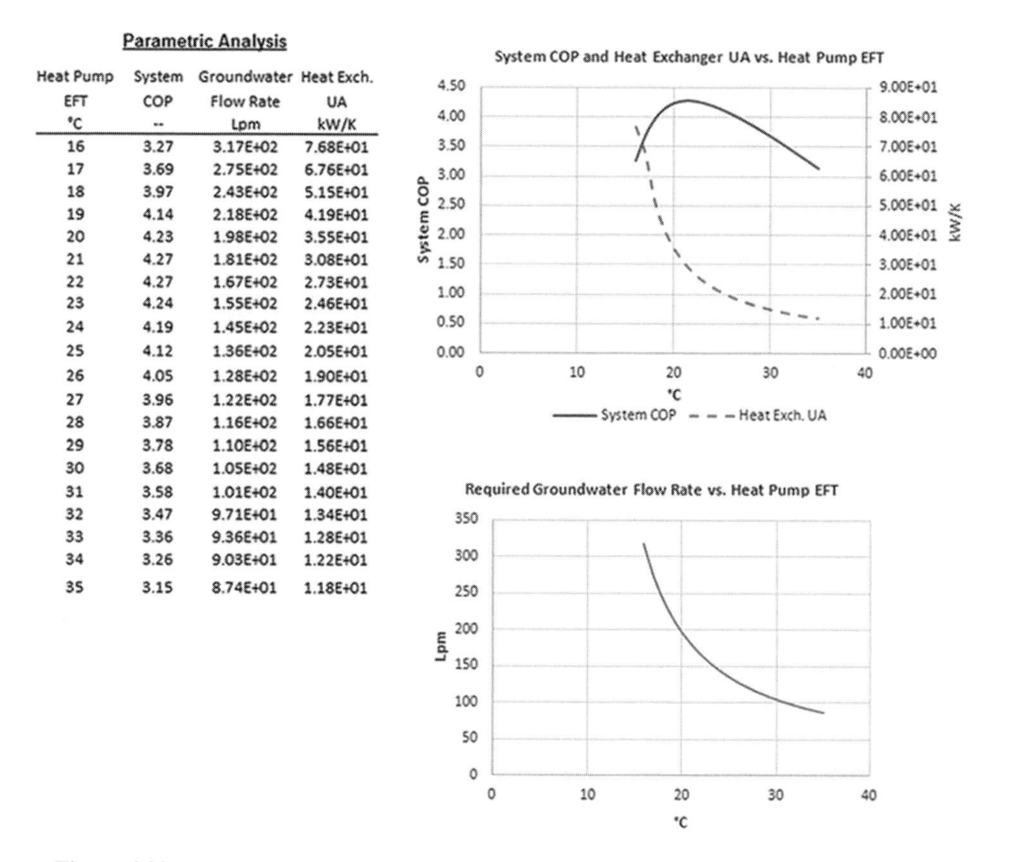

Parametric Analysis

Heat Pump EFT	System COP	Groundwater Flow Rate	Heat Exch. UA
°C	--	Lpm	kW/K
16	3.27	3.17E+02	7.68E+01
17	3.69	2.75E+02	6.76E+01
18	3.97	2.43E+02	5.15E+01
19	4.14	2.18E+02	4.19E+01
20	4.23	1.98E+02	3.55E+01
21	4.27	1.81E+02	3.08E+01
22	4.27	1.67E+02	2.73E+01
23	4.24	1.55E+02	2.46E+01
24	4.19	1.45E+02	2.23E+01
25	4.12	1.36E+02	2.05E+01
26	4.05	1.28E+02	1.90E+01
27	3.96	1.22E+02	1.77E+01
28	3.87	1.16E+02	1.66E+01
29	3.78	1.10E+02	1.56E+01
30	3.68	1.05E+02	1.48E+01
31	3.58	1.01E+02	1.40E+01
32	3.47	9.71E+01	1.34E+01
33	3.36	9.36E+01	1.28E+01
34	3.26	9.03E+01	1.22E+01
35	3.15	8.74E+01	1.18E+01

Figure 4.23 Screen capture of parametric analysis results on the Groundwater HX worksheet

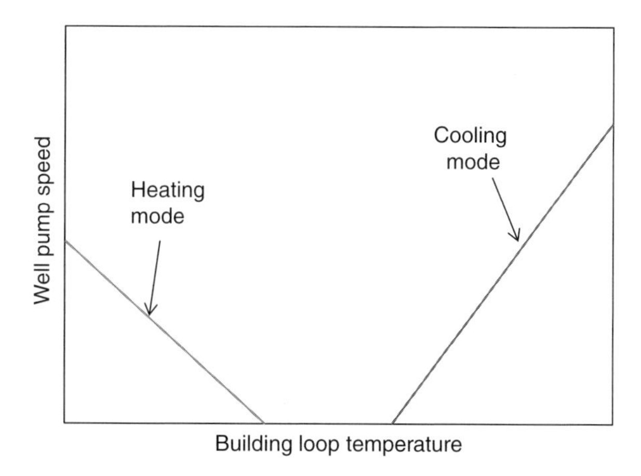

Figure 4.24 Well pump control with variable-speed well pumping

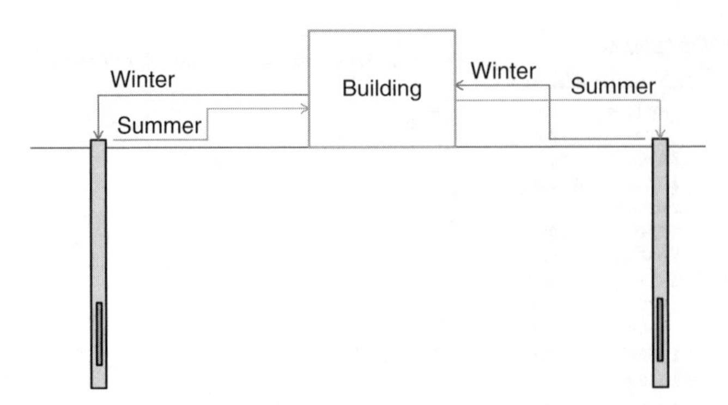

Figure 4.25 Well doublet with seasonal changeover

Another design and well pump control strategy is the well doublet with seasonal changeover, as shown in Figure 4.25. In that design, each well contains a pump and acts as a supply well in one season and as an injection well in the other season (i.e., switching between heating and cooling seasons). There are several benefits to this design, including thermal energy storage from the previous season, enhanced flushing of particulates (as well water flows in a back-and-forth direction seasonally), and availability of a redundant pump, which can extend pump life because each pump is used less.

4.5.5 Single Supply-Return Well Systems

In single supply-return well systems, groundwater is supplied from and returned to the same well. The concept originated with wells of insufficient yield for heat pump application, or in situations where other means of disposal were not viable.

Perhaps the most well-known type of single supply-return well is the so-called *standing-column well* (Figure 4.26). These systems have received a considerable amount of attention in the literature, following an ASHRAE-sponsored research project (Spitler *et al.*, 2002b). The concept purportedly originated in Maine – a New England State comprised mostly of igneous and metamorphic rocks with relatively low-yield wells. As such, residences on well water often needed relatively deep wells (as deep as 500 m or more) to achieve acceptable water quantities for domestic purposes. The crystalline rock geology in New England results in groundwater of good quality and low TDS, and wells are typically uncased.

Well yields in hard, crystalline rock geological environments typically have insufficient yields for geothermal heat pump use, so the concept arose in New England to pump groundwater from the well, run it through a heat pump, and return it to the same well. Heat exchange with the Earth was accomplished as groundwater flowed from the top of the 'standing column' of water, down the well bore, and back into the pump intake. Given the relatively cold New England climate, the well water cools over the winter season, but the well water temperature can be moderated by removal of water (or so-called 'bleed') from the well. Natural 'bleed' occurs in residential systems as water is used and thus removed from the well for domestic purposes. As such, these systems make excellent geothermal systems in residential applications with deep, low-yield wells of good groundwater quality.

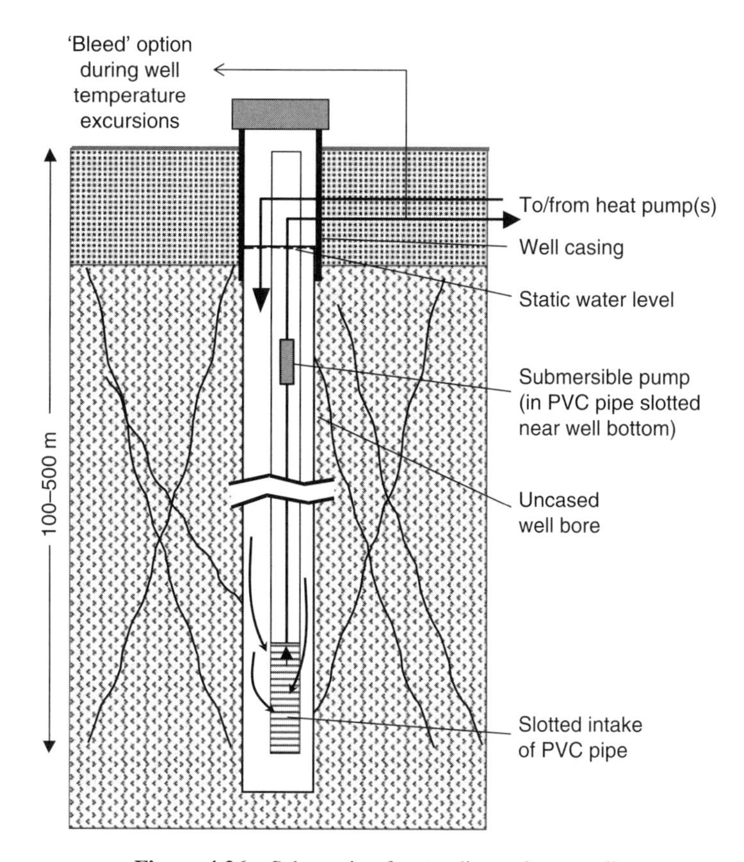

Figure 4.26 Schematic of a standing column well

Standing-column wells have been referred to as semi-closed-loop systems, or open-closed-loop systems, as they have many attributes of the closed-loop systems we shall examine in the next chapter. Thus, we will defer their mathematical analysis to Chapter 5. In theory, a standing column well could be applied in any application where there is groundwater of good quality, and where wells can be feasibly drilled deep enough to allow for proper Earth heat exchange. However, Spitler *et al.* (2002b), in a survey of standing column wells in North America, identified most applications in the New England States, with only sporadic applications in other regions. Most of the installations identified by Spitler *et al.* (2002b) were residential.

In summary, the regional criteria advantageous for standing column wells include: competent rock (which allows uncased well bores), areas of limited well yield, and good groundwater quality. Local criteria include: limited land availability, ability to integrate with domestic well water use ('natural bleed'), and responsible means of bleed water disposal.

Other variants of single supply-return wells have appeared in the literature over the years. These include, for example, a *doublet well* proposed by C.E. Jacob (Wickersham, 1977) and a so-called *sand well* described by Dexheimer (1985). As shown in Figure 4.27, these wells are more conventional, screened wells installed in a single, thick aquifer, or in a sequence of aquifers and confining beds. The difference, as shown in Figure 4.27, is that there are two

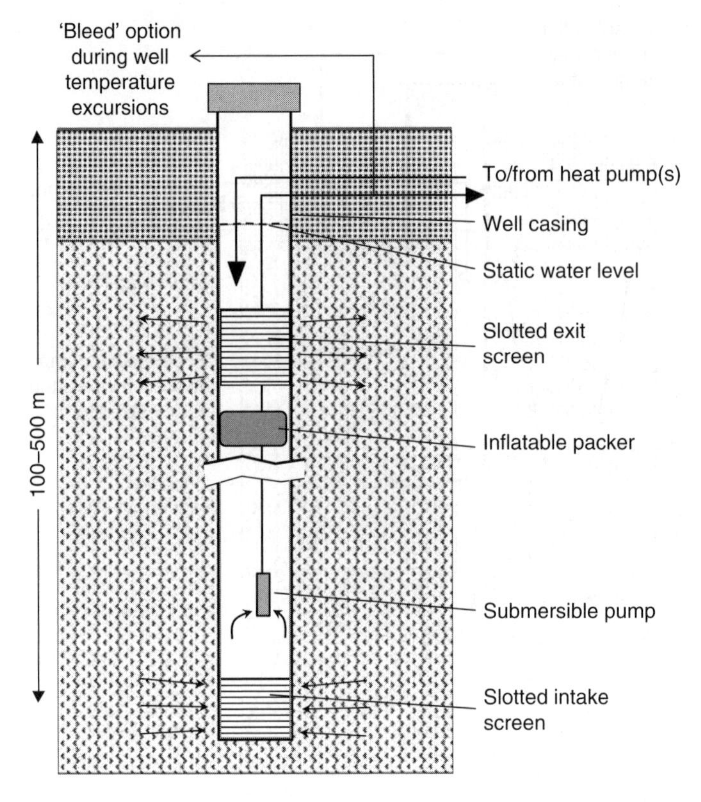

Figure 4.27 Schematic of a single supply-return well with a packer separating the supply and return zones

screened zones, separated by an inflatable packer, such that groundwater is extracted from the lower zone and returned to the upper zone.

4.6 Chapter Summary

This chapter has discussed the numerous theoretical and practical considerations of the use of groundwater in heat exchange applications.

The theory of groundwater flow and the associated governing equations were first examined. Analytical methods were presented for determining groundwater flow direction and velocity, and numerical methods were presented for determining the hydraulic head distribution in an aquifer. The theory of radial flow toward a well and the subject of well hydraulics were discussed, along with methods for determining the well and aquifer parameters from single-well tests. Heat transport in groundwater was also described, and an analytical solution was presented for determining the 2D temperature distribution in an aquifer with a point heat injection source.

Practical considerations of groundwater heat exchange systems were presented, with discussion of the primary equipment used in these systems: wells, pumps, heat exchangers, and piping. Analysis methods were presented for determining pump power requirements and for heat exchanger analysis and design. Groundwater chemical parameters of interest in scaling and corrosion were also discussed.

The theory and practice were combined to analyze the performance of groundwater heat exchange systems. Use of the companion suite of GHX tools was described to determine required groundwater flow rate and system energy consumption. Well pump control methods were described, with emphasis on variable-speed well pumping. Finally, single supply-return wells were described, in particular the standing column well, and a well with dual screens separated with a packer.

Discussion Questions and Exercise Problems

4.1 Conduct some research to determine what government agency has jurisdiction over water well construction where you live. Do they allow groundwater heat pumps?

4.2 Reconstruct the finite difference model shown in Example 4.1. Determine the minimum pumping rate in well 1 that would result in hydraulic breakthrough. That is, what minimum pumping rate results in a reversal of the flow such that all groundwater flows toward the pumping well?

4.3 The following table summarizes the pumping rate and the stabilized drawdown for a single-well step-drawdown test.

Step	Pumping Rate ($L{\cdot}s^{-1}$)	Stabilized Drawdown (m)
—	0	0
1	8.46	6.82
2	12.24	10.75
3	14.64	13.22
4	17.23	16.42
5	19.07	18.62

(a) plot the drawdown vs. pumping rate,
(b) plot the specific drawdown vs. pumping rate,
(c) determine the aquifer and well loss coefficients β and C,
(d) Estimate the drawdown at a pumping rate of 10 L/s, and
(e) Discuss your results.

4.4 Consider the following scenario. A well has a static water level of 100 ft below the ground surface and is pumping at a rate to cause 10 ft of drawdown. If the mechanical room is another 10 ft above the ground surface, and the heat exchanger has 10 ft of head loss (in ft of H_2O equivalent), what is the estimated 'lift' of a well pump in this well?

4.5 If you are the engineer on a project requiring 1000 gpm of groundwater flow, what is your recommended well diameter for this project? Explain.

4.6 Consider a heat exchanger for use in a direct-use geothermal heating application where the heating load is 10 000 kW for a group of buildings. Water is used as the heating fluid in each building. The geothermal resource temperature is 75 °C, and an allowable temperature drop of the geothermal fluid as it passes through the heat exchanger is 20 °C. At temperature decreases of greater than 20 °C, there is concern of mineral precipitation from the groundwater. The heating systems in the buildings are designed for entering water temperatures of 45 °C, and an allowable temperature drop of 5 °C.

(a) sketch the heat exchanger and show the inlet and outlet temperatures on the geothermal and building loop sides,
(b) determine the required flow rate of the geothermal fluid,
(c) determine the heat exchanger approach temperature,
(d) determine the heat exchanger *UA* value, and
(e) determine the heat exchanger effectiveness.

4.7 Consider a groundwater heat pump system to be designed as shown in Figure 4.20. The groundwater temperature is 12 °C. The peak cooling load is 250 kW, and the system is to be designed for maximum heat pump entering water temperatures of 30 °C to minimize groundwater pumping (because the well yield is low). As part of the design team, your task is to determine the following.

(a) the heat to be transferred by the heat exchanger if all the heat pumps have a COP of 4.0 at the peak design condition,
(b) the building loop water temperature entering the heat exchanger,
(c) the required groundwater flow rate, assuming a heat exchanger effectiveness of 0.8,
(d) the groundwater temperature going to the injection well, and
(e) the overall COP of the **whole system** if the electrical power consumption is 5 kW for the circulating pump and 75 kW for the well pump.

4.8 A heating-dominated building has the following design conditions:

- peak heating load = 100 kW
- groundwater temperature = 10 °C
- heat exchanger effectiveness = 0.85
- from well test data:
  - static water level = 10 m below ground surface
  - aquifer loss coefficient = 0.01 min/L·m
  - well loss coefficient = 0.001 $(\text{min/L})^2\cdot\text{m}$
- heat exchanger and associated piping head loss = 6 m-H_2O

(a) What is the optimum groundwater flow rate (in Lpm) and system COP that minimize total system energy consumption? What is the design heat pump entering fluid temperature at this optimum condition?
(b) How does your answer in (a) change if the static water level is 150 m and the groundwater temperature is 5 °C? In other words, what is the new optimum

groundwater flow rate (in Lpm), system COP, and design heat pump entering fluid temperature?

(c) Explain why there is a difference in required groundwater flow rate between questions (a) and (b).

(d) Briefly describe how you would control the well pump.

(e) What materials would you avoid if the groundwater contained 250 ppm chloride?

Hint: You may use the *GHX Tools, Groundwater Heat Exchange* on the companion website.

5

Borehole Heat Exchangers

5.1 Overview of Borehole Heat Exchangers (BHEs)

Predicting the thermal behavior inside and in the vicinity of BHEs is important to establish the required borehole length and to determine the resulting fluid temperature.

In this chapter, we first put borehole heat exchangers into perspective in an overview of their use, followed by a brief discussion of their installation methods. We then turn our focus to the mathematical details of the thermal analysis of single BHEs, with no thermal interaction with the ground surface or other boreholes (those topics will be discussed in subsequent chapters). Analytical solutions of the heat diffusion equation are presented, with applicable boundary conditions to describe the heat transfer in the Earth around a single BHE. Details are presented of the mathematical treatment of the so-called *borehole thermal resistance*, which is described as the resistance to heat transfer between the heat carrier fluid and the surrounding Earth. The chapter concludes with an example use of mathematical methods presented in this chapter to determine subsurface thermal conductivity and borehole thermal resistance from an actual field test experiment, commonly known as a thermal response test.

Learning objectives and goals:

1. Apply models of radial heat conduction around a vertical borehole.
2. Calculate the steady-state borehole thermal resistance for various borehole heat exchanger configurations.
3. Analyze thermal response test data, and apply the so-called line source solution to determine thermal conductivity of geologic materials, and the borehole thermal resistance.

Geothermal Heat Pump and Heat Engine Systems: Theory and Practice, First Edition. Andrew D. Chiasson.

Companion website: www.wiley.com/go/chiasson/geoHPSTP

5.2 What is a Borehole Heat Exchanger?

The term *borehole heat exchanger* (BHE) refers to a closed-loop pipe assembly installed in a vertical borehole with radius r_b over some active depth H for purposes of heat exchange with the Earth, as shown in Figure 5.1. The top of the active BHE is buried at some depth (D) from the ground surface. Typical constructions consist of a single U-tube grouted in a borehole, but other geometries exist (i.e., double U-tube, concentric tube, and groundwater-filled boreholes). These boreholes are designed to extract (or reject) a certain amount of thermal energy ($\dot{q}'$) per unit depth (H) by pumping a fluid, with an average temperature (T_f), through the heat exchanger. Thus, $\dot{q}'$ is the heat extraction or rejection rate divided by the active borehole depth. Heat transfer occurs from the fluid to the ground, the undisturbed 'far-field' temperature of which remains at T_g. T_b represents the average temperature at the borehole wall.

Since the pioneering research and development studies in the 1970s regarding BHEs of the vertical ground-coupled type, the so-called 'U-tube' configuration has emerged as the preferred method of heat exchanger construction. In these types of system, the overall ground heat exchanger commonly consists of an array of vertical boreholes, each containing a high-density polyethylene (HDPE) U-shaped pipe through which a heat exchange fluid is circulated. Depending upon geologic conditions, boreholes for single U-tube heat exchangers are typically drilled to nominal depths ranging from 50 to 100 m (~150 to ~330 ft), with diameters of the order of 5 in (127 mm). Typical U-tubes have a nominal diameter in the range ¾–1¼ in (19.0–38.0 mm). The borehole annulus is generally backfilled with a bentonite-based grout to facilitate the sealing of aquifers and to improve the contact area for heat transfer. In many Scandinavian applications, where most of those countries consist of granite rock, grout is not

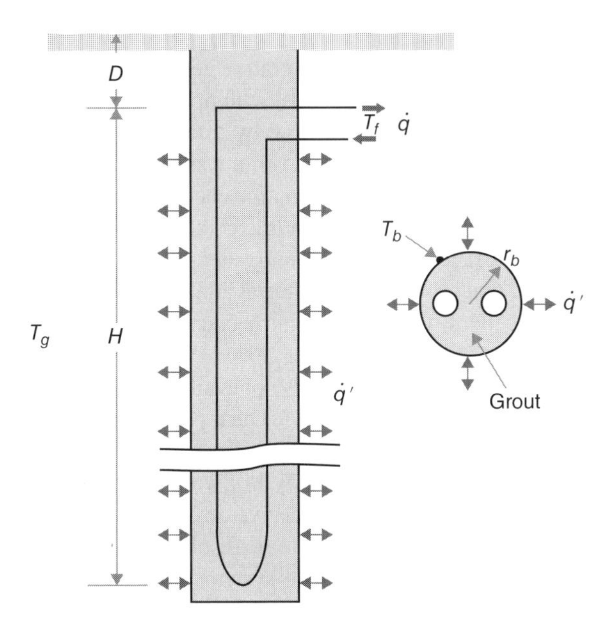

Figure 5.1 Schematic diagram of a typical geothermal borehole heat exchanger (BHE) comprised of a single U-tube grouted in a vertical borehole

used; U-tubes instead are installed in rock boreholes that naturally fill up with groundwater that occurs in rock fractures.

Some other possible types of vertical borehole heat exchanger are the double U-tube and concentric (or coaxial) configuration. These have received much less attention in the literature relative to single U-tube heat exchangers. The basic construction of a double U-tube consists of two single U-tubes installed in one borehole and piped in either a series or parallel flow circuit. The advantage in thermal performance of the double U-tube ground heat exchanger over the single U-tube is that a second channel for the flow of the heat transfer fluid replaces a region of the borehole that would otherwise be filled with grout. Rybach and Sanner (2000) report that the double U-tube heat exchanger is the most common type of BHE installed in Switzerland. Yavuzturk and Chiasson (2002) showed a reduction in overall borehole length of 22% for an office building when double U-tubes were used instead of single U-tubes. This reduction in length is due to the lower borehole thermal resistance with double U-tubes.

5.3 Brief Historical Overview of BHEs

Early uses of 'downhole' heat exchangers (or DHEs) were pioneered for direct-use geothermal applications in the late 1920s in Klamath Falls, Oregon. As described in Chapter 4, these were employed in water wells, and thus were considered a special type of groundwater heat exchange application. However, these laid the 'groundwork' for BHEs to follow. These early DHEs consisted of a black-iron pipe installed in water wells with well water temperatures of the order of 150 °F (65 °C) to over 200 °F (95 °C). Heat was exchanged with groundwater in these wells and was used for directly heating homes, schools, and other buildings. Today, there are over 500 such applications in Klamath Falls alone.

The first commercial applications of low-temperature geothermal exchange for water-source heat pumps is documented as beginning in 1940s, with open-loop systems using groundwater wells in Portland, Oregon. Analytical investigations by Adler *et al.* (1951) provided some of the theoretical basis for later design programs. However, it was the oil crisis of the early 1970s that gave more incentive for the development of alternative energy sources, such as solar, wind, and geothermal. Heat pumps for domestic heating attracted significant interest (and they still do today), using ambient air, surface water, or groundwater as heat sources/sinks. As discussed in Chapter 4, it was realized that groundwater, when available in sufficient quantities with good chemical quality, is an excellent heat source/sink because its temperature is normally quite stable.

As groundwater of good quality and quantity is not available everywhere, research began on the use of the ground itself as a heat source/sink for heat pumps. The ground is obviously globally available, and the temperature (except for the upper few meters) remains stable. Research and development programs began simultaneously in Sweden and in the United States in the 1970s toward the use of closed-loop BHEs. Intensive research on BHEs, in conjunction with developments in water-source heat pump technology, resulted in the emergence of the geothermal heat pump market by the late 1980s. These systems began to be installed in numerous building types, such as homes, schools, and offices, and ongoing research and development resulted in the appearance of design tools and manuals for engineers and other trades. Associations emerged for designers, such as the International Ground Source Heat Pump

Association (IGSHPA) (www.igshpa.okstate.edu) and ASHRAE (www.ashrae.org), and certifications for designers were promoted.

Mathematical methods for BHE thermal analyses are continually revisited by new researchers, and the mid to late 2000s saw a particular research focus on refinements to design and installation procedures, particularly related to hybrid systems and sustainable designs. Heat pump technology continues to improve, with smart controls and variable-speed compressors appearing on the market. Various economic models have come and gone and have had limited success, such as third-party 'loop-leasing', geothermal utilities, and district systems. Heat pump technology continues to improve, impacting how BHEs are designed.

5.4 Installation of BHEs

Successful installation and operation of vertical BHEs require some theoretical considerations, but are always limited by the practical. There is a saying by Albert Einstein that applies here, 'There is no difference between theory and practice, but in practice, there is'. With regard to BHE installation, the designer can specify anything and draw anything in the comfort of the office. But in reality, *borehole constructability* is dictated by field conditions, which are never fully known; when dealing with the subsurface environment, we can't see it beforehand, so there is always some uncertainty about actual conditions. The more effort a designer puts in upfront, the less will be the risk of the unknown.

BHE designers should, at least, have a working knowledge of the soil and rock types that will be encountered during BHE installation. Unexpected groundwater and unstable soils can pose significant problems during drilling if the correct drilling method is not specified. Basic geology and drilling methods have been described in Chapter 3.3.

After the borehole is drilled to the target depth, the BHE is inserted and grouted in place. It is generally not advisable to place soil and rock cuttings back into the borehole without first checking local codes and regulations. Secondly, soil and rock cuttings do not constitute an engineered backfill material, and thus cannot be guaranteed to provide an adequate borehole seal or heat transfer medium around the U-pipe.

Heat transfer is maximized when using a grout that has a relatively high thermal conductivity relative to the surrounding soil/rock, has reasonably low viscosity during placement, is easy to pump, fills annular space and reduces potential for voids, sticks to the pipe wall for good thermal contact, exhibits little or no settling or shrinkage after placement, and retains thermal characteristics over the long term. Grout materials will be discussed in a subsequent section of this chapter.

Figure 5.2 shows a U-tube BHE inserted into a borehole off a loop reel. In boreholes to be grouted, a 'tremie pipe' is inserted simultaneously with the U-tube, and is used to pump grout into the borehole starting at the bottom. Grout is then pumped in as the tremie pipe is removed, ensuring that the borehole is grouted bottom first, which aids in displacing any mud and debris remaining in the borehole. A crew separate from the drilling crew is usually responsible for grouting.

Details on grouting practices and procedures can be found in IGSHPA (2009) and IGSHPA (1991).

Figure 5.2 BHE insertion into a vertical borehole (photo taken by the author)

5.5 Thermal and Mathematical Considerations for BHEs

5.5.1 General BHE Thermal Considerations

Eskilson (1987) and Hellström (1991) provide a detailed thermal analysis of heat extraction boreholes and describe important parameters in their performance. The five most important parameters identified in the performance of a borehole heat exchanger are: (i) the soil/rock thermal conductivity, (ii) the borehole thermal resistance, (iii) the undisturbed Earth temperature, (iv) the heat extraction (and rejection) rates, and (v) the mass flow rate of the heat carrier fluid. Each of these parameters will be discussed in more detail in the paragraphs that follow.

The thermal performance of a borehole heat exchanger is proportional to the *thermal conductivity* of the ground, and a granite with a thermal conductivity of 3.30 $W{\cdot}m^{-1}{\cdot}K^{-1}$ is 3 times better than a clay soil with a thermal conductivity of 1.10 $W{\cdot}m^{-1}{\cdot}K^{-1}$ (Eskilson, 1987). A considerable amount of research has been conducted over the past 20 years regarding *in situ* testing (or thermal response testing) to determine Earth thermal conductivity for use in design and simulation tools. Numerous citations appear in the literature on this topic. Detailed discussions can be found in Austin *et al.* (2000) and Gehlin (2002).

A second very important parameter in borehole heat exchanger performance is the *borehole thermal resistance*. The borehole thermal resistance is defined by a number of design variables, including the composition and flow rate of the heat transfer fluid, the borehole diameter, the heat exchanger pipe material, the arrangement of the flow channels, and the grout material.

The greater the thermal resistance of the individual borehole elements, the less will be the heat transfer rate between the heat carrier fluid and the Earth, thus requiring an increased length of the ground heat exchanger. Therefore, it is most desirable to keep the borehole thermal resistance at a minimum.

A significant portion of recent research efforts regarding U-tube ground heat exchangers has dealt with the development of thermally enhanced grouts (grouts of higher thermal conductivities) for the purpose of improving overall heat exchanger performance, and thereby reducing necessary borehole lengths. Remund and Lund (1993) reported on improving the thermal conductivity of bentonite grouts with the use of quartzite sand. Allan and Kavanaugh (1999) used silica sand, alumina grit, steel grit, and silicon carbide as fillers to increase the thermal conductivity of bentonite grouts to values of 1.70–3.29 $W{\cdot}m^{-1}{\cdot}K^{-1}$, and achieved a theoretical reduction in required bore length of 22–37% for the use of cement–sand grouts. An analysis of the cost savings of borehole fields with thermally enhanced grouts (The University of Alabama, 2000) suggests that an upper practical limit of grout thermal conductivity is approached at values of about 1.47 $W{\cdot}m^{-1}{\cdot}K^{-1}$. This upper limit is attributed to an offset of the reduction in drilling costs by increased labor and handling costs of additives that must be mixed into thermally enhanced grouts. However, recent advancements in grout development have included the introduction of graphite-based grouts, which are lower in density and higher in thermal conductivity than the grouts thermally enhanced with silica sand.

Typical materials for construction of single U-tube ground heat exchangers result in a relatively high borehole thermal resistance. For example, HDPE plastic typically has a thermal conductivity of 0.40 $W{\cdot}m^{-1}{\cdot}K^{-1}$, and bentonite-based grouts have a thermal conductivity of about 0.69 $W{\cdot}m^{-1}{\cdot}K^{-1}$. As the general range of thermal conductivity values for geologic materials is 0.50–3.80 $W{\cdot}m^{-1}{\cdot}K^{-1}$ for soils and 1.00–6.90 $W{\cdot}m^{-1}{\cdot}K^{-1}$ for rocks, it is evident that the components of U-tube systems provide some insulating effect to the ground heat exchanger.

A third important parameter in borehole heat exchanger performance is the undisturbed Earth temperature. Current design and simulation tools take an average of this temperature over the entire borehole depth. Heat extraction and rejection work against this temperature, and the required borehole depth is essentially proportional to the temperature difference between this undisturbed Earth temperature and the minimum (or maximum) design heat pump entering fluid temperature. For example, as described by Eskilson (1987), for a climate where the undisturbed ground temperature is 8.00 °C vs. one with 18.0 °C, the available temperature drop is 3 times less for a minimum heat pump entering fluid temperature of 3.00 °C.

A fourth important parameter in borehole heat exchanger performance is the nature of heat extraction and heat rejection to the ground. At a minimum, current design tools consider peak hour thermal load and its time duration in addition to peak monthly pulse. This is the design approach for residential systems described in IGSHPA (2009). For commercial systems, at a minimum, design tools consider the above hourly and monthly loads in addition to an annual load over a 10 or 20 year period. This is the method described by the ASHRAE Handbook.

A fifth important parameter in borehole heat exchanger performance is the mass flow rate of the heat exchange fluid. This parameter is included in the borehole thermal resistance calculation, but it is important to keep the flow rate large enough to ensure turbulent flow (Eskilson, 1987). The heat transfer fluid is commonly pure water, or in cold climates an aqueous solution of antifreeze such as propylene glycol or methanol.

Eskilson (1987) also identifies negligible parameters and effects on borehole heat exchanger performance. These are:

- deviations from average thermal conductivity owing to stratified ground,
- temperature variations at the ground surface,
- the effect of groundwater flow is negligible if the following criterion is met:

$$\frac{H\rho_w c_{p,w}\dot{q}_w}{2k} < 1 \tag{5.1}$$

  where H is the borehole depth, ρ_w is the groundwater density, $c_{p,w}$ is the groundwater heat capacity, q_w is the Darcy velocity of the groundwater, and k is the ground thermal conductivity. Equation 5.1 is effectively the dimensionless Peclet number (Pe) of the flow. This result has also been verified by Chiasson *et al.* (2000), who utilized a similar definition of the Peclet number, in conjunction with a finite element heat and mass transport model to show that groundwater velocities are negligible in affecting the heat transfer around boreholes in most geologic materials. The exception is for coarse-grained soils and rocks with secondary porosities (i.e., fracturing or solution channels).
- transient thermal effects in the borehole grout and the heat carrier fluid above timescales of t_b:

$$t_b = \frac{5r_b^2}{\alpha} \tag{5.2}$$

  where r_b is the borehole radius and α is the soil/rock thermal diffusivity. Thus, below this timescale, borehole transient thermal effects may be significant. For typical borehole geometries and soil and rock thermal properties, Equation 5.2 implies a timescale of the order 4–8 h.

5.5.2 *Mathematical Models of Heat Transfer around BHEs*

The temperature distribution in the ground may be described by the partial differential heat diffusion equation expressed in cylindrical coordinates:

$$\alpha\left[\frac{\partial^2 T}{\partial r^2} + \frac{1}{r}\frac{\partial T}{\partial r} + \frac{1}{r^2}\frac{\partial^2 T}{\partial \phi^2} + \frac{\partial^2 T}{\partial z^2}\right] + G = \frac{\partial T}{\partial t} \tag{5.3}$$

Note that Equation (5.3) is exactly the same as Equation (4.16), except T is the temperature and α is the thermal diffusivity. As with Equation (4.16), r is the radial coordinate, z is the vertical coordinate, t is the time, and G is a general source/sink term.

Several variations of two analytical models exist for calculating the temperature distribution in the Earth around a BHE, and some have been incorporated into commercially available design software. These two models are known as the *line source model* and the *cylinder source model*, and are solutions to slightly varying forms of Equation (5.3) with slightly different boundary conditions.

It is important to emphasize that there are actually two types of line source models used in BHE design: (1) the infinite line source model and (2) the finite line source model. In both models of radial heat conduction, the term $\frac{1}{r^2}\frac{\partial^2 T}{\partial \phi^2}$ is not applicable. In infinite line source models, a simplifying assumption is made that $\partial^2 T/\partial z^2 = 0$, and therefore only radial heat transfer in the ground is considered. Such a model is valid under certain restrictions of time and borehole depth, but after long periods of time, and/or with short boreholes, the heat transfer around a BHE takes on a significant vertical component and $\partial^2 T/\partial z^2$ must be accounted for; hence, the need for a finite line source model.

Kelvin (1882) originally described an infinite line source model (commonly known as Kelvin's line source model), which is considered to be a classic solution to calculate the temperature distribution around an imaginary vertical line in a semi-infinite solid medium, initially at a uniform temperature. Eskilson (1987) made a further significant contribution by developing mathematical models for describing the temperature distribution around *finite line sources*. The *cylinder source model*, originally developed and evaluated by Carslaw and Jaeger (1947), is generally considered to be a better representation of the cylindrical geometry of a borehole, and was thus proposed by Ingersoll *et al.* (1954) as a method for sizing buried heat exchangers. The solution to the *cylinder source model* contains integrals that can be difficult to evaluate.

Coupled with the complexity of the *cylinder source model*, and the fact that line source models give a closer representation of field data (as we shall see at the end of this chapter), we shall adopt the *line source model* as our standard for modeling the heat transfer around vertical BHEs for practical purposes. We will, however, revisit the *cylinder source model* in our treatment of horizontal ground heat exchangers.

5.5.2.1 Infinite Line Source Analytical Models of Heat Transfer in the Ground

Infinite line source solutions are general solutions to Equation (5.3) when $\partial^2 T/\partial z^2 = 0$ (and $\frac{1}{r^2}\frac{\partial^2 T}{\partial \phi^2}$ is not applicable). The initial conditions are prescribed as

$$T(r,t) = T(r,0) = T_g \tag{5.4}$$

and the boundary conditions are given by

$$T(r,t) = T(\infty,t) = T_g \tag{5.5}$$

$$\lim_{r \to 0}\left(r\frac{\partial T}{\partial r}\right) = \frac{\dot{q}'}{2\pi k} \tag{5.6}$$

where T is temperature, T_g is the undisturbed ground temperature, $\dot{q}'$ is the heat transfer rate per length of line source (W·m^{-1} or Btu/h/ft), and k is the thermal conductivity of the medium (W·m^{-1}·K^{-1} or Btu/h/ft/°F).

Ingersoll and Plass (1948) and Ingersoll *et al.* (1954) provide an adaptation of Kelvin's line source model for practical use with BHEs, exactly analogous to the Theis solution for radial flow to a well [Equation (4.21)]:

$$\Delta T_r = \frac{\dot{q}'}{4\pi k}\int_u^\infty \frac{e^{-u}}{u}du \tag{5.7}$$

where ΔT_r is the temperature change (°C or °F) at radius r, and u is defined as

$$u = \frac{r^2}{4\alpha t} \tag{5.8}$$

where r is the radius from the line source, α is the thermal diffusivity of the medium ($m^2 \cdot s^{-1}$ or ft^2/h), and t is the time duration (s or h) of the $\dot{q}'$ heat input ($W{\cdot}m^{-1}$ or Btu/h/ft).

The infinite integral appearing in Equation (5.7) is well known in mathematics as the *exponential integral*, and was described in Chapter 4. Recall that this integral became known in the field of well hydraulics as the *well function*, $W(u)$. Therefore, Equation (5.7) may be written more simply as

$$\Delta T_r = \frac{\dot{q}'}{4\pi k}W(u) \tag{5.9}$$

For small values of u, Equation (5.9) may be approximated as

$$\Delta T_r = \frac{\dot{q}'}{4\pi k}\left(\ln\left(\frac{4\alpha t}{r^2}\right) - \gamma\right) \tag{5.10}$$

where γ is Euler's constant (0.5772). Equations (5.9) and (5.10) have a lower limit of time at which the heat pulse has not reached the radius r of interest, and an upper limit of time to ensure radial heat flow from the line source. Eskilson (1987) defines these time limits as

$$5r^2/_\alpha < t < {}^{t_s}/_{10} \tag{5.11}$$

where t_s is a characteristic timescale (also known as steady-state time) equivalent to $H^2/(9\alpha)$, where H is the borehole depth (m or ft) and α is the thermal diffusivity ($m^2 \cdot s^{-1}$ or ft^2/h).

For a wider range of u values, approximations for $W(u)$ have been described in Chapter 4 (Section 4.3.2).

Note that Equation (5.9) can also be expressed as

$$\Delta T_r = \dot{q}' R'_g \tag{5.12}$$

where R'_g is the ground thermal resistance per unit length of bore ($m \cdot K \cdot W^{-1}$). Thus, $W(u)/(4\pi k)$ is the ground thermal resistance per unit length of bore.

5.5.2.2 Finite Line Source Analytical Models of Heat Transfer in the Ground

Finite line source solutions are general solutions of Equation (5.3) when $\partial^2 T/\partial z^2 \neq 0$ (and $\frac{1}{r^2}\frac{\partial^2 T}{\partial \phi^2}$ is not applicable). Such solutions are necessary to consider the three-dimensional effects of heat transfer around a BHE when the infinite line source conditions are no longer applicable. These conditions occur after longer times and/or with short boreholes.

The initial conditions for finite line source solutions are prescribed as

$$T(r,z,t) = T(r,z,0) = T_g \tag{5.13}$$

Note that this condition prescribes a geothermal gradient of zero, and assumes a temperature of T_g at a depth D (referring to the notation in Figure 5.1).

The boundary conditions are similar to those of the infinite line source, but are applied over the finite length of the BHE.

Eskilson's Analytical g-Function Approach

Claesson and Eskilson (1987) provide a simple and computationally efficient solution to Equation (5.3) as

$$\Delta T_r = \frac{\dot{q}'}{2\pi k} g \tag{5.14}$$

where g is a dimensionless temperature response factor, referred to as the g-function, and is defined as

$$g\left(\frac{t}{t_s}, \frac{r}{H}\right) = \begin{cases} \ln\left(\frac{H}{2r}\right) + \frac{1}{2}\ln\left(\frac{t}{t_s}\right) & , \frac{5r^2}{\alpha} < t < t_s \quad (5.15a) \\ \ln\left(\frac{H}{2r}\right) & , t > t_s \quad (5.15b) \end{cases}$$

When Equation (5.14) is expressed in terms of a thermal resistance $\dot{q}' = 1/R' \cdot \Delta T$, then it can readily be seen that $g/(2\pi k)$ can be described as the ground thermal resistance per unit length (R').

Eskilson's Numerical g-Function Approach

Eskilson (1987) also developed a two-dimensional finite difference model in radial–axial coordinates to examine heat extraction boreholes. The model was developed in the FORTRAN 77 language for computer application. This model was also used to develop so-called long-time step response factors, which will be discussed in the next chapter. The model uses a variable-size mesh, as shown in Figure 5.3. The boundary conditions are specified as flux conditions equal to zero at all locations except along the finite length of the borehole (H). The initial temperature of the domain is set to zero. Eskilson (1987) spends considerable effort in analyzing the grid dependence on the solution accuracy. The program takes a constant heat flux per unit length of borehole as an input, and calculates the resulting temperature at the borehole wall at

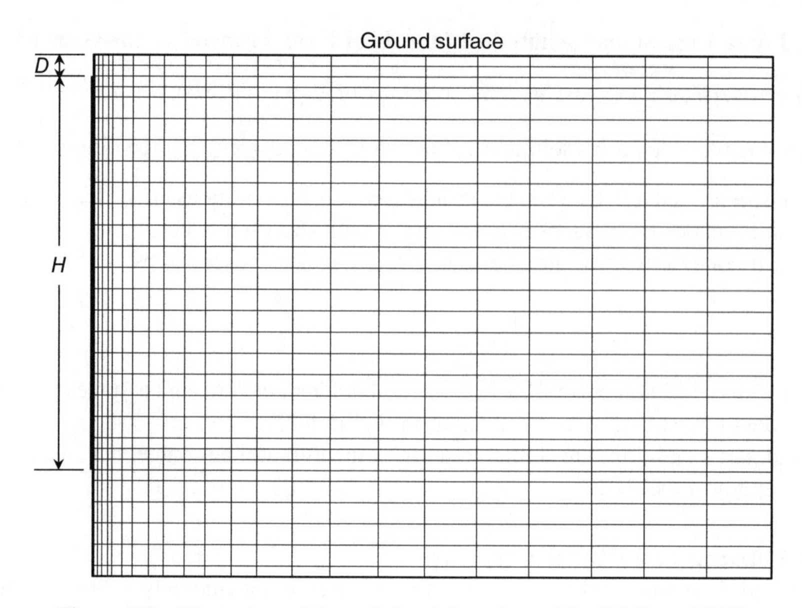

Figure 5.3 Illustration of the radial–axial mesh used by Eskilson (1987)

various times. To examine multiple borehole systems, Eskilson uses the method of superposition, which we shall discuss in Chapter 6.

Incomplete Bessel Function (a.k.a. Leaky Well Function) Approach

Yet another finite-line source solution to Equation (5.3) when $\partial^2 T/\partial z^2 \neq 0$ (and $\frac{1}{r^2}\frac{\partial^2 T}{\partial \phi^2}$ is not applicable) can be developed using the leaky well function ($W(u,B)$), which was described in Chapter 4.3.2. for application to well hydraulics The concept is the same as with 'leaky wells', but now we are referring to the vertical diffusion of heat. Thus, the temperature at radius r can be given by

$$\Delta T_r = \frac{\dot{q}'}{4\pi k} W(u,B) \tag{5.16}$$

where $W(u,B)$ is the incomplete Bessel function described by Equation (4.27). Computationally efficient solution methods were described in Chapter 4 (Section 4.3.2), and Figure 4.4 provides a plot of theoretical curves for the leaky well function for various values of B.

As above with the infinite line source, $u = r^2/(4at)$. For B expressed as

$$B = \frac{r}{\sqrt{\alpha t_s}} = 3\frac{r}{H} \tag{5.17}$$

there are negligible differences when compared with Esklison's g-function for a single borehole. Note that r is the radius from the line source, and B reduces to a simple ratio of $3r/H$ because t_s is equivalent to $H^2/(9\alpha)$. A comparison plot of the g-function of Eskilson (1987)

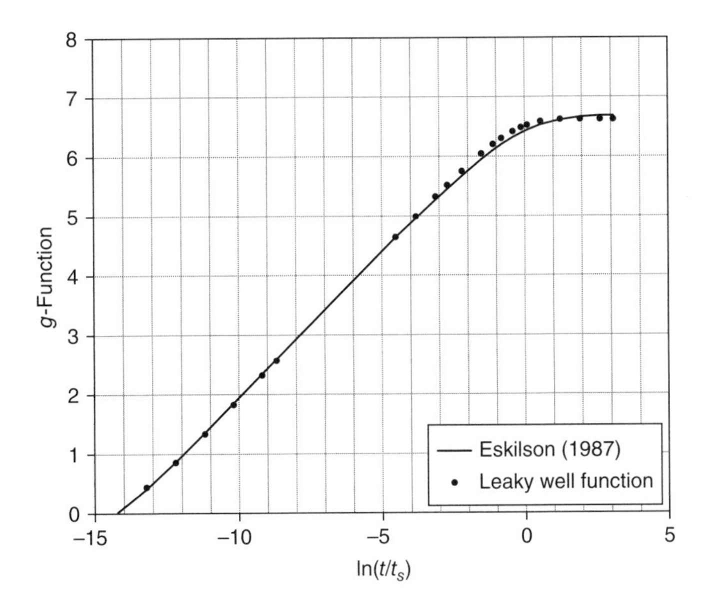

Figure 5.4 Comparison of single-borehole g-functions of Eskilson (1987) with the leaky well function

for a single borehole is compared with that determined by the leaky well function in Figure 5.4. Note the favorable comparison. We shall discuss g-functions in more detail in Chapter 6 for multiborehole arrangements. Thus, the ground thermal resistance per unit length of bore (R'_g) is given by $W(u, B)/(4\pi k)$.

5.5.3 Determining the BHE Fluid Temperature

We are frequently interested in the change in fluid temperature of heat carrier fluid in the borehole heat exchanger, and not necessarily the change in temperature in the ground as would be calculated by the foregoing equations. In order to calculate the change in temperature at a given time in the BHE fluid, we first need to calculate the temperature at the *borehole wall* by substituting the value of r_b in place of r in the above equations, and calculating the ground thermal resistance. Next, we need to account for the thermal resistance of the borehole elements (such as the pipe configuration within the borehole, the borehole grout, and the fluid thermal properties) using the so-called borehole thermal resistance (R'_b):

$$\Delta T_f = \dot{q}' R'_g + \dot{q}' R'_b \tag{5.18}$$

where the subscript f refers to the heat carrier fluid within the BHE, and R'_b is the steady-state borehole thermal resistance per unit length ($\mathrm{m \cdot K \cdot W^{-1}}$). Details and calculation methods of the borehole thermal resistance are described in a later subsection.

The average BHE fluid temperature is then simply calculated by adding the change in the fluid temperature ΔT_f to the undisturbed ground temperature (T_g):

$$T_{f,avg} = \Delta T_f + T_g \tag{5.19}$$

It is emphasized here that the fluid temperature calculated by the above approach is the average temperature of the fluid circulating in the BHE. This is actually a simplifying approximation to the three-dimensional nature of the temperature distribution within the BHE fluid itself. As the BHE is a heat exchanger, the fluid temperature entering the BHE will obviously be different from the fluid temperature exiting the BHE. Calculation of the entering and exiting fluid temperatures from knowledge of the average fluid temperature will be discussed in a subsequent chapter.

Example 5.1 Studying the Thermal Characteristics of a Single BHE

Consider a BHE undergoing continuous heat extraction from the Earth at a rate of 5000 W. The borehole depth (H) is 100 m and the borehole diameter is 125 mm. The initial, undisturbed ground temperature (T_g) was 15 °C. The borehole is completely installed in shale with a thermal conductivity (k) of 2.0 $W \cdot m^{-1} \cdot K^{-1}$, a density ($\rho$) of 2500 $kg \cdot m^{-3}$, and a heat capacity (c_p) of 1000 $J \cdot kg^{-1} \cdot K^{-1}$. The borehole thermal resistance has been estimated to be 0.1000 $m \cdot K \cdot W^{-1}$. Determine the following:

(a) the time (in hours) after which the transient thermal effects in the borehole grout and the heat carrier fluid are negligible,
(b) the time at which heat extraction reaches steady state,
(c) the time (in hours) after which the heat transfer around the BHE is no longer purely radial,
(d) the average temperature of the BHE circulating fluid after 50 h, and
(e) the average temperature of the BHE circulating fluid at steady-state conditions.

Solution

(a) The time after borehole transients can be ignored is given by

$$t_0 = \frac{5r_b^2}{\alpha} = \frac{5(0.125\,\mathrm{m}/2)^2}{\dfrac{2.0\,\mathrm{W \cdot m^{-1} \cdot K^{-1}}}{2500\,\mathrm{kg \cdot m^{-3}} \times 1000\,\mathrm{J \cdot kg^{-1} \cdot K^{-1}}}} = \frac{5(0.0625\,\mathrm{m})^2}{8.0\text{E-}07\,\mathrm{m^2 \cdot s^{-1}}} = 2.44\text{E}+04\,\mathrm{s}$$

note $\alpha = k/(\rho c_p)$.

$$\therefore t_0 = 6.78\,\mathrm{h}$$

(b) The time at which the heat extraction reaches steady state is given by

$$t_s = \frac{H^2}{9\alpha} = \frac{(100\,\mathrm{m})^2}{9 \times 8.0\text{E-}07\,\mathrm{m^2 \cdot s^{-1}}} = 1.39\text{E}+09\,\mathrm{s}$$

$$\therefore t_s = 3.86\text{E}+05\,\mathrm{h} = 44\,\text{years}$$

(c) The time after which the heat transfer around the BHE is no longer purely radial can be estimated by $t_s/10$ [see Equation (5.11)].

$$\therefore t_s = 3.86\text{E}+04\,\text{h} = 4.4\,\text{years}$$

(d) The average temperature of the BHE circulating fluid after 50 h of continuous operation is given by

$$T_f = \dot{q}'R'_g + \dot{q}'R'_b + T_g$$

As $t_0 < 50$ h $< t_s$, then

$$R'_g = \frac{1}{2\pi k}g = \left(\frac{1}{2\pi k}\right)\left(\ln\left(\frac{H}{2r_b}\right) + \frac{1}{2}\ \ln\left(\frac{t}{t_s}\right)\right)$$

$$R''_g = \left(\frac{1}{2\pi \times 2.0\,\text{W}\cdot\text{m}^{-1}\cdot\text{K}^{-1}}\right)\left(\ln\left(\frac{100\,\text{m}}{0.125\,\text{m}}\right) + \frac{1}{2}\ \ln\left(\frac{50\,\text{h}}{3.86\text{E+05 h}}\right)\right) = 0.1758\,\text{m}\cdot\text{K}\cdot\text{W}^{-1}$$

$$\therefore T_f = \frac{-5000\,\text{W}}{100\,\text{m}} \times (0.1758 + 0.1000)\,\text{m}\cdot\text{K}\cdot\text{W}^{-1} + 15\,^\circ\text{C} = 1.21\,^\circ\text{C}$$

$$\therefore T_f = 1.2\,^\circ\text{C}$$

(e) To calculate the average temperature of the BHE circulating fluid at steady state, the approach is the same as in part (c), except that the g-function differs, so R_g' is now

$$R'_g = \frac{1}{2\pi k}g = \left(\frac{1}{2\pi k}\right)\left(\ln\left(\frac{H}{2r_b}\right)\right) = 0.5319\,\text{m}\cdot\text{K}\cdot\text{W}^{-1}$$

$$\therefore T_f = -16.6\,^\circ\text{C}$$

The results of Example 5.1 may be used to illustrate some important aspects related to the design of BHEs for heating and cooling applications in buildings.

First, with the BHE depth and diameter (which are typical, but on the greater end of the depth range, for most US applications), the timescale at which transients within the borehole are negligible occurs after a relatively short period of time (i.e., of the order of 6 h for this example).

Second, the heat extraction process does not reach steady state until after a relatively long period of time (i.e., 44 years for this example), which implies very large thermal time constants for the ground. In other words, the thermally massive nature of the ground results in several important design considerations related to thermal energy storage in the Earth; the time duration at which thermal loads occur strongly impacts the BHE thermal performance over time. Large time constants of the Earth also give rise to BHE designs that rely on seasonal thermal storage (e.g., Drake Landing Solar Community near Okotoks, Alberta, Canada).

Third, the steady-state time has a strong dependence on the borehole depth. If the example BHE were half as deep, then the steady-state time would decrease by a factor of 4 to 11 years – still a relatively long time.

Fourth, the time duration over which thermal loads occur is very important in BHE design. In Example 5.1, let's assume for the moment that −5000 W is the peak design load on the BHE. If that thermal load occurred constantly over a 50 h time duration, the resulting temperature change in the borehole fluid would be −13.8 °C. Similarly, the resulting temperature change in the BHE fluid would be −31.6 °C at steady state. If a designer were trying to size the borehole in Example 5.1 to return an average fluid temperature of 0 °C at steady state, the borehole would need to be more than twice as deep! We therefore never design BHEs to handle steady-state thermal loads. As we have seen in Chapter 2, thermal loads of a building fluctuate on an hourly basis as the building thermal loads respond to fluctuations in weather and occupancy of the building. Further, in buildings with both heating and cooling hours, heat will be both extracted from and rejected to the ground over the annual cycle. Application of the line source equation for BHE design with fluctuating loads is discussed next.

5.5.4 *Fluctuating Thermal Loads*

Once the *g*-function is known, the response to any arbitrary thermal pulse can be determined by superimposing each step pulse as shown in Figure 5.5 for a four-time-step period. The time intervals shown in Figure 5.5 are arbitrary and could represent hours or months or some other time interval. As shown in Figure 5.5, the basic thermal pulse $\dot{q}'_1$ is applied throughout the entire duration of the four time periods, and is denoted as Q'_1. The subsequent pulses are superimposed as $Q'_2 = \dot{q}'_2 - \dot{q}'_1$, effective for three time intervals, as $Q'_3 = \dot{q}'_3 - \dot{q}'_2$, effective for two time intervals, and as $Q'_4 = \dot{q}'_4 - \dot{q}'_3$, effective for one time interval.

The temperature at radius *r* can then be determined by summing the responses of the step functions. In general, at the end of the *n*th time period, the ground temperature is given by

$$\Delta T_r = \frac{1}{2\pi k}\sum_{i=1}^{n}\left(\left(\dot{q}'_i - \dot{q}'_{i-1}\right)\cdot g_i\right) \tag{5.20}$$

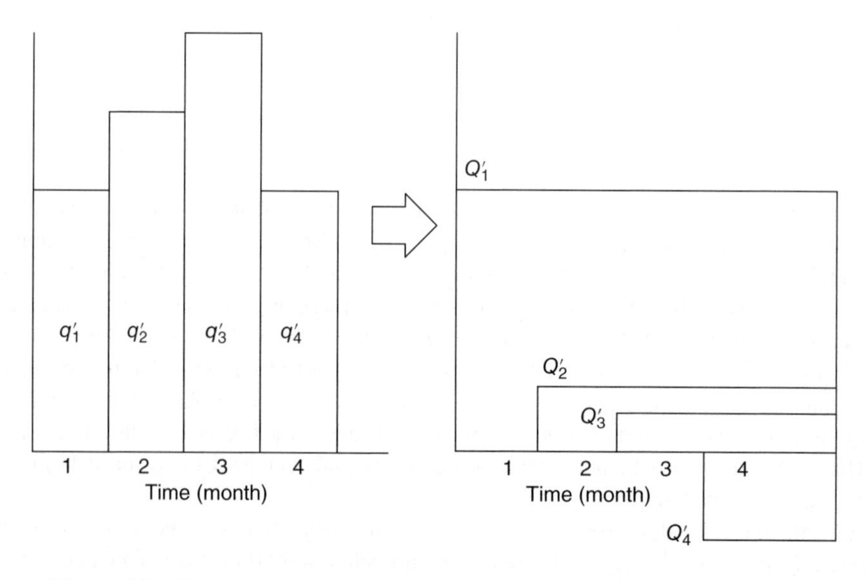

Figure 5.5 Illustration of superpositioning of four evenly spaced thermal pulses

where k is the ground thermal conductivity, T_r is the average ground temperature at radius r, $.q'$ is the step heat injection/extraction pulse per unit length of bore, and i is the index to denote the end of a time step. Note the similarity of Equation (5.20) to Equation (5.14). To determine the actual temperature at the radius r, the ground temperature T_g is added to ΔT_r.

Example 5.2 Examining the Thermal Response of a BHE with Fluctuating Loads
Consider the BHE described in Example 5.1. Recalculate the average temperature of the BHE circulating fluid after 50 h, but this time the heat extraction rate is –5000 W for the first 25 h only, and there is no load applied to the BHE for the second 25 h.

Solution
Here, there are two loads to be superimposed such that the first superimposed load acts over the entire 50 h time duration, and the second superimposed load acts over the time interval of 25–50 h, where $\dot{q}'_0 = 0\text{ W}$, $\dot{q}'_1 = -5000\text{ W}$, and $\dot{q}'_2 = 0\text{ W}$. With r now equal to r_b, the borehole wall temperature (T_b) after 50 h is calculated by

$$T_b = T_g + \sum_{i=1}^{n} \frac{\left(\dot{q}'_i - \dot{q}'_{i-1}\right)}{2\pi k} g = T_g + \frac{1}{2\pi k}\left(\left(\dot{q}'_1 - \dot{q}'_0\right)g_1 + \left(\dot{q}'_2 - \dot{q}'_1\right)g_2\right)$$

$$= T_g + \frac{1}{2\pi k}((-5000\text{ W} - 0\text{ W})g_1 + (0\text{ W} - -5000\text{ W})g_2)$$

$$= T_g + \frac{1}{2\pi k}((-5000\text{ W})g_1 + (5000\text{ W})g_2)$$

$$g_1 = \left(\ln\left(\frac{H}{2r_b}\right) + \frac{1}{2}\ln\left(\frac{t}{t_s}\right)\right) = \left(\ln\left(\frac{100\text{ m}}{0.125\text{ m}}\right) + \frac{1}{2}\ln\left(\frac{50\text{ h}}{3.86\text{E}+05\text{ h}}\right)\right) = 2.2091$$

$$g_2 = \left(\ln\left(\frac{H}{2r_b}\right) + \frac{1}{2}\ln\left(\frac{t}{t_s}\right)\right) = \left(\ln\left(\frac{100\text{ m}}{0.125\text{ m}}\right) + \frac{1}{2}\ln\left(\frac{25\text{ h}}{3.86\text{E}+05\text{ h}}\right)\right) = 1.8625$$

Substituting the values of g_1, g_2, and k into the above equation, we get T_b = 13.62 °C.

To determine the average *fluid* temperature, we must consider the thermal response due to the borehole elements. We have been considering the thermal resistance of the borehole as a steady-state approximation (relative to the large time constants of the ground), and thus the thermal response after 50 h would be given by $q'_2 \times R'_b = (0\text{ W} \div 100\text{ m}) \times 0.100\text{ m}\cdot\text{K}\cdot\text{W}^{-1} = 0$, and $T_f = T_r$ for this case:

$$\therefore T_f = 13.62\,°\text{C}$$

It is important and interesting to note that, even though the thermal loading on the BHE was symmetrical in time (i.e., –5000 W for 25 h and 0 W for 25), the temperature did not recover to the original condition of 15 °C owing to the logarithmic nature of transient heat conduction.

Example 5.3 Further Examination of the Thermal Response of a BHE with Fluctuating Loads
Consider the BHE described in Example 5.2. Recalculate the average temperature of the BHE circulating fluid after 50 h. The heat extraction rate is –5000 W for the first 25 h, as before, but now a heat rejection load of +5000 W is applied to the BHE for the second 25 h.

Solution

Here, there are again two loads to be superimposed such that the first superimposed load acts over the entire 50 h time duration, and the second superimposed load acts over the time interval of 25–50 h, where $\dot{q}'_0 = 0$ W, $\dot{q}'_1 = -5000$ W, and $\dot{q}'_2 = +5000$ W. With r now equal to r_b, the borehole wall temperature (T_b) after 50 h is calculated by

$$T_b = T_g + \sum_{i=1}^{n} \frac{(\dot{q}'_i - \dot{q}'_{i-1})}{2\pi k} g = T_g + \frac{1}{2\pi k}\left((\dot{q}'_1 - \dot{q}'_0)g_1 + (\dot{q}'_2 - \dot{q}'_1)g_2\right)$$

$$= T_g + \frac{1}{2\pi k}((-5000\,\mathrm{W} - 0\,\mathrm{W})g_1 + (5000\,\mathrm{W} - -5000\,\mathrm{W})g_2)$$

$$= T_g + \frac{1}{2\pi k}((-5000\,\mathrm{W})g_1 + (10000\,\mathrm{W})g_2)$$

$$g_1 = \left(\ln\left(\frac{H}{2r_b}\right) + \frac{1}{2}\ln\left(\frac{t}{t_s}\right)\right) = \left(\ln\left(\frac{100\,\mathrm{m}}{0.125\,\mathrm{m}}\right) + \frac{1}{2}\ln\left(\frac{50\,\mathrm{h}}{3.86\mathrm{E}+05\,\mathrm{h}}\right)\right) = 2.2091$$

$$g_2 = \left(\ln\left(\frac{H}{2r_b}\right) + \frac{1}{2}\ln\left(\frac{t}{t_s}\right)\right) = \left(\ln\left(\frac{100\,\mathrm{m}}{0.125\,\mathrm{m}}\right) + \frac{1}{2}\ln\left(\frac{25\,\mathrm{h}}{3.86\mathrm{E}+05\,\mathrm{h}}\right)\right) = 1.8625$$

Substituting the values of g_1, g_2, and k into the above equation, we get $T_b = 21.03$ °C.

To determine the average *fluid* temperature, we must consider the thermal response due to the borehole elements. We have been considering the thermal resistance of the borehole as a steady-state approximation (relative to the large time constants of the ground), and thus the thermal response after 50 h would be given by $q'_2 \times R'_b$ = (5000 W ÷ 100 m) × 0.100 m·K·W^{-1} = 5 °C, and T_f = 21.03 °C + 5.00 °C = 26.03 °C.

$$\therefore T_f = 26.03\,°\mathrm{C}.$$

5.5.5 *Effects of Groundwater Flow on BHEs*

The partial differential equation describing heat transport in groundwater flow was discussed in Chapter 4 (Section 4.3.3). Claesson and Hellström (2000) present a solution for the borehole wall temperature of a single vertical borehole in a uniform regional groundwater flow field that extends well below the borehole depth. The method is based on the g-function temperature response factor approach described above, and is described as the groundwater g-function.

The groundwater g-function (g_{gw}) represents the change in the average borehole wall temperature as a result of groundwater flow. It does not include the effects of thermal dispersion, and depends on two parameters defined by Claesson and Hellström (2000) as dimensionless time (τ) and dimensionless groundwater flow rate (h):

$$\tau = \frac{4\alpha t}{H^2} \tag{5.21}$$

and

$$h = \frac{H\rho_l c_l \dot{q}_w}{2k_{eff}} \tag{5.22}$$

where q_w is the Darcy velocity, and recall from Chapter 4 that k_{eff} is defined as $\phi k_l + (1-\phi)k_s$.

After considerable mathematical manipulation, the groundwater g-function (g_{gw}) can be approximated with the following simple expression:

$$g_{gw}(\tau, h) = \frac{h^2\tau}{8}\left(1 - \sqrt{\frac{4\tau}{9\alpha}}\right), \ \tau < 1, \ h\tau < \tag{5.23}$$

Finally, the average borehole wall temperature (T_b) is determined by

$$T_b = \frac{\dot{q}'}{2\pi k_{eff}} g_{total} \tag{5.24}$$

where $g_{total} = g(t/t_s, r_b/H) - g_{gw}(\tau, h)$.

Chiasson and O'Connell (2011) present a solution using a mass-heat analogy that accounts for thermal dispersion. The solution is for temperature at distance (x, y) resulting from a continuous injection of heat (located at the origin, x = 0, y = 0) into a two-dimensional flow field with uniform groundwater flow velocity (v_x) parallel to the x-axis. The solution will not be derived here as it is exactly analogous to Equation (4.49), with slight modification as follows:

$$\Delta T(x, y, t) = \frac{\dot{q}'}{4\pi(\rho c)_{eff}\left(\frac{D_L D_T}{R^2}\right)^{1/2}} e^{\frac{v_x x}{2D_L}}\left[W(0, B) - W(t_D, B)\right] \tag{5.25}$$

where $(\rho c)_{eff}$ is defined as $\phi(\rho c)_l + (1-\phi)(\rho c)_s$, t_D is given by Equation (4.50), and B is given by Equation (4.51). Equation (5.25) produces the same results as the groundwater g-function when thermal dispersion is negligible.

The groundwater g-function has been implemented in the GHX Tools software on the companion website for use with Eskilson's line source solution for a single borehole. The groundwater g-function will also be applied in Chapter 6 to size borehole arrays under conditions with groundwater flow.

5.5.6 *Mathematical Models of the Borehole Thermal Resistance*

As described above, the current BHE design approaches approximate the borehole thermal resistance as a steady-state value, thus ignoring the thermal storage effects of the heat carrier fluid, pipe, and grout. This has been shown to be a reasonable approximation where r_b/H ratios are small, of the order of 0.0005 as determined by Eskilson (1987). We will spend considerable time dealing with the borehole thermal resistance because it is perhaps one of the main features over which BHE designers have control (i.e., we cannot engineer the ground). The main goal in BHE design is to minimize the borehole thermal resistance within practical limits.

Simply put, the borehole thermal resistance relates the heat rate applied to the average fluid temperature circulating in a BHE to the average borehole wall temperature:

$$R'_b = \frac{T_f - T_b}{\dot{q}'} \tag{5.26}$$

where R'_b is the borehole thermal resistance per unit length, T_f is the *average* fluid temperature, T_b is the *average* temperature at the borehole wall, and $\dot{q}'$ is the thermal pulse per unit length.

The concept of the borehole thermal resistance is shown schematically in Figure 5.6 for a case of heat rejection to the ground. Recall that the purpose of a BHE is to transfer heat to (or from) the Earth, and the borehole construction materials pose a resistance to heat transfer; heat first must be transferred by fluid convection, and then by conduction through the pipe, and then through the borehole filling material. Figure 5.6 depicts a borehole 'filling material', which may be either grout or natural groundwater. In grouted boreholes, the thermal resistance of the grout is essentially a geometry problem. In groundwater-filled boreholes where no grout is present, the thermal resistance becomes somewhat more complicated by natural convection of the groundwater in the borehole. Note that the concept of the borehole resistance is exactly analogous to the hydraulic efficiency of a well, as discussed in Chapter 4.

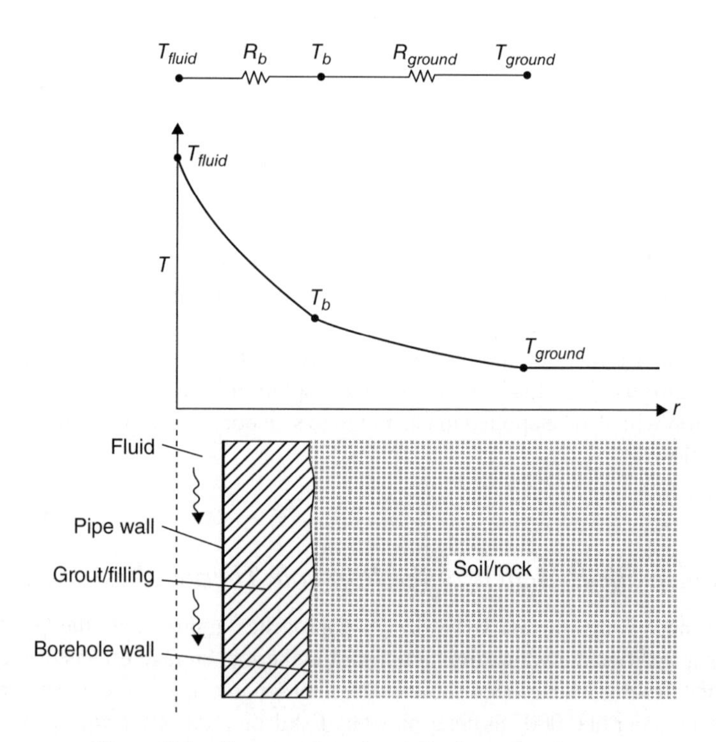

Figure 5.6 Concept of the borehole thermal resistance

The borehole thermal resistance is also dependent on the type and arrangement of flow channels in the borehole. Recall as described previously, there are many configurations of BHEs, but each configuration is generally some type of U-tube or concentric tube (coaxial tube) configuration as shown in Figure 5.7.

Although appearing simple to calculate, the borehole thermal resistance is quite complex, as shown in the accompanying delta thermal circuit (Figure 5.8) for a single U-tube in a grouted borehole after Hellström (1991). The heat fluxes associated with each leg of the U-tube are not equal (except at the borehole bottom) and vary with depth. Further, each leg thermally interacts with the surrounding ground, as well as with each other. Calculation of the borehole thermal resistance is further complicated with multiple U-tubes (i.e., two U-tubes or three U-tubes).

Calculation of the steady-state borehole thermal resistance for BHEs, regardless of the method of calculation, essentially consists of two parts: (i) the pipe resistance and (ii) the grout or filling resistance. These are each described in what follows in the context of single U-tube, double U-tube, concentric (or coaxial) tube, and groundwater-filled boreholes.

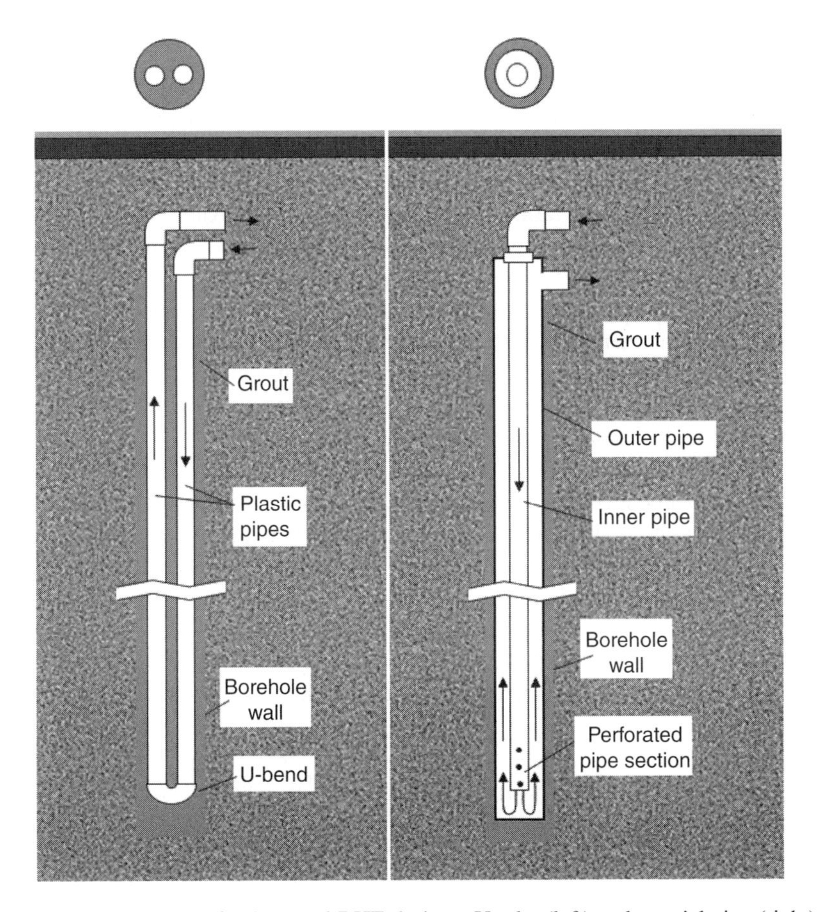

Figure 5.7 The two fundamental BHE designs: U-tube (left) and coaxial pipe (right)

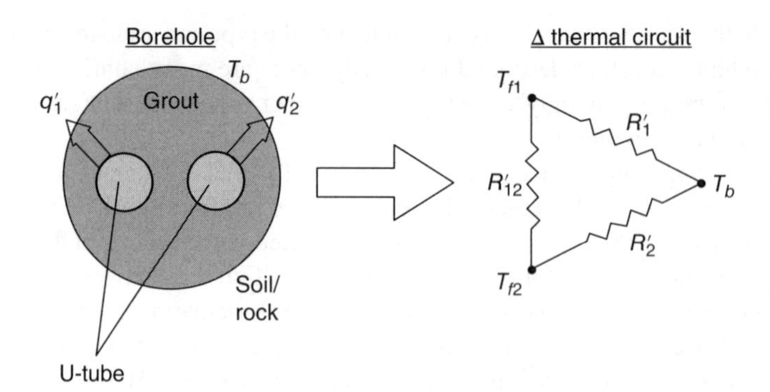

Figure 5.8 Borehole thermal resistance represented by a delta thermal circuit

5.5.6.1 The Pipe Thermal Resistance

Prior to presenting pipe thermal resistance calculations, some discussion of BHE pipes and associated terminology is warranted here. As previously mentioned, the most common material of construction for BHEs is high-density polyethylene (HDPE) or polyethylene high-density (PEHD) thermoplastic made from the refining process of petroleum. Known for its large strength to density ratio, HDPE is commonly used in the production of plastic bottles, *corrosion-resistant piping*, geomembranes, and plastic lumber.

Many HDPE pipe manufacturers use the standard dimension ratio (SDR) method of rating pressure piping. The SDR is the ratio of pipe diameter to wall thickness. Thus, the SDR is also used to describe the wall thickness of the pipe. SDR 11 is a common pipe wall thickness for BHE, and is similar to Schedule 40. SDR 11 means that the outside diameter of the pipe is 11 times the thickness of the wall. With a high SDR ratio, the pipe wall is relatively thin compared with the pipe diameter, and conversely, pipe with a low SDR ratio has a pipe wall that is relatively thick compared with the pipe diameter. As a consequence, a high-SDR pipe has a lower pressure rating, and low-SDR pipe has a higher pressure rating. Thus, SDR 9 is sometimes specified in higher pressure rating environments.

Thermal conductivity values of various pipe materials currently and historically used in BHEs are listed in Appendix B. HDPE survives as the industry standard owing to its ability to be thermally fused in the field, resulting in a strong, reliable joint. Crosslinked polyethylene (PEX) is also gaining market strength, but joints are mechanical fittings. Note the very low thermal conductivity of PVC pipe.

Calculation of the thermal resistance of the BHE pipe essentially consists of application of classic 'textbook' heat transfer equations, and is comprised of two subparts: the convective resistance of the fluid and the conductive resistance of the pipe:

$$R_{total} = \frac{1}{2\pi r_{p,in} L h_{in}} + \frac{1}{2\pi k_p L} \ln\left(\frac{r_{p,out}}{r_{p,in}}\right) \qquad (5.27)$$

where L is the pipe length (m or ft), k is thermal conductivity ($W \cdot m^{-1} \cdot K^{-1}$ or Btu/h·ft·°F), r is radius (m or ft), h is the convection coefficient ($W \cdot m^{-2} \cdot K^{-1}$ or Btu/h·ft^2·°F), subscripts *in* and *out* refer to the inside and outside pipe radius, and subscript p refers to the pipe.

Recall that we have been using the borehole resistance per unit length of bore (H), and Equation (5.27) is expressed in terms of total pipe area relative to the inside pipe area. Thus, resistance per unit length of pipe becomes

$$R'_p = \frac{1}{\pi D_{p,in} h_{in}} + \frac{\ln\left(D_{p,out} / D_{p,in}\right)}{2\pi k_p} \tag{5.28}$$

Recall relationships for the convection coefficient (h) for internal flow in pipes:

$$h = \frac{Nu k_f}{D_{p,in}} \tag{5.29}$$

where Nu is the Nusselt number, k_f is the thermal conductivity of the fluid, and $D_{p,in}$ is the inside diameter of the pipe. Under laminar flow conditions, Nu = constant = 4.36 (for a constant heat flux condition, which is the typical case considered here), but under turbulent flow conditions (i.e., Re > ~2300), $Nu = f(Re, Pr)$.

Before proceeding, let us recall some definitions of dimensionless numbers commonly used in heat transfer:

- ***Nu*** (Nusselt number) is the dimensionless temperature gradient at a surface: $Nu = hL/k$ (where L is some characteristic length and k is the fluid thermal conductivity).
- ***Re*** (Reynolds number) is the ratio of inertial to viscous forces: $Re = \rho VL/\mu$ or VL/ν (where ρ is the density in $kg \cdot m^{-3}$ or lb/ft^3), V is the fluid velocity ($m \cdot s^{-1}$ or ft/s), L is some characteristic length (m or ft), μ is the viscosity ($kg \cdot m^{-1} \cdot s^{-1}$ or lbm/ft·s), and ν is the kinematic viscosity ($m^2 \cdot s^{-1}$ or ft^2/s)).
  For circular pipes, the characteristic length is taken as the inside pipe diameter, and calculation of Re simplifies to $Re = 4\dot{m}/(\pi D_{p,in}\mu)$, where $\dot{m}$ is the mass flow rate of the fluid ($kg \cdot s^{-1}$ or lb/s).
- **Pr** (Prandtl number) is the ratio of momentum to thermal diffusivities: $Pr = c_p\mu/k$ or ν/α, where c_p is the heat capacity ($J \cdot kg^{-1} \cdot K^{-1}$ or Btu/lb·°F) and α is the thermal diffusivity ($m^2 \cdot s^{-1}$ or ft^2/h).

There are numerous correlations for Nu as a function of Re and Pr in the turbulent range. We shall use that of Gnielinski (1976), which is valid over a wide range of $0.5 < Pr < 2000$ and $3000 < Re < 5 \times 10^6$:

$$Nu = \frac{(f/8)(Re-1000)Pr}{1+12.7(f/8)^{1/2}(Pr^{2/3}-1)} \tag{5.30}$$

where f is the Moody (or Darcy) friction factor for smooth pipes after Petukhov for a large range of Reynolds numbers ($3000 < Re < 5 \times 10^6$):

$$f = (0.790 \ln Re - 1.64)^{-2} \tag{5.31}$$

The heat transfer fluid used in BHEs ranges from pure water to an aqueous antifreeze mixture with various proportions of, most commonly, propylene glycol. The antifreeze proportion of the mixture should be chosen such that the heat transfer fluid remains liquid at the minimum expected fluid temperature plus some margin. Common designs specify an aqueous mixture of 15% propylene glycol by volume, which has a freeze point of approximately −5.5 °C (22 °F). More conservative designs specify an aqueous mixture of 20% propylene glycol by volume, which has a freeze point of approximately −7.5 °C (18.5 °F).

5.5.6.2 The Thermal Resistance of Grouted Single U-Tube BHEs

Prior to presenting BHE thermal resistance calculations, some discussion of grouting materials and associated terminology is warranted here.

An ideal BHE bore grout would protect groundwater from contamination, promote heat transfer, be easy to install, and have a reasonable cost. Conventional grouts that are used to seal the annular region around U-tubes may protect groundwater at the expense of effective heat transfer. Grouting materials that are most commonly used in the GHP (as well as the water-well industry) are either bentonite based or cement based. The choice represents a trade-off in installation cost, which is generally reflected by its 'pumpability' into the borehole. Appendix B lists thermal conductivity values of various BHE grouting materials. The trade-off is that the lower-thermal-conductivity grouts (bentonite) are easier, and therefore cheaper, to install. The higher-thermal-conductivity grouts (i.e., those with sand mixtures) are grittier, and therefore more difficult to pump and more abrasive on the pumping equipment, and often must be mixed with bentonite in the field, giving rise to their costlier installation. As we will see, the higher-thermal-conductivity materials result in lower borehole thermal resistance, which translates to less drilling required. Designers must weigh these trade-offs of more drilling at a lower cost rate per bore against less drilling at a higher cost rate per bore. Generally speaking, a practical upper limit of achievable grout thermal conductivity is of the order of 1.7 $W{\cdot}m^{-1}{\cdot}K^{-1}$ or 1.0 Btu/hr·ft·°F.

A considerable amount of research on methods for calculating the steady-state borehole thermal resistance in grouted boreholes has been conducted since the late 1990s, and the so-called *multipole method* developed by Bennet *et al.* (1987) has proven to be the most accurate. In the multipole concept, cylinders of varying diameter (representing U-tubes) are arbitrarily placed in a homogeneous cylindrical medium, which is surrounded by another medium. Thus, any number of U-tubes of any diameter can be modeled. They are called multipoles because for each source there is an imaginary mirror sink outside the borehole. The multipole method then solves the steady-state, two-dimensional heat conduction equation. A computer program for solution of the multipole method is given by Bennet *et al.* (1987), and Liu and Hellström (2006) present a simplified set of expressions related to single U-tubes only.

The simplified approximations by Liu and Hellström (2006) for a single U-tube take the arithmetic average of the two cases of a uniform borehole wall temperature ($R'_{1,effective}$) and a uniform heat flux on the borehole wall ($R'_{2,effective}$):

$$R'_{b,single\ U\text{-}tube} = \left(R'_{1,effective} + R'_{2,effective}\right)/2 \tag{5.32}$$

where

$$R'_{1,effective} = R'_2 + \frac{1}{3R'_{1,2}}\left(\frac{H}{\dot{m}c_p}\right)^2 + \frac{1}{12R'_2}\left(\frac{H}{\dot{m}c_p}\right)^2 \text{ and } R'_{2,effective} = R'_2 + \frac{1}{3R'_1}\left(\frac{H}{\dot{m}c_p}\right)^2 \quad (5.33a,b)$$

where H is the borehole depth and $\dot{m}c_p$ is the heat capacity rate of heat transfer fluid.

The intermediate resistances appearing in Equation (5.33) are given by

$$R'_1 = \frac{1}{\pi k_{grout}}\left[\beta + \ln\left(\frac{2S}{r_p}\right) + \sigma ln\left(\frac{r_b^2+S^2}{r_b^2-S^2}\right)\right] - \frac{1}{\pi k_b}\frac{\frac{r_p^2}{4S^2}\left[1+\sigma\frac{4r_b^2S^2}{(r_b^4-S^4)}\right]^2}{\left\{\frac{1+\beta}{1-\beta}+\frac{r_p^2}{4S^2}+\sigma\frac{2r_p^2r_b^2(r_b^4+S^4)}{(r_b^4-S^4)^2}\right\}} \quad (5.34)$$

$$R'_2 = \frac{1}{4\pi k_{grout}}\left[\beta + \ln\left(\frac{r_b}{r_p}\right) + \ln\left(\frac{r_b}{2S}\right)\right] + \sigma\ln\left(\frac{r_b^4}{r_b^4-S^4}\right) - \frac{1}{4\pi k_b}\frac{\frac{r_p^2}{4S^2}\left[1-\sigma\frac{4S^4}{(r_b^4-S^4)}\right]^2}{\left\{\frac{1+\beta}{1-\beta}+\frac{r_p^2}{4S^2}\left[1+\sigma\frac{16r_b^4S^4}{(r_b^4-S^4)^2}\right]\right\}} \quad (5.35)$$

and

$$R'_{1,2} = \frac{R'_1R'_2}{R'_2-0.25R'_1} \quad (5.36)$$

where

$$\beta = 2\pi k_{grout}R'_p \quad (5.37)$$

and

$$\sigma = \frac{k_{grout}-k_{ground}}{k_{grout}+k_{ground}} \quad (5.38)$$

where r is radius, R'_p is the pipe thermal resistance described previously by Equation (5.28a), subscripts b and p refer to the borehole and pipe respectively, S is the distance between the center of the borehole and one leg of the U-tube, and k is the thermal conductivity.

5.5.6.3 The Thermal Resistance of Grouted Double U-Tube BHEs

Solution for the borehole thermal resistance by the multipole method with more than one U-tube requires a computer program. However, simplified solutions given by Hellström (1991) provide practical approximations for the borehole resistance of double U-tube BHEs in a circular region. Accounting for counterflow in the up and down legs of the U-tubes, we have

$$R'_{b,\ double\ u-tube} = R'_{sf} + \frac{1}{3}\frac{1}{R'_a}\left(\frac{H}{C}\right)^2 \tag{5.39}$$

where R'_{sf} is an effective steady-flux thermal resistance, R'_a is an internal resistance, H is the borehole length, and C is the heat capacity rate of the borehole heat transfer fluid. The intermediate resistances are given by:

$$R'_{sf} = \frac{1}{2\pi k_{grout}}\left[\ln\left(\frac{r_b}{r_{p,out}}\right) - \frac{3}{4} + b^2 - \frac{1}{4}\ln\left(1-b^8\right) - \frac{1}{2}\ln\left(\frac{\sqrt{2}br_b}{r_{p,out}}\right) - \frac{1}{4}\ln\left(\frac{2br_b}{r_{p,out}}\right)\right] + \frac{R'_p}{4} \tag{5.40}$$

$$R'_a = \frac{1}{\pi k_{grout}}\left[\ln\left(\frac{\sqrt{2}br_b}{r_{p,out}}\right) - \frac{1}{2}\ln\left(\frac{2br_b}{r_{p,out}}\right) - \frac{1}{2}\ln\left(\frac{1-b^4}{1+b^4}\right)\right] + R'_p \tag{5.41}$$

where r_b is the borehole radius, $r_{p,out}$ is the pipe outer radius, k_{grout} is the thermal conductivity of the grout, b is an eccentricity parameter equivalent to the U-tube shank spacing (distance between pipe edges) divided by the borehole diameter, R'_p is the pipe thermal resistance described previously by Equation (5.28), and subscripts b and p refer to the borehole and pipe respectively.

5.5.6.4 The Thermal Resistance of Concentric Pipe (or Coaxial) BHEs

The borehole thermal resistance for the concentric pipe arrangement consists of two subparts: (1) the resistance between the inner and outer flow channels and (2) the resistance between the outer flow channel and the borehole wall. A schematic of a concentric pipe BHE is shown in Figure 5.9.

The resistance between the inner and outer flow channels consists of three parts: (i) convective thermal resistance between the fluid in the inner flow channel and the inside surface of the inner pipe, (ii) conductive resistance through the inner pipe, and (iii) convective thermal resistance between the outside surface of the inner pipe and the fluid in the outer flow channel. The inner flow channel is not in direct contact with the ground.

The resistance between the outer flow channel and the borehole wall also consists of three parts: (i) convective thermal resistance between the fluid in the outer flow channel and the inside surface of the outer pipe, (ii) conductive resistance through the outer pipe, and (iii) conductive thermal resistance through the grout, between the outside surface of the outer pipe and the borehole wall.

Accounting for counterflow in the inner and outer flow channels, we have

$$R'_{b,\ concentric\ pipe} = R'_{sf} + \frac{1}{3}\frac{1}{R'_a}\left(\frac{H}{C}\right)^2 \tag{5.42}$$

where the terms are as described above. The intermediate resistances are simplified from Hellström (1991) for a BHE in a circular region:

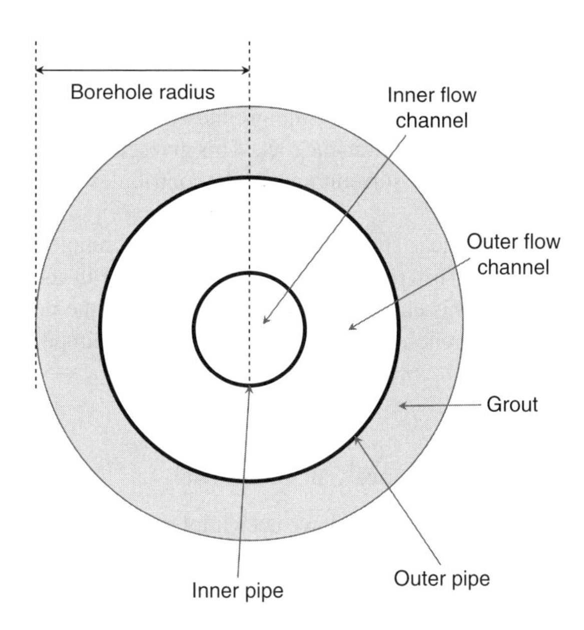

Figure 5.9 Schematic of a concentric pipe BHE

$$R'_{sf} = \frac{1}{2\pi k_{ground}} \left[\frac{k_{ground}}{k_{grout}} \left[(1+\xi\pi r_p)^2 \ln\left(\frac{r_b}{r_p}\right) - \xi\pi(1+\xi\pi r_p)^2 \left(r_b^2 - r_p^2\right) + \frac{1}{4}\left(\xi\pi r_b^2\right)^2 - \frac{1}{4}\left(\xi\pi r_p^2\right)^2 \right] \right] + R'_p \tag{5.43}$$

where $\xi = 1/\left(r_b^2 - r_p^2\right)$ and R'_p is the resistance per unit length between the outer flow channel and the borehole wall. R'_a is the sum of the resistance between the inner and outer flow channels and the resistance between the outer flow channel and the borehole wall.

5.5.6.5 The Thermal Resistance of Groundwater-Filled Boreholes with U-Tubes

As mentioned earlier in this chapter, it is common in Sweden and Norway to install U-tube BHEs in ungrouted boreholes, and let the borehole fill up with groundwater. The crystalline rock geology of these locations allows for a fracture network that transmits groundwater. Boreholes drilled in this type of geology may require minutes to days for the groundwater levels to equilibrate to static levels.

The thermal resistance of these so-called *groundwater-filled boreholes* is more difficult to quantify owing to natural convection of groundwater in the bore. The process of natural convection is the result of density differences caused by the temperature gradients in the groundwater during operation of the BHE. These gradients vary, depending on whether heat is being rejected to or extracted from the ground. Further, the temperature gradients, and therefore the rate of natural convection, will increase with increasing thermal load on the BHE. Another complicating factor in determining borehole thermal resistance with natural convection in groundwater-filled boreholes is that some of the groundwater will leave the borehole and be replenished by surrounding groundwater in the rock. This gives rise to the concept of a borehole 'skin' effect, where some of the surrounding rock contributes to the borehole thermal resistance.

In short, there are no analytical solutions describing the borehole thermal resistance of groundwater-filled boreholes. This parameter must be measured in the field using methods described in a later section of this chapter. Experience shows that the borehole thermal resistance of groundwater-filled boreholes can be one order of magnitude lower than a similar grouted BHE.

5.5.6.6 Borehole Thermal Resistance Calculator Tool

The foregoing thermal resistance equations have been implemented into the companion suite of GHX tools. The calculator also contains thermal property data for pure water as well as aqueous propylene glycol mixtures. Use of this tool is demonstrated in the following examples. Both SI and US customary units are allowed.

Example 5.4a Borehole Thermal Resistance Calculation – IP Units

Consider a BHE design consisting of a nominal 1 in diameter U-tube in a 5 in diameter borehole, 250 ft deep. The borehole is grouted with standard bentonite. The heat transfer fluid is pure water, and the design flow rate is 3 gpm. The minimum temperature expected is 40 °F.

Solution

The BHE design in this example is typical of basic US installations. Without information given about the spacing of the legs of the U-tube, it is typical to assume that they are evenly spaced in the borehole (i.e., each leg is equidistant from each other and from the borehole wall). However, in reality, the legs may be touching each other at some places in the borehole, and be against the borehole wall at other locations. The thermal properties of 'standard bentonite grout' may be found in Appendix B.

Results from the software tool are shown in Figure E.5.4a. Input data are entered in the yellow boxes, and calculations are done to the right of the input fields. Dropdown boxes allow options for: the U-tube spacing, the nominal pipe size, the pipe SDR, and the borehole heat transfer fluid. The ground thermal conductivity is assumed to be 1.0 Btu/hr·ft·°F.

Example 5.4b Borehole Thermal Resistance Calculation – SI Units

Consider a BHE design consisting of a nominal 25 mm diameter U-tube in a 125 mm diameter borehole, 75 m deep. The borehole is grouted with standard bentonite. The heat transfer fluid is pure water, and the design flow rate is 10 Lpm. The minimum temperature expected is 5 °C.

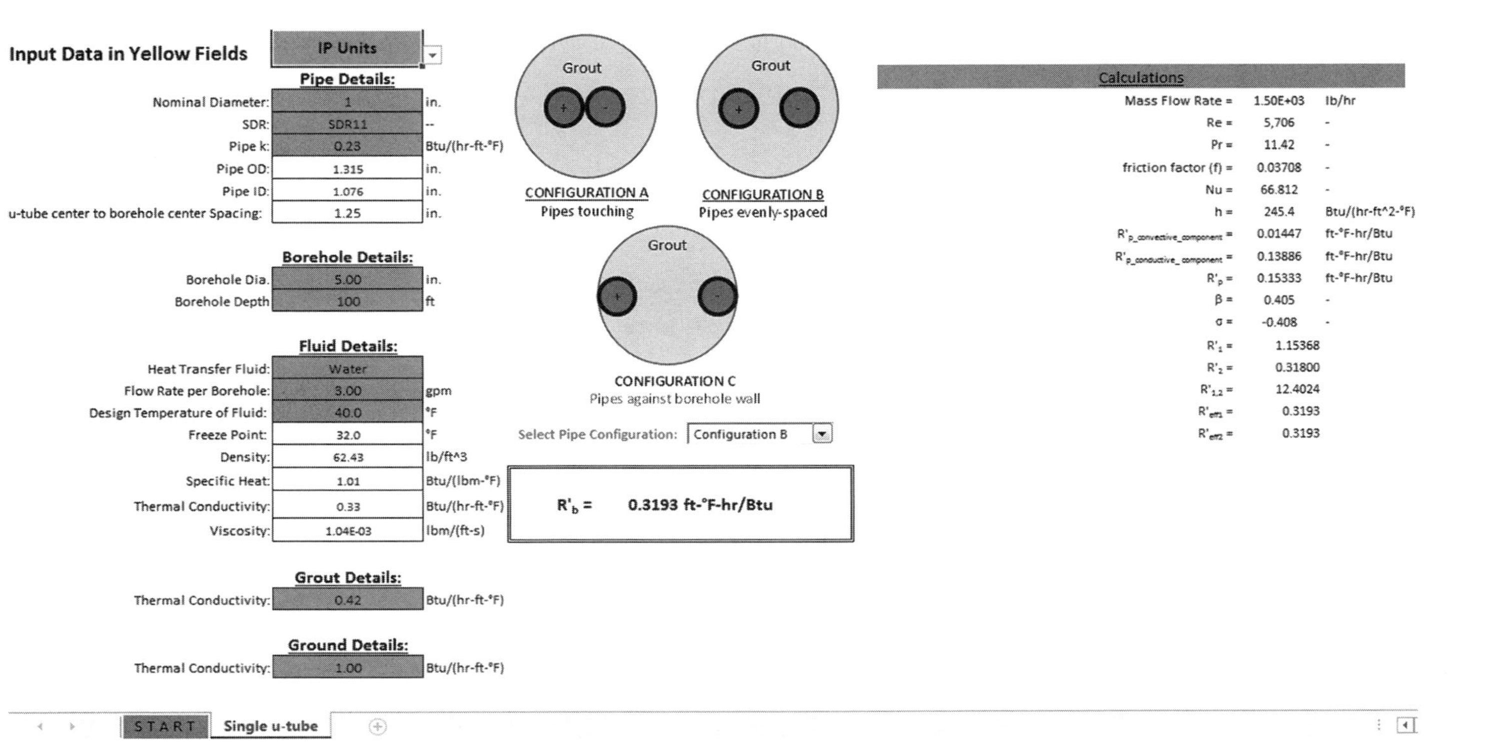

Figure E.5.4a Screen capture of the single U-tube borehole thermal resistance calculator tool for Example 5.4a (IP units)

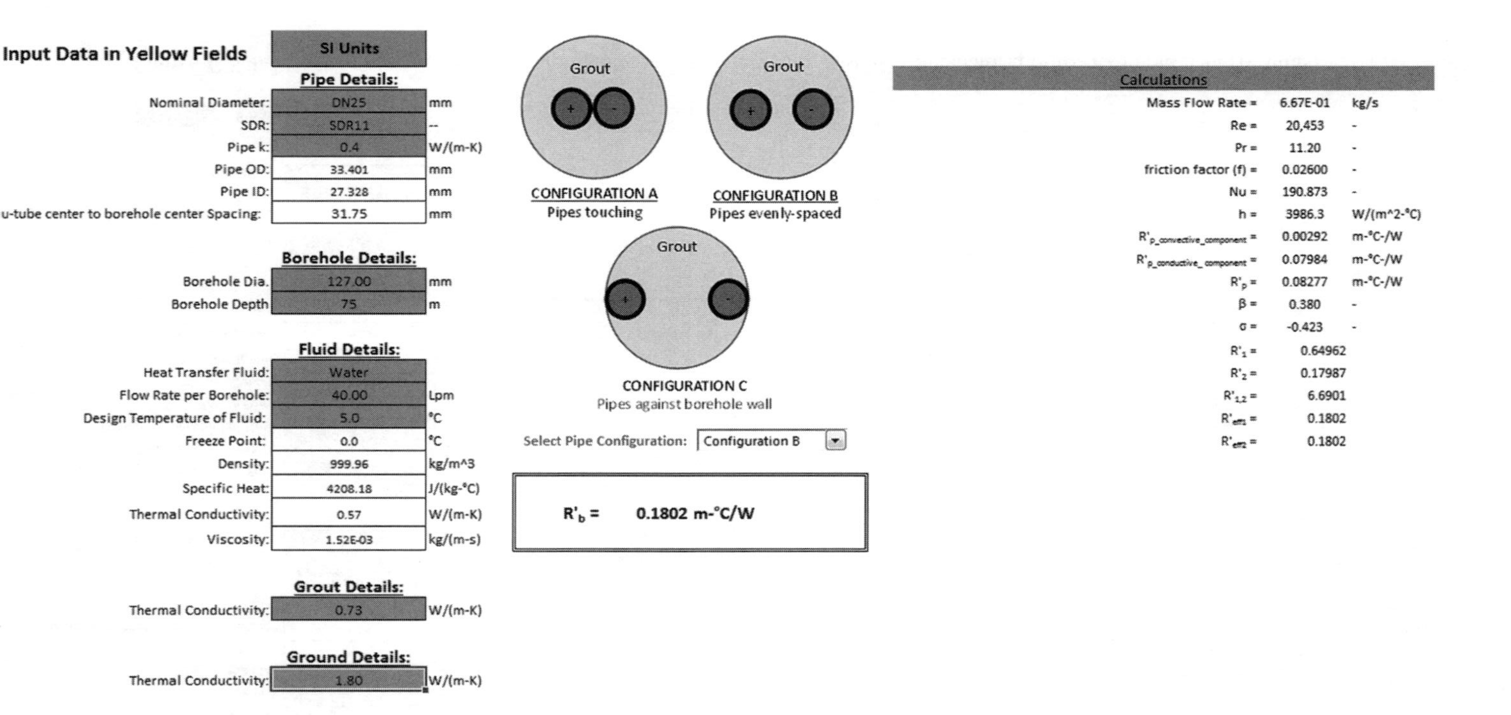

Figure E.5.4b Screen capture of the single U-tube borehole thermal resistance calculator tool for Example 5.4b (SI units)

Solution

The BHE design in this example is typical of basic US installations. Without information given about the spacing of the legs of the U-tube, it is typical to assume that they are evenly spaced in the borehole (i.e., each leg is equidistant from each other and from the borehole wall). However, in reality, the legs may be touching each other at some places in the borehole, and be against the borehole wall at other locations. The thermal properties of 'standard bentonite grout' may be found in Appendix B.

Results from the software tool are shown in Figure E.5.4b. Input data are entered in the yellow boxes, and calculations are done to the right of the input fields. Dropdown boxes allow options for: the U-tube spacing, the nominal pipe size, the pipe SDR, and the borehole heat transfer fluid. The ground thermal conductivity is assumed to be 1.8 $W \cdot m^{-1} \cdot K^{-1}$.

Example 5.5 Studying Effects of Borehole Details on the Thermal Resistance

We are concerned with reducing the thermal resistance of the borehole described in Example 5.4b. Let's try (a) enhanced thermal conductivity grout (k = 1.50 $W \cdot m^{-1} \cdot K^{-1}$), (b) adding spacers on the U-tube legs to push them against the borehole wall, and (c) both (a) and (b). Comment on the results.

Solution

(a) Here the only change is replacing the grout thermal conductivity in the spreadsheet to 1.50 $W \cdot m^{-1} \cdot K^{-1}$. The resulting borehole thermal resistance is 0.1120 $m \cdot K \cdot W^{-1}$, or about 38% lower.

(b) Here the only change involves selecting *Configuration C* from the dropdown box. The resulting borehole thermal resistance is 0.1242 $m \cdot K \cdot W^{-1}$, or about 33% lower.

(c) Combining the effects of both (a) and (b) into the *borehole thermal resistance calculator*, the resulting borehole thermal resistance is 0.0902 $m \cdot K \cdot W^{-1}$, or about 50% lower.

Discussion: The borehole thermal resistance is reduced significantly by adding either enhanced thermal conductivity grout, U-tube spacers, or both. The reduction in borehole thermal resistance using thermally enhanced grout is of the same order as adding U-tube spacers, but using both can reduce the borehole thermal resistance by over half. Many designers will focus on the grout thermal conductivity and miss opportunities to reduce the borehole thermal resistance by careful spacing of the u-tube legs. The choice of which design to select would depend on other factors and the required size of the borehole array. The most economic choice could then be selected. Borehole array design is the subject of the next chapter.

Example 5.6: Calculating the Thermal Resistance of a Double U-tube BHE

Let's reconsider the BHE design in Example 5.4b but now for a double U-tube BHE. A larger-diameter borehole is typically required to facilitate insertion of the BHE, so let's design for a 150 mm diameter borehole. All other input data can remain the same.

Solution and Discussion

Results from the calculation are shown in Figure E.5.6. The borehole thermal resistance is calculated to be 0.086 $m \cdot K \cdot W^{-1}$, or about half that relative to the single U-tube case. Note

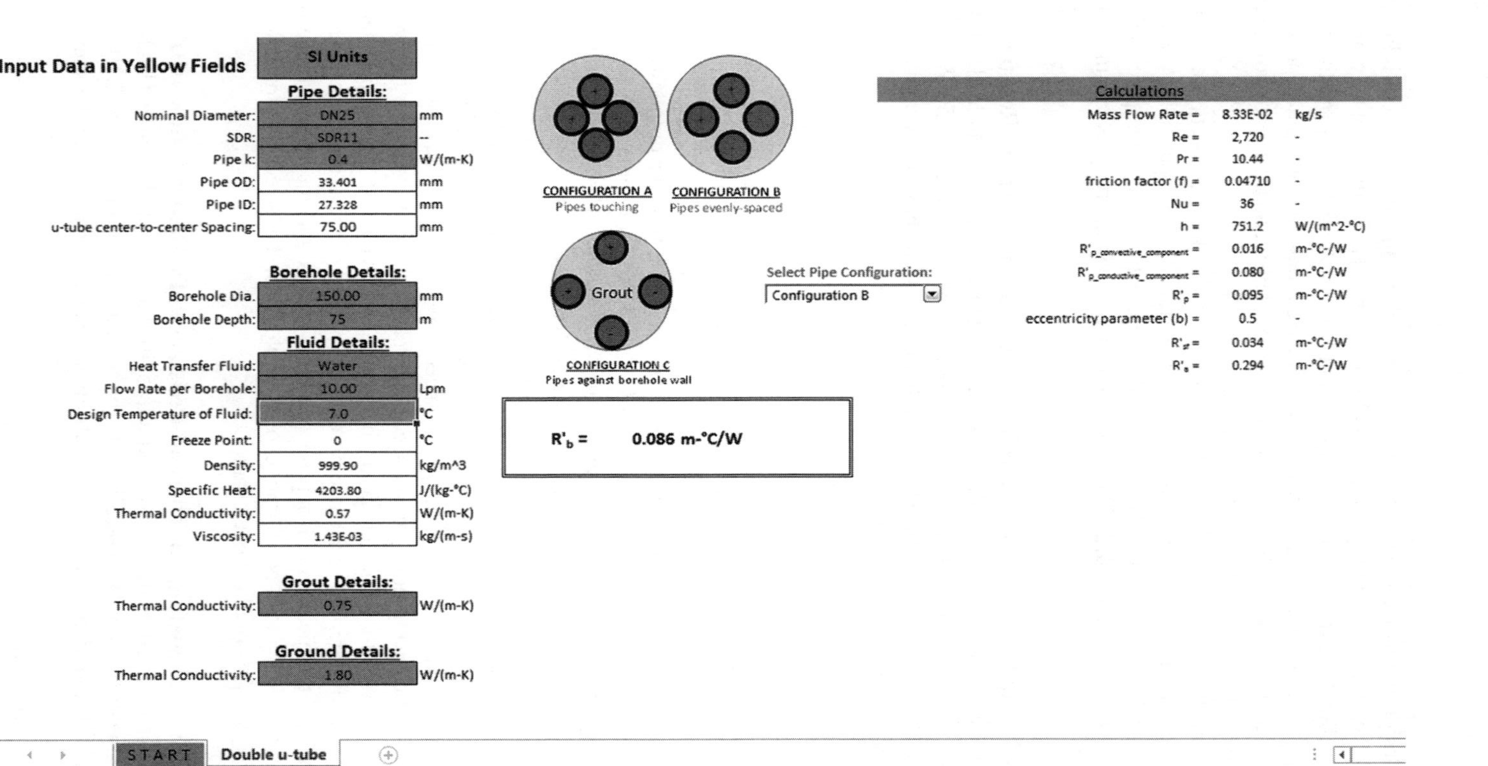

Figure E.5.6 Screen capture of the double U-tube borehole thermal resistance calculator tool for Example 5.6 (SI units)

that this thermal resistance of 0.086 m·K·W^{-1} is of the same order as the single U-tube BHE with U-tube spacers and thermally enhanced grout. It is puzzling why double U-tube BHEs are not more common in the United States. As we shall see in Chapter 6, a half-reduction in the borehole thermal resistance does not constitute a direct reduction in required borehole lengths of a GHX; the ground thermal resistances must be considered.

5.6 Thermal Response Testing

Thermal response testing (or *in situ* thermal conductivity testing) is a field method for determining the thermal properties of subsurface materials, mainly the thermal conductivity and subsurface temperature. Intensive research on this topic was conducted in the 1990s, and was aimed both at defining field methods for thermal response testing and analysis methods of the test data. An excellent compilation of worldwide research on the topic of thermal response testing can be found in Gehlin (2002).

5.6.1 Field Methods

The basic concept of thermal response testing is shown schematically in Figure 5.10. The test procedure involves application of a constant heat rate to the fluid flowing through a BHE.

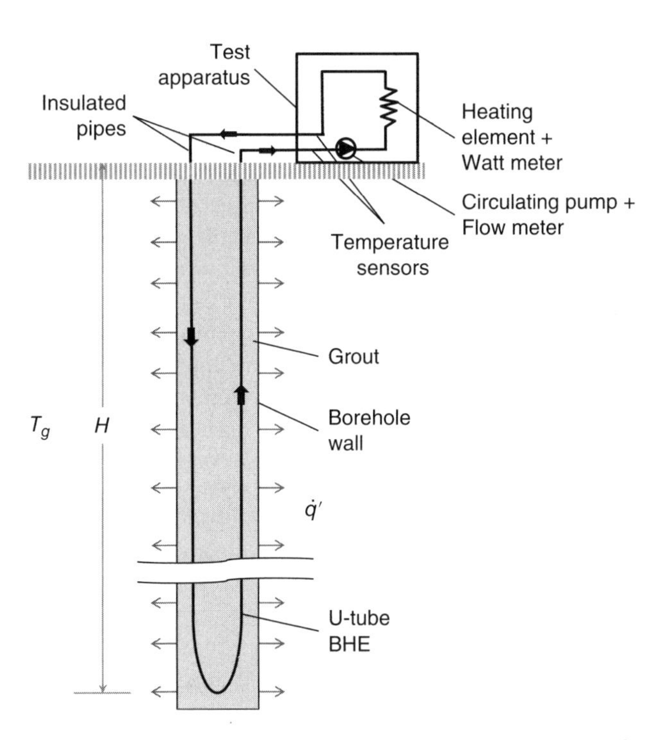

Figure 5.10 Schematic of a thermal response test apparatus and test set-up for heat rejection

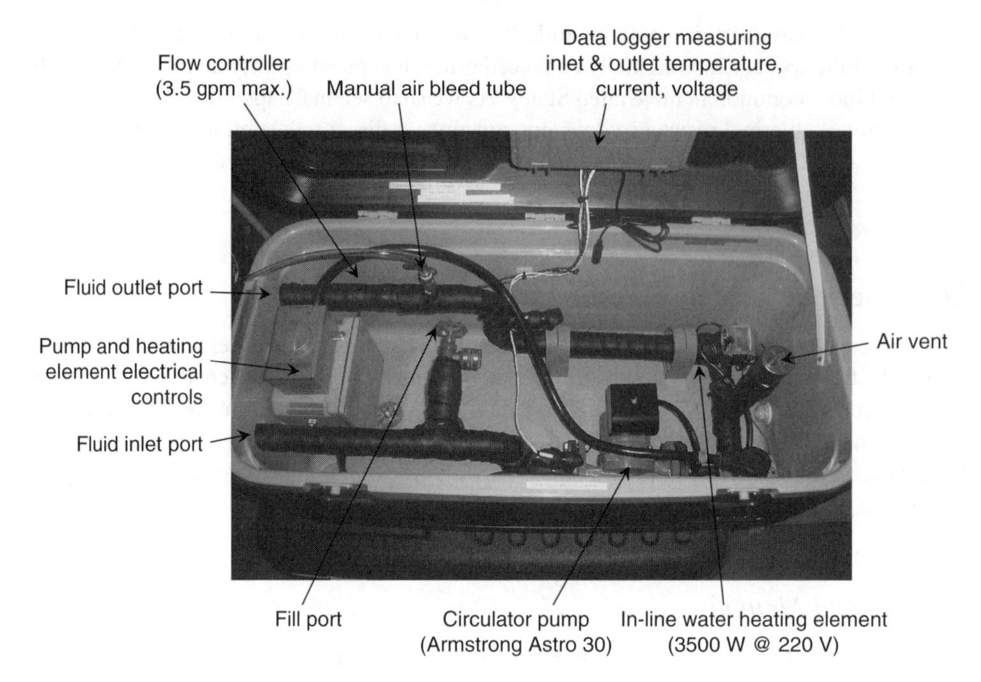

Figure 5.11 Photo of a portable thermal response test apparatus in a ~120 L cooler(photo taken by the author)

A data logger records the inlet and outlet fluid temperatures at a minimum, and some tests are also designed to record fluid flow rate and power added to the fluid stream. The test originated as a heat rejection test, but in cold climates some tests are designed as heat extraction tests. In either case, the heat rejection or extraction rate must be known in order to evaluate the test data. Research suggests that test duration should be of the order of 40 h.

Early prototype thermal response test apparatus in the 1990s was housed in trailers that were towed by vehicles. Those were convenient in cases were power generators were needed at the site. Today, most are built into portable, durable boxes that can be easily transported or shipped to a project site. Figure 5.11 is a photo of a portable thermal response testing apparatus built into a ~120 L cooler. In heat rejection tests, a 220 V power supply is needed.

Conducting a thermal response test also includes accurate measurement of the average subsurface Earth temperature prior to the start of the test. This is a crucial piece of information for proper analysis of the test because it represents the initial condition of the subsurface temperature field. Acceptable methods to do this include: (i) using a temperature probe to measure the temperature of the standing fluid in the BHE with depth, and/or (ii) circulating the BHE fluid with the test pump with no heat addition, and recording the stabilized fluid temperature.

5.6.2 Analysis Methods of Field Test Data

Two analysis methods are described here for determining the ground thermal conductivity from thermal response test data: (1) a graphical method and (2) an inverse modeling method. Each

Table 5.1 Summary of Borehole Construction and Thermal Response Test Details

Parameter	Units	Value
	Borehole Heat Exchanger Construction	
Borehole diameter	in (mm)	5.75 (146.0)
Borehole depth	ft (m)	300 (91)
U-tube diameter (nominal)	in (mm)	1.25 (32)
Grout thermal conductivity	Btu/h·ft·°F (W·m^{-1}·K^{-1})	1.0 (1.73)
Borehole thermal resistance calculated from multipole method	h·ft·°F/Btu (m·K·W^{-1})	0.1413 (0.0816)
	Thermal Response Test Details	
Undisturbed ground temperature	°F (°C)	56.9 (13.8)
Heat rejection rate	Btu/h/ft (W·m^{-1})	53.9 (51.8)
Test duration	h	47.8
Estimated ρc_p	Btu/ft^3·°F (J·m^{-3}·°C^{-1})	30 (2.01E+6)

method is based on application of the infinite line source solution as described in Section 5.5.2. Thermal response tests are not conducted for a long enough time duration for end effects to be important. The first analysis method, independent of the borehole thermal resistance, employs a graphical technique and can be accomplished manually. The second analysis method determines the borehole thermal resistance and requires a computer solution with optimization.

The analysis methods are described in what follows in the context of a thermal response test conducted at a site where the geology consists of 75 ft (23 m) of clay underlain by dense limestone. Details of the test set-up are summarized below in Table 5.1, and the inlet and outlet BHE temperatures are plotted with time in Figure 5.12.

5.6.2.1 Graphical Analysis

In the graphical method, the average BHE fluid temperature is plotted versus the natural logarithm of time, as shown in Figure 5.13 (IP units) and Figure 5.14 (SI units). Note that the average BHE fluid temperature is simply the arithmetic average of the inlet and outlet fluid temperature at each time measurement.

As can be seen in Figures 5.13 and 5.14, when the test data points are plotted versus the natural logarithm of time, a straight line results of the form

$$T_{fluid}(t) = m \cdot lnt + b \tag{5.44}$$

where m represents the slope of the line and is equivalent to

$$m = \frac{\dot{q}'}{4\pi k} \tag{5.45}$$

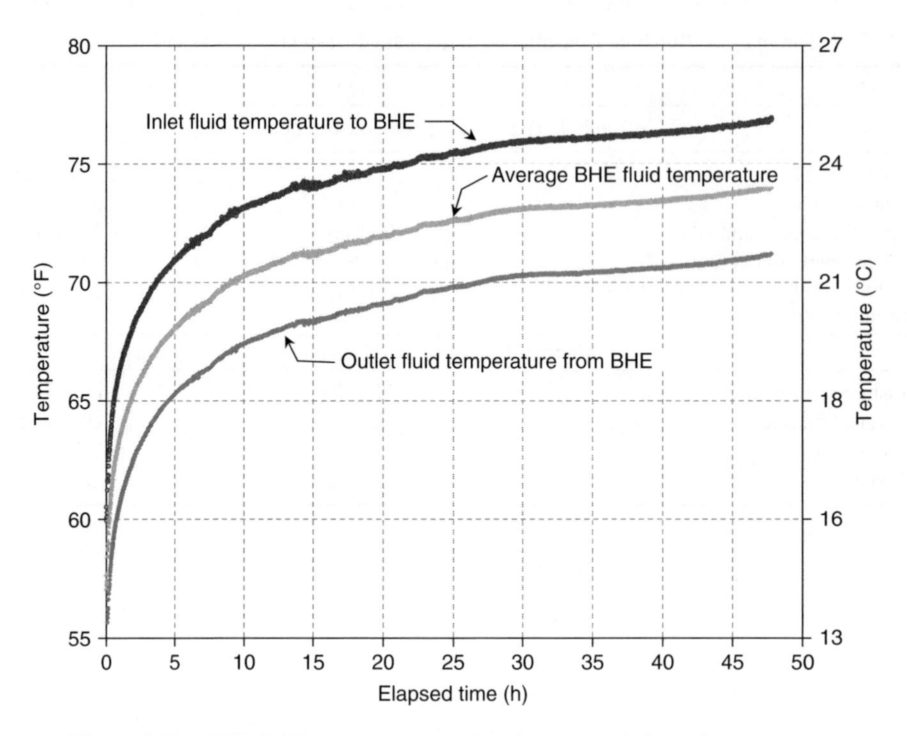

Figure 5.12 BHE fluid temperatures vs. time for an actual thermal response test

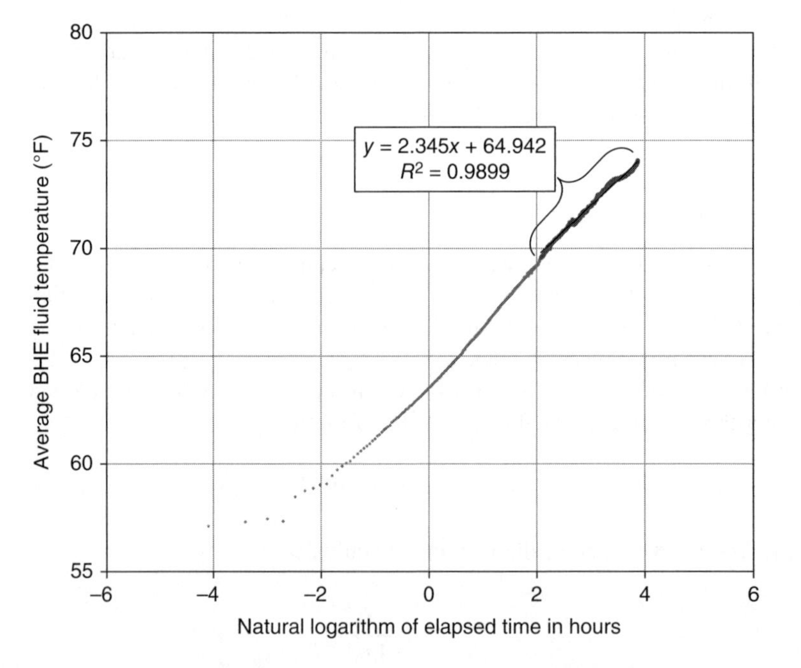

Figure 5.13 Average BHE fluid temperatures vs. natural logarithm of time in IP units

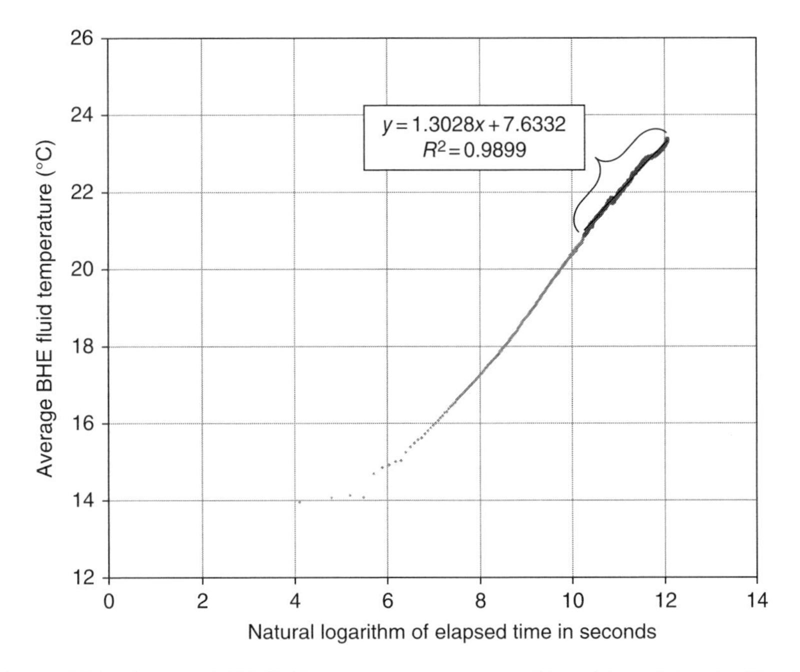

Figure 5.14 Average BHE fluid temperatures vs. natural logarithm of time in SI units

Note that the early-time data are non-linear, and are related to the transient effects of the borehole materials (i.e., fluid, pipe, and grout). As the object of this analysis is to determine the thermal conductivity of the ground, early-time data in the analysis are ignored, and only the late-time data are of interest, after the borehole thermal resistance has reached steady state. The line of best fit is determined from the data points shown by the brace bracket in Figures 5.13 and 5.14, and include data after ~8 h only.

Result in IP Units: Rearranging Equation (5.45) to solve for k, and substituting the value of m (from Figure 5.13) and the value of $\dot{q}'$ (from Table 5.1), we obtain

$$k = \frac{\dot{q}'}{4\pi m} = \frac{53.9\text{Btu/h} \cdot \text{ft}}{4\pi(2.345)}$$

$$k = 1.83 \frac{\text{Btu}}{\text{h} \cdot \text{ft} \cdot {}^{\circ}\text{F}}$$

Result in SI Units: Rearranging Equation (5.45) to solve for k, and substituting the value of m (from Figure 5.14) and the value of $\dot{q}'$ (from Table 5.1), we obtain

$$k = \frac{\dot{q}'}{4\pi m} = \frac{51.8\,\text{W} \cdot \text{m}^{-1}}{4\pi(1.3028)}$$

$$k = 3.16\,\text{W} \cdot \text{m}^{-1} \cdot \text{K}^{-1}$$

This thermal conductivity value is typical of that expected from a dense rock such as limestone.

5.6.2.2 Inverse Modeling

Inverse modelling (also known as *parameter estimation*) involves minimizing the difference between experimentally obtained results and results predicted by a mathematical model by adjusting inputs to the model. Therefore, any adequate mathematical model that describes heat transfer in BHEs coupled to an optimization routine is suitable. A two-variable optimization is needed to solve for the thermal conductivity of the ground and the borehole resistance.

The mathematical model used here is the infinite line source analytical solution described previously in Section 5.5.2.1. Implemented into an Excel spreadsheet, the model fluid temperature can be calculated at each time step and compared with thermal response test results. Comparisons are made by calculating the squared error between the mathematical model calculation and the experimental measurement at each time of measurement. The error (or residual) is squared in these types of optimization problem because large errors become magnified, but more importantly, negative residuals become positive and do not cancel positive errors when all of the N individual error calculations are summed. Thus, the objective function for the optimization is the sum of the squared error (SSE), given by

$$\text{SSE} = \sum_{1}^{N} \left(T_{experimental} - T_{model}\right)^2 \tag{5.46}$$

The optimization is performed conveniently using the Excel Solver. The Excel Solver employs the mathematics of optimization systematically to vary relevant parameters to minimize the objective function [Equation (5.46)]. In this case, when the objective function is minimized (note that other types of optimization problem might maximize an objective function), the best values of the varied parameters are said to be found. Here, the parameters of interest to be varied include the average ground thermal conductivity (k) and the borehole thermal resistance (R_b'). For those not familiar with using the Excel Solver, a discussion is included in Appendix A.

The following steps summarize the optimization when using a spreadsheet:

(a) First, make some initial guesses of the thermal conductivity and borehole thermal resistance. Enter the remaining values needed to calculate the BHE fluid temperature (i.e., ρc_p, undisturbed temperature, average heat input during the test (in W or Btu/h), borehole depth, borehole radius).

(b) At each time interval, calculate the average *measured* borehole fluid temperature $T_{avg} = (T_{in} + T_{out})/2$ and plot on a graph,

(c) At each time interval, calculate the ground thermal resistance (R_g').
- If using the well function, R_g' is given by $W(u)/(4\pi k)$.
- If using Eskilson's analytical g-function, R_g' is given by $g/(2\pi k)$.

(d) At each time interval, calculate the BHE fluid temperature ($T_{f,\,avg}$) using equations (5.18) and (5.19). Thus, $T_{f,avg} = \dot{q}'R_g' + \dot{q}'R_b' + T_g$. Plot the calculated $T_{f,avg}$ on the same graph as the measured data.

(e) At each time interval, calculate the squared error between the experimental and modeled fluid temperatures. Sum the squared errors for times greater than $5r_b^2/\alpha$.

(f) Use the Excel Solver to minimize the sum of all the squared errors, with the Solver being set up to adjust the thermal conductivity and the borehole thermal resistance. Report the thermal conductivity and borehole thermal resistance when the sum of squared error is minimized.

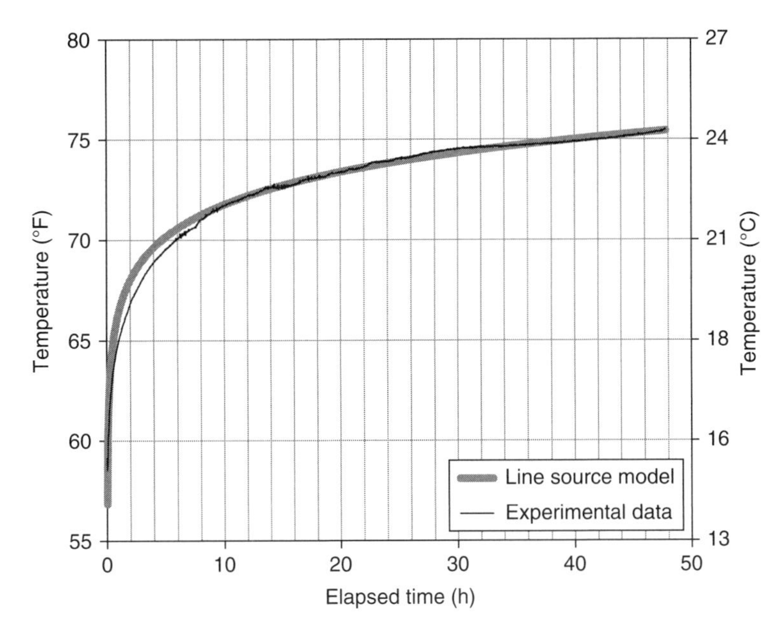

Figure 5.15 Comparison of model and experimental results for a thermal response test

Figure 5.15 shows the results of the optimization. A review of Figure 5.15 shows that the model produces an excellent match to the experimental data. The optimized values of the parameters were $k = 1.84$ Btu/h·ft·°F (3.18 $W \cdot m^{-1} \cdot K^{-1}$) and $R_b' = 0.1379$ h·ft·°F/Btu (0.0797 $m \cdot K \cdot W^{-1}$). The k value is in excellent agreement (i.e., <1%) with that determined by the graphical method, and the R_b' value is also in excellent agreement (i.e., ~2.5%) with that calculated by the multipole method (see Table 5.1).

Note that in Figure 5.15 the early *line source model* results do not match the experimental data exactly owing to the thermal storage effects of the borehole elements as discussed above. For this dataset, that minimum time is 4.8 h for line source model validity, and therefore calculations before that time were not included in the sum of squared errors to be minimized.

5.7 Pressure Considerations for Deep Vertical Boreholes

ASHRAE (2015) provides brief guidance on pressure considerations for boreholes deeper than 120 m. This author has been involved with at least one case where collapse of an HDPE BHE occurred in a borehole of the order of 120 m. With plastic pipe, circumferentially applied external pressure or internal vacuum, or a combination, will tend to flatten the pipe. Thus, pipe manufacturers report external pressure ratings as a function of temperature and ovality (typically 3%). As external pressure (or internal vacuum) increases the pipe ovality, stress cracking occurs, leading to pipe failure. It should be noted that the BHE pipe may not be exactly round when it is unrolled from a reel and inserted into a borehole.

The main reason for pipe collapse is attributed to the borehole grout, which has a specific gravity (*SG*) of 1.5–2.5 (relative to the density of water). The higher-thermal-conductivity (or thermally enhanced) grouts are those with the higher *SG*. Thus, it is easy to see that, with increasing depth, the external pressure progressively exceeds the internal pressure of the BHE pipe.

A common specification for BHEs in deeper boreholes is to use SDR 9 pipe. The external pressure rating for SDR 11 HDPE pipe at 23 °C is about 410 kPa, and it is about twice that value (820 kPa) for SDR 9 pipe. These pressure ratings pertain to a load duration of 12 h. External pressure ratings for load durations of 1 year are about 30% lower than those for a load duration of 12 h. In theory, the fluid added to grouts is designed to gel or harden the grout, but nobody has been able to go down a borehole and verify this. Thus, recent advancements in grout development have included the introduction of graphite-based grouts, which are lower in density and higher in thermal conductivity than the grouts thermally enhanced with silica sand.

To determine the depth of a BHE where collapse is possible, we may consider basic fluid statics:

$$P(z) = gz\left(SG_{grout} - SG_{BHE\,fluid}\right) \tag{5.47}$$

where g is gravitational acceleration, z is depth, and SG is the specific gravity.

Example 5.7 Calculating the External Pressure on a BHE Pipe

Calculate the external pressure on a BHE at the base of a 120 m deep borehole. The pipe is SDR 11 with an external pressure rating of 400 kPa at the design temperature. The BHE fluid is water, and a thermally enhanced grout with $SG = 2.0$ is being considered. Would you recommend this design?

Solution

$$P(120\,\text{m}) = 9.81\frac{\text{m}}{\text{s}^2} \times 120\,\text{m} \times (2-1) \times 1000\frac{\text{kg}}{\text{m}^3} = 1177\,\text{kPa}$$

This design would not be recommended, as the calculated pressure is almost 3× the allowable. SDR 9 pipe should be specified with lower-*SG* grout in order to achieve these depths.

5.8 Special Cases

In this section, we address some borehole heat exchangers that don't conform to the foregoing discussion. These are: (i) standing column wells and (ii) heat pipes.

5.8.1 Standing Column Wells Revisited

Standing column wells were discussed in Chapter 4 (Section 4.5.5) because of their reliance on groundwater and because of the fact that they are water wells. However, the thermal characteristics of standing column wells are more akin to those of closed-loop borehole heat exchangers, as the heat transfer fluid (groundwater) is recirculated in the borehole.

One main difference between the thermal characteristics of standing column wells and other types of BHE is the borehole thermal resistance. With standing column wells, the heat exchange fluid is in direct contact with the borehole wall, thus eliminating pipe and grout thermal resistances. Further, there is a 'well bore skin effect', where some of the groundwater may leave and re-enter the borehole through fractures. Methods described in Section 5.5.6.4 for calculation of the thermal resistance of concentric pipes can be applied to standing columns wells, but with no outer pipe and no grout resistance. The borehole thermal resistance of standing column wells is typically an order of magnitude lower than that of single U-tube BHEs. The heat transfer around standing column wells can be described by the line source model.

The other main difference between the thermal characteristics of standing column wells and other types of BHE is the ability to 'bleed' fluid from the system. As mentioned in Chapter 4 (Section 4.5.5), bleed occurs naturally in residential systems, as well water is used for domestic purposes. In many jurisdictions, groundwater disposal to the sewer is not allowed, and thus in commercial applications, bleed should only be used in extreme situations if there is no other responsible disposal option. Bleed complicates the mathematics because mass is removed from the system. Hydraulically, as discussed in Chapter 4, pumping induces hydraulic gradients toward the well. Thermally, groundwater flowing toward the well moderates the temperature of the rock around the well bore, and also the well water temperature.

To a first approximation, we can account for the thermal effects of bleed by combining Equations (5.18) and (5.19), and adding mass flow of groundwater into the well to calculate the average water temperature in the well ($T_{f,avg}$):

$$T_{f,avg} = \dot{q}'R'_g + \dot{q}'R'_b + \dot{m}c_p\left(T_{gw} - T_{f,avg}\right) + T_g \tag{5.48}$$

where T_{gw} is the groundwater temperature entering the well bore, which is assumed to be at the temperature of the borehole wall. Recall the borehole wall temperature is calculated by the line source model. Typical bleed rates are ~10% of the flow rate of the water circulating in the borehole. Note that the solution of Equation (5.48) is iterative because $T_{f,avg}$ is unknown. Also note sign convention; heat extraction is negative and heat rejection is positive.

Deng *et al.* (2005) developed a one-dimensional (1D) numerical model of standing column wells with bleed. The model was validated with experimental data and a more detailed reference numerical model. The 1D model is intended for use in hourly simulation programs or design tools.

5.8.2 Heat Pipes

A heat pipe is a passive means of transporting thermal energy from one location to another. As shown in Figure 5.16, in a vertical orientation, heat pipes are closed-loop structures containing a two-phase fluid. Heat is transported from the lower portion of the heat pipe as the fluid boils and travels upward. The latent heat of vaporization is released at the top of the structure, and then the liquid fraction travels back downward via a wick.

Heat pipes have seen geothermal applications in a few ways. The most common application is soil freezing for permafrost, railway, roadway, and pipeline stabilization. In these applications, typically in permafrost regions, the warmer Earth is used to boil a working fluid (typically CO_2) in a stainless steel tube. When the heat sink is atmospheric air, cold winter temperatures result in cooling of the subsurface to temperatures well below freezing. Heat pipes are installed on support

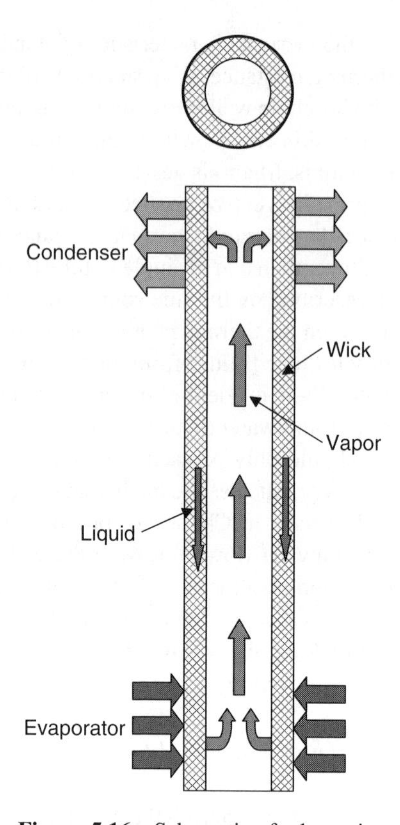

Figure 5.16 Schematic of a heat pipe

pillars of the Alaska Pipeline. In some cases, heat pipes may be used to freeze groundwater for the purpose of minimizing dewatering during excavations, or stabilizing excavations in soft soils.

Nydahl *et al.* (1984) report on the use of ammonia heat pipes in bridge deck de-icing and snow-melting in Wyoming. That system used ~16 m deep heat pipes, constructed of black iron, with ammonia as the working fluid. Over the temperature range to which the heat pipes were exposed in operation, part of the ammonia working fluid resided as liquid in a pool at the bottom of the pipe, while the remaining ammonia was in a vapor phase filling the rest of the tube. At any time the bridge deck temperature fell below the temperature of the ground in contact with the heat pipe, the vapor in the upper section condensed and flowed down toward the bottom of the pipe, creating the boiling–condensing cycle.

5.9 Chapter Summary

This chapter presented fundamental heat transfer theory of single-borehole heat exchangers, in addition to a practical overview of their installation. Infinite and finite line source mathematical models were reviewed for radial heat conduction around vertical boreholes. Similarity of these solutions to those for radial groundwater flow toward a well was also presented. Mathematical models for determining the effects of groundwater flow on the heat transfer characteristics of vertical closed-loop BHEs were discussed.

Mathematical models for borehole thermal resistance were presented for single U-tube, double U-tube, and concentric pipe BHEs. Designers have control over the construction specifics of BHEs (i.e., grout type, pipe size, pipe spacing within the borehole), and the effects of these design choices can be quantified with the borehole thermal resistance.

Thermal response testing of single BHEs was also described. Field methods and data analysis methods were illustrated through an example. Pressure considerations in deep vertical boreholes were also discussed, demonstrating that designers should be cognizant of possible pipe collapse in boreholes deeper than 100 m in certain conditions. Finally, the special single BHE cases of standing column wells and heat pipes were briefly described.

Discussion Questions and Exercise Problems

5.1 An actual thermal response test dataset is provided on this book's companion website. Your assignment is to estimate the average thermal conductivity of the subsurface materials and the borehole thermal resistance using inverse modeling methods.

The borehole details are as follows:

IP Units.

Borehole diameter = 5 in
Borehole depth = 245 ft
Assumed ρc_p of geologic materials from drilling = 40 Btu/ft^3·°F

(a) Determine the thermal conductivity of the subsurface materials using the inverse modeling approach.
(b) The value of ρc_p is an assumed value. What happens to the estimates of the thermal conductivity and borehole thermal resistance if ρc_p is assumed to be 35 Btu/ft^3·°F? What if it is assumed to be 45 Btu/ft^3·°F? What can you conclude about the effect of ρc_p on the thermal conductivity and borehole thermal resistance estimates?

The borehole details are as follows:

SI Units.

Borehole diameter = 125 mm
Borehole depth = 75 m
Assumed ρc_p of geologic materials from drilling = 2700 kJ·m^{-3}·°C

(a) Determine the thermal conductivity of the subsurface materials using the inverse modeling approach.
(b) The value of ρc_p is an assumed value. What happens to the estimates of the thermal conductivity and borehole thermal resistance if ρc_p is assumed to be 2100 kJ·m^{-3}·°C? What if it is assumed to be 3500 kJ·m^{-3}·°C? What can you conclude about the effect of ρc_p on the thermal conductivity and borehole thermal resistance estimates?

5.2 Determine the thermal conductivity of the subsurface materials described in problem 5.1 using the graphical method.

5.3 **IP Units.** Use the borehole thermal resistance calculator tool on the companion website to determine the thermal resistance per unit length of a borehole with the following construction details:

- 5 in diameter
- 250 ft borehole depth
- Flow rate: 3 gpm pure water
- Borehole grout: standard bentonite grout

(a) a 1 in U-tube with tubes approximately evenly spaced in the borehole,
(b) a 1 in U-tube with tubes touching each other,
(c) a 1 in U-tube with tubes touching the borehole wall,
(d) repeat problems (a), (b), and (c) but with thermally enhanced grout with a thermal conductivity of 1.0 Btu/h/ft/°F.

SI Units. Use the borehole thermal resistance calculator tool on the companion website to determine the thermal resistance per unit length of a borehole with the following construction details:

- 125 mm diameter
- 75 m borehole depth
- Flow rate: 10 Lpm pure water
- Borehole grout: standard bentonite grout

(e) a DN25 U-tube with tubes approximately evenly spaced in the borehole,
(f) a DN25 U-tube with tubes touching each other,
(g) a DN25 U-tube with tubes touching the borehole wall.

Repeat problems (a), (b), and (c) but with thermally enhanced grout with a thermal conductivity of 1.75 $W{\cdot}m^{-1}{\cdot}K^{-1}$.

5.4 Boreholes are being planned with pure water as the heat transfer fluid and grout with $SG = 1.5$. Determine the maximum allowable depth of the borehole to avoid collapse of the pipe if the external pressure rating is 400 kPa.

6

Multi-Borehole Heat Exchanger Arrays

6.1 Overview

The focus of this chapter is the design and dimensioning of vertical bore ground heat exchangers (GHXs) coupled to heat pump applications. GHXs consist of multiple BHEs connected together. For example, residential buildings may require 3–5 vertical BHEs, while larger buildings such as schools may require tens to hundreds. In this chapter, we will use the term *vertical ground heat exchanger* (or GHX) to refer to the entire array of BHEs, meaning that the GHX length (*L*) will now be defined to mean the total number of boreholes (*NB*) × the borehole depth (*H*).

One of the fundamental tasks in the design of a GHX is the proper sizing of the total number, depth, and spacing of the BHEs so that they provide fluid temperatures to the heat pump(s) within design limits over its lifetime. The thermal loads of a building are time dependent, but the Earth does not respond to these loads instantaneously. Therefore, unlike with conventional heating and cooling systems, design of GHXs requires consideration of transient functions owing to the thermal storage effects of the Earth. If improperly designed, the GHX can quickly become cost prohibitive if oversized, or unable to meet the intended heating and/or cooling loads if undersized. A more advanced design involves a so-called hybrid system, where a supplemental component is designed to handle some portion of the thermal loads on the GHX to reduce its required size.

In this chapter, we will first review the configuration of vertical GHXs, and then examine each design parameter in the context of the so-called *GHX design length equation*, which is implemented into the suite of GHX tools provided with this book. We will then examine simulation of vertical GHXs, followed by a discussion of hybrid GHP systems. The chapter concludes with the topic of vertical GHX modeling with software applications. The focus is on use of the companion suite of GHX tools provided with this book.

Geothermal Heat Pump and Heat Engine Systems: Theory and Practice, First Edition. Andrew D. Chiasson.

Companion website: www.wiley.com/go/chiasson/geoHPSTP

Learning objectives and goals:

1. Appreciate the complexities in the heat transfer of ground heat exchangers coupled to buildings.
2. Quantify all necessary parameters required properly to size a vertical ground heat exchanger for a building.
3. Appreciate the role of simulation in the study of vertical GHXs.
4. Appreciate the role of software solutions for vertical GHX design and simulation.
5. Calculate the required size of a vertical ground heat exchanger for a building.
6. Calculate the required size of a hybrid vertical ground heat exchanger for a building.

6.1.1 Introduction

Perhaps the main complicating factor of GHXs as compared with single BHEs is the thermal interaction between boreholes. It is a common misconception that the Earth remains at a constant temperature; the *undisturbed* Earth temperature is at some equilibrium temperature based on heat fluxes at the ground surface and geothermal processes from below, but periodic BHE heat extraction and heat rejection over time lead to changing temperatures in the Earth volume in which the BHEs are installed. As we have seen, and will continue to see, heat transfer rates in the Earth are not rapid, and the Earth is characterized by large time constants. In applications with highly unbalanced loads over the annual cycle (i.e., cases where the total annual heating demand greatly exceeds the total annual cooling demand, or vice versa), the required total length of the GHX may become prohibitively large. In such cases, hybrid geothermal heat pump (GHP) systems may be considered. In hybrid GHP systems, a supplemental component (i.e., a dry fluid cooler, solar collectors) is used to handle some portion of the ground loads.

The geothermal heat pump industry, especially in the residential sector, is fraught with *rules of thumb* for GHX design length. Such rules of thumb are frequently reported as ft/ton or W/m or m/W. As we will see, the GHX design length depends on several variables, and it can be very difficult to apply rules of thumb from one situation to another. So where did these rules of thumb originate? Their origin, at least in the United States, was in the south-central region, and it applied to residential buildings with distinct heating and cooling seasons. The rule of thumb was meant to relate total GHX length to the installed capacity of the heat pump, which was usually closely matched to the cooling load. The rule of thumb was adequate for similar buildings in that area with similar occupancy schedules, similar geologic conditions, and similar heat pump design. Extrapolation of the rule of thumb outside the assumptions upon which the rule was developed may lead to significant under- or oversizing of the GHX. Designers that use rules of thumb should first verify them through mathematical and engineering calculations and/or field evidence, and make sure that the assumptions built into the rule are applicable to the situation at hand.

6.1.2 Vertical GHX Configurations

A schematic of the configuration of a vertical GHX coupled to an office building is shown in Figure 6.1. A photograph of completion of a GHX is shown in Figure 6.2, after completion of

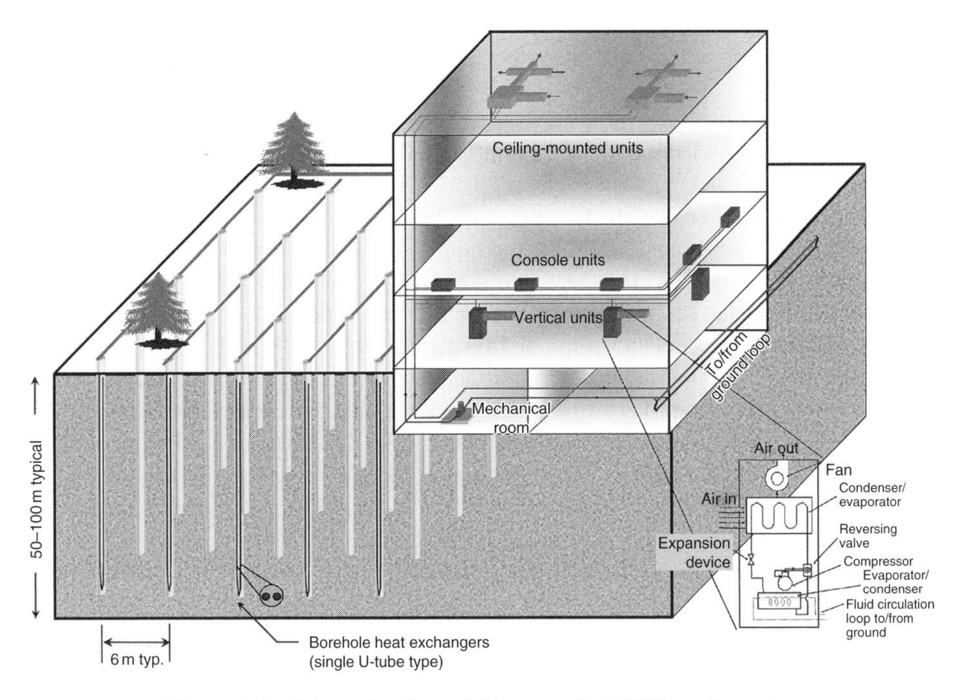

Figure 6.1 Schematic of a multi-bore vertical GHX configuration

Figure 6.2 Photo of a GHX after completion of BHE drilling and installation. Note the U-tubes sticking out of the ground, prior to excavating and connecting them all together (photo taken by the author)

BHE drilling and installation, prior to excavating and connecting the U-tubes together. Laying out and designing the horizontal transfer piping to and from each BHE will be discussed in a subsequent chapter.

As discussed in the previous chapter, the typical depth of BHEs ranges from 50 to 100 m. BHEs are typically constructed of HDPE thermoplastic, with all joints thermally fused, and with nominal pipe diameters ranging from ¾ to 1¼ in or DN20 to DN32, with 1 in or DN25 nominal diameter being the most common. Other options exist, including PEX pipe U-tube assemblies and concentric pipe arrangements. Each BHE is typically designed as one parallel flow circuit in a reverse-return arrangement with other BHEs. Groups of BHEs are sometimes connected in series if BHEs are relatively shallow in depth. These considerations are dictated by fluid management constraints that will be discussed in a subsequent chapter. Reverse-return and direct-return piping systems will also be discussed in a later chapter.

6.2 Vertical GHX Design Length Equation and Design Parameters

6.2.1 *The Vertical GHX Design Length Equation*

If we combine Equations (5.18) and (5.19) to arrive at an expression of the borehole average fluid temperature, we have

$$T_{f,avg} = \dot{q}' R'_g + \dot{q}' R'_b + T_g \tag{6.1}$$

Recall that R'_g could be calculated with a suitable finite line source model for a single borehole.

For multi-BHE systems, Equation (6.1) can be rearranged to solve for length (i.e., instead of using $\dot{q}'$), and accounting for monthly and annual thermal superimposed pulses, to obtain

$$L = \frac{\dot{q}_h R'_b + \dot{q}_a R'_{g,a} + \hat{\dot{q}} R'_{g,m} + \hat{\dot{q}} R'_{g,d}}{T_f - T_g} \tag{6.2}$$

where $\dot{q}$ is the ground load, R' is the effective thermal resistance per unit length of borehole, T_f is the average fluid temperature in the BHE, and T_g is the undisturbed ground temperature. Subscripts *g, b, a, m, h,* and *d* refer to ground, borehole, annual, monthly, hourly, and daily. The ^ symbol denotes superimposed loads. NOTE: positive values of $\dot{q}$ are associated with heat rejection to the ground, so cooling loads are positive and heating loads are negative.

Each of the terms in Equation (6.2) represents a vertical GHX design parameter, and each is described in the sections that follow. Note that Equation (6.2) gives the total length (number of boreholes *NB* × borehole depth *H*) of the GHX. As we shall see, the long-term (monthly and annual) thermal resistances of the ground depend on *NB* and *H*, but *H* is unknown. Thus, *Equation (6.2) must be solved by iteration*, and hence the necessity to employ a software tool for its solution.

6.2.2 *The Undisturbed Ground Temperature* (T_g)

Proper consideration of the undisturbed ground temperature is perhaps one of the most overlooked parameters in GHX design. An inspection of Equation (6.2) shows that its relative difference to the average BHE fluid temperature has a direct impact on the required GHX length. In fact, Equation (6.2) shows that to have T_f equal to T_g would result in an infinitely sized GHX. Recall that the GHX is a heat exchanger that exchanges heat with a solid material having a far-field temperature of T_g. Therefore, the operating temperature in the GHX (i.e., T_f) can only *approach* T_g. That approach temperature is a function of the GHX effectiveness, which is a function of its total length.

The variation in Earth temperature over the depth interval of interest to vertical bore GHXs has been discussed in Chapter 3.5. The actual value of T_g chosen for a GHX design is site-specific, and can be thought of as an average equilibrium temperature resulting from surface processes (i.e., atmospheric and solar) and deep geothermal processes (i.e., magmatic activity, plate tectonics, and hydrothermal activity).

There are a few methods available to estimate the undisturbed ground temperature (T_g) for vertical closed-loop GHX systems: (i) field measurement during thermal response tests as discussed in Chapter 5, (ii) from maps, and (iii) from surface meteorological data.

The first method for determining T_g has been discussed in Chapter 5, and involves measurement of the BHE fluid prior to conducting a thermal response test. Recall that this can be accomplished by (a) measuring the vertical temperature profile of the standing fluid in the U-tube, and/or (b) circulating the BHE fluid with the pump to be used in the thermal response test, and recording the stabilized fluid temperature. Gehlin (2002) found that method (a) is more reliable because method (b) can result in the addition of unnecessary pump heat to the fluid and disturb the temperature significantly.

The second method for determining T_g is by estimating it through published maps of the Earth temperature. Such maps are available through many various agencies. While seemingly simple to obtain Earth temperatures from maps, one should be cautious of the depth interval over which the temperature is applicable. Soil science groups publish several map series, but their interest usually lies in the topmost soil layer only, which has significant seasonal effects. At the other end of the spectrum, geothermal power plant engineers and exploration geologists may only be interested in the Earth temperature at a depth of hundreds or thousands of meters, and would consequently be interested in yet a different set of maps.

Finally, a third method for estimating T_g is from surface meteorological data as described in Chapter 2.2. Alternatively, Signorelli (2004) suggests that the average underground temperature equivalent over the typical geoexchange borehole depth (~100 m) is approximated by the average annual air temperature plus about 1.5 °C. This estimate captures the subsurface temperature due to heat fluxes from above and below, and is applicable only in areas of normal geothermal gradient.

6.2.3 Soil/Rock Thermal Properties

Aside from the ground temperature, the main thermal properties of interest for GHX design are the thermal conductivity (k), the density (ρ), and the heat capacity (c_p) of the ground. The ρc_p product is commonly referred to as volumetric heat capacity. Recall that the thermal diffusivity (α) is defined as $k/(\rho c_p)$. Thermal conductivity values of common soils and rocks are listed in Appendix B.

Soils and rocks are complex materials consisting of three phases: solids (minerals and organic matter), liquids (water), and gases. Therefore, the thermal properties are usually reported as macroscopic (or bulk) values. In general, the thermal conductivity is directly proportional to the density of the material. Thermal conductivity of rocks and soils is also related to mineral content and water content. Crystalline rocks tend to have higher k values than materials with clay minerals. Soils and rocks that are saturated have higher thermal conductivity values than the same soil or rock that is dry, not because water has a high thermal conductivity (in fact it does not – it is only ~0.62 $W{\cdot}m^{-1}{\cdot}K^{-1}$), but because the thermal conductivity of water is more than an order of magnitude greater than that of air (k_{air} = 0.026 $W{\cdot}m^{-1}{\cdot}K^{-1}$). Therefore, it is better to have soils and rocks with pore spaces saturated with water, rather than unsaturated with some volume of air in the pores.

The volumetric heat capacity of water is 62.4 $lb/ft^3{\cdot}°F$ or 4180 $kJ{\cdot}m^{-3}{\cdot}K^{-1}$, and the value for most soils and rocks is about one-half that of water. The actual ρc_p varies according to mineral content and water content, and ranges from about 20 $lb/ft^3{\cdot}°F$ (1340 $kJ{\cdot}m^{-3}{\cdot}K^{-1}$) for light, dry soils to over 40 $lb/ft^3{\cdot}°F$ (2680 $kJ{\cdot}m^{-3}{\cdot}K^{-1}$) for dense, wet soils and rocks. Minor variations in the ρc_p value in the context of the GHX design length equation do not significantly impact calculation results. Therefore, a reasonable estimate of its value usually suffices.

6.2.4 The Ground Loads

Geothermal process loads are the subject of Chapter 2. Recall that the term *load*, as used here, refers to the time-dependent thermal energy needs to be met by the energy system. Also, recall that the ground load in geothermal heat pump systems (GHPs) is not the same as the building load. The ground loads are related to the heat pump loads through the heat pump coefficient of performance (COP) by multiplying the building load by the appropriate factor, denoted here as F:

$$F_{htg} = \frac{\mathrm{COP}_{htg} - 1}{\mathrm{COP}_{htg}} \qquad F_{clg} = \frac{\mathrm{COP}_{clg} + 1}{\mathrm{COP}_{clg}} \tag{6.3a, b}$$

where subscripts *htg* and *clg* refer to heating and cooling respectively. Note that we have switched here from using the term *building loads* to the term *heat pump loads*. This distinction is made here because the two loads may not be the same. As we shall see, some of the building loads may be met with a heat pump(s) plus some other piece of equipment (for example, electric resistance heating). Therefore, the term *building loads* will be used henceforth to refer to the thermal load required by the building, while the term *heat pump load* will be used to refer only to the building loads met by the heat pump(s). *The ground load, therefore, is related to the heat pump load.*

The GHX design length equation requires the following *ground* loads: the peak hour load ($\dot{q}_h$), the average monthly load ($\dot{q}_m$), and the average annual load ($\dot{q}_a$). However, there are actually six loads needed for GHX design, because the peak hour, peak monthly loads, and annual loads are required for both heating and cooling. Therefore, Equation (6.2) is actually applied twice to a GHX design, to determine the GHX length required for the heating load and the GHX length required for the cooling load. The designer can then either choose to design for the greater length (in order to meet all loads) or choose to design for the smaller length but meet the imbalance with a supplemental piece of equipment that offsets the building loads (i.e., an electric heating element) or a hybrid GHX with a supplemental system that offsets the ground loads (i.e., a dry fluid cooler or solar thermal array).

In the design tool described subsequently in this chapter, found on the accompanying website, $\dot{q}_m$ and $\dot{q}_a$ are not used directly. Rather, the concept of the *load factor* (*LF*) is used, which is defined as the ratio of the average heat pump load to the peak heat pump load during a given time and can be expressed for a month and a year as

$$LF_m = \frac{Q_{m,HP}}{\dot{q}_{h,HP} \times (N_m)}, \quad LF_a = \frac{Q_{a,HP}}{\dot{q}_{h,HP} \times (N_a)} \tag{6.4a,b}$$

where Q is an energy quantity (in Btu or kWh), $\dot{q}$ is an energy rate (Btu/h or kW), N is the number of hours, and subscripts HP, h, m, and a refer to heat pump, hour, month, and annual respectively. Therefore, Q_m represents the total monthly heating (or cooling) energy load, and Q_a represents the total annual heating (or cooling) energy load. The peak heating load (in the northern hemisphere) typically occurs in January, while the peak cooling load typically occurs in July. Therefore, the number of hours is generally: 31 days/month × 24 h/day = 744 h. There are 8760 h per year.

The *ground loads* necessary for Equation (6.2) are then computed as follows. The peak hour ground load is determined by

$$\dot{q}_h = \dot{q}_{h,HP} \times F \tag{6.5}$$

where the appropriate factor F is used as described by Equation (6.3a) or Equation (6.3b), depending on whether $\dot{q}_h$ is the heating or cooling load.

The average ground load acting over the peak month ($\dot{q}_m$) is given by

$$\dot{q}_m = \dot{q}_h \times LF_m \times F \tag{6.6}$$

where $\dot{q}_h$ is the peak hour ground load (heating or cooling), LF_m is the monthly load factor (heating or cooling), and F is the appropriate COP factor, depending on whether loads are heating or cooling.

The average ground load acting over the year $\dot{q}_a$ is given by

$$q_a = q_{h,htg} \times LF_{a,htg} \times F_{htg} + q_{h,clg} \times LF_{a,clg} \times F_{clg} \tag{6.7}$$

Recall our sign convention: heat rejection loads to the ground (i.e., building cooling loads) are positive, and heat extraction loads from the ground (i.e., building heating loads) are negative. Therefore, $\dot{q}_a$ will be positive or negative, depending on whether the building is cooling or heating dominated.

While building load calculations are beyond the scope of this course, they are necessary for use of the GHX design length equation. The peak hour load for a building is that used to size the heat pump equipment (or any piece of heating or cooling equipment), but determination of monthly and annual loads is usually an extra step for conventional HVAC designers. Load calculation methods have been discussed in Chapter 2.

6.2.4.1 Loads from a Building Simulation Software Program

Building simulation software programs have been discussed in Chapter 2. Recall, as described here, these refer to whole building energy software tools that calculate energy use in buildings on an hourly basis. The general approach to the use of these tools is similar: users must input details of the building envelope, occupancy, HVAC systems and controls, and all other factors that impact the thermal behavior of the building. These building simulation programs are mainly driven by weather files and other time-dependent functions. The output of interest to GHX design include: hourly heating loads, hourly cooling loads, and hourly domestic hot water loads. Time and budget permitting, application of building simulation programs provides handy output for use in GHX design, as all hourly loads throughout the year are calculated. However, the effort of constructing a building model is not warranted in all projects.

Example 6.1 Calculation of the 'Ground Loads'

Revisiting Example 2.3, the hourly loads from a building energy simulation program are summarized in Table E.6.1. You are involved in GHX design for this building, and the heat pump will be handling all of the loads. Thus, the heat pump loads will be the same as the building loads. Assuming that the heat pump COP under design conditions is 4.0 for heating and 5.0 for cooling, determine the following for use in your subsequent GHX design length calculations:

(a) the peak hourly ground loads for both heating and cooling,
(b) the monthly load factors for both heating and cooling,
(c) the average ground load over the peak heating and cooling months,
(d) the annual load factors for both heating and cooling, and
(e) the average ground load over the year.

Table E.6.1 Monthly Load Summaries for Example 6.1

A	B	C	D	E	F
	h/month	Total Heating (1000 Btu)	Total Cooling (1000 Btu)	Peak Heating (1000 Btu/h)	Peak Cooling (1000 Btu/h)
Jan	744	18 963	0	47	0
Feb	672	15 782	5	46	2
Mar	744	9640	596	36	16
Apr	720	4352	772	19	17
May	744	0	3730	0	21
Jun	720	0	3987	0	23
Jul	744	0	5545	0	24
Aug	744	67	4748	6	21
Sep	720	4453	2211	30	20
Oct	744	4460	411	22	9
Nov	720	9711	89	30	11
Dec	744	15 745	0	47	0

Solution

By inspection of the above table, we see that the peak heating loads occur in January, and the peak cooling loads occur in July.

(a) The peak hourly ground loads for both heating and cooling are given by:

$$\dot{q}_h = \dot{q}_{h,HP} \times F$$

For heating: $\dot{q}_{h,htg} = -47000\frac{\text{Btu}}{\text{h}} \times \frac{\text{COP}_{htg} - 1}{\text{COP}_{htg}} = -47000\frac{\text{Btu}}{\text{h}} \times \frac{3.0}{4.0} = -35250\frac{\text{Btu}}{\text{h}}$

For cooling: $\dot{q}_{h,clg} = 24000\frac{\text{Btu}}{\text{h}} \times \frac{\text{COP}_{htg} + 1}{\text{COP}_{htg}} = 24000\frac{\text{Btu}}{\text{h}} \times \frac{5.0}{4.0} = 30000\frac{\text{Btu}}{\text{h}}$

(b) The monthly load factors for both heating and cooling are given by:

$$LF_m = \frac{Q_{m,HP}}{\dot{q}_{h,HP} \times (N_m)}$$

For heating: $LF_{m,htg} = \frac{-18963\,\text{kBtu/month}}{-47\,\text{kBtu/h} \times (744\,\text{h/month})} = 0.54$

For cooling: $LF_{m,clg} = \frac{5545\,\text{kBtu/month}}{24\,\text{kBtu/h} \times (744\,\text{h/month})} = 0.31$

(c) The average ground load over the peak heating and cooling months is given by

$$\dot{q}_m = \dot{q}_h \times LF_m \times F$$

The ground loads were calculated in part (a) and the load factors were calculated in part (b). Therefore,

For heating: $\dot{q}_{m,htg} = -35250\frac{\text{Btu}}{\text{h}} \times 0.54 = -19035\frac{\text{Btu}}{\text{h}}$

For cooling: $\dot{q}_{m,clg} = 30000\frac{\text{Btu}}{\text{h}} \times 0.31 = 9300\frac{\text{Btu}}{\text{h}}$

(d) The annual load factor for both heating and cooling is given by

$$LF_a = \frac{Q_{a,bldg}}{\dot{q}_{h,bldg} \times (N_a)}$$

For heating: $LF_{a,htg} = \frac{-83173\,\text{kBtu/year}}{-47\,\text{kBtu/h} \times (8760\,\text{h/year})} = 0.20$

For cooling: $LF_{a,clg} = \frac{22094\,\text{kBtu/year}}{24\,\text{kBtu/h} \times (8760\,\text{h/year})} = 0.11$

(e) The average ground load over the year is given by

$$\dot{q}_a = \dot{q}_{h,htg} \times LF_{a,htg} \times F_{htg} + \dot{q}_{h,clg} \times LF_{a,clg} \times F_{clg}$$

The ground loads were calculated in part (a) and the load factors were calculated in part (d). Therefore

$$\dot{q}_a = -35250\frac{\text{Btu}}{\text{h}} \times 0.20 + 30000\frac{\text{Btu}}{\text{h}} \times 0.11 = -3750\frac{\text{Btu}}{\text{h}}$$

6.2.4.2 Dealing with Auxiliary or Supplemental Heating

The advantages of supplemental heating, or baseload-peaking arrangements, has been discussed in Chapter 2. The building in the previous example demonstrates a typical case where the peak hour loads are mismatched (i.e., the peak heating load is about twice the peak cooling load). As described in Chapter 2, one method of dealing with such mismatches is to add a supplemental heating element in the heat pump discharge duct in water-to-air GHP systems. In fact, this a common design approach in the residential sector of the United States: select the heat pump for the cooling load, and meet the balance of the heating load with electric resistance heat. Such an approach results in better heat pump performance and occupant comfort in cooling, because the heat pump will run more continuously, thereby providing better humidity control. If sized for the heating load, the heat pump would be grossly oversized for the cooling load in this case, and frequently short cycle.

We were careful to point out in the foregoing, that the heat pump loads may not be the same as the building loads, and that would be the case here if an electric heating element were to be added to the heat pump discharge duct. The heat pump would be configured to meet the base heating loads, and the electric heating element would come on during times when the heat pump could not meet the load. Hourly heating loads from a building simulation program could then be *post-processed* such that the heat pump load was the lesser of the heat pump heating capacity or the actual building load.

Example 6.2 Calculating Ground Loads with Supplemental Heating

The post-processed hourly loads from a building energy simulation program with a 50% peak heating element are shown in Table E.6.2.

Recompute the following from Example 6.1:

(a) the peak hourly ground heating load,
(b) the monthly load factors for heating,
(c) the average ground load over the peak month,
(d) the annual load factors heating, and
(e) the average ground load over the year.

Table E.6.2 Monthly Load Summaries for Example 6.2

A	B	C	D	E	F
	h/month	Total Heating (1000 Btu)	Total Cooling (1000 Btu)	Peak Heating (1000 Btu/h)	Peak Cooling (1000 Btu/h)
Jan	744	16 107	0	23.5	0
Feb	672	13 548	5	23.5	2
Mar	744	9129	596	23.5	16
Apr	720	4352	772	19.0	17
May	744	0	3730	0.0	21
Jun	720	0	3987	0.0	23
Jul	744	0	5545	0.0	24
Aug	744	67	4748	6.0	21
Sep	720	4376	2211	23.5	20
Oct	744	4460	411	22.0	9
Nov	720	9483	89	23.5	11
Dec	744	13 480	0	23.5	0

Solution

By inspection of the above table, we see that (as in Example 6.1) the peak heating loads occur in January, and the peak cooling loads occur in July. Note that the cooling loads have not changed.

(a) The peak hourly ground load for heating is given by

$$\dot{q}_h = \dot{q}_{h,HP} \times F$$

$$\dot{q}_{h,htg} = -23500\frac{\text{Btu}}{\text{h}} \times \frac{\text{COP}_{htg}-1}{\text{COP}_{htg}} = -23500\frac{\text{Btu}}{\text{h}} \times \frac{3.0}{4.0} = -17625\frac{\text{Btu}}{\text{h}}$$

(b) The monthly load factor for heating is given by

$$LF_m = \frac{Q_{m,HP}}{\dot{q}_{h,HP} \times (N_m)}$$

$$LF_{m,htg} = \frac{-16107\,\text{kBtu/month}}{-23.5\,\text{kBtu/h} \times (744\,\text{h/month})} = 0.92$$

(c) The average ground load over the peak heating month is given by

$$\dot{q}_m = \dot{q}_h \times LF_m \times F$$

The ground load was calculated in part (a) and the load factor was calculated in part (b). Therefore,

$$\dot{q}_{m,htg} = -17625\frac{\text{Btu}}{\text{h}} \times 0.92 = -16215\frac{\text{Btu}}{\text{h}}$$

(d) The annual load factor for heating is given by

$$LF_a = \frac{Q_{a,bldg}}{\dot{q}_{h,bldg} \times (N_a)}$$

$$LF_{a,htg} = \frac{-75001\,\text{kBtu/year}}{-23.5\,\text{kBtu/h} \times (8760\,\text{h/year})} = 0.36$$

(e) The average ground load over the year is given by

$$\dot{q}_a = \dot{q}_{h,htg} \times LF_{a,htg} \times F_{htg} + \dot{q}_{h,clg} \times LF_{a,clg} \times F_{clg}$$

The ground loads were calculated in part (a) and the load factors were calculated in part (d). Therefore

$$\dot{q}_a = -17625\frac{\text{Btu}}{\text{h}} \times 0.36 + 30000\frac{\text{Btu}}{\text{h}} \times 0.11 = -3045\frac{\text{Btu}}{\text{h}}$$

Discussion: This example further illustrates the concept developed in Chapter 2, which is that, even though the heating element is sized to meet 50% of the peak load, the heat pump still meets 90% of the annual load (i.e., the total annual heating load from Example 6.1 is 83 173 kBtu, and is 75 001 kBtu in this example). This may not seem intuitive, but this phenomenon occurs because peak loads are experienced very rarely over the year, but we must design for them. In capital-intensive systems such as GHP systems, it is important to balance first costs with operating cost. As we shall see, a 50% peak heating element considerably reduces the necessary size of the GHX, in addition to reducing the size of the heat pump itself, while still meeting 90% of the annual load. This may result in significant capital cost savings.

6.2.4.3 Monthly and Annual Load Estimation

In early project design stages, or in the case of small projects, use of building simulation programs may not be warranted or available. Peak hour loads may be the only loads available, for reasons mentioned above: peak hour loads are the more familiar set of loads to HVAC engineers, and are calculated for purposes of equipment sizing. As discussed in Chapter 2, in addition to the peak load, other energy loads (month and year) are needed for sizing the GHX. If these loads are unavailable from a building simulation program, they must be estimated. Methods of estimating these loads have been described in Chapter 2.

6.2.4.4 Temporal (Time) Superpositioning of the Ground Loads

Dealing with time-varying loads has been discussed in the previous chapter, and the same principles are applied here. The GHX design length equation makes use of three superpositioned

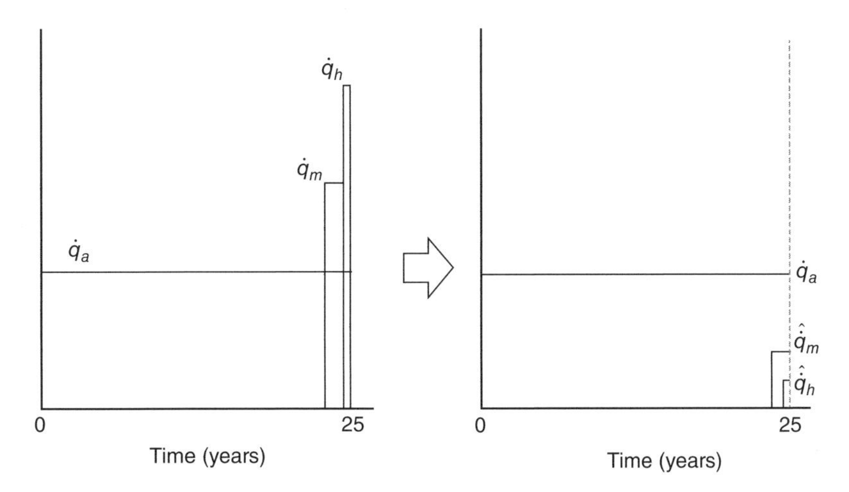

Figure 6.3 Superposition of piecewise linear step heat inputs in time. The monthly and peak hourly pulses are superimposed in time onto the basic annual pulse (*not necessarily to scale*). Heat rejection loads are shown here

loads as piecewise linear steps illustrated in Figure 6.3. For a GHX design life of 20 years, the subsequent pulses are superimposed as:

- $\dot{q}_a$ effective for 20 years,
- $\hat{\dot{q}}_m = \dot{q}_m - \dot{q}_a$, effective for 1 month,
- $\hat{\dot{q}}_h = \dot{q}_h - \dot{q}_m$, effective for n hours (typically 4–8 h).

Note that the peak hour load is superimposed at the end of the design life to account for a 'worst case' scenario. We will discuss impacts of the design life on the GHX size in further detail later in this chapter. In short for now, it takes about 10 years for most GHX systems to stabilize owing to the thermal storage effects of the Earth. In systems that are about 3 times or more out of annual thermal balance (in favor of either heating or cooling), it may take of the order of 30 years for Earth temperatures to stabilize. Hence the potential for hybrid systems, to be discussed at the end of this chapter.

Example 6.3

Referring to Example 6.1, calculate the superpositioned ground loads for heating if the GHX is being designed for 20 years.

Solution

From Example 6.1, we have:

$$\dot{q}_h = -35\,250\,\text{Btu/h};\ \dot{q}_m = -19\,035\,\text{Btu/h};\ \dot{q}_a = -3750\,\text{Btu/h}$$

$$\therefore \dot{q}_a = -3750\,\text{Btu/h effective for 20 years,}$$

$$\hat{\dot{q}}_m = \dot{q}_m - \dot{q}_a = -19\,035\,\text{Btu/h} - (-3750\,\text{Btu/h}) = -15\,285\,\text{Btu/h effective for 1 month,}$$

$$\left[\hat{\dot{q}}_h = \dot{q}_h - \dot{q}_m\right] = -35\,250\,\text{Btu/h} - (-19\,035\,\text{Btu/h}) = -16\,215\,\text{Btu/h effective for 6 h.}$$

6.2.5 *The Average BHE Fluid Temperature* (T_f)

As previously mentioned, T_f is an important parameter in dictating the GHX length. It appears in the denominator of the GHX design length equation as a difference term with T_g, and as such the calculated GHX length is quite sensitive to its value. It is perhaps the term in the GHX design length equation that designers have the most freedom to control (within heat pump limits, of course); the T_f term is related to the design heat pump entering fluid temperature that designers must choose to strike a good balance between capital cost of the GHX and the heat pump operating energy cost. As will be discussed in further detail in a later chapter, geothermal heat pumps have good COPs in heating down to 0 °C and in cooling up to 35 °C, and thus these represent good limits for GHX sizing; life-cycle economic optimization of the GHX size as a function of heat pump entering fluid temperatures is rarely done. In short, T_f, which is dictated by the heat pump entering fluid temperature, controls the GHX length while also affecting the heat pump energy consumption.

IGSHPA (2009) provides some guidance for selecting the design heat pump entering fluid temperatures for heating and cooling modes. As a starting point, the design entering fluid temperature for heating should be 15–20 °F (8–11 °C) below the undisturbed Earth temperature, or 25 °F (–4 °C), whichever is greater. The design entering fluid temperature for cooling should be 30–40 °F (16–22 °C) above the undisturbed Earth temperature, or 95 °F (35 °C), whichever is less.

As we shall see later in this chapter, the vertical GHX design tool provided on the book companion website (as well as commercially available software) requires minimum and maximum heat pump entering fluid temperatures and the design flow rate of the GHX heat transfer fluid. T_f is then related to these input parameters by an energy balance on the heat transfer fluid:

$$\begin{aligned}
\dot{q}_h &= \dot{m}c_p\left(T_{in,GHX} - T_{out,GHX}\right) \\
T_f &= \frac{T_{in,GHX} + T_{out,GHX}}{2} \\
T_{in,GHX} &= 2T_f - T_{out,GHX} \\
\dot{q}_h &= \dot{m}c_p\left(2T_f - T_{out,GHX} - T_{out,GHX}\right) \\
\dot{q}_h &= \dot{m}c_p\left(2T_f - 2T_{out,GHX}\right) \\
\dot{q}_h &= 2\dot{m}c_p\left(T_f - T_{out,GHX}\right) \\
T_f &= T_{out,GHX} + \frac{\dot{q}_h}{2\dot{m}c_p} = T_{in,HeatPump} + \frac{\dot{q}_h}{2\dot{m}c_p}
\end{aligned} \tag{6.8}$$

where $\dot{q}_h$ is the total hourly heat transfer rate of the GHX fluid, and $\dot{m}$ is the mass flow rate of the heat transfer fluid. Recalling our sign convention on the ground load, positive ground loads (heat rejection or building cooling loads) will result in a GHX inlet fluid temperature greater than the outlet, and vice versa for negative ground loads (heat extraction or building heating loads).

Example 6.4 Calculating the Average BHE Fluid Temperature from an Energy Balance
Calculate the average fluid temperature in a GHX under the design condition of supplying 5 °C fluid to heat pumps at a mass flow rate of 1.0 kg·s^{-1} when the peak hour ground load is 20 kW (extraction). Assume the heat transfer fluid in the GHX is pure water.

Solution

Assuming the heat capacity of water is 4.2 kJ·kg^{-1}·K^{-1}, from Equation (4.10) we have

$$\begin{aligned} T_f &= T_{in,HeatPump} + \frac{\dot{q}_h}{2\dot{m}c_p} \\ &= 5\,^\circ\text{C} + \frac{-20\,\text{kW}}{2\cdot(1.0\,\text{kg}\cdot\text{s}^{-1})\cdot(4.2\,\text{kJ}\cdot\text{kg}^{-1}\cdot\text{K}^{-1})} \\ &= 2.62\,^\circ\text{C} \end{aligned}$$

6.2.6 The Ground Thermal Resistances

The final pieces to the GHX design length equation are the ground thermal resistances associated with the peak hourly (well, actually daily, because we assume the building peaks near the peak hour load for a block of 4–6 h), peak monthly, and annual thermal loads on the ground.

We've already dealt with calculating the transient ground thermal resistance for a single borehole using the dimensionless g-function, and our task now is to calculate transient ground thermal resistances for multiple, thermally interacting boreholes. Fortunately, there is a straightforward way to do this. We will again make use of the principle of superposition, but this time in space. In other words, as we can obtain the solution for the temperature response in the ground due to a single BHE at a particular time, we can superimpose multiple solutions in space (i.e., at various locations in the ground). Once the temperature response can be calculated for the BHE array at a particular time, it can be simply multiplied by the ground load to get the ground thermal resistance.

6.2.6.1 Determining Borehole Thermal Interaction with Eskilson's Classic *g*-functions

First, some more background. Eskilson (1987) was the first, and most well known, to develop dimensionless temperature response factors for various BHE field patterns. It is important to realize that Eskilson's g-functions pertain to the ground thermal resistance only; elements of BHEs, such as fluid, pipe, and grout, were not considered. Eskilson used a numerical (finite difference) computer model (briefly described in Chapter 5) to superimpose solutions in space for numerous rectangular grid patterns, circular patterns, open rectangular patterns, and other miscellaneous patterns. For example, he began with a 2 × 1 multi-borehole pattern, then 2 × 2, 2 × 3, 2 × 4, and so on, up to a 10 × 12 borehole grid pattern. For each borehole field pattern, there is actually a family of curves representing varying borehole spacing to depth ratio (B/H ratio, where B is the borehole-to-borehole spacing and H is the borehole depth). The borehole-to-borehole spacing (B) is fixed for a particular curve, and the borehole radius to borehole depth ratio (r_b/H) is 0.0005.

An example family of B/H curves for a large 10 × 10 borehole grid is shown in Figure 6.4. Notice that the g-function is reported as a function of $\ln(t/t_s)$. The B/H curves may be interpolated. For example, a GHX with 75 m deep boreholes spaced at 7.5 m apart would result in a B/H ratio of 0.1. The thermal interaction among boreholes as a function of their spacing is evident from analysis of the curves shown in Figure 4.8. For example, the temperature response

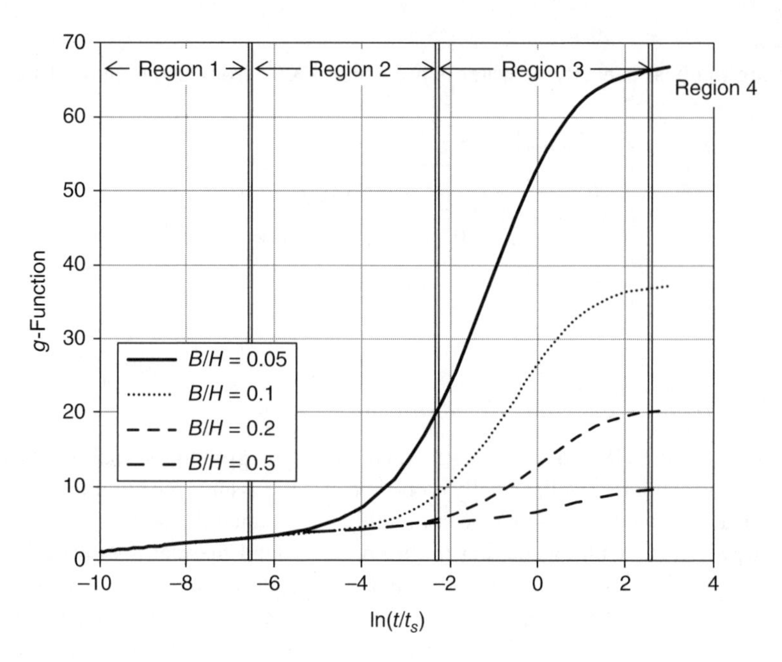

Figure 6.4 Dimensionless temperature response factors (g-functions) for a 10 × 10 borehole field at various borehole spacing/depth ratios (B/H ratios). These have been developed for $r_b/H = 0.00009$

of a 10 × 10 borehole field at $\ln(t/t_s)$ would be approximately doubled for a B/H ratio of 0.05 compared with a B/H ratio of 0.1.

It is important to note that four different thermal regions can be identified on a *g-function* curve. Referring to Figure 6.4, these four regions are described as follows:

- *Region 1* occurs at the lowest values of $\ln(t/t_s)$, and is where all curves are linear and fall on top of one another. Here, the heat transfer can be considered to be one-dimensional in the radial direction around single boreholes, and boreholes have not started thermally to interact with one another. Note this region would also describe the temperature response of a single borehole.
- *Region 2* demarks the time at which boreholes begin thermally to interact with one another. The actual time is specific to a particular *B/H* ratio and borehole field configuration. As can be seen in Figure 6.4, the second region starts at $\ln(t/t_s) \approx -6.5$, where the specific g-function curves diverge from the g-function of a single borehole. In the second region, borehole-to-borehole thermal interaction increases with increasing time.
- *Region 3* starts when $t > 0.1t_s$ (i.e., $\ln(t/t_s) > -2.3$), and represents the time when heat transfer near boreholes becomes two-dimensional (radial and axial). As can be seen in Figure 6.4, the onset of this region is marked by a change in slope of the g-function curve; the volume of Earth in which the BHEs are placed has changed temperature enough for heat conduction to proceed in the vertical direction down below the storage volume. At $\ln(t/t_s) = 0$, t is equivalent to t_s, meaning that the system is approaching steady state. As can be seen in Figure 6.4, $\ln(t/t_s) = 0$ represents the approximate time where the g-function is at 90% of its true steady-state value.

- Finally, *Region 4* is characterized by a plateau where g-functions reach their 'steady-state' values at $\ln(t/t_s) \approx 2.5$.

To summarize, Eskilson (1987) converted the temperature response of specific borehole field patterns to a set of non-dimensional temperature response factors, called g-functions. The g-function allows the calculation of the temperature change at the borehole wall in response to a step heat input for a time step. Once the response of the borehole field to a single step heat pulse is represented with a g-function, the response to any arbitrary heat rejection/extraction function can be determined by devolving the heat rejection/extraction into a series of step functions, and superimposing the response to each step function.

6.2.6.2 Determining Borehole Thermal Interaction by Spatial Superpositioning of Analytical Finite Line Sources Using the Leaky Well Function

We are now at the final step before we can use the GHX length equation. By space-superpositioning of finite line sources calculated with the leaky well function (see Chapter 5.5.2.2.3.), we have the power to approximate Eskilson's g-functions for any arbitrary borehole field pattern (i.e., we are not necessarily limited to the prescribed geometries of Eskilson). The leaky well function has been incorporated into the suite of GHX Design Tools found on the book companion website, and provides a computationally efficient means of determining temperature response factors in thermally interacting boreholes.

Spatial superpositioning involves calculation of the dimensionless temperature response (g) in each borehole due to all the others at a particular time. Thus, the dimensionless temperature response of a particular borehole (g_b) is the sum of the influence from all other boreholes:

$$g_{bx,y} = \sum_{i=1}^{xbores} \sum_{j=1}^{ybores} g_{i,j} \tag{6.9}$$

where i and j are indices, and *xbores* and *ybores* are the number of boreholes in the x and y direction respectively. The formulation for g is now expressed with the leaky well function notation as

$$g\left(\frac{t}{t_s}, \frac{r}{H}\right) = \frac{1}{2} W(u, B) \tag{6.10}$$

With Equation (6.10), g can readily be calculated for a particular borehole as a function of distance from other boreholes, H, t, and t_s. Note that the main thermal effect on a particular borehole will be due to its immediate neighbors. If r is too large, the thermal response is simply zero.

The spatial superpositioning method as applied here is illustrated in Figure 6.5. For the case of the upper left corner borehole indexed at $(i, j) = (1, 1)$, Figure 6.5 shows the radial distances from all other boreholes in a 4×4 grid pattern for the borehole at position $(i, j) = (2, 3)$. As you can see, this would not be easy to calculate by hand for large borehole fields. We could take advantage of borehole field symmetries and reduce the number of calculations considerably,

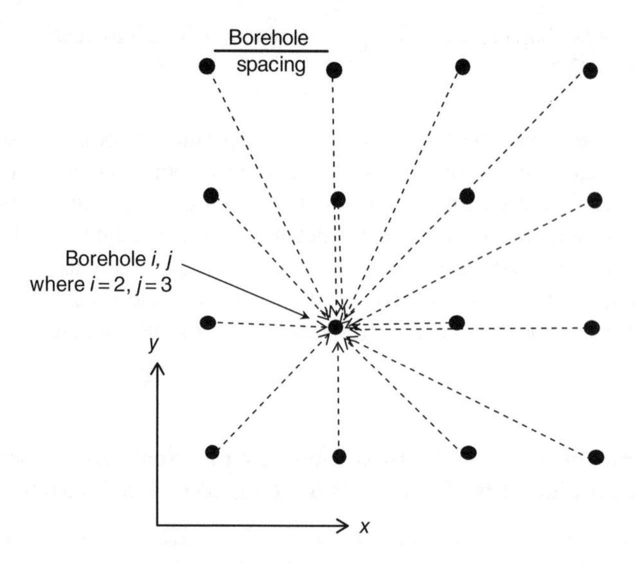

Figure 6.5 Illustration of spatial superpositioning, where the temperature response at a particular borehole is influenced by all others. Sometimes that influence is zero. The particular borehole pointed out here is at position $(i, j) = (2, 3)$ in a 4×4 grid pattern, with $(i, j) = (1, 1)$ at the upper left-hand corner

but it's much easier to set these calculations up in a nested looping structure in a computer program, and let a computer do the work.

Finally, the g-function for the entire borehole field at a particular time is determined by summing the calculated g values for all individual boreholes ($g_{b\ x,y}$) as described above, and dividing by the number of boreholes (NB):

$$g_{field\ pattern} = \frac{\sum g_{b\ x,y}}{NB} \tag{6.11}$$

6.3 Vertical GHX Simulation

The process of *simulation* refers to modeling the performance of a real-world process over time. Regarding GHXs, *simulation* refers to the calculation of the outlet fluid temperature of the GHX and the heat pump energy consumption as a function of time. Some approaches examine GHX performance at monthly intervals, others on an hourly basis. Thus, monthly or hourly building loads are needed for these simulations. Knowledge of the GHX length is also required.

There are two basic approaches that could be applied to the modeling of multi-borehole vertical arrays: (i) numerical and (ii) temperature-response factor.

A fully numerical method, where an entire borehole field is discretized in space and time, is computationally prohibitive owing to wide-ranging time and space scales that need to be modeled. In other words, such a model would have to capture transients on the scale of hours to years, on space scales of millimeters up to hundreds of meters. The TRNSYS energy simulation program uses the so-called duct ground storage (DST) model (Hellström, 1991), which incorporates both numerical and analytical solutions, to calculate the outlet fluid temperature of a GHX as a function of time. The DST model assumes that boreholes are placed uniformly within a cylindrical storage volume of Earth. Convective heat transfer within the pipes and conductive heat transfer to the storage volume are modeled, and the temperature of the surrounding ground volume is calculated from three parts: a global temperature, a local solution, and a steady-flux solution. The global and local problems are discretized using a radial-axial mesh similar to that described above for Eskilson's (1987) model, and then solved with the use of an explicit finite difference method. The steady-flux solution is obtained analytically, and the resulting temperature is then calculated using the method of superposition. The exact locations of boreholes are not considered in the DST model.

A temperature-response factor approach essentially is a precalculated, non-dimensional set of temperature responses of a borehole field. Yavuzturk and Spitler (1999) have extended the work of Eskilson (1987) to develop a g-function-based routine that has been incorporated into TRNSYS and other building energy simulation programs. The previously described calculation of T_f can be generalized for n time steps as follows:

$$T_f = T_g - \sum_{i=1}^{n} \frac{\left(Q'_i - Q'_{i-1}\right)}{2\pi k_g} g\left(\frac{t_n - t_{i-1}}{t_s}, \frac{r_b}{H}, \frac{B}{H}\right) - Q_n R_b \tag{6.12}$$

where Q' is the step heat pulse (in W·m^{-1}), and subscript g denotes ground. Note the syntax for the g-function: it is a function of t_s, r_b/H, and B/H, as before, and it multiplies the superimposed load over the associated time interval.

Equation (6.12) can be implemented on a monthly or hourly basis. A mathematical model of heat pump performance must be included in order to track the heat pump COP and ground load with time-varying fluid temperature. Such approaches typically involve curve-fit correlations with heat pump COP vs. entering fluid temperature.

Solving Equation (6.12) can be computationally intensive if the number of time steps is large, which would be true for hourly simulations. This is because at each new time step n, as time marches forward, the ground loading history needs to be updated. Thus, as the simulation progresses, the computational speed becomes progressively slower. To improve computational speed while maintaining acceptable accuracy, load aggregation algorithms are typically employed, particularly in hourly simulations. These algorithms are based on the premise that more recent ground loads have a more significant impact on the current mean fluid temperature than less recent ground loads.

6.4 Hybrid Geothermal Heat Pump Systems

We saw the benefit of adding a supplemental electric resistance element to a GHP in Chapter 6 (Section 6.2.4.2) in reducing the heat pump load. Now, we will look at offsetting loads on the 'other side' of the heat pump, namely the ground loads, in what we will refer to as hybrid GHP systems.

Hybrid GHP systems have received considerable attention in recent years (e.g., Hackel, 2008) because they have been shown significantly to improve the economics of GHP systems. Hybrid GHP systems couple a supplemental heat extraction or rejection subsystem to a conventional GHP system to handle some portion of the building or the ground loads, and as such, permit the use of a smaller, lower-cost GHX. Hybrid GHPs are especially effective in applications that have large peak loads and/or have highly imbalanced loads over the year (i.e., heavily heating-dominated or heavily cooling-dominated buildings). In cooling-dominated buildings, the most common type of supplemental component to offset ground loads is a closed-loop, dry fluid cooler. In heating-dominated buildings, boilers or solar thermal collectors are commonly used as the supplemental component to offset ground loads. In both cooling- and heating-dominant applications, other supplemental components are possible, such as shallow ponds or pools, or shallow horizontal GHX systems.

Hybrid GHP systems are more complex in their design than conventional GHP systems owing to the transient nature of the supplemental component. Further, recent research on hybrid GHPs identifies more than one method to design a hybrid GHP. For example, Chiasson and Yavuzturk (2009a, 2009b) describe a method for designing hybrid GHP systems based on annual ground load balancing. Xu (2007) and Hackel *et al.* (2009) describe hybrid GHP system design based on lowest life-cycle cost, while Kavanaugh (1998) describes a method based on designing the GHX for the non-dominant load, and the hybrid component for the balance of the load. Cullin and Spitler (2010) describe yet another method based on minimizing first cost of the system, while designing the GHX to supply both the minimum and maximum design entering heat pump fluid temperature over the life cycle of the system.

Here, we will adopt an approach to hybrid GHP system design such that ground loads are balanced, or in other words $\dot{q}_a \cong 0$. This approach is based on the premise that the supplemental component can operate year round, either loading or unloading thermal energy to/from the ground when the differential temperature between the GHX and the hybrid source/sink is favorable. Thus, the GHX is used for diurnal and seasonal thermal energy storage. With this approach, applications that make good candidates for hybridization are those with unbalanced annual ground loads. How ‘unbalanced’ do the ground loads need to be to consider a hybrid design? The answer to that question is an economic one. Generally speaking, when one annual load relative to the other is at least 2 to 3 (i.e., the annual heating load relative to the annual cooling load, or vice versa), then the size (and therefore cost) of the GHX can become prohibitively large, and hybrid options become more viable.

The approach of hybrid GHX design to balance ground loads is essentially a surrogate for an economic optimization, based on the premise that the life-cycle cost is mostly wrapped up in the capital cost of the GHX.

6.5 Modeling Vertical GHXs with Software Tools

Manual solution of the foregoing equations is not practical owing to the iterative and tedious nature of the process. Thus, a number of commercially available software packages have been developed for GHX design.

The foregoing algorithms and equations have been implemented into the suite of GHX tools (found on the book companion website) for use in GHX design problems. Two basic options are available for vertical GHX modeling based on the available loads, as shown in Figure 6.6.

The most basic option for vertical GHX modeling is if the user only has peak hour loads available and estimates of the monthly and annual load factors for heating and cooling. This option is suitable for single-zone residential applications, or for preliminary GHX sizing in other applications, where the user is only interested in the GHX size.

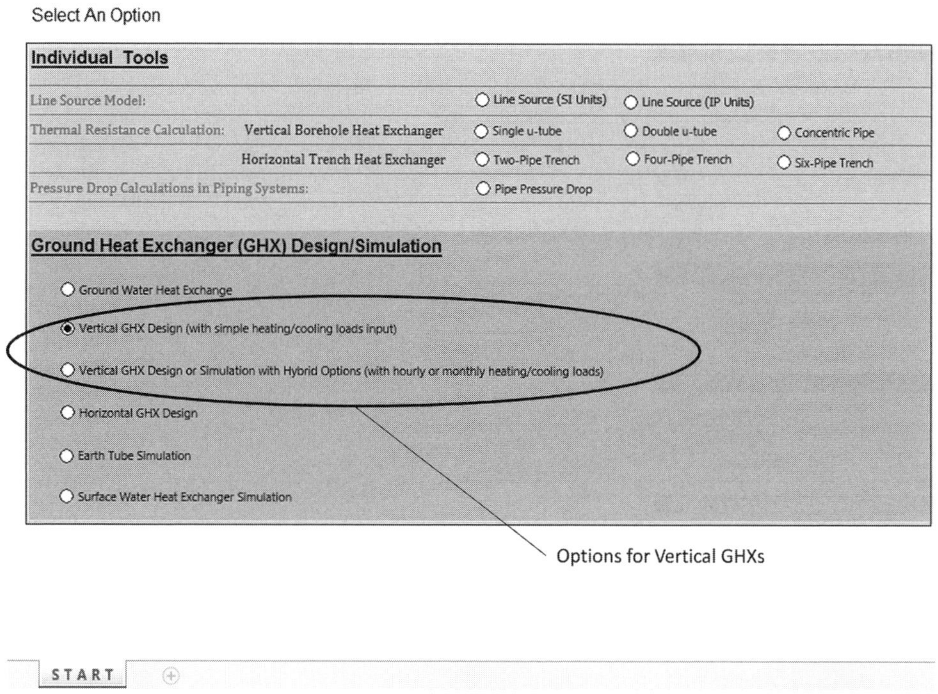

Select An Option

Individual Tools

Line Source Model:		Line Source (SI Units)	Line Source (IP Units)	
Thermal Resistance Calculation:	Vertical Borehole Heat Exchanger	Single u-tube	Double u-tube	Concentric Pipe
	Horizontal Trench Heat Exchanger	Two-Pipe Trench	Four-Pipe Trench	Six-Pipe Trench
Pressure Drop Calculations in Piping Systems:		Pipe Pressure Drop		

Ground Heat Exchanger (GHX) Design/Simulation

- Ground Water Heat Exchange
- Vertical GHX Design (with simple heating/cooling loads input)
- Vertical GHX Design or Simulation with Hybrid Options (with hourly or monthly heating/cooling loads)
- Horizontal GHX Design
- Earth Tube Simulation
- Surface Water Heat Exchanger Simulation

Options for Vertical GHXs

START

Figure 6.6 Screen capture of the START worksheet of the companion GHX tools highlighting vertical GHX modeling options

The more advanced option is if the user has hourly or monthly loads available. This allows more refinement in modeling the transient loading on the ground, and thus allows more accurate GHX design length calculation. In addition, GHX performance simulation is possible for both standalone GHX systems or for hybrid solar or fluid cooler GHP systems. The output results include monthly heat pump entering fluid temperatures and monthly heat pump energy use. If design options are selected, GHX size and hybrid component size (if applicable) are calculated. Use of the vertical GHX software tools is described in the subsections that follow.

6.5.1 Vertical GHX Modeling with Basic Load Input

Constructing the Model. Selection of the option button *Vertical GHX Design (with simple heating/cooling loads input)* launches two worksheets, the first of which is shown in Figure 6.7. The input fields on the *Vertical GHX* worksheet shown in Figure 6.7 are divided into four groups: (1) ground parameters, (2) heat pump details, (3) load information, and (4) supplemental ground load information for application to hybrid GHX systems.

In the *Ground Parameters* section of the *Vertical GHX* worksheet, all information is entered regarding the ground thermal properties, in addition to the specific heat of the GHX fluid. A dropdown box allows options for entering the borehole thermal resistance; this

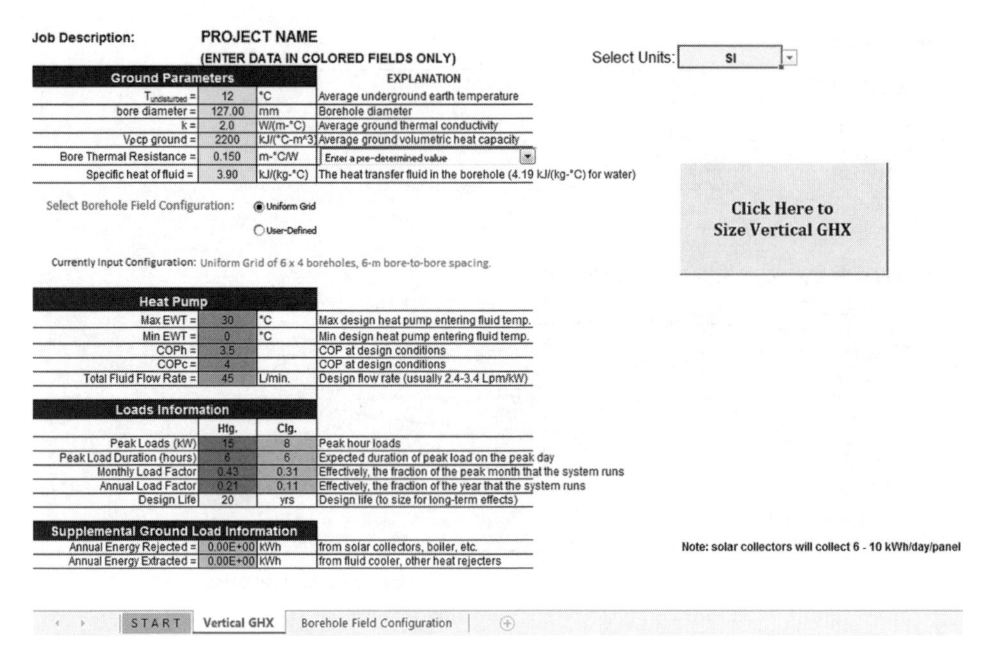

Figure 6.7 Screen capture of the *Vertical GHX* worksheet of the companion GHX tools. This *Vertical GHX* worksheet is for use with basic load input

value can be entered either directly or calculated by selecting *single U-tube*, *double U-tube*, or *concentric pipe* options from the dropdown box, which launches a Visual Basic macro to display the appropriate worksheet for borehole thermal resistance calculation as described in Chapter 5.

Details of the borehole field configuration are also handled in the *Ground Parameters* section of the *Vertical GHX* worksheet. Two suboptions are available: (i) a uniform grid and (ii) a user-defined grid. Selection of *Uniform Grid* launches a Visual Basic macro to provide the display shown in Figure 6.8. Here, input data are simply the number of bores in the x- and y-directions and the borehole-to-borehole spacing. Thus, the input data shown in Figure 6.8 would be for a 2 × 2 grid pattern with a borehole-to-borehole spacing of 6 m.

Selection of the *User-Defined* borehole field configuration launches a Visual Basic macro to provide the display shown in Figure 6.9. An interactive grid is displayed with 100 × 100 cells in the x–y direction, where users can place boreholes in any arbitrary pattern. Input data include the number of boreholes to be entered, along with the grid spacing in the x- and y-directions. Boreholes can be denoted with any text character. Thus, the input data shown in Figure 6.9 would be for six boreholes in an open-circular pattern. Note the cell size is 2 m. Internally, the model determines the radial distance from one borehole to another for use in the g-function calculation for any arbitrary borehole field.

Note that in the *Borehole Field Configuration* worksheet, input data for groundwater flow are also included. Selecting the option button *INCLUDE effects of groundwater flow* launches a Visual Basic macro to display groundwater flow input parameters as shown in Figure 6.10. Internally, these data are used to calculate the groundwater g-function.

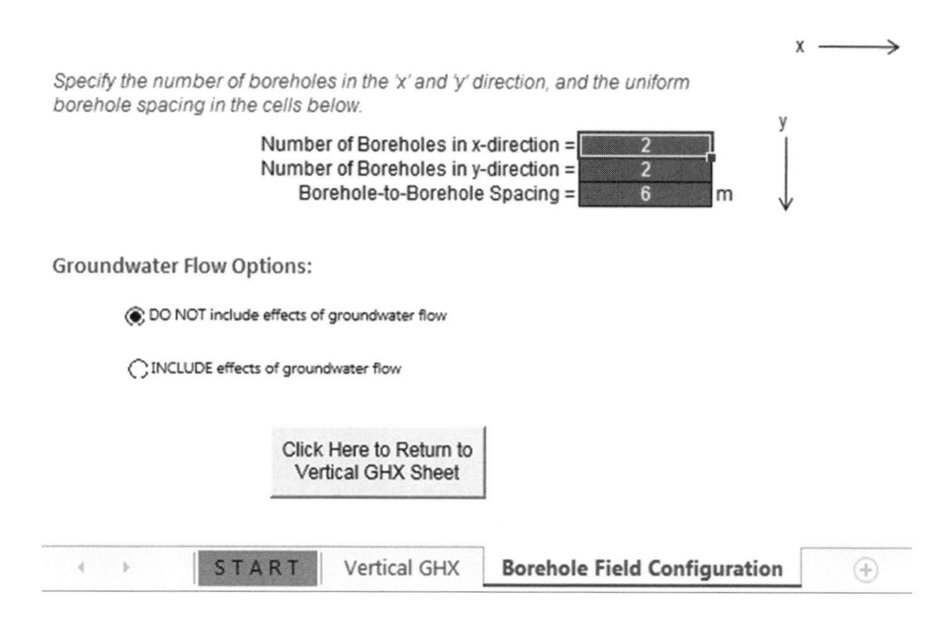

Figure 6.8 Screen capture of uniform borehole field grid input data on the *Borehole Field Configuration* worksheet of the companion GHX tools. This *Borehole Field Configuration* worksheet is for use with vertical GHX modeling

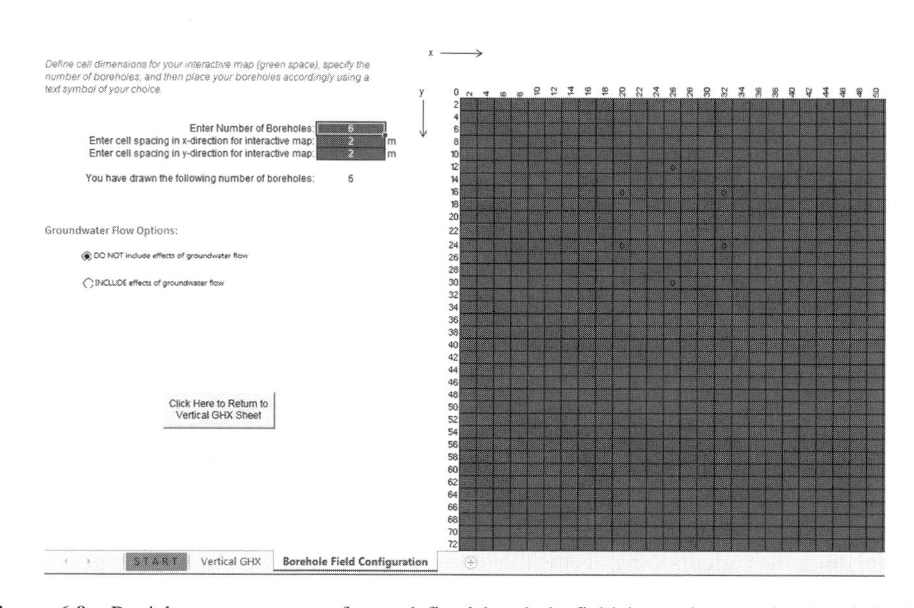

Figure 6.9 Partial screen capture of user-defined borehole field input data on the *Borehole Field Configuration* worksheet of the companion GHX tools. This *Borehole Field Configuration* worksheet is for use with vertical GHX modeling

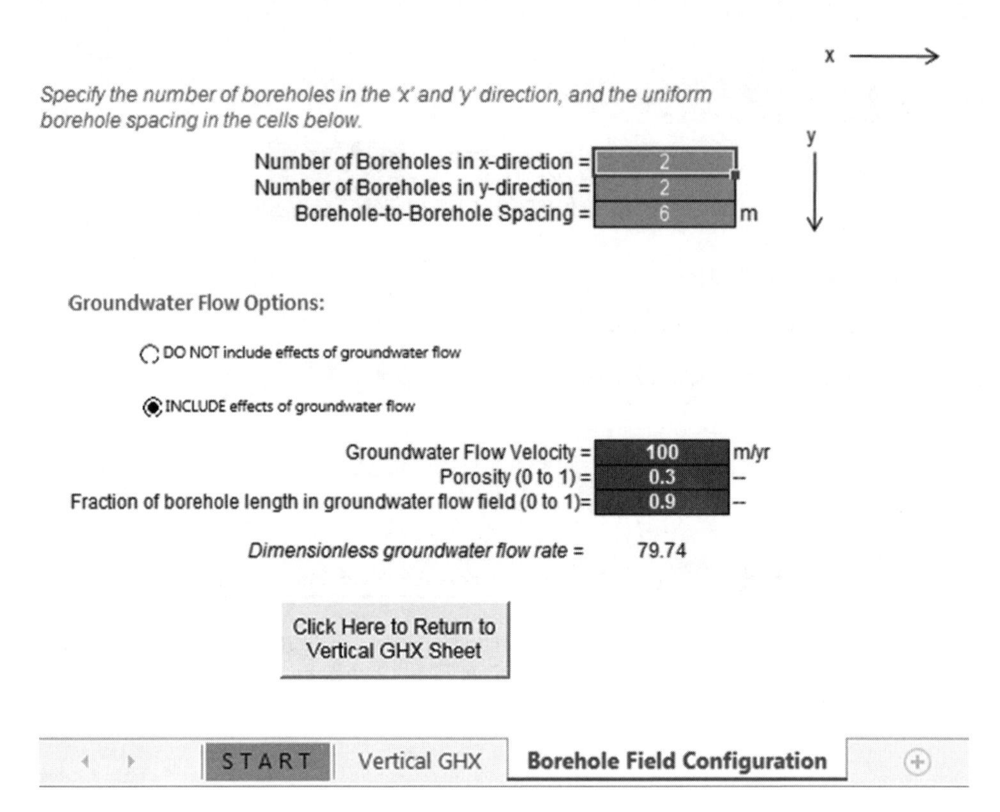

Figure 6.10 Screen capture of uniform borehole field grid input data with the effects of groundwater flow included on the *Borehole Field Configuration* worksheet of the companion GHX tools. This *Borehole Field Configuration* worksheet is for use with vertical GHX modeling

Returning to the *Vertical GHX* worksheet, in the *Heat Pump* section, the design limits of the heat pump fluid entering temperature are entered, along with the associated heat pump COP values at these conditions. As mentioned, these are important constraints in controlling the size of the GHX. The heat pump design flow rate is also entered here. Typical design flow rates are suggested in the worksheet, but users may enter the value appropriate for their design.

In the *Loads Information* section of the *Vertical GHX* worksheet, users enter the peak loads and the monthly and annual load factors. Methods for determining these values have been discussed in Chapter 2. The *Design Life* is based on the life cycle of the GHP project.

In the *Supplemental Ground Loads Information* section of the *Vertical GHX* worksheet, users enter estimates of annual thermal energy handled by a supplemental component. This allows a crude means to size hybrid GHXs. A more detailed approach is described in the following section of this chapter.

Performing the Calculations. As mentioned at the beginning of our discussion of the GHX design length equation, its solution is iterative; the ground thermal resistances depend on the borehole depth *(H)*, but *H* is unknown, and the main goal of the GHX design length equation is to determine the total GHX length. Also recall that the GHX design length equation is solved twice: once for the length required to meet the heating loads, and once for the length required to meet the cooling loads. It is rare that the two calculated lengths will be equal owing to the numerous factors that affect the ground loads. Therefore, users should select the greater length as the design choice. If the lengths are unacceptably large, the GHX can be redesigned by changing some parameters within engineering reason, or by designing a hybrid GHX.

Mouse-clicking on the command button *Click Here to Size Vertical GHX* executes a Visual Basic for Applications (VBA) code that makes use of a single-variable optimization method (i.e., a golden section search) for iterative solution of the vertical GHX design length equation [Equation (6.2)]. The VBA optimization method automatically adjusts the borehole depth in order to minimize the objective function, which is defined as the difference between the input length (determined by $NB \times H$) and the maximum calculated length (whether for heating or cooling). The results of a vertical GHX size calculation are printed to a message box on the screen. Figure 6.11 is a screen capture of the results related to the input data shown in Figure 6.7.

As shown in Figure 6.11, a summary of the input borehole field configuration is echoed in the results box. The calculated depth of each borehole is 69 m, resulting in a total length requirement of 276 m for boreholes. Other useful information is provided, such as the length driven by the non-dominant load. For this example, cooling is the non-dominant load, and the required length gives users some sense of the load imbalance on the GHX. In this example, the required borehole length for cooling is about 61% of that required for heating. This imbalance suggests possible consideration of a supplemental heating element to balance the ground loads.

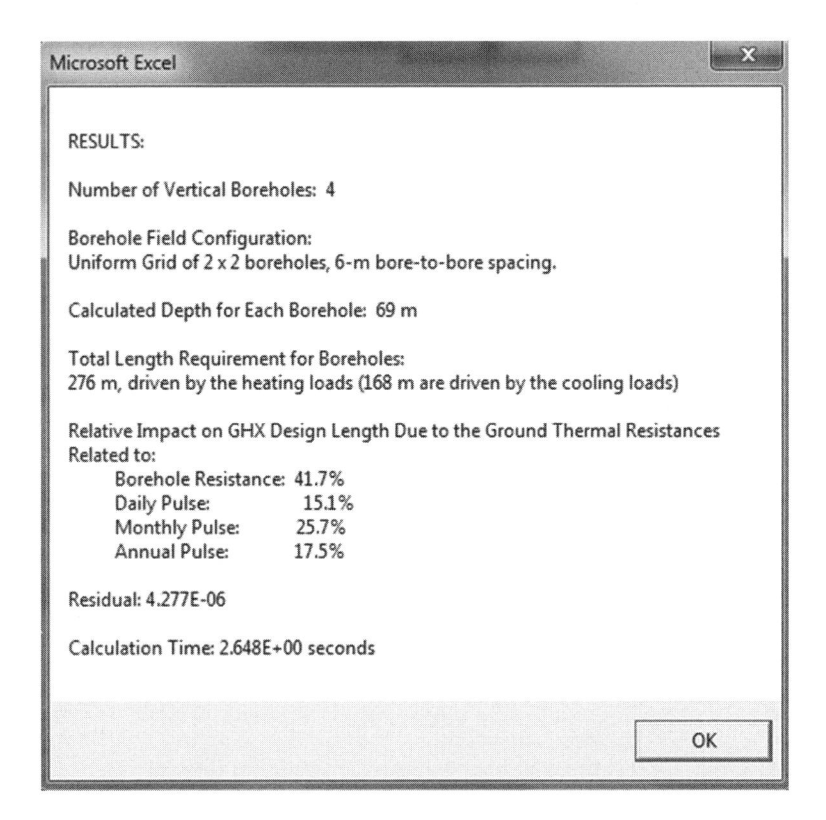

Figure 6.11 Screen capture of the calculated results related to the input screen capture shown in Figure 6.7

Also as shown in Figure 6.11, the results box displays the relative thermal resistances of the borehole, along with the ground thermal resistances due to the daily, monthly, and annual loads. As seen in this example, the largest thermal resistance is due to the borehole. Thus, some reconsideration of the borehole completion is warranted.

6.5.2 *Vertical GHX Modeling with Hourly or Monthly Load Input*

Constructing the Model. Selection of the option button *Vertical GHX Design or Simulation with Hybrid Options (with hourly or monthly heating/cooling loads)* launches initially four worksheets, the first of which is shown in Figure 6.12. A fifth worksheet for entering load information is launched when the user selects 'paste hourly loads' or 'paste monthly loads'. Additional worksheets are launched if the user selects hybrid options.

A review of Figure 6.12 reveals that the input fields on the *Vertical GHX* worksheet for use with hourly of monthly loads are divided into four similar groups, as before in Figure 6.7, for use with basic loads. These groups are: (1) ground parameters, (2) heat pump performance data, (3) load information, and (4) hybrid options. In addition, the currently computed *g*-function curve is shown.

The *Ground Parameters* section of the *Vertical GHX* worksheet is the same as that described previously in Section 6.5.1, and thus will not be repeated here. One notable difference is that the borehole depth is entered on the *Borehole Field Configuration* worksheet for use in GHX simulations.

In the *Heat Pump Performance Data* section, a command button brings the user to the *Heat Pump Input* worksheet. In that worksheet, users enter the heat pump entering fluid temperature and the corresponding COP as shown in Figure 4.22. The heat pump design flow rate is also

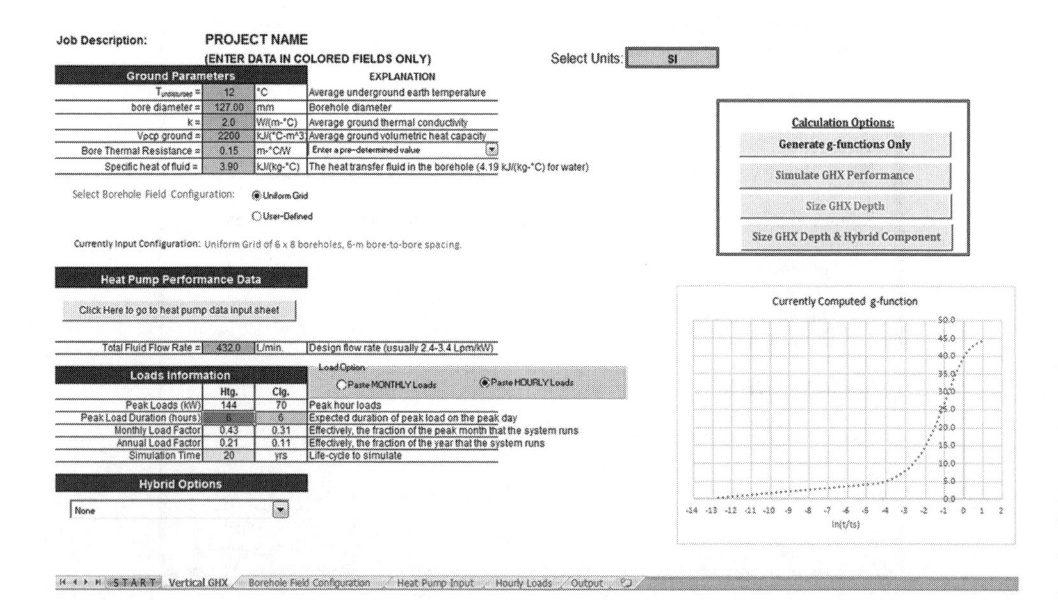

Figure 6.12 Screen capture of the *Vertical GHX* worksheet of the companion GHX tools. This *Vertical GHX* worksheet is for use with hourly or monthly load input

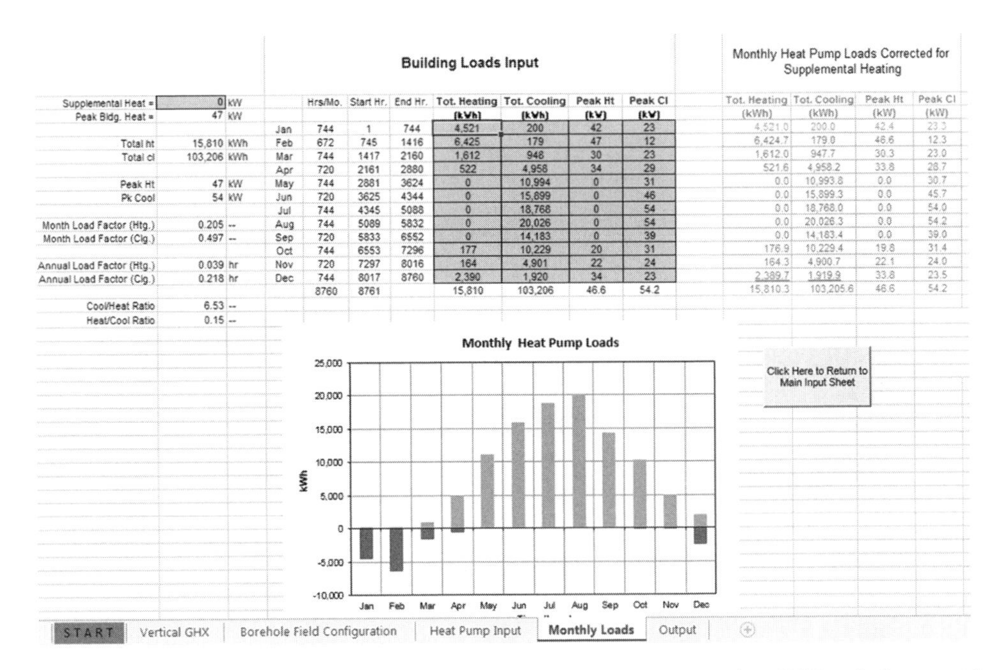

Building Loads Input

	Hrs/Mo.	Start Hr.	End Hr.	Tot. Heating (kWh)	Tot. Cooling (kWh)	Peak Ht (kW)	Peak Cl (kW)
Jan	744	1	744	4,521	200	42	23
Feb	672	745	1416	6,425	179	47	12
Mar	744	1417	2160	1,612	948	30	23
Apr	720	2161	2880	522	4,958	34	29
May	744	2881	3624	0	10,994	0	31
Jun	720	3625	4344	0	15,899	0	46
Jul	744	4345	5088	0	18,768	0	54
Aug	744	5089	5832	0	20,026	0	54
Sep	720	5833	6552	0	14,183	0	39
Oct	744	6553	7296	177	10,229	20	31
Nov	720	7297	8016	164	4,901	22	24
Dec	744	8017	8760	2,390	1,920	34	23
	8760	8761		15,810	103,206	46.6	54.2

Monthly Heat Pump Loads Corrected for Supplemental Heating

Tot. Heating (kWh)	Tot. Cooling (kWh)	Peak Ht (kW)	Peak Cl (kW)
4,521.0	200.0	42.4	23.3
6,424.7	179.0	46.6	12.3
1,612.0	947.7	30.3	23.0
521.6	4,958.2	33.8	28.7
0.0	10,993.8	0.0	30.7
0.0	15,899.3	0.0	45.7
0.0	18,768.0	0.0	54.0
0.0	20,026.3	0.0	54.2
0.0	14,183.4	0.0	39.0
176.9	10,229.4	19.8	31.4
164.3	4,900.7	22.1	24.0
2,389.7	1,919.9	33.8	23.5
15,810.3	103,205.6	46.6	54.2

Figure 6.13 Screen capture of the *Monthly Loads* worksheet of the companion GHX tools for use with vertical GHX modeling

entered in this section. Typical design flow rates are suggested in the worksheet, but users may enter any value appropriate for their design.

In the *Loads Information* section of the *Vertical GHX* worksheet, users enter the duration of the peak load on the peak day. The *Design Life* is also entered, which is based on the life cycle of the GHP project. Selecting the option button *Paste MONTHLY Loads* launches the worksheet shown in Figure 6.13, while selecting the option button *Paste HOURLY Loads* launches the worksheet shown in Figure 6.14. Internally, the model works with monthly loads. Supplemental heating calculations may be done.

In the *Hybrid Options* section of the *Vertical GHX* worksheet, users have four modeling choices of hybridization of the GHP system:

- none,
- solar thermal recharging of the GHX,
- dry fluid cooler,
- nocturnal solar panel cooling.

Screen captures of the input data worksheets for solar hybrids and fluid cooler hybrids are shown in Figures 6.15 and 6.16 respectively.

Selection of any of the hybrid options also loads the *Weather Data* worksheet. The model expects average monthly weather data, such as the data described in Chapter 2.2. Weather data of importance to solar hybrids include: average daily solar radiation on horizontal, the ambient air temperature, and the wind speed. Weather data of importance to dry fluid cooler hybrids are limited to the ambient air temperature.

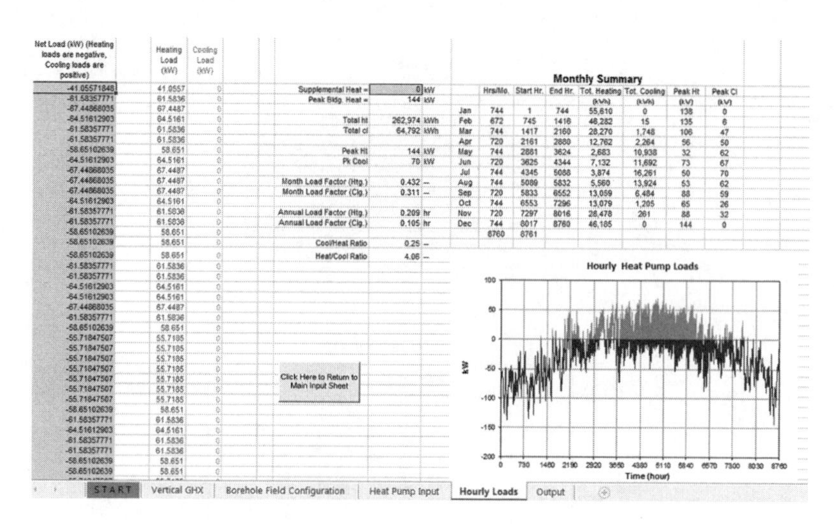

Figure 6.14 Partial screen capture of the *Hourly Loads* worksheet of the companion GHX tools for use with vertical GHX modeling

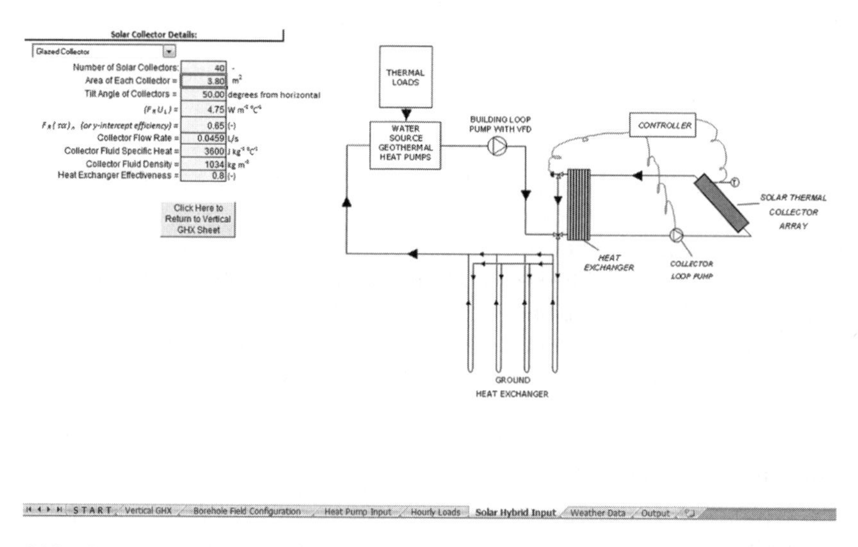

Figure 6.15 Screen capture of the *Solar Hybrid Input* worksheet of the companion GHX tools for use with vertical GHX modeling

Selection of *Solar thermal recharging of the GHX* as a hybrid option models solar energy storage in the GHX, and is applicable to heating-dominated buildings. As depicted in Figure 6.15, the solar collector loop is isolated from the GHX with a plate-type heat exchanger, modeled with a constant effectiveness, input by the user. The user also inputs properties of the collector fluid and the fluid flow rate. In practice, the collector fluid must be an aqueous fluid, typically with ~40–50% propylene glycol by volume, for adequate freeze protection. Also shown in Figure 6.15 is a controller, which is actually a differential temperature controller similar to that used in conventional thermal systems. When the solar collector temperature exceeds

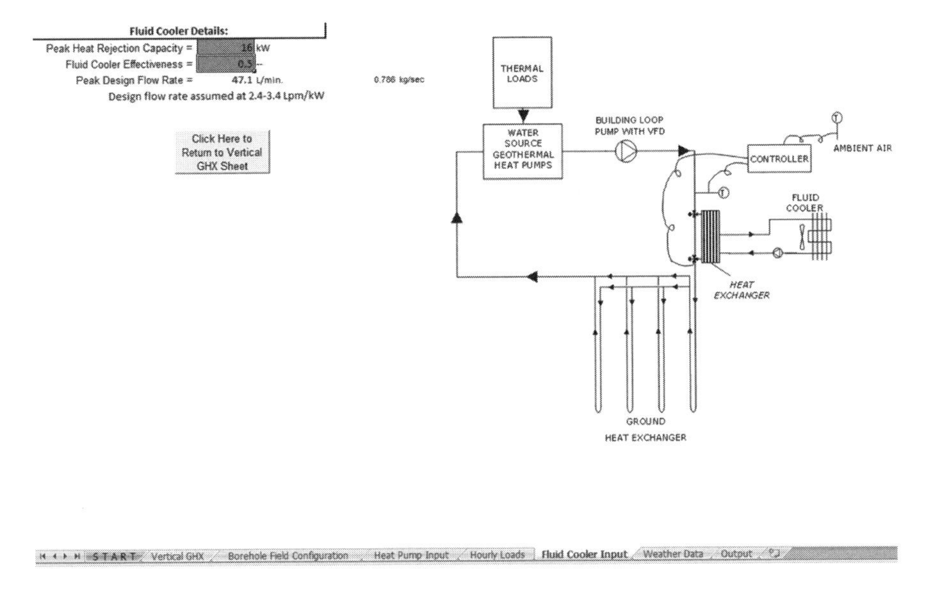

Figure 6.16 Screen capture of the *Fluid Cooler Input* worksheet of the companion GHX tools for use with vertical GHX modeling

the heat pump fluid exiting temperature by a differential set point, the solar collector loop pump is energized, and the control valve on the heat pump side of the heat exchanger is opened to allow the fluid exiting the heat pump to flow through the heat exchanger. The model assumes a differential set point of 5 °C.

The user also inputs the number of solar thermal collectors, the area of each collector, the tilt angle from the horizontal, and the efficiency parameters ($F_R U_L$ and $F_R(\tau\alpha)_n$), where F_R is the heat removal factor, U_L is the overall heat transfer coefficient for the collector (W·m^{-2}·K^{-1}), and $\tau\alpha$ is the transmittance-absorptance product. The parameters $F_R U_L$ and $F_R(\tau\alpha)_n$ are readily determined from collector efficiency tests, where the subscript n denotes normal incidence. The worksheet contains a dropdown box for users to select default values for unglazed, glazed, and evacuated tube collectors. More specifically, solar collectors are tested, rated, and certified by the Solar Rating and Certification Corporation (SRCC) according to industry-accepted standards, and test data for specific collectors may be found at: http://www.solar-rating.org/ratings/index.html

The collector heat removal factor (F_R) is corrected for the heat exchanger effectiveness and fluid mass flow rate after Beckman *et al.* (1977):

$$\frac{F'_R}{F_R} = \left[1 + \left(\frac{F_R U_L}{\dot{m} c_p}\right)\left(\frac{1}{\varepsilon} - 1\right)\right]^{-1} \tag{6.13}$$

where $\dot{m}$ is the collector fluid mass flow rate per unit area (kg·s^{-1}·m^{-2}), c_p is the specific heat of the solar collector fluid (J·kg^{-1}·K^{-1}), A is the collector area (m^2), and ε is the constant heat exchanger effectiveness (—). Note that Equation (6.13) is applicable for the smaller flow rate being on the solar collector side of the heat exchanger, which is a reasonable assumption for practical purposes.

Useful solar energy added to the GHX per month is determined using the solar utilizability method described by Duffie and Beckman (2013). That method is summarized in Appendix C.

Selection of *Dry fluid cooler* as a hybrid option models monthly heat removal from the GHX via rejection to the atmosphere, and is applicable to cooling-dominated buildings. As depicted in Figure 6.16, the fluid cooler loop is isolated from the GHX with a plate-type heat exchanger. The heat exchange effectiveness is modeled with a combined effectiveness of the plate heat exchanger and the fluid cooler liquid-to-air effectiveness. A differential control set point is used as described above, and is assumed to be 5 °C. Thus, heat rejection to the atmosphere will not take place unless the air temperature is 5 °C less than the exiting heat pump fluid temperature.

Selection of *Nocturnal solar panel cooling* as a hybrid option models heat removal from the GHX via convection losses to the atmosphere and nocturnal radiation to the sky from an unglazed solar collector, and is applicable to cooling-dominated buildings. All input data are the same as the data described above for solar thermal recharging of the GHX. Heat losses to the atmosphere and to the sky are assumed to occur after sunset hours only. The average monthly sky temperature is calculated in the *Weather Data* worksheet. The number of night-time hours per month is calculated from the sunset angle. A differential control set point is used as described above, and is assumed to be 5 °C. Thus, heat rejection to the atmosphere and sky will not take place unless the radiant panel temperature is 5 °C less than the exiting heat pump fluid temperature.

The efficiency (η) of the unglazed solar panel is modeled after Burch *et al.* (2004) as

$$\eta = \left(F_R(\tau\alpha)_n - 0.029\frac{s}{m}v_{wind}\right)\times IAM - \left(F_R U_L + 4.69\frac{W\cdot s}{m^3\cdot K}v_{wind}\right)\frac{(T_{in}-T_{air})}{q_{rad}} \tag{6.14}$$

where v_{wind} is the wind speed (in m·s^{-1}), *IAM* is an incidence angle modifier assumed at 0.95, q_{rad} is the long-wave thermal radiation (in W) from the panel to the sky, determined by Stefan–Boltzmann's law for an assumed panel emissivity of 0.90, T_{in} is the inlet fluid temperature to the collector (°C), and T_{air} is the air temperature (°C).

Performing the Calculations. Four calculation options are available as displayed in Figure 6.12 and shown in more detail in Figure 6.17. The first calculation option determines *g*-functions only for the borehole field geometry described in the input worksheets. The second calculation option simulates thermal performance for GHX and hybrid data entered. The third calculation option sizes the GHX depth, and the fourth calculation option sizes GHX depth and hybrid component size. In the second, third, and fourth calculation options, the calculations are done on a monthly basis, where the GHX fluid temperature is computed using Equation (6.12). The peak maximum and minimum heat pump entering fluid temperatures are determined by superimposing (over the average monthly load) the line source model with the peak load for the user-entered time duration. The ground loads and the heat pump energy consumption are determined using curve fits to the heat pump input data of COP vs. entering fluid temperature. Hybrid loads are added/subtracted to/from the ground loads.

Selection of the third and fourth calculation options (i.e., the sizing options) brings up a control sheet as shown in Figure 6.18. Users can start the calculations at any month of the year, based on the time of start-up of the GHP system. The target minimum and maximum heat pump entering fluid temperatures are also entered in this control sheet.

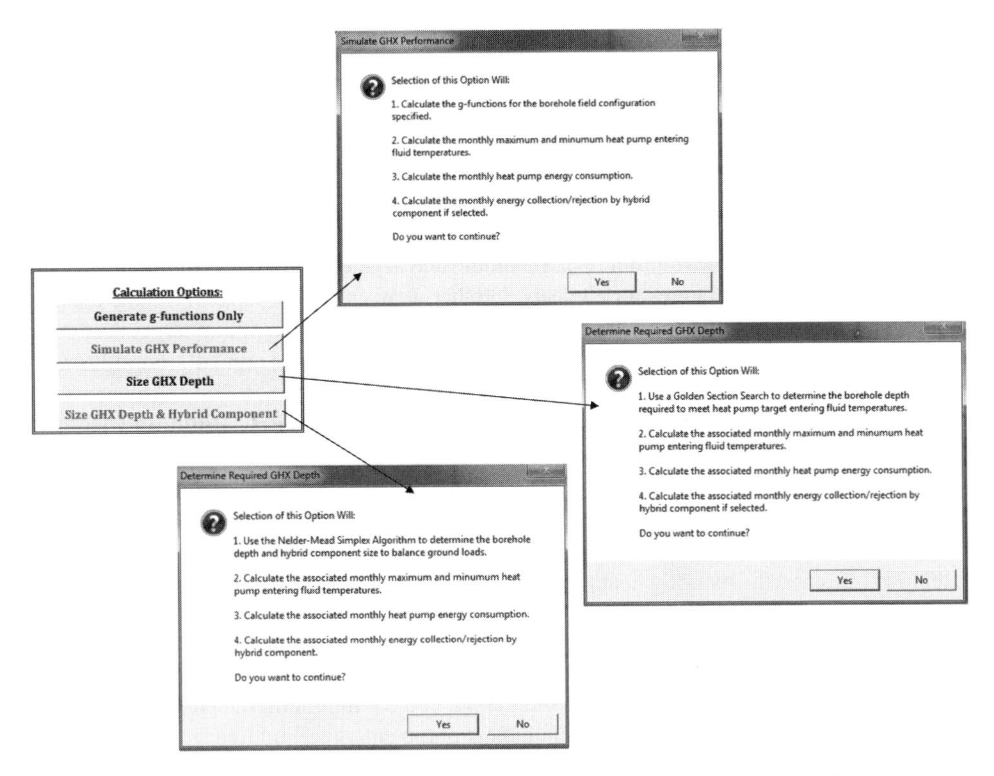

Figure 6.17 Screen capture of calculation options on the *Vertical GHX* worksheet of the companion GHX tools. This *Vertical GHX* worksheet is for use with hourly or monthly load input

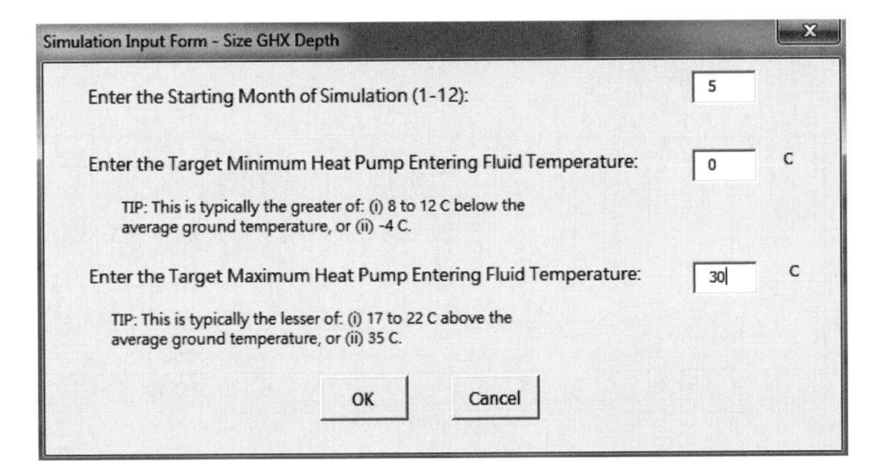

Figure 6.18 Screen capture of the input form for GHX and Hybrid GHX sizing options on the *Vertical GHX* worksheet of the companion GHX tools. This *Vertical GHX* worksheet is for use with hourly or monthly load input

With the third calculation option *Size GHX Depth*, a single-variable optimization routine (i.e., a golden section search) is used automatically to adjust the GHX depth such that the squared error is minimized between the calculated and the target heat pump minimum and maximum entering fluid temperatures at any time in the simulation.

With the fourth calculation option *Size GHX Depth & Hybrid Component*, a two-variable optimization routine (i.e., the simplex method of Nelder and Mead (1965)) is used automatically to adjust the GHX depth and hybrid component size such that the sum of squared error is minimized between the calculated annual minimum or maximum and the target heat pump minimum or maximum entering fluid temperatures for all years. In other words, the optimization routine attempts to size the GHX and hybrid components such that the minimum or maximum target heat pump entering fluid temperature is met each year, thus resulting in balanced annual ground loads.

Example 6.5 Sizing a Vertical GHX with Precalculated Hourly Building Loads
Determine the depth of boreholes for a GHX with input data as shown in Figure 6.12 for the hourly loads shown in Figure 6.14. Heat pump entering fluid temperature and the corresponding COP are as shown in Figure 4.22. The design minimum and maximum heat pump entering fluid temperatures are 0 °C and 30 °C respectively. Plot the monthly heat pump entering fluid temperatures and the monthly heat pump energy consumption, starting in January of year 1.

Solution
With the data entered in the Vertical GHX model, we select *Size GHX Depth* in the calculation options, and specify month 1 to start the simulation, a target minimum heat pump entering fluid temperature of 0 °C, and a target maximum heat pump entering fluid temperature of 30 °C.

The calculated depth of each borehole is 97.1 m, resulting in a total drilling length of 6 × 8 × 97.1 m = 4661 m. Plots of monthly heat pump temperatures and heat pump energy consumption are shown in Figures E.6.5a and E.6.5b.

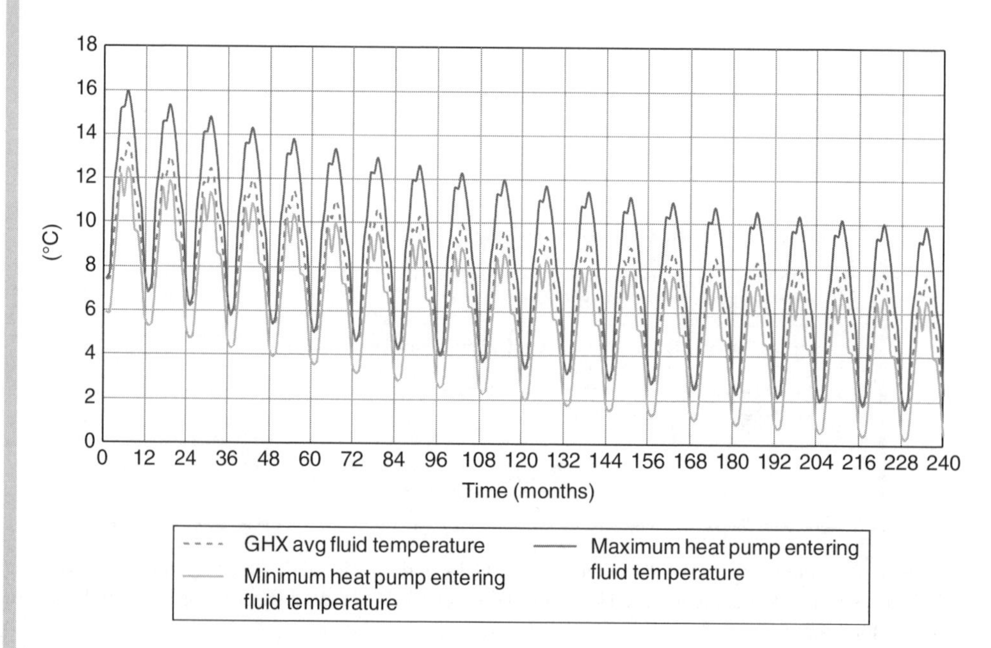

Figure E.6.5a Plot of monthly heat pump entering fluid temperatures for Example 6.5

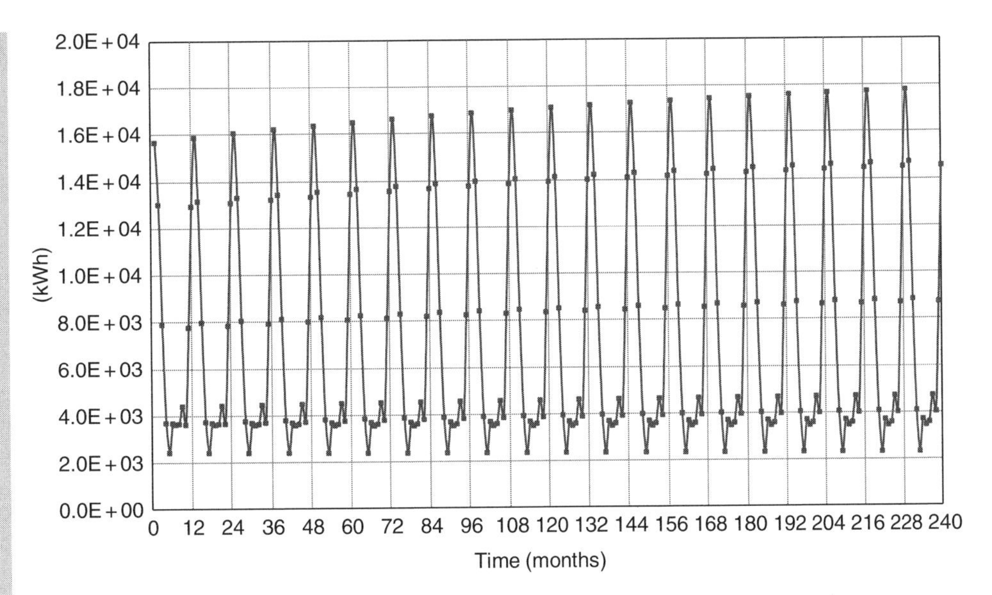

Figure E.6.5b Plot of monthly heat pump energy consumption for Example 6.5

Discussion: As seen in Figure E.6.5a, the optimizer sized the borehole depths to meet the-heat pump target temperatures. A review of Figure E.6.5a reveals a trend of decreasing heat pump entering temperatures over the years, typical of a heating-dominant building. A review of Figure E.6.5b shows a corresponding increase in monthly heat pump energy consumption as time progresses. Over the 20 year life cycle, the total energy consumption is 1.750E+6 kWh. Thus, the GHX size in this case is constrained by the heating loads; the heat pump temperatures do not even approach the target maximum temperature of 30 °C. This is due to the fact that the annual heating loads are 4× the annual cooling loads (see Figure 6.14), making this application a good candidate for hybrid design consideration.

Example 6.6 Simulating a Vertical GHX Performance with Precalculated Hourly Building Loads

Upon drilling and installation of the design in Example 6.5, the driller informs you that, owing to unforeseen drilling circumstances, the boreholes can only be drilled to 90 m – i.e., 7 m short of the target depth. Simulate the performance of this design change and recommend action for proceeding or not.

Solution

With the same Vertical GHX model as in Example 6.5, the borehole depth is changed to 90 m in the *Borehole Field Configuration* worksheet. We select *Simulate GHX Performance* and choose month 1 as the starting month.

The monthly heat pump temperatures are shown in Figure E.6.6.

Discussion: A review of Figure E.6.6 shows that the minimum heat pump entering fluid temperature drops slightly below the design target of 0 °C toward the end of the 20 year life cycle. Also, over the 20 year life cycle, the total energy consumption increases slightly to 1.764E+6 kWh. On their own, these changes negligibly impact the thermal performance

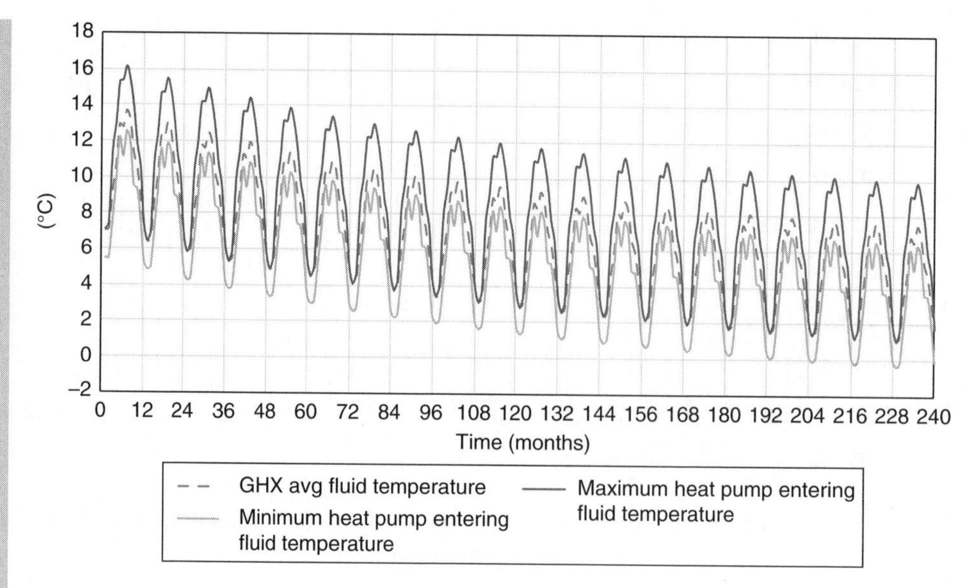

Figure E.6.6 Plot of monthly heat pump entering fluid temperatures for Example 6.6

of the GHX. However, of more concern is the continued downward trend of the temperature profile with time. Thus, in our next example, we shall consider a hybrid solar design.

Example 6.7 Sizing a Vertical Hybrid Solar GHX with Precalculated Hourly Building Loads

Determine the size (number of boreholes and depth) of a vertical GHX and the number of solar collectors for Example 6.5. Use the solar input data as shown in Figure 6.15. Monthly weather data are the data for Dayton, Ohio.

Solution

First, let's decrease the number of boreholes to 4 × 6 = 24, and observe the depth that the optimizer determines. Note that the optimizer will only adjust the borehole depth, not the number of boreholes, so we are thus assuming a reduction in GHX size of about 50%. If the resultant depth of boreholes is unacceptable, we will repeat the optimization process. We select *Size GHX Depth and Hybrid Component* and specify month 1 to start the simulation, a target minimum heat pump entering fluid temperature of 0 °C, and a target maximum heat pump entering fluid temperature of 30 °C.

The results box is shown in Figure E.6.7a. Plots of the monthly heat pump temperatures and energy consumption are shown in Figures E.6.7b and E.6.7c.

Discussion: Review of the results box shown in Figure E.6.7a reveals a quite acceptable drilling depth of 78.3 m, resulting in a total drilling length of 4 × 6 × 78.3 m = 1879 m. In Example 6.5, we had calculated a drilling length of 4661 m. Thus, this solar hybrid design results in a reduction in drilling length of about 60%! However, this design comes at a trade-off of the addition of 36 solar thermal collectors. We will evaluate the economics of these systems in a subsequent chapter.

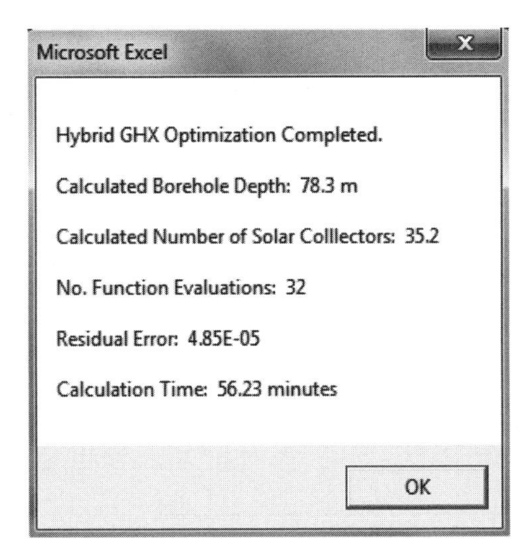
Microsoft Excel

Hybrid GHX Optimization Completed.

Calculated Borehole Depth: 78.3 m

Calculated Number of Solar Colllectors: 35.2

No. Function Evaluations: 32

Residual Error: 4.85E-05

Calculation Time: 56.23 minutes

OK

Figure E.6.7a Results message box for Example 6.7

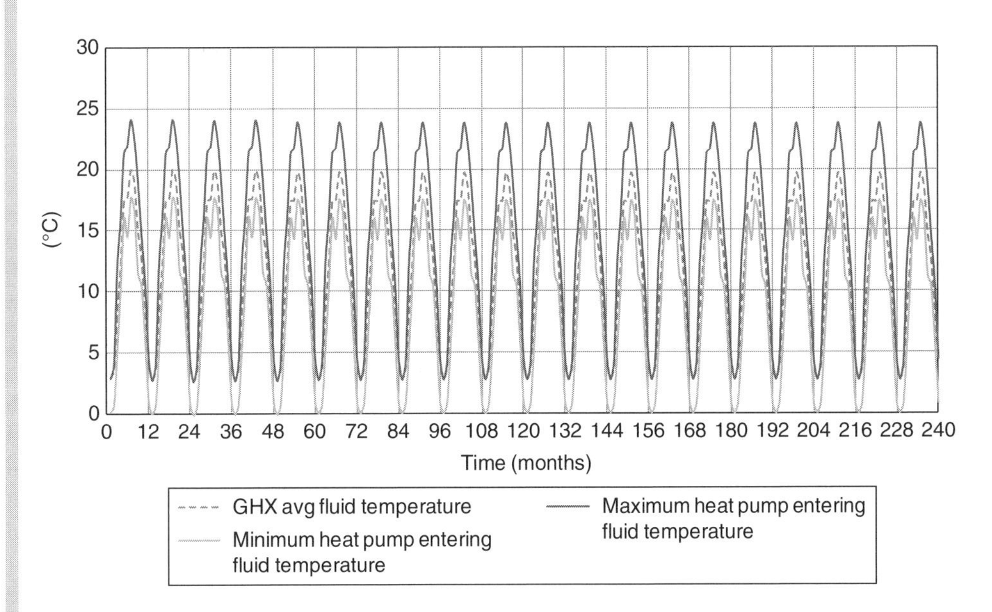

Figure E.6.7b Plot of monthly heat pump entering fluid temperatures for Example 6.7

Further analysis of this design can be seen by review of Figures E.6.7b and E.6.7c. Note the stable, periodic temperature profile shown in Figure E.6.7b, in contrast to the declining temperature profile shown in Figure E.6.5a. This stable, periodic profile suggests balanced ground loads and a 'right-sized' GHX to meet both the heating and cooling loads with the aid of solar thermal recharging. The stable, periodic temperature profile is also reflected in the monthly

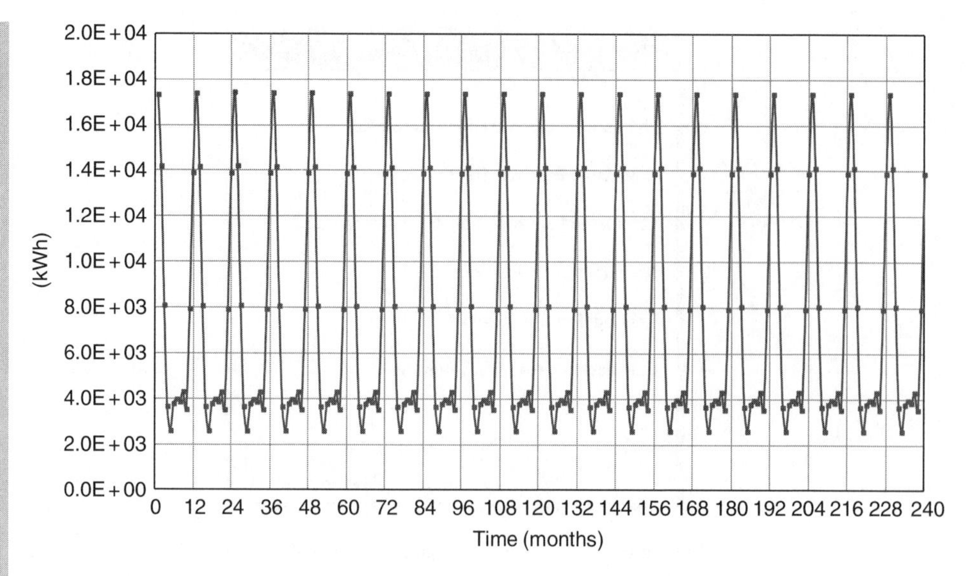

Figure E.6.7c Plot of monthly heat pump energy consumption for Example 6.7

heat pump energy consumption plot shown in Figure E.6.7c. Over the 20 year life cycle, the total energy consumption decreases relative to that in Example 6-5 to 1.741E+6 kWh.

6.6 Chapter Summary

This chapter extended the heat transfer theory of single vertical borehole heat exchangers to multiple vertical borehole arrays. In real applications with multiple vertical boreholes, design complexity arises owing to the thermal interaction among the boreholes and the periodic nature of the thermal loading.

This chapter addressed the fundamental task in the design of a vertical borehole GHX, which is properly to size the total number, depth, and spacing of the BHEs so that they provide fluid temperatures to the heat pump(s) within design limits over its lifetime. The so-called design length equation, and each parameter in the equation, was analyzed in detail. Further discussion of the *g*-function was presented as a computationally efficient means to determine the ground thermal resistance of multiple, finite-length, thermally interacting boreholes. The role of *g*-functions, as well as numerical methods, in the simulation of GHX thermal performance was described.

Hybrid GHP systems were described in this chapter as a means of reducing the necessary size of a vertical GHX. It was emphasized that there is more than one means to design a hybrid GHP system. In this chapter, we adopted the design approach of balancing annual ground loads.

Finally, this chapter discussed vertical GHX design and simulation with software tools. Use of the software tools provided on the book companion website for use in vertical GHX design problems was illustrated.

Discussion Questions and Exercise Problems

6.1 A school building in a mixed humid climate (climate zone 4A) is in the preliminary stages of design and is considering a vertical GHX. The estimated peak heating and cooling loads are approximately equal at 300 kW. The undisturbed Earth temperature is 16 °C, and the subsurface material is primarily sandstone. Choose a reasonable borehole construction and associated thermal resistance, heat pump design minimum and maximum entering fluid temperatures and the associated COPs, flow rate through the GHX, and design life. If the maximum drilling depth of boreholes is limited to 100 m, determine:

(a) the number of boreholes and the required depth of the GHX with no groundwater flow,
(b) the number of boreholes and the required depth of the GHX with a groundwater flow rate of 10 m/year, and
(c) the number of boreholes and required depth of the GHX with a groundwater flow rate of 100 m/yr.

Hint: Use Table 2.1 to determine the monthly and annual load factors.

6.2 Calculate the design GHX length for a residence in a cold to very cold climate where the Earth temperature is 7.5 °C. The loads file may be found on the book companion site.

The driller you've contacted can comfortably drill to 75 m. The driller has experience in the area and tells you to expect heavy clay at the site. Assume for this exercise that you have ample room to drill and install the GHX in the backyard of the house.

Your analysis should consist of the following steps:

(a) First, calculate the borehole thermal resistance. This is a heating-dominated climate, so you should pick an adequate antifreeze solution as the heat transfer fluid. Assume that the individual U-tube pipes will be evenly spaced within the borehole, and that the driller only works with standard bentonite grout.
(b) Complete the input data sheet in the GHX sizing spreadsheet and list your sources of information and describe any assumptions you make.
(c) Size the GHX. Remember that the driller cannot drill to depths greater than 250 ft, but you want to take advantage of deep boreholes as much as possible to minimize land area that the GHX occupies, so you'll have to adjust the number of boreholes and the spacing to get the calculated depth as close as possible to 75 m.
(d) Now, examine how the subsurface thermal conductivity impacts the total GHX length.
    (aa) Resize the GHX with *half* the value of the thermal conductivity you have in part (d). How does the GHX size change? What type of rock/soil would this new thermal conductivity value be representative of?
    (ab) Resize the GHX with *double* the value of the thermal conductivity you have in part (d). How does the GHX size change? What type of rock/soil would this new thermal conductivity value be representative of?

6.3 You've just completed your design for the residence in problem 6.2, and the architect notifies you that the homeowner has decided to install an underground pool and a patio in the backyard, and has decided to expand the size of the garage. This decision eliminates the backyard as an option for the GHX as originally planned. The only available space to drill and install the GHX is a linear strip of land that measures 4 m × 20 m along the depth of the lot.

Your task now is to fit a vertical GHX as described in problem 6.2 into the available space. Note that you have to use the same driller, so the vertical depth is limited to 75 m. Also, assume that the original information from the driller is correct, which is to expect heavy clay at the site.

There are two things to focus on:

(a) First, recalculate the borehole thermal resistance with the U-tubes against the borehole wall. In practice, this would be accomplished with spacers that physically separate the U-tube pipes. Recalculate the GHX length and determine whether it fits in the available land area.
(b) Second, reduce the peak heating load on the heat pump by including supplemental electric resistance heat. Recalculate the GHX length and determine whether it fits in the available land area.
(c) Comment on the kW/m of heating that you calculated for problem 6.2 and problem 6.3(b). What do these differences mean for a design ‘rule of thumb’?
(d) The homeowner has heard about swimming pool heating with geothermal and asks you about this possibility for the new pool being planned for this new home. Given the space constraints, what is your recommendation about adding swimming pool heating loads to the GHX?

6.4 Calculate the design GHX length for a 1000 m^2 office building in a cold to very cold climate. The loads file may be found on the book companion site.

(a) Starting with your input data from problem 6.2, enter the proper load information for the office building. Make initial guesses of the number of boreholes. Calculate the design GHX length.
(b) Comment on how the kW/m values compare with the residential case in problem 6.2. Explain why there are differences, if any.

6.5 The same 1000 m^2 office building as in problem 6.4 has been constructed in a hot-humid climate where the Earth temperature is 23 °C. The loads file may be found on the book companion site. Assume that limestone bedrock underlies the site. Calculate the design GHX length. Comment on how the kW/m values compare with your results for problem 6.4. Explain why there are differences, if any.

6.6 Design a hybrid GHX for the building in problem 6.5. You will need monthly weather data for a hot-humid climate (e.g., Miami, Florida).

7

Horizontal Ground Heat Exchangers

7.1 Overview

We will now extend our knowledge of vertical GHXs to horizontal ones. As we shall see, there are many similarities, but also some significant differences. Thus, the focus of this chapter is the design and dimensioning of horizontal ground heat exchangers (GHXs) coupled to heat pump applications in buildings. We will also examine so-called *Earth tubes*, which is an old idea receiving new attention. Earth tubes are buried, open-ended pipes used to introduce tempered ventilation air into buildings, specifically low-energy and passive buildings.

The most obvious difference between horizontal and vertical GHXs is their orientation. Thus, the horizontal GHX configuration is a prudent choice for buildings where sufficient land area is available. Relative to their vertical GHX counterparts, they are generally known for a simpler, lower-cost installation. As they are buried at relatively shallow depths (typically less than 2 m or 6 ft), drilling equipment is not necessary; installation is typically in shallow excavations or trenches. Ironically, it is these practical simplicities that lead to more mathematical complexities as compared with vertical GHXs. The proximity of horizontal GHXs to the ground surface leads to violation of our constant ground temperature assumption because the shallow Earth temperature changes seasonally. Further, the ground surface itself presents a physical as well as a mathematical boundary condition.

In this chapter, we will first review the numerous configurations of horizontal GHXs, and then jump right into analysis of the horizontal *GHX design length equation*. As with vertical GHXs, we will examine each design parameter in the context of the horizontal *GHX design length equation*, and then look at implementation of this equation into the companion suite of GHX tools provided with this book for purposes of determining the horizontal GHX length for buildings. The chapter will conclude with a discussion and examination of Earth tubes.

Geothermal Heat Pump and Heat Engine Systems: Theory and Practice, First Edition. Andrew D. Chiasson.

Companion website: www.wiley.com/go/chiasson/geoHPSTP

Learning objectives and goals:

1. Appreciate the complexities in the heat transfer of shallow ground heat exchangers coupled to buildings.
2. Quantify all necessary parameters required properly to size a horizontal ground heat exchanger for a building.
3. Calculate the required size of a horizontal ground heat exchanger for a building.
4. Appreciate the role of software solutions for horizontal GHX design and simulation.
5. Calculate heat transfer rates from Earth tubes.

7.1.1 Horizontal GHX Configurations

The prevalence (and limitations) of rules of thumb for vertical GHX length design was discussed in Chapters 5 and 6. Similar rules of thumb, as you can imagine, also exist with respect to horizontal GHXs, but can get even more confusing owing to the numerous possible field configurations, as described in the paragraphs that follow.

The GHX can be laid in large, open excavation if digging is easy and plenty of land is available. In these cases, the heat exchange pipe may be laid in a serpentine arrangement, or in coiled arrangements on the floor of the excavation. As available land and cost and/or ease of digging become constraints, pipe can be laid into trenches just wide enough to fit the pipe with adequate pipe spacing. Such common designs are known as two-, four-, or six-pipe GHXs (Figure 7.1). There are a few variants on the designs shown in Figure 7.1; pipes in the GHX can be positioned in a different orientation such as in a *stacked* orientation. Notice that these types of GHX design are referred to in multiples of two pipes, even though the heat transfer fluid flows out to the end of the trench and back in the same pipe, similarly to the U-tube we discussed in the last chapter. Thus, a six-pipe trench GHX typically consists of three pipes or three flow circuits. Also notice that each of these designs progressing from Figure 7.1a through to 7.1c has successively more pipe per unit length of trench (or GHX circuit), resulting in more heat transfer per unit length of GHX. Thus, one configuration is not necessarily better than another – the best one really depends on how much land area you have, the cost of digging, and contractor experience.

Another type of GHX is a very compact one, invented in Oklahoma, and is known as a slinky. These may be installed in a horizontal orientation in open excavations or in trenches, as shown in Figure 7.2. Slinky GHXs may also be oriented vertically. These types of GHX offer even more pipe per unit length of trench than six-pipe horizontal GHXs, and are thus a prudent choice in areas of limited space. A typical *pitch* (i.e., the separation distance between successive loops) is about 450 mm or 18 in. Smaller or larger pitches are possible, depending on land area availability. Figure 7.3 shows a photo of a compact slinky GHX installed in the backyard of an urban residence. The GHX footprint is evident by the unmelted snow area. Continual heat extraction over the winter heating season caused the subsurface temperature to decrease, thus resulting in the snow taking a longer time to melt.

Yet another type of horizontal GHX consists of horizontally drilled boreholes, similar to those drilled by underground utility companies. Such a GHX resembles a vertical GHX, and U-tubes can be installed as borehole heat exchangers. Now routine practice in utility line and natural gas line installations under roadways, drilling contractors drill down to target depth,

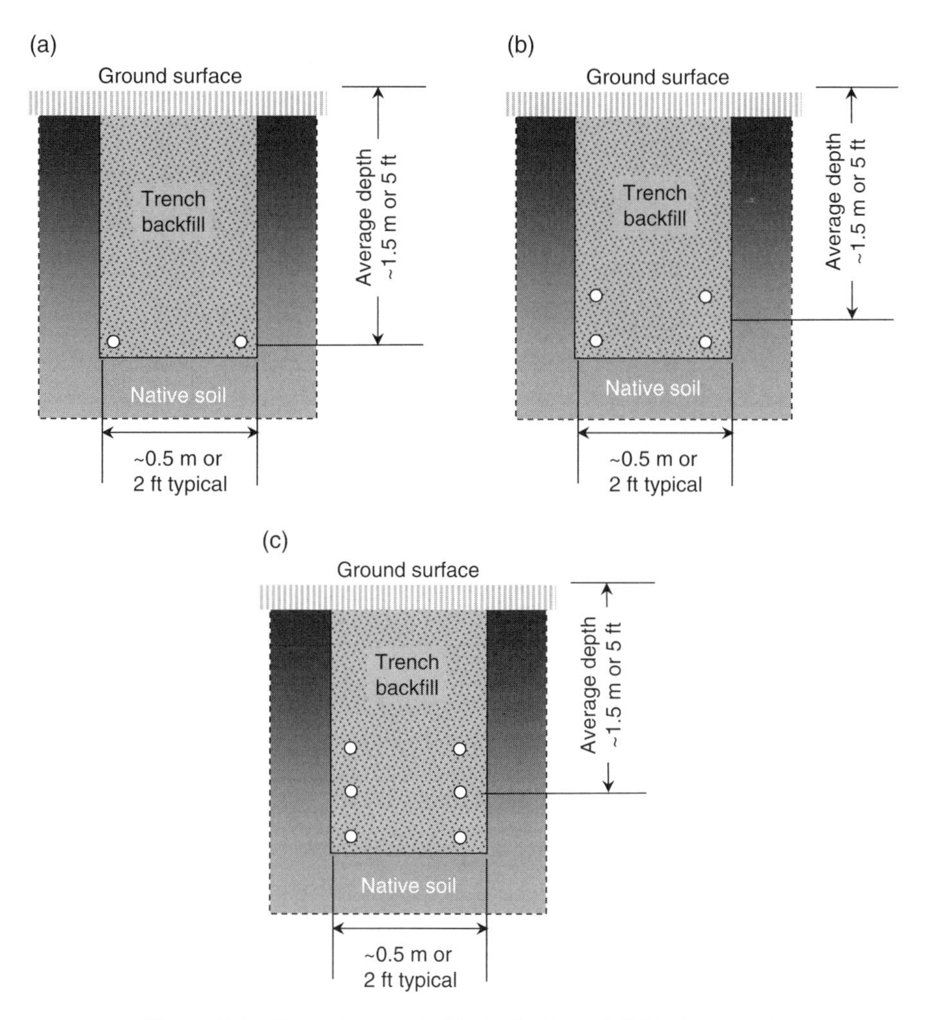

Figure 7.1 Two-, four-, and six-pipe horizontal GHXs in a trench

and then horizontally for several hundreds of meters. Soil boring must 'daylight' (penetrate the surface) and then pull pipe back through the hole as drilling tools are retracted. If soil conditions are favorable, these horizontal boreholes may be placed at depths of tens of meters.

7.2 Horizontal GHX Design Length Equation and Design Parameters

7.2.1 Mathematical Contrasts Between Vertical and Horizontal GHXs

As mentioned above, two main assumptions we employed in the line source model for vertical GHX modeling are violated with application to horizontal GHXs. The undisturbed or 'far-field' temperature is not a constant (it varies seasonally), and we no longer have semi-infinite heat

Figure 7.2 Compact *slinky* GHX installed in a trench (photo by J. Bohrer; permission obtained)

Figure 7.3 Footprint of a compact *slinky* GHX marked by the unmelted snow area in an urban backyard setting (photo by J. Bohrer; permission obtained)

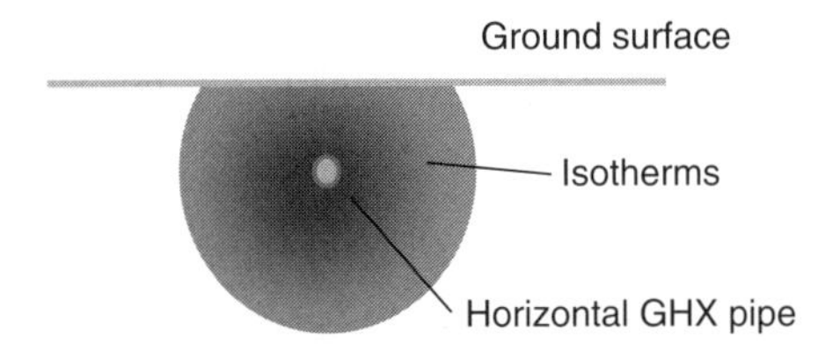

Figure 7.4 Isotherms around a horizontal GHX

transfer (the ground surface interferes with radial heat transfer). Both of these violations are due to horizontal GHXs being buried close to the ground surface, as illustrated in Figure 7.4.

However, we can still apply the same approach to horizontal GHX design length calculation that we used for vertical GHX length calculations, but we have to make some significant modifications. These are related to: (1) the seasonal effect on the ground temperature, (2) geometry considerations and the fact that trenches are not appropriately modeled as line sources, and (3) thermal interference at the ground surface. These modifications are accomplished by (1) introducing a new equation to calculate the seasonal ground temperature, (2) replacing the line source model with a cylinder source model, and (3) implementing 'mirror image' cylinder sources to account for the thermal interference at the ground surface.

7.2.2 *The Horizontal GHX Design Length Equation*

Equation (6.2), which we used to size vertical GHXs, can be slightly reconfigured to solve for total *trench* length (instead of total borehole length), and distinguishing between the seasonal ground temperatures, to obtain

$$L=\frac{\dot{q}_h R'_T+\dot{q}_a R'_{g,a}+\hat{\dot{q}}_m R'_{g,m}+\hat{\dot{q}}_h R'_{g,d}}{T_f-T_{g,\,winter}},\qquad L=\frac{\dot{q}_h R'_T+\dot{q}_a R'_{g,a}+\hat{\dot{q}}_m R'_{g,m}+\hat{\dot{q}}_h R'_{g,d}}{T_f-T_{g,\,summer}} \tag{7.1a,b}$$

where $\dot{q}$ is the ground load as before, R' is the effective thermal resistance per unit length, T_f is the average fluid temperature in the GHX, and T_g is the undisturbed ground temperature distinguished for summer and winter. Subscripts *g, T, a, m, h,* and *d* refer to ground, trench, annual, monthly, hourly, and daily respectively. The ^ symbol denotes superimposed loads. NOTE: as before, positive values of $\dot{q}$ are associated with heat rejection to the ground, so cooling loads are positive and heating loads are negative.

Notice the similarities between Equations (7.1) and (6.2). Recall with Equation (6.2), we solved it twice to get the required GHX length for heating and the required length for cooling. With Equation (7.1), we have two separate equations because of the seasonally varying ground temperature. Each of the terms in Equation (7.1) represents a GHX design parameter, and each will be described in the sections that follow. As with vertical GHXs, note that Equation (7.1) gives the total length (number of trenches *NB* × trench length *H*) of the GHX. Even though we're talking about trenches now, let's still use *NB* and *H*, to be consistent with our previous notation for vertical bores. Also, just as before, the long-term (monthly and annual) thermal resistances of the ground depend on *NB* and *H*, but *H* is unknown and Equation (7.1) must also be solved iteratively. Thus, we will employ a new software tool for its solution.

7.2.3 *The Seasonal Ground Temperature* ($T_{g,winter}$ *and* $T_{g,summer}$)

An equation for calculating the shallow Earth temperature as a function of depth and day of the year has been given by Kasuda and Achenbach (1965):

$$T(d,t) = T_M - A_S \cdot e^{\left[-d \cdot \left(\frac{\pi}{365\alpha}\right)^{1/2}\right]} \cdot \cos\left[\frac{2\pi}{365}\left(t - t_o - \frac{d}{2} \cdot \left(\frac{365}{\pi\alpha}\right)^{1/2}\right)\right] \tag{7.2}$$

where $T(d, t)$ is the Earth temperature (in °F or °C) at soil depth (d) after t days from 1 January, T_M is the mean Earth temperature (°F or °C), A_S is the Earth surface temperature amplitude above/below T_M (°F or °C) and is a measured weather parameter, typically 18 °F or 10 °C, t is the number of days after 1 January (days), t_o is the number of days after 1 January where the minimum Earth surface temperature occurs (days), typically 30–35 days, d is the depth (ft or m), and α is the soil thermal diffusivity (ft^2/day or $m^2 \cdot s^{-1}$).

Recall, soil temperature data are available at: http://eosweb.larc.nasa.gov/sse/RETScreen/. NOTE: this website gives the surface amplitude as the *difference of the maximum and minimum surface temperature*, not the difference about the mean. Therefore, when using the amplitude from this data source, it must be divided by 2 to get the appropriate value of A_s for Equation (7.2).

Equation (7.2) can be applied to illustrate the soil temperature as a function of depth and time at a location where the mean soil temperature is 50 °F or 10 °C, the surface amplitude is 18 °F or 10 °C, t_o is 33 days, and α = 0.6 ft^2/day or 6.45E-7 $m^2 \cdot s^{-1}$. Substituting these values into Equation (7.2), temperature profiles for various depths have been constructed as shown in Figure 7.5.

Review of the temperature profiles in Figure 7.5 reveals two important features with respect to horizontal GHX design. First, the damping of the soil temperature amplitude increases with depth, and rapidly diminishes by 5 m depth. Thus, coupled with practical reasons, most horizontal GHXs are installed at depths less than about 2 m; deeper excavations must be braced or shored, and can become cost-prohibitive. Second, and more importantly, there is a time lag, increasing with depth, when the peak minimum and maximum temperatures occur. This is extremely important because, for example, a GHX with a 2 m burial depth does not experience the minimum and maximum soil temperatures until mid-March and mid-September – at times far removed from the times of peak heating and cooling load. Therefore, we customarily calculate the ground temperatures at times that more closely coincide with the peak heating and cooling load – typically at day 35 and day 215.

7.2.4 *Ground Thermal Properties*

You may have noticed that the foregoing discussion started using the word 'soil' quite a bit. This is because horizontal GHXs are installed almost always in soil. It is neither easy nor practical to excavate rock; rock excavations are typically blasted and are therefore cost prohibitive for GHXs

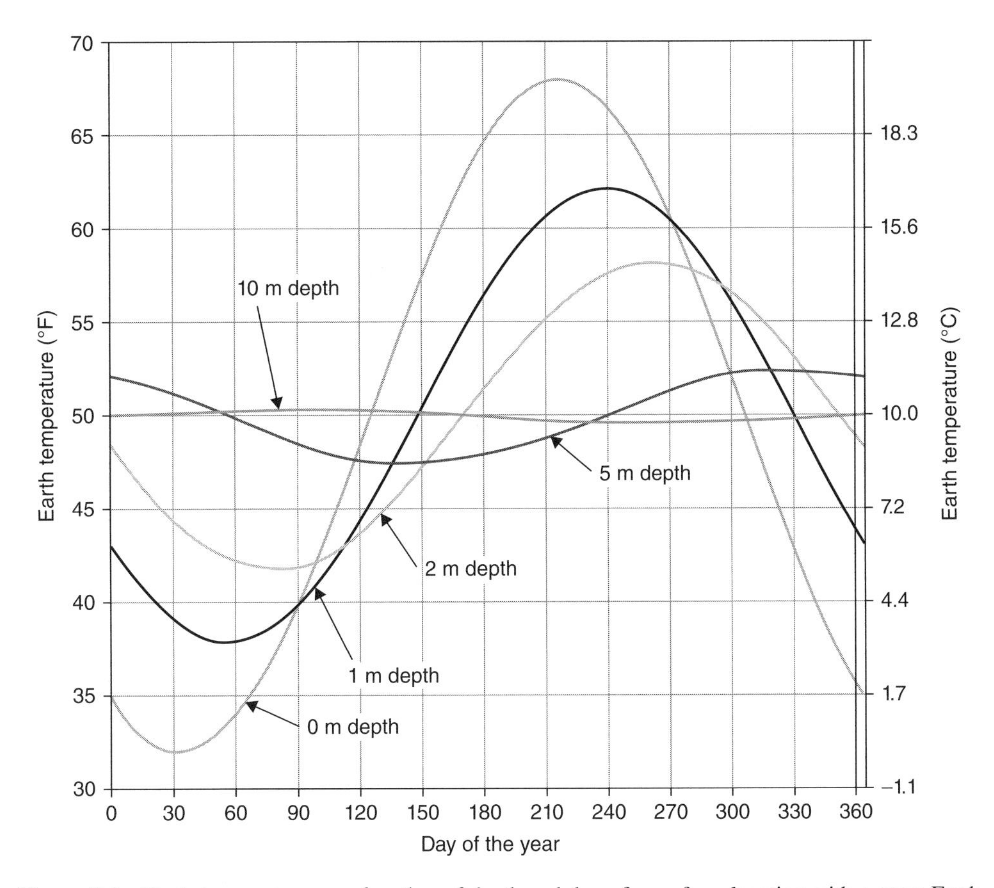

Figure 7.5 Earth temperature as a function of depth and day of year for a location with a mean Earth temperature of 50 °F or 10 °C

alone, unless rock excavations are done for some other reason (such as a building parking garage foundation). In such cases, horizontal piping could be laid into these excavations if practical. Thus, for horizontal GHXs, the soil thermal property data listed in Appendix B will be more commonly used. As before, the main thermal properties of interest for horizontal GHX design are the thermal conductivity (k), the density (ρ), and the heat capacity (c_p) of the soil.

We spent a considerable amount of time studying thermal response tests for determining the thermal conductivity of the Earth surrounding vertical GHXs. This kind of test is not warranted for horizontal GHXs; shallow soils can be trenched and examined, and thus the thermal properties are estimated on the basis of visual inspection or prior knowledge of the site. Soil temperatures and thermal conductivity can be measured with the use of field probes.

7.2.5 The Ground Loads

Determination of the ground loads for horizontal GHXs is no different from that for vertical GHXs. Therefore, no further discussion is warranted here.

7.2.6 *The Average GHX Fluid Temperature* (T_f)

Determination of the average GHX fluid temperature for horizontal GHXs is no different from that for vertical GHXs. Therefore, no further discussion is warranted here.

7.2.7 *The Trench Thermal Resistance* (R'_T)

The trench thermal resistance is exactly analogous to the borehole thermal resistance discussed in Chapter 5. As with the borehole thermal resistance, we will treat the trench thermal resistance as a steady-state approximation, thus ignoring the thermal storage in the trench materials themselves.

Approximations for two-, four-, and six-pipe trench thermal resistances are available through analytical solutions by Hellström (1991), where the pipes are approximated as line sources in a circular region. Accounting for counterflow in the pipe legs, we have

$$R'_{trench} = R'_{sf} + \frac{1}{3}\frac{1}{R'_a}\left(\frac{H}{C}\right)^2 \tag{7.3}$$

where R'_{sf} is an effective steady-flux thermal resistance, R'_a is an internal resistance, H is the trench length, and C is the heat capacity rate of the heat transfer fluid. The intermediate resistances are given for a two-pipe, four-pipe, and six-pipe trench as follows:

$$R'_{sf,\,two\text{-}pipe} = \frac{1}{2\pi k_{backfill}}\left[\ln\left(\frac{r_{trench}}{r_{p,out}}\right) - \frac{3}{4} + b^2 - \frac{1}{2}ln\left(1-b^4\right) - \frac{1}{2}\ln\left(\frac{B_u}{r_{p,out}}\right)\right] + \frac{R'_p}{2} \tag{7.4}$$

$$R'_{sf,\,two\text{-}pipe} = \frac{1}{\pi k_{backfill}}\left[\ln\left(\frac{2br_{trench}}{r_{p,out}}\right) - \ln\left(\frac{1-b^2}{1+b^2}\right)\right] + 2R'_p \tag{7.5}$$

$$R'_{sf,four\text{-}pipe} = \frac{1}{2\pi k_{backfill}}\left[\ln\left(\frac{r_{trench}}{r_{p,out}}\right) - \frac{3}{4} + b^2 - \frac{1}{4}\ln\left(1-b^8\right) - \frac{1}{2}\ln\left(\frac{\sqrt{2}br_{trench}}{r_{p,out}}\right) - \frac{1}{4}\ln\left(\frac{2br_{trench}}{r_{p,out}}\right)\right] + \frac{R'_p}{4} \tag{7.6}$$

$$R'_{a,four\text{-}pipe} = \frac{1}{\pi k_{backfill}}\left[\ln\left(\frac{\sqrt{2}br_{trench}}{r_{p,out}}\right) - \frac{1}{2}\ln\left(\frac{2br_{trench}}{r_{p,out}}\right) - \frac{1}{2}\ln\left(\frac{1-b^4}{1+b^4}\right)\right] + R'_p \tag{7.7}$$

$$\begin{aligned} R'_{sf,\,six\text{-}pipe} = \frac{1}{2\pi k_{backfill}}\Bigg[&\ln\left(\frac{r_{trench}}{r_{p,out}}\right) - \frac{3}{4} + b^2 - \frac{1}{6}\ln\left(1-b^{12}\right) - \frac{1}{3}\ln\left(\frac{br_{trench}}{r_{p,out}}\right) \\ &- \frac{1}{3}\ln\left(\frac{\sqrt{3}br_{trench}}{r_{p,out}}\right) - \frac{1}{6}\ln\left(\frac{2br_{trench}}{r_{p,out}}\right)\Bigg] + \frac{R'_p}{6} \end{aligned} \tag{7.8}$$

$$R'_{a,\,six\text{-}pipe} = \frac{1}{\pi k_{backfill}}\left[\frac{2}{3}\ln\left(\frac{br_{trench}}{r_{p,out}}\right) + \frac{1}{3}\ln\left(\frac{2br_{trench}}{r_{p,out}}\right) - \frac{2}{3}\ln\left(\frac{\sqrt{3}br_{trench}}{r_{p,out}}\right) - \frac{1}{3}\ln\left(\frac{1-b^6}{1+b^6}\right)\right] + \frac{2}{3}R'_p \tag{7.9}$$

where B_u is the shank spacing between pipes, r_{trench} is the trench radius, $r_{p,out}$ is the pipe outer radius, $k_{backfill}$ is the thermal conductivity of the trench backfill material, b is an eccentricity parameter equivalent to the pipe center-to-center distance, S, divided by the trench diameter, and R'_p is the pipe thermal resistance given by Equation (5.28). Subscript p refers to pipe.

7.2.7.1 The Trench Thermal Resistance Calculator Tool

The thermal resistance equations for two-, four-, and six-pipe trenches have been implemented into the companion suite of GHX tools (found on the book companion website) for use in horizontal GHX design problems requiring the trench thermal resistance. The calculator also contains thermal property data for pure water as well as aqueous propylene glycol mixtures. Use of this trench thermal resistance calculator tool is demonstrated in the following example.

Example 7.1a Trench Thermal Resistance Calculation – IP Units

Consider a six-pipe horizontal trenched GHX design consisting of nominal 1 in diameter HDPE pipes installed in a 24 in wide trench, 5 ft deep, and 50 ft long. The average spacing of the pipes is 18 in, and the trench is backfilled with sand. The heat transfer fluid is a 20% aqueous mixture of propylene glycol with a design flow rate of 3 gallons per minute (gpm) per flow circuit. The minimum temperature expected is 30 °F.

Solution

The trench design in this example is typical of basic US installations.

Results from the software tool are shown in Figure E.7.1a. Input data are entered in the yellow boxes, and calculations are done to the right of the input fields. Dropdown boxes allow options for: the nominal pipe size, the pipe SDR, and the borehole heat transfer fluid. The ground thermal conductivity is assumed to be 1.0 Btu/hr·ft·°F.

Example 7.1b Trench Thermal Resistance Calculation – SI Units

Consider a six-pipe horizontal trenched GHX design consisting of nominal 25 mm diameter HDPE pipes installed in a 600 mm wide trench, 2 m deep, and 17 m long. The average spacing of the pipes is 450 mm, and the trench is backfilled with sand. The heat transfer fluid is a 20% aqueous mixture of propylene glycol with a design flow rate of 10 liters per minute (Lpm) per flow circuit. The minimum temperature expected is −1.0 °C.

Solution

The trench design in this example is typical of basic US installations.

Results from the software tool are shown in Figure E.7.1b. Input data are entered in the yellow boxes, and calculations are done to the right of the input fields. Dropdown boxes allow options for: the nominal pipe size, the pipe SDR, and the borehole heat transfer fluid. The ground thermal conductivity is assumed to be 1.8 $W \cdot m^{-1} \cdot K^{-1}$.

Discussion: A review of the trench thermal resistance values reveals that they are of similar order to that of a well-designed vertical BHE (see Example 5.5).

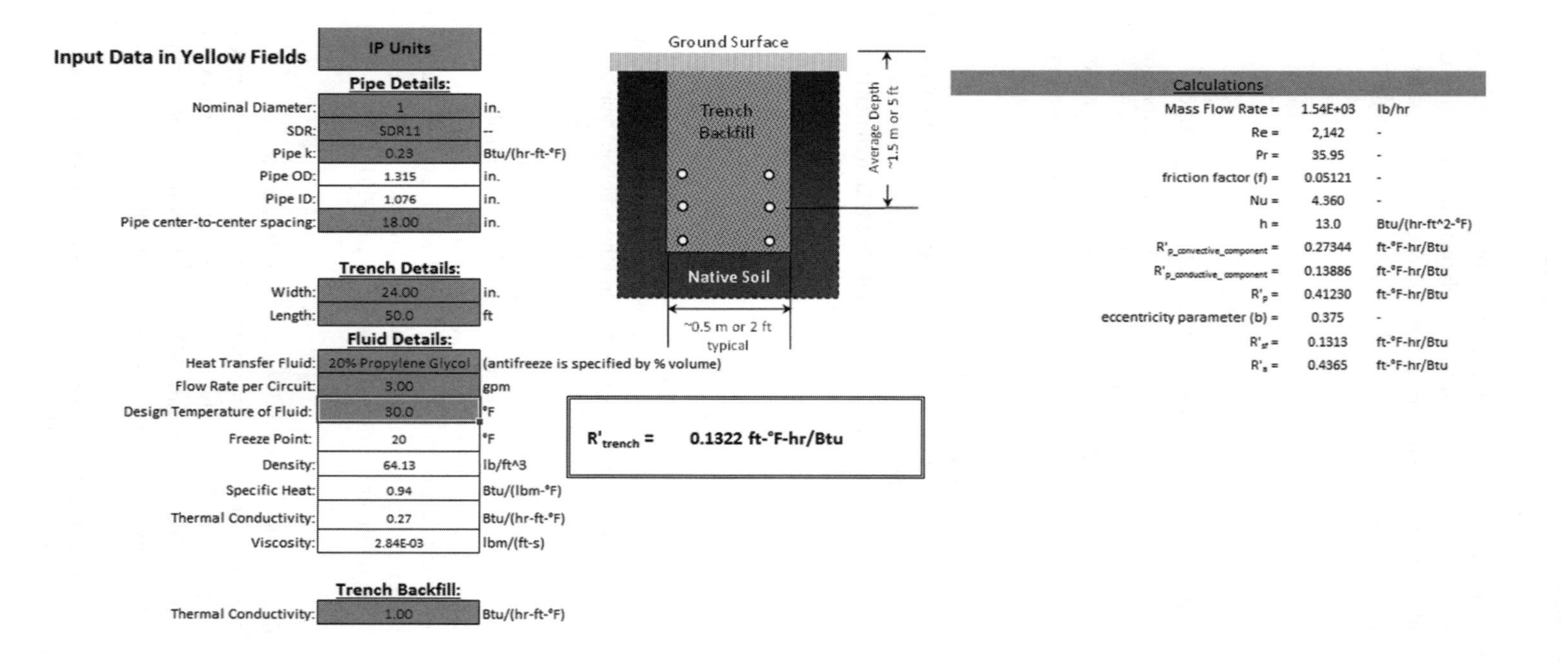

Figure E.7.1a Screen capture of six-pipe trench thermal resistance calculator tool for Example 7.1a (IP units)

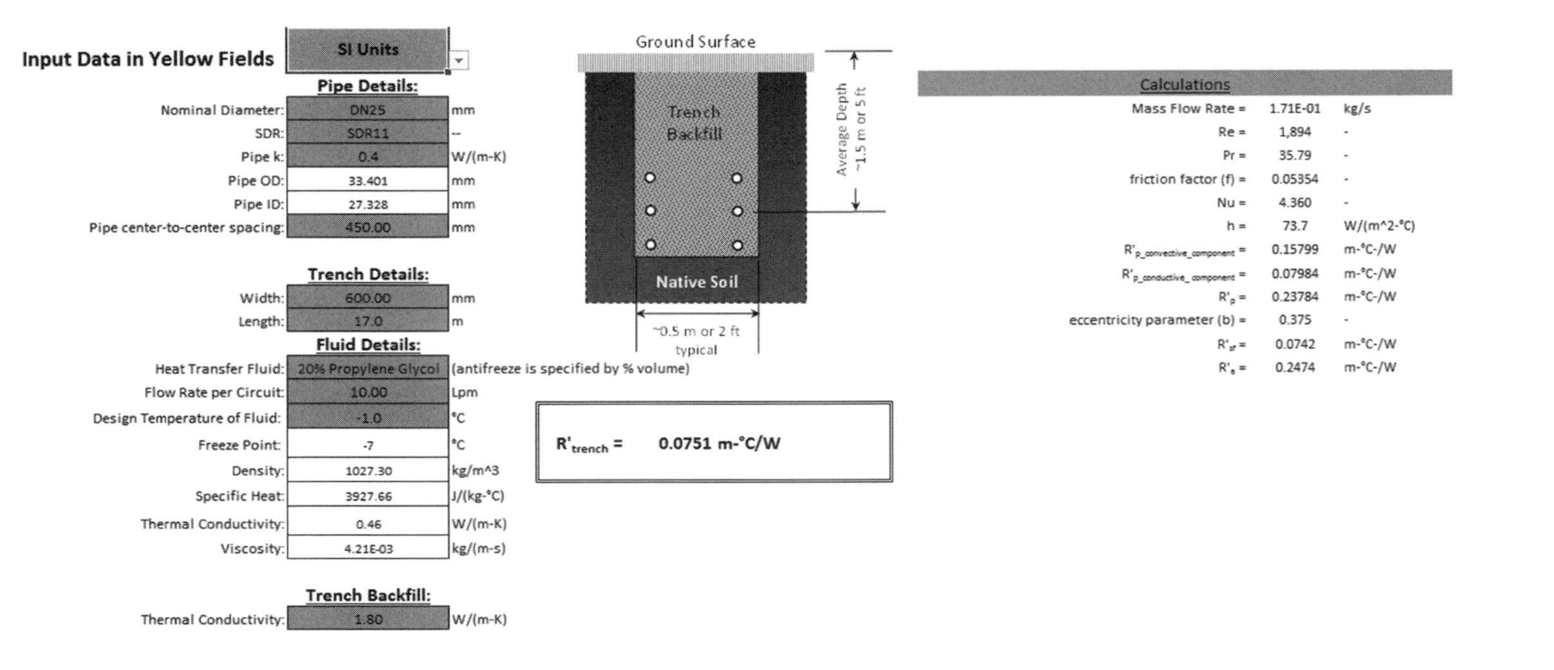

Figure E.7.1b Screen capture of six-pipe trench thermal resistance calculator tool for Example 7.1b (SI units)

7.2.8 The Ground Thermal Resistances

In Chapter 5, recall that we mentioned the existence of two analytical models for calculating the temperature distribution in the ground owing to radial heat conduction. These were the *line source model* and the *cylinder source model*. We also saw that the line source model gave a good representation of the average fluid temperature when used in conjunction with the steady-state borehole thermal resistance, and thus we used it solely in calculating the design length of vertical GHXs. However, in horizontal trenches, the ratio of trench radius to length is not nearly as small as the r_b/H ratio for vertical boreholes, and thus the *cylinder source model* gives a better approximation to heat transfer around a single trench. The line source model is still adequate for describing the thermal interference between trenches after long times of operation.

Thus, we now have two models that we will use to calculate the ground thermal resistances related to horizontal GHXs: (i) the *cylinder source model* for calculating the daily ground thermal resistance around an individual trench, and (ii) the *finite line source model* for calculating the monthly and annual ground thermal resistances for multiple, thermally interacting trenches.

7.2.8.1 Cylinder Source Model for Determining the Temperature at the Trench Wall

Assuming a horizontal trench GHX approximates a cylindrical shape around the trench, the solution to the so-called *cylinder source model* is an exact solution to the temperature on the cylinder periphery. This model was developed and evaluated by Carslaw and Jaeger (1947), and was proposed by Ingersoll *et al.* (1954) as a method for sizing buried heat exchangers. The interior boundary condition for this solution is a constant heat flux, representative of a borehole or trench heat exchanger.

The cylinder source model solution to the temperature change (ΔT) at radius r for a constant heat rate per unit length ($\dot{q}'$) applied at a cylinder of radius r_o is given by

$$\Delta T = \frac{\dot{q}'}{k_g} G\left(Fo, \frac{r}{r_o}\right) \tag{7.10}$$

where k_g is the thermal conductivity of the ground, Fo is the dimensionless Fourier number ($\alpha t/r^2$), and G is the so-called G-factor (not to be confused with the g-function), given by

$$G\left(Fo, \frac{r}{r_o}\right) = \frac{1}{\pi^2} \int_0^\infty \frac{e^{-\beta^2 Fo} - 1}{J_1^2(\beta) + Y_1^2(\beta)} [J_0(p\beta)Y_1(\beta) - J_1(\beta)Y_0(p\beta)] \frac{1}{\beta^2} d\beta \tag{7.11}$$

where J_0, J_1, Y_0, Y_1 are Bessel functions of zero and first orders.

As you might imagine, the integral in Equation (7.11) is not easy to evaluate. Fortunately, for our purposes, we are only interested in using the *cylinder source model* to calculate the temperature at the periphery of the cylinder, where $r = r_o$. We will use the *finite line source model* to evaluate temperatures at other radial distances from the trench, as we did with vertical boreholes. Bernier (2001) provides an efficient solution to Equation (7.11) for $r/r_o = 1$ (i.e., the trench wall):

$$G(Fo, 1) = 10^{\left[-0.89129 + 0.36081\log Fo - 0.05508\log^2 Fo + 0.003596171\log^3 Fo\right]} \tag{7.12}$$

Now we have the change in temperature at the trench periphery (ΔT_T), given by

$$\Delta T_T = \frac{\dot{q}'}{k} G(Fo, 1) \tag{7.13}$$

We can now calculate the short-term, daily ground resistance (per unit length of trench) as

$$R'_{g,d} = \frac{G(Fo, 1)}{k_g} \tag{7.14}$$

7.2.8.2 Determining Trench Thermal Interaction by Spatial Superpositioning of Finite Line Sources

Here, we will use the same approach as we did for vertical boreholes, which made use of spatial superpositioning of the dimensionless temperature response (g) in each trench due to all the others at a particular time. The configuration of most horizontal GHXs can be considered to be a linear array, unlike vertical GHXs, which are configured as rectangular (or other) arrays. Therefore, with respect to spatial superpositioning, horizontal GHXs are mostly configured in $x \times 1$ arrays, where x here denotes the number of trenches. However, as mentioned previously, we make use of 'mirror image' trenches to account for thermal interference imposed by the proximity of the ground surface, as shown in Figure 7.6. Thus, linear horizontal trenches are actually modeled mathematically as $x \times 2$ arrays to account for the ground surface.

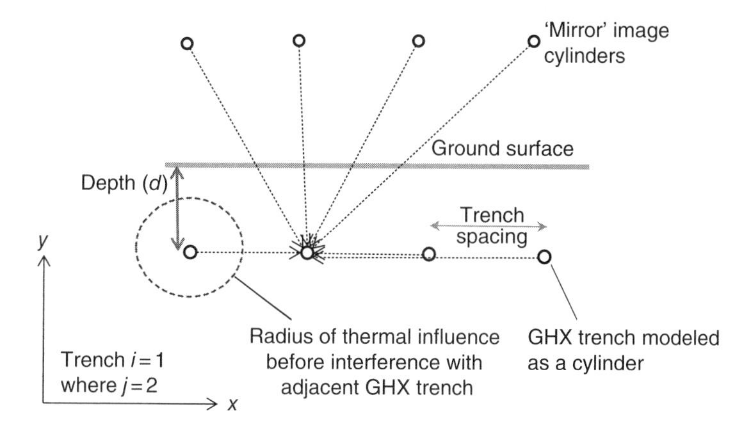

Figure 7.6 Cross-sectional illustration of spatial superpositioning of linear trenches, where the temperature response at a particular trench is influenced by all others, plus the mirror image trenches that act to model the ground surface

In a horizontal GHX, thermal interference will occur between the ground surface or adjacent pipes, depending on the radius of thermal influence relative to the burial depth. Therefore, there is no reason to space GHX pipes or trenches at a spacing greater than twice the depth.

Now we can proceed with application of the *g*-function (as described in Chapter 6) to determine the monthly and annual ground thermal resistances. The software tool described in the following subsection determines trench thermal interaction by spatial superpositioning of analytical finite line sources using the leaky well function as described in Chapter 6 (Section 6.2.6.1).

7.3 Modeling Horizontal GHXs with Software Tools

As we saw with vertical GHX design, similarly with horizontal GHXs, manual solution of the foregoing equations is not practical owing to the iterative and tedious nature of the process. Thus, a number of commercially available software packages have been developed for GHX design.

The foregoing algorithms and equations have been implemented into the suite of GHX tools found on the book companion website for use in horizontal GHX design problems (Figure 7.7). The model has been developed for basic load input only: peak hour heating and cooling loads and monthly and annual load factors for heating and cooling. More advanced simulation

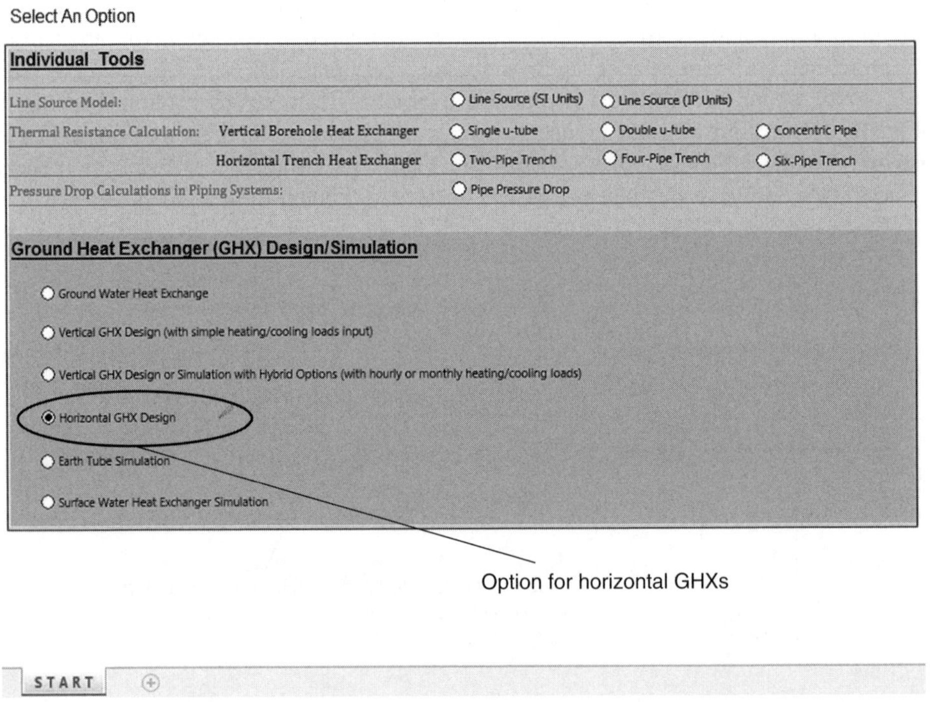

Figure 7.7 Screen capture of the START worksheet of the companion GHX tools highlighting the horizontal GHX design option

modeling of horizontal GHXs would require the use of a numerical model, as described in a subsequent section.

Use of the companion horizontal GHX software tool is described in the paragraphs that follow.

Constructing the Model. Selection of the option button *Horizontal GHX Design* launches one worksheet which is shown in Figure 7.8. The input fields on the *Horizontal GHX* worksheet shown in Figure 7.8 are divided into four groups similar to those for vertical GHXs with basic load input: (1) ground parameters, (2) heat pump details, (3) load information, and (4) supplemental ground load information for application to hybrid GHX systems.

In the *Ground Parameters* section of the *Horizontal GHX* worksheet, all information is entered regarding the ground thermal properties, the trench or horizontal borehole details, in addition to the specific heat of the GHX fluid. A dropdown box allows options for entering the trench thermal resistance; this value can either be entered directly or calculated by selecting *two-pipe trench, four-pipe trench,* or *six-pipe trench* options from the dropdown box, which launches a Visual Basic macro to display the appropriate worksheet for trench thermal resistance calculation as described in Section 7.2.7. Users enter the Earth temperature details as described above in Section 7.2.3. The winter and summer temperatures are calculated at day 35 and day 215, and these temperatures are used in the calculations on the Horizontal GHX design equation. As seen in Figure 7.8, the average daily Earth temperature at the average GHX depth is displayed on an *x–y* plot. Users also enter the average depth of the horizontal GHX, the number of trenches or horizontal boreholes, and the trench or borehole spacing.

In the *Heat Pump* section of the *Horizontal GHX* worksheet, the design limits of the heat pump fluid entering temperature are entered, along with the associated heat pump COP values at these conditions. As mentioned, these are important constraints in controlling the size of the GHX. The heat pump design flow rate is also entered here. Typical design flow rates are suggested in the worksheet, but users may enter the value appropriate for their design.

In the *Loads Information* section of the *Horizontal GHX* worksheet, users enter the peak loads and the monthly and annual load factors. Methods for determining these values have been discussed in Chapter 2. The *Design Life* is based on the life cycle of the GHP project.

In the *Supplemental Ground Loads Information* section of the *Vertical GHX* worksheet, users enter estimates of annual thermal energy handled by a supplemental component. This provides a crude means to size hybrid horizontal GHXs.

Performing the Calculations. As mentioned at the beginning of our discussion of the GHX design length equation, its solution is iterative; the ground thermal resistances depend on the trench or horizontal borehole length (H), but H is unknown, and the main goal of the GHX design length equation is to determine the total GHX length. Also recall that for horizontal GHXs there are two GHX design length equations to solve owing to the seasonally changing ground temperature: one for the winter season and one for the summer season. It is rare for the two calculated lengths to be equal, given the numerous factors that affect the ground loads. Therefore, users should select the greater length as the design choice. If the lengths are unacceptably large, the GHX can be redesigned by changing some parameters, within engineering reason, by adding supplemental electric heat, or by designing a hybrid GHX.

Mouse-clicking on the command button *Click Here to Size Horizontal GHX* executes a Visual Basic for Applications (VBA) code that makes use of a single-variable optimization method (i.e., a golden section search) iteratively to solve the horizontal GHX design length equations [Equations (7.1a) and (7.1b)]. The VBA optimization method automatically adjusts the trench

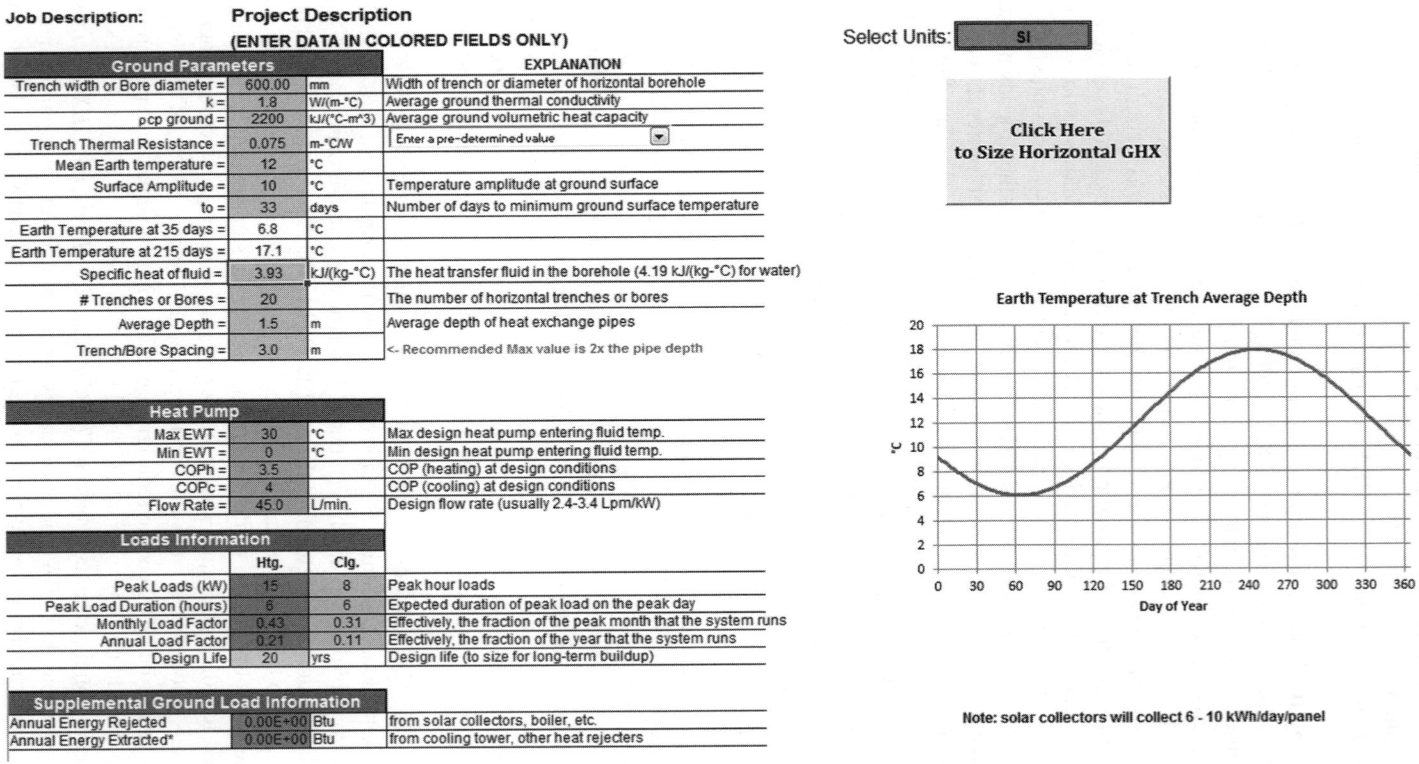

Job Description: **Project Description**

(ENTER DATA IN COLORED FIELDS ONLY)

Select Units: SI

Ground Parameters			EXPLANATION
Trench width or Bore diameter =	600.00	mm	Width of trench or diameter of horizontal borehole
k =	1.8	W/(m-°C)	Average ground thermal conductivity
ρcp ground =	2200	kJ/(°C-m^3)	Average ground volumetric heat capacity
Trench Thermal Resistance =	0.075	m-°C/W	Enter a pre-determined value
Mean Earth temperature =	12	°C	
Surface Amplitude =	10	°C	Temperature amplitude at ground surface
to =	33	days	Number of days to minimum ground surface temperature
Earth Temperature at 35 days =	6.8	°C	
Earth Temperature at 215 days =	17.1	°C	
Specific heat of fluid =	3.93	kJ/(kg-°C)	The heat transfer fluid in the borehole (4.19 kJ/(kg-°C) for water)
# Trenches or Bores =	20		The number of horizontal trenches or bores
Average Depth =	1.5	m	Average depth of heat exchange pipes
Trench/Bore Spacing =	3.0	m	<- Recommended Max value is 2x the pipe depth

Click Here to Size Horizontal GHX

Heat Pump			
Max EWT =	30	°C	Max design heat pump entering fluid temp.
Min EWT =	0	°C	Min design heat pump entering fluid temp.
COPh =	3.5		COP (heating) at design conditions
COPc =	4		COP (cooling) at design conditions
Flow Rate =	45.0	L/min.	Design flow rate (usually 2.4-3.4 Lpm/kW)

Loads Information	Htg.	Clg.	
Peak Loads (kW)	15	8	Peak hour loads
Peak Load Duration (hours)	6	6	Expected duration of peak load on the peak day
Monthly Load Factor	0.43	0.31	Effectively, the fraction of the peak month that the system runs
Annual Load Factor	0.21	0.11	Effectively, the fraction of the year that the system runs
Design Life	20	yrs	Design life (to size for long-term buildup)

Supplemental Ground Load Information			
Annual Energy Rejected	0.00E+00	Btu	from solar collectors, boiler, etc.
Annual Energy Extracted*	0.00E+00	Btu	from cooling tower, other heat rejecters

Note: solar collectors will collect 6 - 10 kWh/day/panel

Figure 7.8 Screen capture of the *Horizontal GHX* worksheet of the companion GHX tools

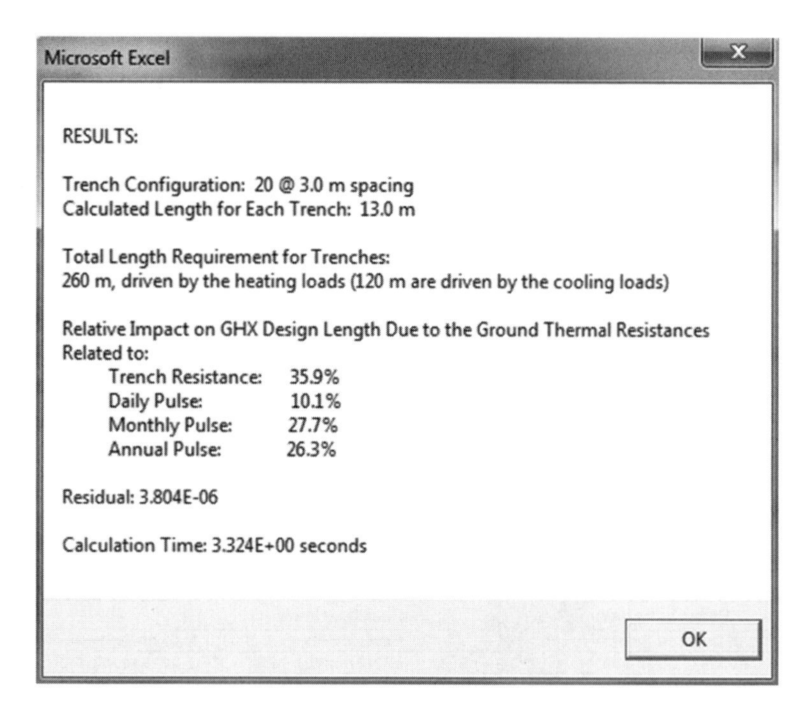
RESULTS:

Trench Configuration: 20 @ 3.0 m spacing
Calculated Length for Each Trench: 13.0 m

Total Length Requirement for Trenches:
260 m, driven by the heating loads (120 m are driven by the cooling loads)

Relative Impact on GHX Design Length Due to the Ground Thermal Resistances
Related to:
Trench Resistance: 35.9%
Daily Pulse: 10.1%
Monthly Pulse: 27.7%
Annual Pulse: 26.3%

Residual: 3.804E-06

Calculation Time: 3.324E+00 seconds

OK

Figure 7.9 Screen capture of the calculated results related to the input screen capture shown in Figure 7.8

or horizontal borehole length to minimize the objective function, which is defined as the difference between the input length (determined by $NB \times H$) and the maximum calculated length (whether for heating or cooling). The results of a horizontal GHX size calculation are printed to a message box on the screen. Figure 7.9 is a screen capture of the results related to the input data shown in Figure 7.8.

As shown in Figure 7.9, a summary of the input GHX configuration is echoed in the results box. The calculated length of each trench is 13.0 m, resulting in a total length requirement of 260 m for the GHX. Other useful information is provided, such as the length driven by the non-dominant load. For this example, cooling is the non-dominant load, and the required length gives users some sense of the load imbalance on the GHX. In this example, the required trench length for cooling is about 46% of that required for heating. This imbalance suggests possible consideration of a supplemental heating element to balance the ground loads.

Also as shown in Figure 7.9, the results box displays the relative thermal resistances of the trench, along with the ground thermal resistances due to the daily, monthly, and annual loads. As seen in this example, the largest thermal resistance is due to the trench.

Example 7.2 Supplemental Electric Resistance Heating with Horizontal GHXs
The horizontal GHX described in the foregoing discussion is quite heating dominated. Add a supplemental electric heating element to handle 50% of the peak heating load and resize the GHX.

Solution

As we don't have the hourly load output, let's use the correlations provided in Chapter 2. Recalling Equations (2.8) and (2.9):

$$\text{Monthly load factor multiplier} = 0.561x^2 + 1.1366x + 0.9954, \mathbf{valid\ for\ 0 \le x \le 0.5} \quad (2.8)$$

$$\text{Annual load factor multiplier} = 1.3281x^2 + 0.941x + 1.0007, \mathbf{valid\ for\ 0 \le x \le 0.5} \quad (2.9)$$

Thus, for $x = 0.50$, we have

$$\text{Monthly load factor multiplier} = 1.704$$

$$\text{Annual load factor multiplier} = 1.803$$

The peak heating load and the monthly and annual heating load factors are adjusted as shown in Figure E.7.2a.

The results are shown in the screen capture of the results box in Figure E.7.2b.

Loads Information			
	Htg.	Clg.	
Peak Loads (kW)	8	8	Peak hour loads
Peak Load Duration (hours)	6	6	Expected duration of peak load on the peak day
Monthly Load Factor	0.73	0.31	Effectively, the fraction of the peak month that the system runs
Annual Load Factor	0.38	0.11	Effectively, the fraction of the year that the system runs
Design Life	20	yrs	Design life (to size for long-term buildup)

Figure E.7.2a Screen capture of adjusted load information for Example 7.2

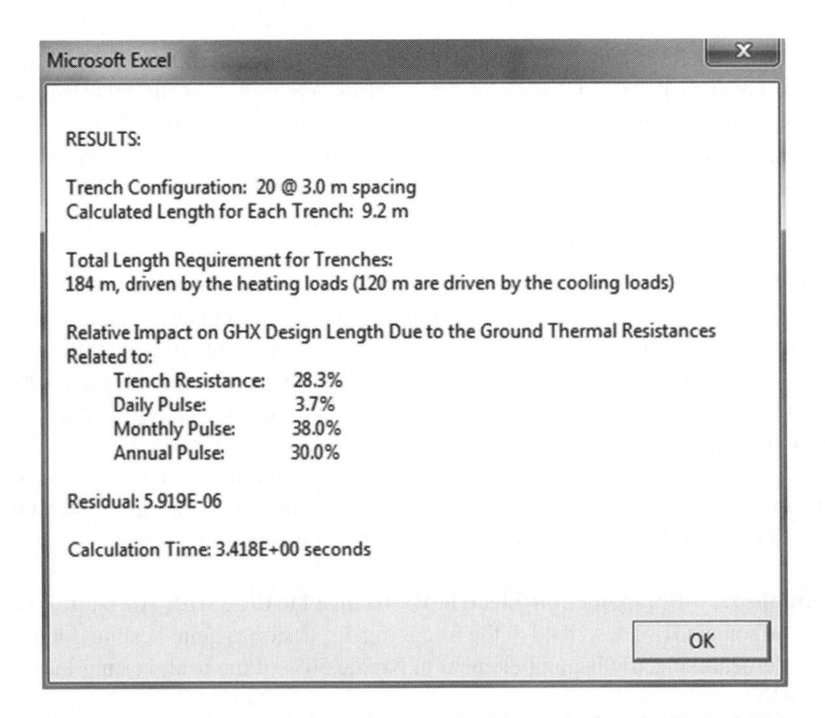

Figure E.7.2b Screen capture of horizontal GHX results box for Example 7.2

Discussion: In comparing the total trench length of 184 m as shown in Figure E.7.2b (i.e., the horizontal GHX design with the 50% supplemental heating element) with that of 260 m as shown in Figure 7.9 (i.e., the horizontal GHX design without the 50% supplemental heating element), we see that there is a 30% reduction in the GHX size when the heating element is included with the design. This reduction in size would result in a significant capital cost savings. There would be a trade-off in increased operating cost of the electric heating element, but as discussed in Chapter 2, such a design would still meet over 90% of the annual heating loads.

7.4 Simulation of Horizontal GHXs

Interest in simulation of the thermal performance of horizontal GHXs is for the same reasons as for vertical GHXs: we may wish to examine system temperatures over the operating cycle of the GHX and examine hourly or monthly heat pump electrical energy consumption. However, simulation of the thermal performance of horizontal GHXs is not as straightforward (and thus not as common) as simulation of vertical GHXs. Again, this has to do with the proximity of the GHX to the ground surface; proper simulation of the thermal performance of these systems over time involves accounting for transient weather conditions at the ground surface.

Chiasson (2010) describes simulation of a horizontal GHX using Comsol Multiphysics, a finite element software package. Chiasson (2010) constructed finite element models of horizontal GHXs for various geometries in 2D cross-section, with focus on a six-pipe horizontal GHX. The 2D approach was considered reasonable, as long as the cross-section was taken through the mid-section of the trench along the major direction of fluid flow in the pipe system.

Thus, the governing partial differential equation was

$$\frac{\partial^2 T}{\partial x^2} + \frac{\partial^2 T}{\partial z^2} = \frac{1}{\alpha}\frac{dT}{dt} \tag{7.15}$$

where T is temperature (K), x is the lateral space coordinate (m), z is the depth space coordinate (m), α is the thermal diffusivity ($m^2 \cdot s^{-1}$), and t is time (s).

Owing to symmetry conditions, the x-dimension of the model domain consisted of half the trench section plus the distance midway to the adjacent trench. Thus, the left- and right-hand boundaries are adiabatic by definition of the symmetry condition. The model domain in the z-direction was extended to a depth of 10 m, and the bottom boundary was specified as a constant temperature (Dirichlet condition). The top boundary represented the ground surface where heat fluxes due to varying weather conditions were modeled as a Neumann condition:

$$n \cdot (k\nabla T) = q''_{solar} + h\left(T_{inf} - T\right) + \varepsilon\sigma\left(T_{sky}^4 - T^4\right) \tag{7.16}$$

where k is the thermal conductivity of the soil ($W \cdot m^{-1} \cdot K^{-1}$), T is ground surface temperature (K), q''_{solar} is the solar heat flux on horizontal ($W \cdot m^{-2}$), h is the convection coefficient ($W \cdot m^{-2} \cdot K^{-1}$) due to wind, T_{inf} is the ambient air temperature (K), α is the thermal emissivity of the ground surface (—), σ is the Stefan–Boltzmann constant ($W \cdot m^{-2} \cdot K^{-4}$), and T_{sky} is the sky temperature (K). The boundary condition at internal pipe surfaces was modeled as a

Neumann condition, representing the heating and cooling loads on the GHX. Hourly loads were computed for an actual 140 m^2 (1500 ft^2) home, and were then converted to thermal loads on the GHX assuming a heat pump coefficient of performance (COP) of 4.0 for cooling mode and 3.0 for heating mode at design conditions.

For improvement in computational speed, Chiasson (2010) averaged the hourly weather data and the hourly building loads per month, with the monthly peak load being imposed for a 6 h time period. Parameters defined as time-varying functions included: ambient air temperature, wind speed, solar radiation on horizontal, sky temperature, and building thermal loads.

COMSOL simulations were conducted for 20 year simulation times, and the GHX was iteratively adjusted such that the calculated fluid temperatures remained within the target design criteria. Thus, the main model output was the integral temperature over the internal surface of the heat exchange pipes. The 20 year simulation times were completed in less than 1 min on a typical desktop computer of the day, showing that numerical methods for horizontal GHX simulation could be a viable option for design of these systems.

7.5 Earth Tubes

As mentioned at the beginning of this chapter, Earth tubes are an old idea receiving new attention. Earth tubes, also known as Earth-to-air heat exchangers, are buried, open-ended pipes used to introduce tempered ventilation air into buildings, specifically low-energy and passive buildings.

7.5.1 Introduction

Earth tubes are a special case of horizontal GHXs. They differ from the horizontal ground loops coupled to geothermal heat pumps that we have been studying so far in two main respects: (1) they are open-loop systems (i.e., the working fluid is *not* recirculated through the loop), and (ii) the working fluid is air that is introduced directly into buildings.

As you might expect, heat transfer rates in Earth tube systems are not outstanding, but are useful when combined with other HVAC systems. Air is generally not a great heat transfer fluid when run through the Earth, but the advantages of Earth tubes are that they provide controlled fresh air ventilation that is filtered for dust and allergens. Thus, they have the potential significantly to reduce the amount of additional energy required to heat and cool a building, especially when combined with a heat or energy recovery ventilator (HRV or ERV). Also, Earth tubes may eliminate the need for defrost cycles in an HRV, and typically require only a small amount of electrical power to operate their air intake fan. Use of HRVs will be examined in more detail in Chapter 15. In some climates and/or in low-energy buildings such as passive houses, Earth tubes may supply all of a building's cooling needs.

7.5.2 Practical Considerations

Earth tubes can be constructed in many different field configurations, a couple of which are shown in Figure 7.10. In residential buildings, the pipe can be wrapped around the house,

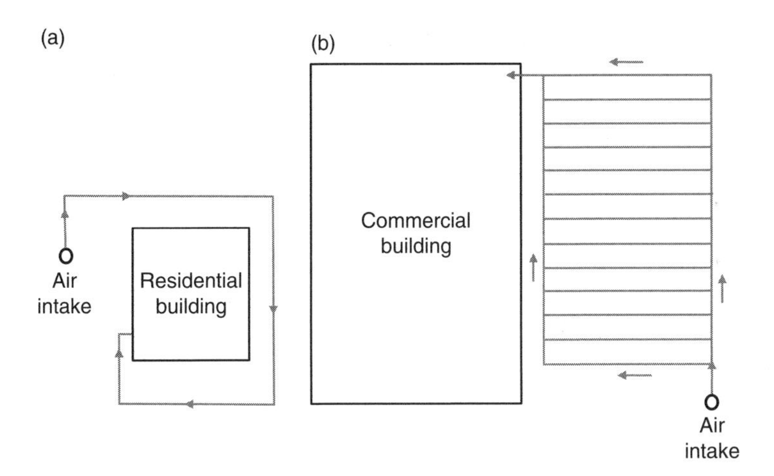

Figure 7.10 Various Earth tube field configurations

or installed in a serpentine configuration. In larger commercial buildings, parallel patterns are common in order to handle more airflow.

There are some new considerations with Earth tube design that we did not encounter with horizontal GHXs for heat pumps. As Earth tubes are open-loop systems and convey air that can be directly introduced into buildings, considerable care must be taken in their design to inhibit microbial growth in the Earth tube, in addition to managing moisture that condenses out of the air. Thus, the Earth tube system consists of an intake tower on a stable base with filters to remove dust and allergens from the air, and screens to keep animals and insects out of the pipe. The Earth tube itself is constructed of specialized pipe (PVC or HDPE) treated with a microbial growth inhibitor. To allow for proper condensate management, the pipe system is typically sloped, and a gravity drain system is included.

Thus, to summarize the operation of an Earth tube system, a fan inside the building draws fresh air in through an air inlet tower with an integrated air filter. The air becomes tempered as it flows through the buried Earth pipe(s), and exchanges heat with the Earth. Upon entering the building, the air ideally enters a heat recovery ventilator, and exchanges heat with air being exhausted from the building. Alternately, the tempered air could be directly introduced into the HVAC system of the building.

7.5.3 Mathematical Considerations

Recall how we have been handling the design of GHXs for geothermal heat pumps: we have been calculating the heating and cooling loads for the building, and then sizing the GHX to meet those loads. Here, we will need to take an alternative approach because the air is being used *directly* in the building, and therefore the energy available from the ground will be more limiting than with heat pump applications owing to the Earth temperature. Now, for Earth tubes, we will choose a pipe length first, and then calculate the outlet air temperature and associated energy savings. Thus, we are, in effect, designing by simulation.

To restate the above, we can easily calculate a ventilation air load, but we can't necessarily design the Earth tube to meet that load, because the load that the Earth tube can actually meet is limited by the Earth temperature. Note that this is different to before with geothermal heat pump applications because we were not using the heat transfer fluid in the GHX directly; it was being used as a source/sink for the heat pump. Now if we, for example, calculated our ventilation air load based on a desired outlet air temperature of 25 °C in winter in a cold-climate location, that is an impossible condition for the Earth tube to meet because we know that the ground temperature is less than 10 °C in winter. Thus, the ground temperature represents an upper (or lower) limit to which an Earth tube could warm up (or cool down) air at any given time of the year. We will need to revisit our mathematical approach.

Let's recall some heat transfer principles here. Heat transfer due to fluid flowing through a pipe can either be described by a constant heat flux case or a constant pipe surface temperature case. Neither of these cases truly occurs in nature, but solutions to engineering problems can be reasonably approximated as one case or the other. Note the main differences between the two cases: in the constant heat flux case, the medium surrounding the pipe is free to change temperature along with the fluid, but in the constant surface temperature case, the fluid temperature is constrained by the pipe surface (or surrounding medium) temperature.

Recalling how we have been modeling the heat transfer in the Earth around GHXs, we have been assuming a *constant heat flux case* around boreholes and trenches. This is a reasonable approximation for those situations because the borehole or trench is an assemblage of pipes with fluids that flow back and forth, transferring heat to each other; the pipes in boreholes and trenches contained the same fluid, but at two different temperatures. However, this is not the case with Earth tubes. They are open-loop, single pipes with single flow-through systems, and thus will not necessarily be characterized by a constant heat flux through their entire length. Therefore, to a more reasonable approximation, we will adopt the constant pipe surface temperature case as described mathematically below and illustrated in Figure 7.11.

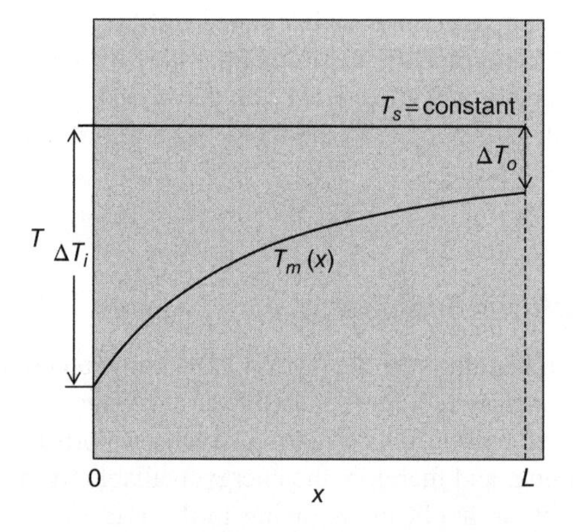

Figure 7.11 Schematic of temperature variation in a pipe with a constant exterior temperature

On a monthly basis, we can relate the pipe surface temperature (approximated as the monthly soil temperature) (T_s) to the monthly average inlet air temperature (T_i) as

$$\frac{\Delta T_o}{\Delta T_i} = \frac{T_s - T_o}{T_s - T_i} = \exp\left(-\frac{1}{\dot{m}c_p R_{tot}}\right) \quad (7.17)$$

where T_s is the soil temperature (°C or °F), T_o is the outlet air temperature (°C or °F), T_i is the inlet air temperature (°C or °F), R_{tot} is the total thermal resistance of the pipe (°F·h/Btu or °C/W), c_p is the heat capacity of air ($\text{J} \cdot \text{kg}^{-1} \cdot \text{K}^{-1}$ or Btu/lb·°F), and $\dot{m}$ is the mass flow rate of the air (lb/h or kg·s^{-1}). T_s can be calculated using Equation (7.2), and R_{tot} is the total pipe thermal resistance and can be calculated using Equation (5.27) and methods described in Chapter 5 (Section 5.5.6.1). With the outlet air temperature known from Equation (7.17), the average hourly heat transfer rate can be calculated from an energy balance on the fluid stream:

$$\dot{q} = \dot{m}c_p(T_o - T_i) \quad (7.18)$$

It is also useful to calculate the pressure drop (ΔP) in the pipe and the associated fan power ($\dot{W}_{fan}$) requirements:

$$\Delta P = f\frac{\rho u_m^2}{2D}L \quad (7.19)$$

$$\dot{W}_{fan} = \Delta P\dot{Q} \quad (7.20)$$

where ΔP is the pressure drop (Pa), $\dot{W}_{fan}$ is the fan power (W), f is the Darcy friction factor [see Equation (5.31)], ρ is the density of the air (kg·m^{-3}), u_m is the mean air velocity (m·s^{-1}), D is the pipe inside diameter (m), L is the pipe inside diameter (m), and $\dot{Q}$ is the volumetric air flow rate ($\text{m}^2\text{·s}^{-1}$).

7.5.4 Earth Tube Analysis with Software Tools

The foregoing algorithms and equations have been implemented into the suite of GHX tools found on the book companion website for use in Earth tube performance analysis (Figure 7.12). The model has been developed for average monthly analysis of thermal performance of an Earth tube. Thermal interference of pipes is not considered. Use of the companion Earth tube simulation software tool is described in the paragraphs that follow.

Constructing the Model. Selection of the option button *Earth Tube Simulation* launches three worksheets, the first of which is the *Earth Tube Analysis* worksheet, shown in Figure 7.13. The second worksheet *PipeThermalResistanceCalcs* shows the monthly calculations for determining the total pipe thermal resistance, with no input data required. The third worksheet *Weather Data* contains monthly average weather data, where the only required input is the monthly average air temperature.

The input fields on the *Earth Tube Analysis* worksheet shown in Figure 7.13. are divided into three simple sections: (1) pipe details, (2) soil details, and (3) analysis.

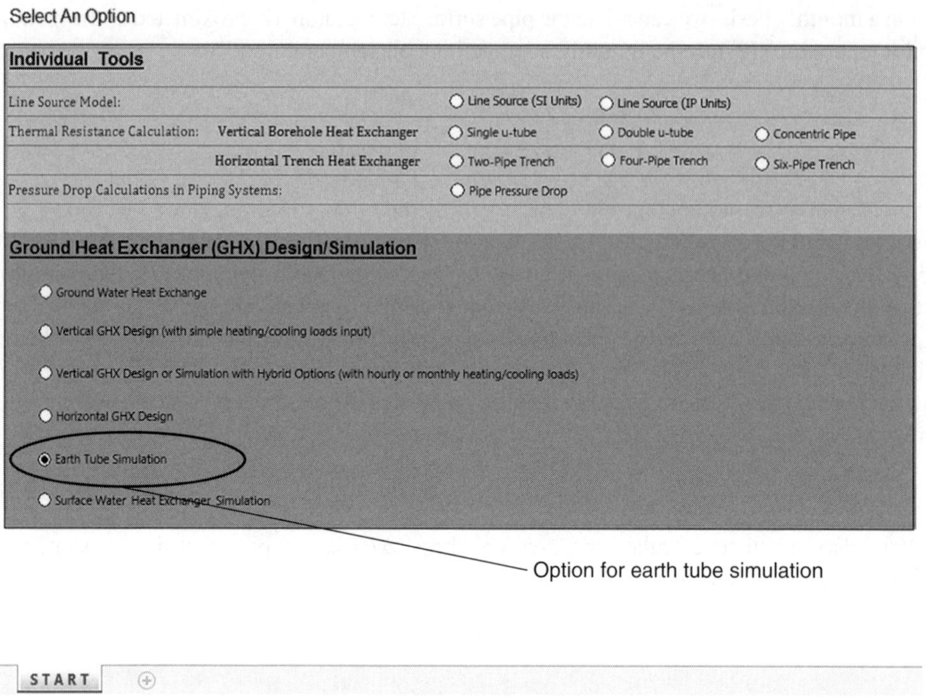

Figure 7.12 Screen capture of the START worksheet of the companion GHX tools highlighting the Earth tube simulation option

In the *Pipe Details* section of the *Earth Tube Analysis* worksheet, all information is entered regarding the Earth tube pipe. The nominal diameter and SDR are entered through dropdown boxes. The pipe thermal conductivity and length are also entered.

In the *Soil Details* section of the *Earth Tube Analysis* worksheet, the same basic information is entered as described above for horizontal GHX calculations. That is, the Earth temperature details as described above in Section 7.2.3, in addition to the pipe burial depth.

In the *Analysis* section of the *Earth Tube Analysis* worksheet, users enter the monthly average air intake volumetric flow rate. All other data in this section are calculated. Values in the *Average Tair In* column are the monthly air temperatures translated from the *Weather Data* worksheet. Values in the *Average Pipe T* column are the monthly average pipe surface temperatures calculated with Equation (7.2). Values in the *Pipe Resistance* column are translated from calculations made on the *PipeThermalResistanceCalcs* worksheet. Values in the *Average Tair Out* column correspond to T_o calculated with Equation (7.17). Values in the *Average Heat Transfer* column are determined by an energy balance on the air stream [Equation (7.18)], and values in the *Monthly Heat Transfer* column multiply the corresponding values in the *Average Heat Transfer* column by the number of hours per month. Finally, the values in the *Pressure Drop in Pipe* and *Fan Power* columns are calculated by Equations (7.19) and (7.20) respectively. Monthly fan energy would be determined by multiplying the fan power by the number of hours per month. The *Earth Tube Analysis* worksheet also displays a plot of monthly average air temperature (the inlet air), the soil temperature, and the outlet air temperature.

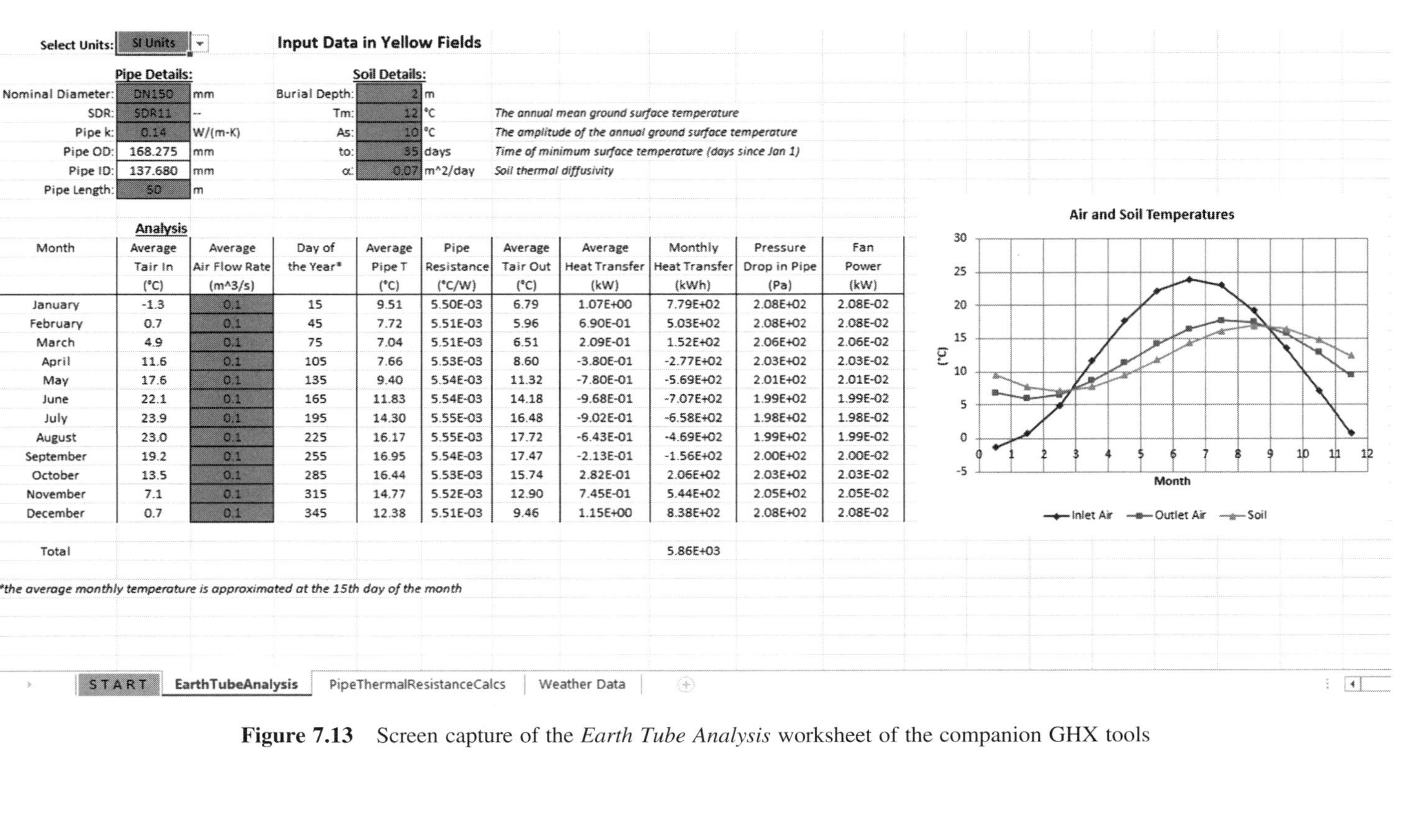

Select Units: SI Units

Input Data in Yellow Fields

Pipe Details:

Nominal Diameter:	DN150	mm
SDR:	SDR11	--
Pipe k:	0.14	W/(m-K)
Pipe OD:	168.275	mm
Pipe ID:	137.680	mm
Pipe Length:	50	m

Soil Details:

Burial Depth:	2	m	
Tm:	12	°C	*The annual mean ground surface temperature*
As:	10	°C	*The amplitude of the annual ground surface temperature*
to:	35	days	*Time of minimum surface temperature (days since Jan 1)*
α:	0.07	m^2/day	*Soil thermal diffusivity*

Analysis

Month	Average Tair In (°C)	Average Air Flow Rate (m^3/s)	Day of the Year*	Average Pipe T (°C)	Pipe Resistance (°C/W)	Average Tair Out (°C)	Average Heat Transfer (kW)	Monthly Heat Transfer (kWh)	Pressure Drop in Pipe (Pa)	Fan Power (kW)
January	-1.3	0.1	15	9.51	5.50E-03	6.79	1.07E+00	7.79E+02	2.08E+02	2.08E-02
February	0.7	0.1	45	7.72	5.51E-03	5.96	6.90E-01	5.03E+02	2.08E+02	2.08E-02
March	4.9	0.1	75	7.04	5.51E-03	6.51	2.09E-01	1.52E+02	2.06E+02	2.06E-02
April	11.6	0.1	105	7.66	5.53E-03	8.60	-3.80E-01	-2.77E+02	2.03E+02	2.03E-02
May	17.6	0.1	135	9.40	5.54E-03	11.32	-7.80E-01	-5.69E+02	2.01E+02	2.01E-02
June	22.1	0.1	165	11.83	5.54E-03	14.18	-9.68E-01	-7.07E+02	1.99E+02	1.99E-02
July	23.9	0.1	195	14.30	5.55E-03	16.48	-9.02E-01	-6.58E+02	1.98E+02	1.98E-02
August	23.0	0.1	225	16.17	5.55E-03	17.72	-6.43E-01	-4.69E+02	1.99E+02	1.99E-02
September	19.2	0.1	255	16.95	5.54E-03	17.47	-2.13E-01	-1.56E+02	2.00E+02	2.00E-02
October	13.5	0.1	285	16.44	5.53E-03	15.74	2.82E-01	2.06E+02	2.03E+02	2.03E-02
November	7.1	0.1	315	14.77	5.52E-03	12.90	7.45E-01	5.44E+02	2.05E+02	2.05E-02
December	0.7	0.1	345	12.38	5.51E-03	9.46	1.15E+00	8.38E+02	2.08E+02	2.08E-02
Total								5.86E+03		

**the average monthly temperature is approximated at the 15th day of the month*

Figure 7.13 Screen capture of the *Earth Tube Analysis* worksheet of the companion GHX tools

The input data shown in Figure 7.13 correspond to a cold climate in the US Midwest. The Earth tube is a DN150 pipe, 50 m long. The average monthly air flow is 0.1 $m^3 \cdot s^{-1}$. The calculation results show that about 5860 kWh of energy is saved with air tempering by the Earth tube.

7.6 Chapter Summary

This chapter extended our knowledge of multiple vertical borehole arrays to horizontal GHXs. While simpler to install than their vertical counterparts, mathematical modeling of horizontal GHXs is complicated by the fact that the 'undisturbed' ground temperature changes seasonally, and heat transfer around trenches is not semi-infinite.

In Chapter 6, the so-called design length equation for vertical GHXs was introduced to address the fundamental task in the design of a vertical borehole GHX, which is properly to size the total number, depth, and spacing of the BHEs so that they provide fluid temperatures to the heat pump(s) within design limits over its lifetime. Even with the heat transfer complexities in horizontal GHXs, the same design length equation was applied with some modification. The main modifications were: (i) the introduction of a summer and winter ground temperature, resulting in two design length equations, and (ii) revised methods for calculation of the ground thermal resistances, which included the use of the cylinder source model and the use of 'mirror-image' trenches to account for the ground surface. The analytical *g*-function is still used to determine the longer-term ground thermal resistance due to thermally interfering finite line sources. The software tools provided on the book companion website were illustrated for use in horizontal GHX design problems.

Simulation of horizontal GHXs using a finite element software package was discussed. The main benefit of a numerical method includes explicit modeling of the heat transfer between pipes in the trench, and modeling of transient boundary conditions at the ground surface. Heat transfer due to soil moisture migration and freeze/thaw cycles could also be included.

Finally, this chapter discussed practical and mathematical considerations of Earth tubes or Earth-to-air heat exchangers. The software tools provided on the book companion website were illustrated for use in Earth tube analysis for ventilation air tempering.

Discussion Questions and Exercise Problems

7.1 A school building in a mixed humid climate (climate zone 4A) is in the preliminary stages of design and is considering a horizontal GHX in the playing field. The estimated peak heating and cooling loads are approximately equal at 300 kW. The undisturbed Earth temperature is 16 °C, and the subsurface soils primarily consist of sandy soils. Choose a reasonable horizontal trench GHX configuration and associated thermal resistance, heat pump design minimum and maximum entering fluid temperatures and the associated COPs, flow rate through the GHX, and design life. Determine:

(a) the number of trenches and required length, and
(b) the land area required by your GHX.

Hint: use Table 2.1 to determine the monthly and annual load factors.

7.2 Design an Earth tube for preconditioning of the ventilation air for a residential building where you live. Assume a constant air flow rate of 0.05 $m^3 \cdot s^{-1}$. The pipe is DN150 PVC, 20 m long, and buried at a depth of 2 m. Report on the following:

(a) The annual amount of energy provided by the Earth tube for preconditioning the ventilation air.
(b) How do your results in part (a) change for a 40 m long pipe?
(c) How do your results in part (a) change for a 10 m long pipe?

8

Surface Water Heat Exchange Systems

8.1 Overview

Surface water is generally defined as a body of water that is exposed to the atmosphere, and includes ponds, lakes, oceans, and rivers. Use of surface water as a heat exchange medium is not a new idea, but has seen increasing popularity in recent decades in parallel with the growth of geothermal heat pumps. As such, surface water heat exchange falls under the umbrella of low-temperature geothermal energy utilization.

Surface water heat exchange systems are of two general configurations: (i) open-loop systems and (ii) closed-loop systems. In open-loop systems, surface water is used directly for heating or cooling purposes (usually cooling). Closed-loop systems are generally integrated with a heat pump, and are sometimes known as surface water heat pump (SWHP) systems. These are similar in principle to the systems we have already examined (i.e., the ground-coupled systems), except the heat transfer processes are much different.

In this chapter, we will first review the practical and theoretical considerations related to simple open-loop surface water heat exchange systems. We will then extend this knowledge to closed-loop systems coupled to heat pumps (i.e., SWHP systems), and examine the thermal performance of these systems using a monthly system simulation software tool.

Learning objectives and goals:

1. Appreciate the complexities in the heat transfer in surface water bodies.
2. Calculate heat transfer rates in surface water bodies.
3. Calculate the required size of a surface water heat exchanger for a building.

Geothermal Heat Pump and Heat Engine Systems: Theory and Practice, First Edition. Andrew D. Chiasson.

Companion website: www.wiley.com/go/chiasson/geoHPSTP

8.2 Thermal Processes in Surface Water Bodies

Thermal processes in surface water bodies have been discussed in Chapter 2 (Section 2.5), so we will not elaborate further here. The subsections that follow summarize the modes of heat transfer and describe the seasonal dynamics in deeper lakes.

8.2.1 Governing Modes of Heat Transfer

Recall from Chapter 2 (Section 2.5) that the main heating mechanism of surface water bodies is by solar radiation. The exception is when the water surface is ice covered. Evaporation and convection are the main cooling mechanisms. Conduction heat transfer to/from the ground is a weak process when the surface is not frozen, but may become a highly significant process if the surface is frozen. Inflows due to surface water run-off and from tributaries may also be significant when present. The processes are summarized graphically in Figure 8.1.

Mathematical equations describing each process are described in Chapter 2 (Section 2.5), and the resultant steady-state energy balance is given by Equation (2.27) [now relabeled as Equation (8.1)]:

$$\dot{q}_{net} = A\left(\dot{q}''_{solar} + \dot{q}''_{thermal\ radiation} + \dot{q}''_{convection} + \dot{q}''_{evaporation} + \dot{q}''_{ground}\right) + \dot{q}_{make-up} \tag{8.1}$$

where A is the surface area of the water body (m^2), and $\dot{q}_{make\text{-}up}$ refers to the thermal load on the surface water body owing to water entering the surface water body from any arbitrary source.

Surface water bodies have a large thermal mass (or thermal inertia), and therefore their temperature does not change quickly. Thus, daily or monthly average energy balances are usually sufficient. Ignoring vertical temperature gradients, the bulk average temperature change of a

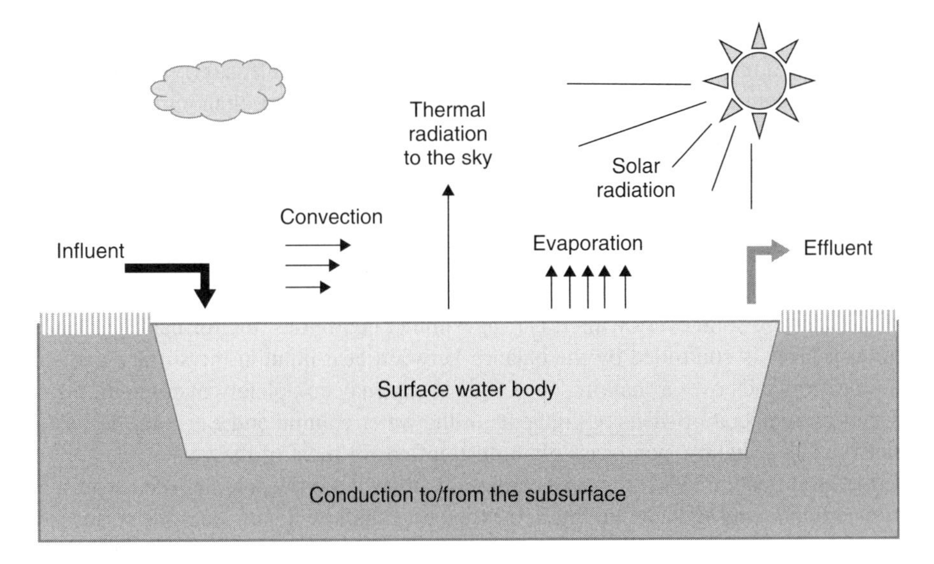

Figure 8.1 Heat transfer processes acting on an unfrozen surface water body

surface water body with time (dT/dt) may be described by the so-called lumped capacitance or lumped parameter approach:

$$\frac{dT}{dt} = \frac{\dot{q}_{net}}{\rho c_p V} \tag{8.2}$$

where ρ, c_p, and V are the density, specific heat, and volume of the surface water body respectively.

The overall energy balance equation described by Equation (8.2) may be rearranged into the form of a linear, first-order, ordinary differential equation:

$$\frac{dT}{dt} = x_1 T + x_2 \tag{8.3}$$

where T represents the surface water temperature, x_1 contains all terms in Equations (2.13) to (2.26) that multiply T, and x_2 contains all terms in Equations (2.13) to (2.26) that are independent of T. Note that in Equations (2.13) to (2.26), T is represented by T_{pool}. Equation (8.3) may be solved at an arbitrary time step (Δt) by

$$T_t = \left(T_{t-\Delta t} + \frac{x_2}{x_1}\right) e^{x_1 \Delta t} - \frac{x_2}{x_1} \tag{8.4}$$

8.2.2 Seasonal Dynamics

The dynamic distribution of mass, energy, and momentum in a surface water body is due to the hydrodynamic transport processes of advection and diffusion. Advection refers to the transport of mass, momentum, and energy by the fluid's bulk motion. Diffusion in surface water bodies takes place by: (i) turbulent mixing (turbulent diffusion or eddy diffusion), and (ii) molecular diffusion. Turbulent diffusion, typically orders of magnitude greater than molecular diffusion, is characterized by irregular, random fluctuations caused by turbulent eddies in the water body. Other mechanisms include surface wind stress, internal waves, shear waves, inflows, and outflows.

Idealized seasonal dynamics of deeper lakes is shown in Figure 8.2. Owing to the unique property of water, where its density is greatest at about 4 °C, seasonal dynamics results in stratification and spring and fall 'overturn'. Without this unique property of water, surface water bodies could freeze solid every winter. During summer conditions, the formation of stratification in deep lakes is controlled by the balance between heat input at the surface and surface winds. Surface winds over a shallow, open water body may completely overwhelm the naturally occurring temperature–density gradients in the water column and cause the water to mix completely. Thus, shallow ponds are often unstratified for most of the year.

Thermal stratification in lakes is most commonly observed in deeper ones (i.e., at least 5 m in depth) where the wind-induced mixing is restricted to shallow depths near the surface. Well-stratified lakes have three distinct thermal regions: (i) the epilimnion or upper mixed region, (ii) the metalimnion or thermocline, and (iii) the hypolimnion or lower mixed region. The

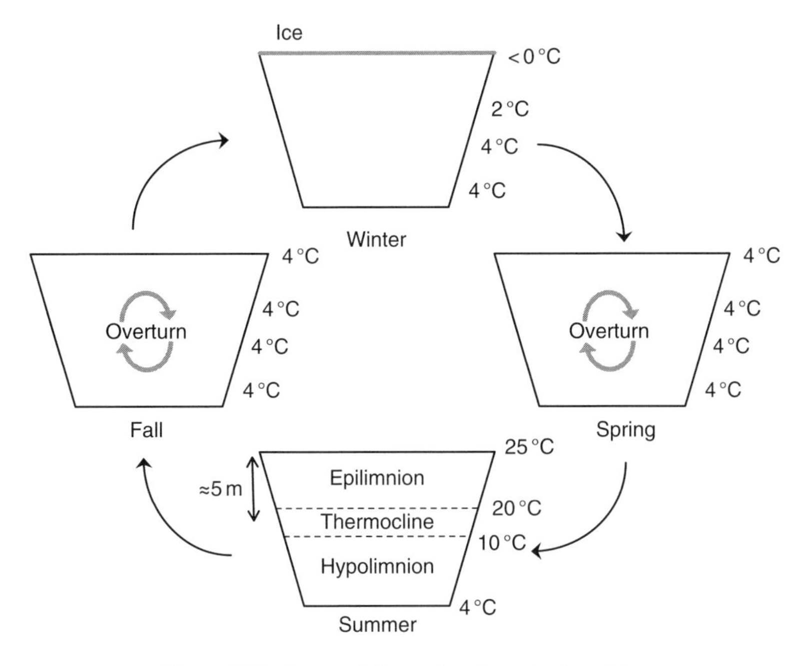

Figure 8.2 Seasonal thermal patterns in deep lakes

epilimnion or upper mixed region is characterized by well-mixed and relatively high water temperatures. The *metalimnion* or *thermocline* region is a transition region between the warm surface water and cooler bottom water, and is characterized by large temperature and density gradients. The *hypolimnion* or lower mixed region is the relatively cold region below the thermocline that extends to the bottom of the surface water body. The hypolimnion region is well mixed, and generally contains the densest water. It tends to be unaffected by surface winds.

By late fall, the lake surface temperature cools, and the density gradient in the water column is slowly destroyed. This is coupled with turbulent energy from surface winds, which combine to destabilize or mix the entire water column. This phenomenon is called *turnover*, and the temperature of the lake becomes approximately uniform from the surface to the bottom. During winter, the lake bottom temperature approaches 4 °C, the maximum density of water. The surface may freeze, and the coldest water will remain below the ice. Lakes that develop surface ice cover exhibit the same overturn phenomenon in the spring when the surface ice melts, and 4 °C water sinks to the bottom.

Numerous mathematical models of lake dynamics have been developed over the years, many of which have been reviewed by Spitler *et al.* (2012), who build upon a model by Saloranta and Andersen (2007). Their model predicts the temperature distribution for a horizontally mixed and vertically stratified lake, and is written as

$$A\frac{\partial T}{\partial t} = \frac{\partial}{\partial z}\left(k_z A\frac{\partial T}{\partial z}\right) + A\frac{\dot{q}}{\rho c_p} \tag{8.5}$$

where A is the horizontal area of the surface water body, which varies with depth (z), T is the temperature, t is time, k_z is the vertical eddy diffusion coefficient, which may be estimated with a range of models, $\dot{q}$ is the volumetric rate of heat transfer at depth z, and ρ and c_p are the density and specific heat respectively. The mixing dynamics of a lake is controlled by the eddy diffusion, of which there are numerous correlations reviewed and summarized by Spitler *et al.* (2012).

8.3 Open-Loop Systems

A thorough review of open-loop surface water heat exchange systems is given by Mitchell and Spitler (2013). Although they identified a number of system configurations, each one falls into one of three categories: (i) direct surface water cooling (DSWC) systems, (ii) surface water heat pump (SWHP) systems, and (iii) hybrid surface water heat pump (HSWHP) systems. A generalized schematic is shown in Figure 8.3.

Direct surface water cooling (DSWC) systems are those that use seawater or lake water to provide cooling without the use of heat pumps or chillers. Systems may optionally incorporate an intermediate heat exchanger to isolate surface water from the building fan-coil units. Mitchell and Spitler (2013) describe several operating systems worldwide. For example, those at Cornell University and Ithaca High School in Ithaca, New York, began providing direct cooling in 2000. That system draws cold water from Lake Cayuga, and is capable of delivering up to 70 MW (20 000 tons) at peak capacity at a pumping rate of $2\,m^3 \cdot s^{-1}$ (32 000 gpm) through roughly 3.2 km (2 miles) of 1600 mm (63 in) diameter HDPE pipe. The system COP as of 2010 was stated to be 25.8.

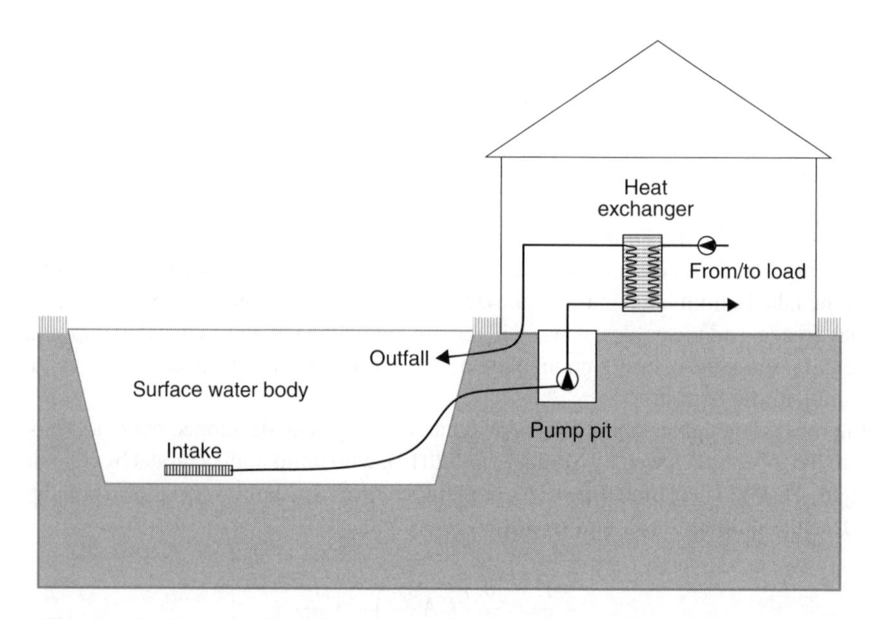

Figure 8.3 General schematic of an open-loop surface water heat exchange system

Surface water heat pump (SWHP) systems are those that use surface water as a heat source or sink for heat pumps or chillers. Depending on location and application, the system may provide heating only, cooling only, or both. Mitchell and Spitler (2013) cite numerous operating systems worldwide. For example, a district heating system in Stockholm, Sweden, utilizes six seawater heat pumps at the Värtan Ropsten plant with a total heating capacity of 180 MW. The heat pumps supply hot water at 80 °C. During the winter, water is taken from the Baltic Sea at 3 °C and returned at 0.5 °C. Under these conditions, a heat pump COP of 3.75 is reported.

Hybrid surface water heat pump (HSWHP) systems are those that use heat pumps or chillers to provide heating and/or cooling, and under favorable conditions use the same surface water body directly to provide cooling. Mitchell and Spitler (2013) cite some operating systems worldwide. For example, the City of Toronto employs an HSWHP system to draw 4 °C water from Lake Ontario at a depth of 83 m. The lake water is then filtered to be used for the city's potable water supply and then passed through the isolation heat exchangers at the heat exchange facility. The direct cooling portion of this system alone is capable of producing 141 MW (40 000 tons) of cooling, while bottoming chillers provide an additional 58 MW (16 500 tons) for a rated capacity of 185 MW (52 200 tons).

Major system components and subsystems of DSWC, SWHP, and HSWHP include:

- Piping with intake screen (similar in configuration to a groundwater well) to bring the surface water source to the on-shore equipment. Intake piping is often the largest capital cost of a surface water heat exchange system, and thus the most feasible systems are very close to the surface water body.
- Pumps and pump sumps to deliver the surface water to the load. Sumps act as pumping pits, and can be either wet or dry sumps.
- Isolation heat exchanger to isolate potentially contaminated or corrosive lake water or seawater from the distribution piping system, heat pump condenser/evaporator, or building fan-coil units.
- Heat pumps or chillers for SWHP and HSWHP systems, and fan coil units for DSWC systems.
- Return piping and mixing devices designed to mix the system return water with the original water source.

8.4 Closed-Loop Systems

Several configurations of closed-loop surface water heat exchanger (SWHX) layouts are possible. Of the numerous types, they generally exist as two broad configurations: (i) coil types and (ii) plate types.

Coil-type SWHXs are commonly installed as bundle coils (or bundle spools), with HDPE being the most common pipe material. A generalized schematic is shown in Figure 8.4. Copper coils have been common in parts of the southern United States. Slinky coil (or 'mat' types) also exist, similar to the slinky configurations installed in horizontal GHXs in soils. Typical construction of bundle coils consists of 100 m (300 ft) or 150 m (500 ft) rolls of DN19 or DN25 (nominal ¾ or 1 in) HDPE, with the pipes either loosely tied or installed with spacers to allow contact with the pond or lake water. The bundles are usually submerged with a weight, which also helps to prevent the loop from becoming buoyant. Buoyancy forces may become significant when cold, dense water sinks to the bottom of the water body.

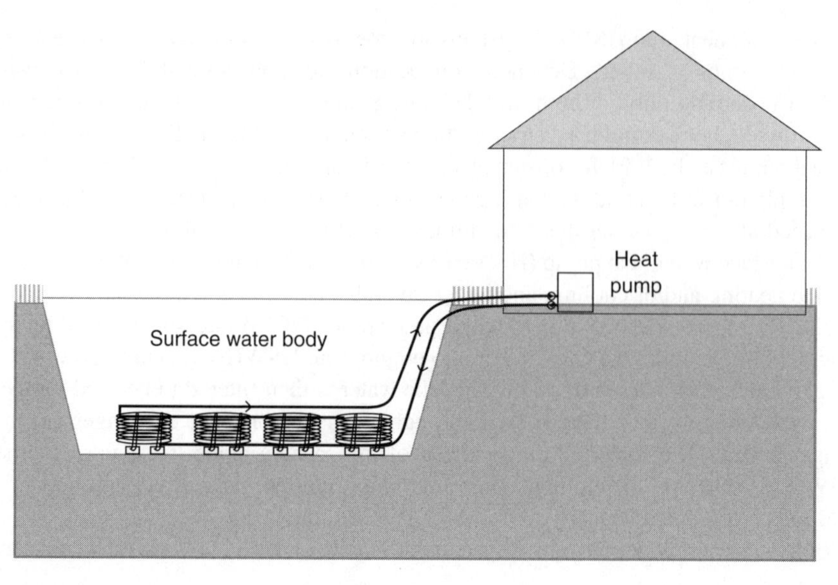

Figure 8.4 General schematic of a closed-loop surface water heat exchange system consisting of a bundle coil configuration

The other main configuration of SWHX is the plate type, which resembles plate-type stainless steel heat exchangers or flat-plate solar collectors. These are compact, and can achieve high heat transfer rates in a small surface area.

8.4.1 Mathematical Models of Closed-Loop Heat Exchangers

Mathematical modeling of closed-loop heat exchangers involves calculating the overall heat transfer coefficient (*UA*) of the heat exchanger, which is a function of the inside and outside convective resistance and conductive resistance through the heat exchanger material. The total thermal resistance is given by

$$R_{total} = R_i + R_p + R_o \tag{8.6}$$

where R_i, R_p, and R_o represent the convective thermal resistance inside the heat exchanger flow channel, the conductive resistance of the pipe or plate heat exchanger material, and the convective thermal resistance outside the heat exchanger. The overall heat transfer coefficient (*UA*) is then given by

$$UA = \frac{1}{R_{total}} \tag{8.7}$$

Assuming a large thermal inertia of the surface water body over the time step of interest, the number of transfer units (NTU) of the surface water heat exchanger can be approximated as

$$\mathrm{NTU} = \frac{UA}{\dot{m}c_p} \tag{8.8}$$

where $\dot{m}$ is the mass flow rate of the heat exchange fluid, and c_p is the specific heat of the heat exchange fluid evaluated at the average fluid temperature. The heat exchanger effectiveness (ε_{HX}) is then given by

$$\varepsilon_{HX} = 1 - \mathrm{e}^{-NTU} \tag{8.9}$$

The heat transfer rate ($\dot{q}$) between the heat exchanger and the surface water body is then given by

$$\dot{q} = \varepsilon_{HX}\dot{m}c_p\left(T_{in} - T_{pond}\right) \tag{8.10}$$

where T_{in} is the inlet fluid temperature to the heat exchanger (which is also the fluid temperature exiting the heat pump) and T_{pond} is the average temperature of the surface water body. If $\dot{q}$ is known from the building loads and heat pump performance (i.e., the surface water loads analogous to the ground loads), T_{in} can be readily calculated, thus allowing calculation of the outlet fluid temperature of the heat exchanger (or the heat pump entering fluid temperature).

8.4.1.1 Thermal Resistance Formulas for Bundle Spool Heat Exchangers

The convective resistance due to fluid flow inside the pipe (R_i) is given by

$$R_i = \frac{1}{\pi D_i L h_i} \tag{8.11}$$

where D_i is the pipe inside diameter, L is the pipe length, and h_i is the inside convection coefficient, given by

$$h_i = \frac{Nu_i k_{f,i}}{D_i} \tag{8.12}$$

where $k_{f,i}$ is the thermal conductivity of the heat exchange fluid, and Nu_i is the Nusselt number of the internal flow. An appropriate correlation for the internal Nusselt number for flow in spiraled pipes (Nu_i) is given by Salimpour (2009):

$$Nu_i = 0.152 De^{0.431} Pr^{1.06} PitchRatio^{-0.277} \tag{8.13}$$

where De is the Dean number, defined as

$$De = Re\sqrt{\frac{D_i}{D_{coil,o}}} \tag{8.14}$$

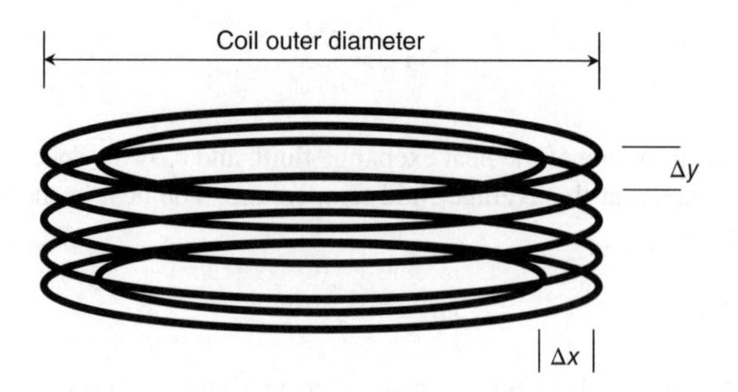

Figure 8.5 Coil heat exchanger geometric parameters for determining convective thermal resistances

and the pitch ratio is defined as

$$PitchRatio = \frac{\Delta y}{\pi D_{coil,o}} \tag{8.15}$$

where $D_{coil,o}$ is the coil outer diameter, and Δy is the vertical center-to-center distance between adjacent heat exchanger pipes, as shown in Figure 8.5.

The conductive resistance of the pipe (R_p) is given by

$$R_p = \frac{\ln\left({}^{D_o}/_{D_i}\right)}{2\pi k_p L} \tag{8.16}$$

where D_i and D_o are the pipe inside and outside diameter respectively, L is the pipe length, and k_p is the pipe thermal conductivity.

The convective resistance due to fluid flow on the outside surface of the pipe (R_o) is given by

$$R_o = \frac{1}{\pi D_o L h_o} \tag{8.17}$$

where D_o is the pipe outside diameter, L is the pipe length, and h_o is the outside convection coefficient, given by

$$h_o = \frac{Nu_o k_{f,o}}{D_o} \tag{8.18}$$

where $k_{f,o}$ is the thermal conductivity of the surface water, and Nu_o is the Nusselt number of the external flow. Assuming natural convection on the pipe outside surface, an appropriate correlation for the external Nusselt number is given by Hansen (2011):

$$Nu_o = 0.16 Ra^{0.264}\left(\frac{\Delta y}{D_o}\right)^{0.078}\left(\frac{\Delta x}{D_o}\right)^{0.223} \tag{8.19}$$

where Δx is the horizontal center-to-center distance between adjacent heat exchanger pipes, and Ra is the Rayleigh number, defined as

$$Ra = \frac{g\beta\left(T_{p,s} - T_{pond}\right)D_o^3}{\nu\alpha} \tag{8.20}$$

where $T_{p,s}$ is the pipe surface temperature, and all properties are evaluated at the pipe outside film temperature.

8.4.1.2 Thermal Resistance Formulas for Lake Plate-Type Heat Exchangers

Figure 8.6 shows some of the geometric parameters for determining the convective resistances of plate-type heat exchangers in surface water bodies.

The convective resistance due to fluid flow inside the flow channels (R_i) is given by

$$R_i = \frac{1}{PLh_i} \tag{8.21}$$

where P and L are the perimeter and length of the flow channel, respectively, which are determined by the width and height of the plate, the thickness of the plate heat exchanger assembly (i.e., the dimension between the plates), and the number of passes, and h_i is the inside convection coefficient, given by

$$h_i = \frac{Nu_i k_{f,i}}{D_i} \tag{8.22}$$

where D_i is the hydraulic diameter of the flow channel (i.e., 4 × cross-sectional area ÷ perimeter), $k_{f,i}$ is the thermal conductivity of the heat exchange fluid, and Nu_i is the Nusselt number of

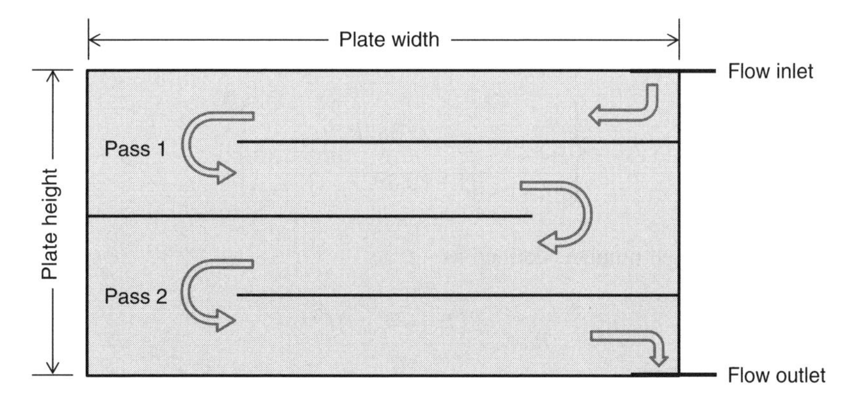

Figure 8.6 Lake plate heat exchanger geometric parameters for determining convective thermal resistances

the internal flow. An appropriate correlation for the Nusselt number for internal flow is given by the Dittus–Boelter equation:

$$Nu_i = 0.0243 Re^{0.8} Pr^{0.4} \quad (8.23a)$$

when the heat exchange fluid is heating, and

$$Nu_i = 0.0265 Re^{0.8} Pr^{0.3} \quad (8.23b)$$

when the heat exchange fluid is cooling.

The conductive resistance of the plate wall (R_p) is given by

$$R_p = \frac{x}{k_p A} \quad (8.24)$$

where x is the thickness of the plate wall (typically 1.5 mm or 1/16 in), A is the plate exterior area (i.e., approximately $2 \times \text{height} \times \text{width}$), and k_p is the thermal conductivity of the plate material.

The convective resistance due to fluid flow on the outside surface of the plate (R_o) is given by

$$R_o = \frac{1}{A h_o} \quad (8.25)$$

where A is the exterior surface area of the plate heat exchanger, and h_o is the outside convection coefficient, given by

$$h_o = \frac{Nu_o k_{f,o}}{L} \quad (8.26)$$

where L is a characteristic length (taken as the height of the plate), $k_{f,o}$ is the thermal conductivity of the surface water, and Nu_o is the Nusselt number of the external flow. Assuming natural convection on the plate outside surface, an appropriate correlation for the external Nusselt number is given by the Churchill–Chu correlation for vertical plates:

$$Nu = \left\{ 0.825 + \frac{0.387 Ra^{1/6}}{\left[1 + \left(0.492/_{\mathrm{Pr}}\right)^{9/16}\right]^{8/27}} \right\}^2 \quad (8.27)$$

where Ra is the Rayleigh number, defined as

$$Ra = \frac{g\beta\left(T_{p,s} - T_{pond}\right)L^3}{\nu\alpha} \quad (8.28)$$

where $T_{p,s}$ is the plate surface temperature, L is a characteristic length (taken as the height of the plate), and all properties are evaluated at the plate outside film temperature.

8.4.2 *Modeling Closed-Loop Surface Water Heat Pump Systems with Software Tools*

As we saw with other GHX design procedures, manual solution of the governing equations for heat transfer in surface water bodies is not practical owing to the iterative and tedious nature of the process. Thus, commercially available software packages have been developed for surface water heat exchanger design.

The foregoing algorithms and equations have been implemented into the suite of GHX tools found on the book companion website for use in closed-loop surface water HX design problems (Figure 8.7). The model has been developed for simulating bundle spool or plate-type heat exchanger performance in unstratified, well-mixed surface water bodies with known building loads.

Selection of the option button *Surface Water Heat Exchanger Simulation* launches two worksheets: (i) *SurfaceWaterHX* and (ii) *Weather Data*. A screen capture of the input data on the *SurfaceWaterHX* spreadsheet is shown in Figure 8.8. The model allows simulation of two basic options: with or without building loads. In the first option, only the average monthly temperature of the surface water body is calculated. In the second option, the average monthly temperature of the surface water body is calculated, in addition to monthly heat pump

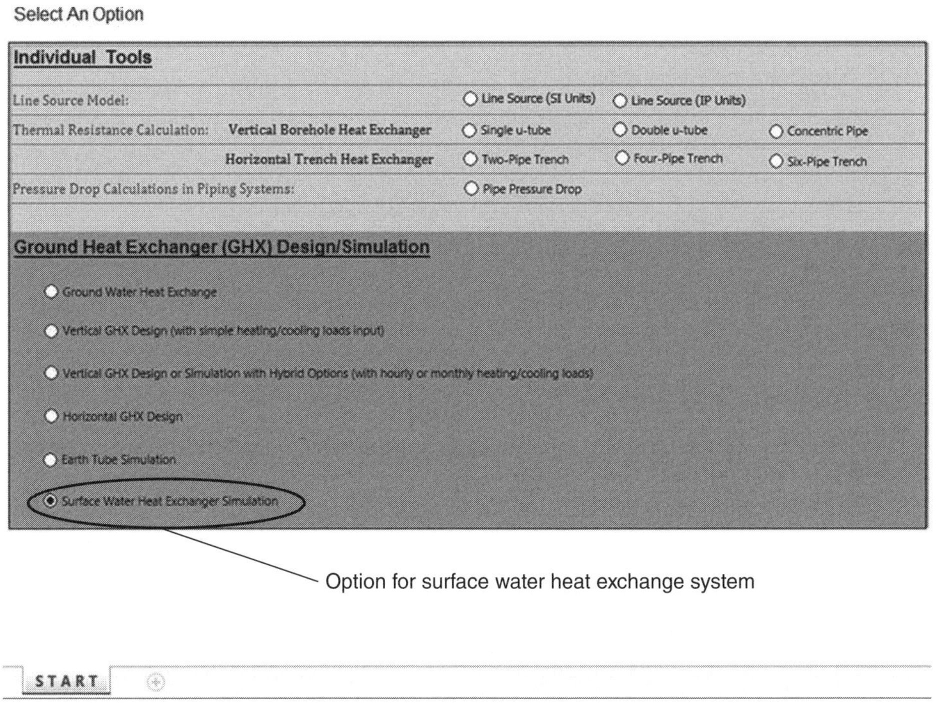

Figure 8.7 Screen capture of the START worksheet of the companion GHX tools highlighting the horizontal GHX design option

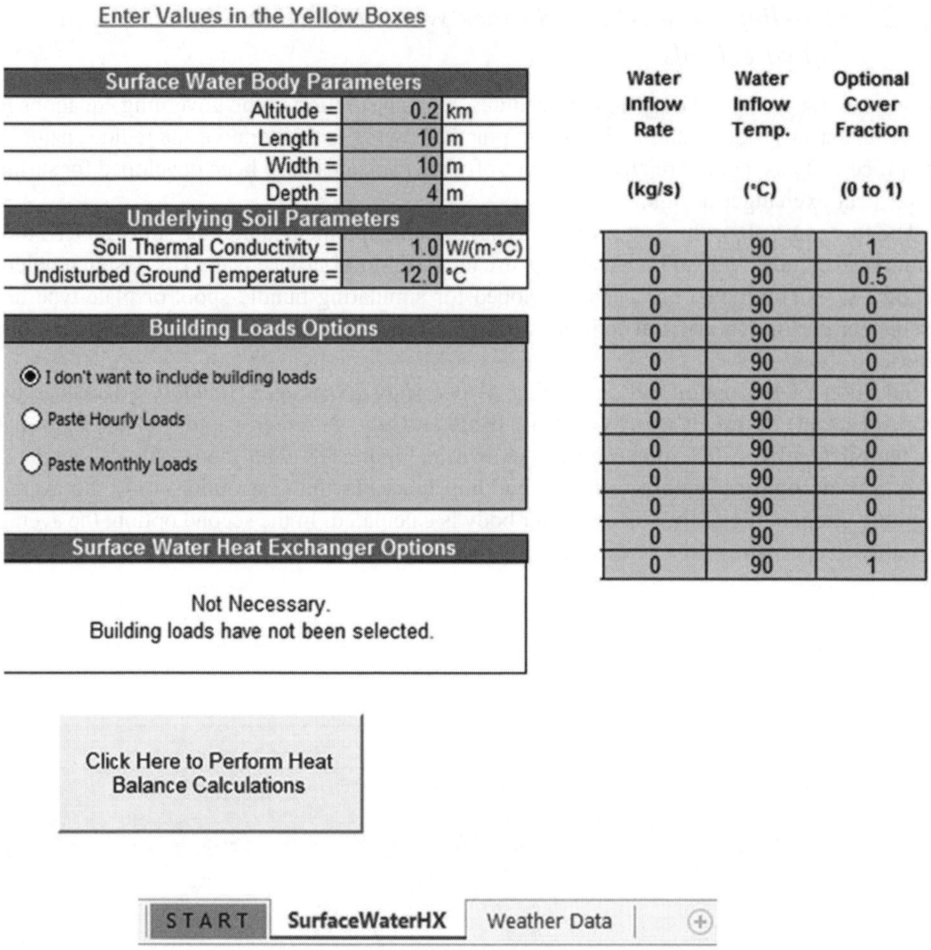

Enter Values in the Yellow Boxes

Surface Water Body Parameters		
Altitude =	0.2	km
Length =	10	m
Width =	10	m
Depth =	4	m
Underlying Soil Parameters		
Soil Thermal Conductivity =	1.0	W/(m·°C)
Undisturbed Ground Temperature =	12.0	°C

Building Loads Options

◉ I don't want to include building loads

○ Paste Hourly Loads

○ Paste Monthly Loads

Surface Water Heat Exchanger Options

Not Necessary.
Building loads have not been selected.

Click Here to Perform Heat Balance Calculations

Water Inflow Rate (kg/s)	Water Inflow Temp. (°C)	Optional Cover Fraction (0 to 1)
0	90	1
0	90	0.5
0	90	0
0	90	0
0	90	0
0	90	0
0	90	0
0	90	0
0	90	0
0	90	0
0	90	0
0	90	1

START | SurfaceWaterHX | Weather Data

Figure 8.8 Partial screen capture of the *SurfaceWaterHX* worksheet of the companion GHX tools with no loads

fluid temperatures and monthly heat pump energy consumption. These options are described in the subsections that follow.

On the *Weather Data* worksheet, the model expects average monthly weather data, such as the data described in Chapter 2.2. Weather data of importance to the surface water HX model include: ambient air temperature, relative humidity, average daily solar radiation on horizontal, atmospheric pressure, and wind speed.

Further elaboration on the use of relative humidity data is warranted here. Recall that the evaporation heat balance equation uses humidity ratios [Equation (2.20)]. Thus, correlations of ASHRAE (2009, 2013) are used to determine the humidity ratios of moist air and the saturated surface of the water body as a function of relative humidity and atmospheric pressure. The humidity ratio of moist air (w_{air}) in units of kg H_2O/kg dry air is given by

$$w_{air} = 0.62198\left(\frac{p_w}{p_{air} - p_{w,sat}}\right) \tag{8.29}$$

where pressures are in Pa, and p_w is the partial pressure of water, given by

$$p_w = p_{w,sat} \times RH \tag{8.30}$$

where RH is the relative humidity of the air, and $p_{w,sat}$ is the saturation pressure of water for T_{air} (in K) less than 273.15 K, given by

$$p_{w,sat} = \exp\left(\frac{c_1}{T_{air}} + c_2 + c_3 T_{air} + c_4 T_{air}^2 + c_5 T_{air}^3 + c_6 T_{air}^4 + c_7 \ln(T_{air})\right) \tag{8.31a}$$

For T_{air} (in K) greater than or equal to 273.15 K:

$$p_{w,sat} = \exp\left(\frac{c_8}{T_{air}} + c_9 + c_{10} T_{air} + c_{11} T_{air}^2 + c_{12} T_{air}^3 + c_{13} \ln(T_{air})\right) \tag{8.31b}$$

The values of the constants in Equations (8.31a) and (8.31b) are: $c_1 = -5674.5359$, $c_2 = 6.3925247$, $c_3 = -0.009677843$, $c_4 = 0.00000062215701$, $c_5 = 2.0747825\text{E-}09$, $c_6 = -9.484024\text{E-}13$, $c_7 = 4.1635019$, $c_8 = -5800.2206$, $c_9 = 1.3914993$, $c_{10} = -0.048640239$, $c_{11} = 0.000041764768$, $c_{12} = -0.000000014452093$, and $c_{13} = 6.5459673$.

The humidity ratio at the surface of the water body (w_{surf}) in units of kg H_2O/kg dry air is given by

$$w_{surf} = \frac{(2501000 + (1805 - 4186)T_w)w^*}{2501000 + 1805T_w - 4186T_w} \tag{8.32}$$

where T_w is the water surface temperature (K), and w^* is given by

$$w^* = 0.62198\left(\frac{p_w}{p_{air} - p_w}\right) \tag{8.33}$$

where p_w is calculated with Equation (8.31b).

8.4.2.1 Surface Water Temperature Calculation with No Loads

Constructing the Model. The input fields on the *SurfaceWaterHX* worksheet shown in Figure 8.8 are divided into three groups (excluding *Building Loads Options*): (1) surface water body parameters, (2) underlying soil parameters, and (3) monthly inflows (mass flow rate and temperature) and cover fraction (due to ice or an engineered cover).

In the *Surface Water Body Parameters* section of the *SurfaceWaterHX* worksheet, the basic geometry parameters of length, width, and depth are entered. In the *Underlying Soil Parameters* section of the *SurfaceWaterHX* worksheet, the soil thermal conductivity and the underground undisturbed temperature are entered. The model also includes monthly average inflow rates and the associated temperature. These inflows could be attributed to storms, surface runoff, tributaries, or engineered tempering of the surface water temperature with groundwater wells. An optional cover fraction is also included to model surface ice or an engineered cover. Evaporation enhancements such as fountains or bubblers are not explicitly included in the model. As shown in Figure 8.8, 100% ice cover is assumed in December and January, and 50% ice cover is assumed in February.

Performing the Calculations. The calculation of the average temperature of the surface water body is iterative; thermal properties depend on the surface water temperature, which is unknown. Also, the long-wave thermal radiation energy balance equation involves temperatures to the fourth power.

Mouse-clicking on the command button *Click Here to Perform Heat Balance Equations* executes a Visual Basic for Applications (VBA) code that makes use of a half-interval method iteratively to solve for the average surface water temperature for each month. Energy balance equations as described in Chapter 2 are assembled into the form of a linear, first-order, ordinary differential equation described by Equation (8.3). Heat transfer due to evaporation is determined using methods described above. Equation (8.3) is solved using monthly time steps as described by Equation (8.4). A convergence criterion of the surface water temperature of 0.01 °C is used. The model also iterates on an annual basis under the assumption that the initial average surface water temperature should match the average temperature in December.

A partial screen capture of the *SurfaceWaterHX* worksheet displaying the model results is shown in Figure 8.9, based on the input data shown in Figure 8.8 and weather data representative of Dayton, Ohio. Monthly temperatures are plotted, and monthly heat fluxes are tabulated.

8.4.2.2 Adding Building Loads and Closed-Loop Heat Exchangers

Constructing the Model. Selecting the option button *Paste Hourly Loads* or *Paste Monthly Loads* launches the *Heat Pump Input* worksheet, and either the *Hourly Loads* worksheet or the *Monthly Loads* worksheet. In addition, option buttons for selecting the type of surface water heat exchanger become active. Depending on which surface water heat exchanger option button is selected, either the *BundleSpoolHXDetails* worksheet or the *PlateHXDetails* worksheet becomes active. Figure 8.10 shows a partial screen capture of the *SurfaceWaterHX* worksheet where monthly load input for a bundle/spool coil heat exchanger has been selected.

The only new worksheets here are the *BundleSpoolHXDetails* and the *PlateHXDetails* worksheets. The *Heat Pump Input* worksheet has been described in Chapters 4 and 6 for simulation of groundwater and closed-loop vertical borehole GHXs, and a screen capture is shown in Figure 4.22. Similarly, the *Monthly Loads* and the *Hourly Loads* worksheets have been

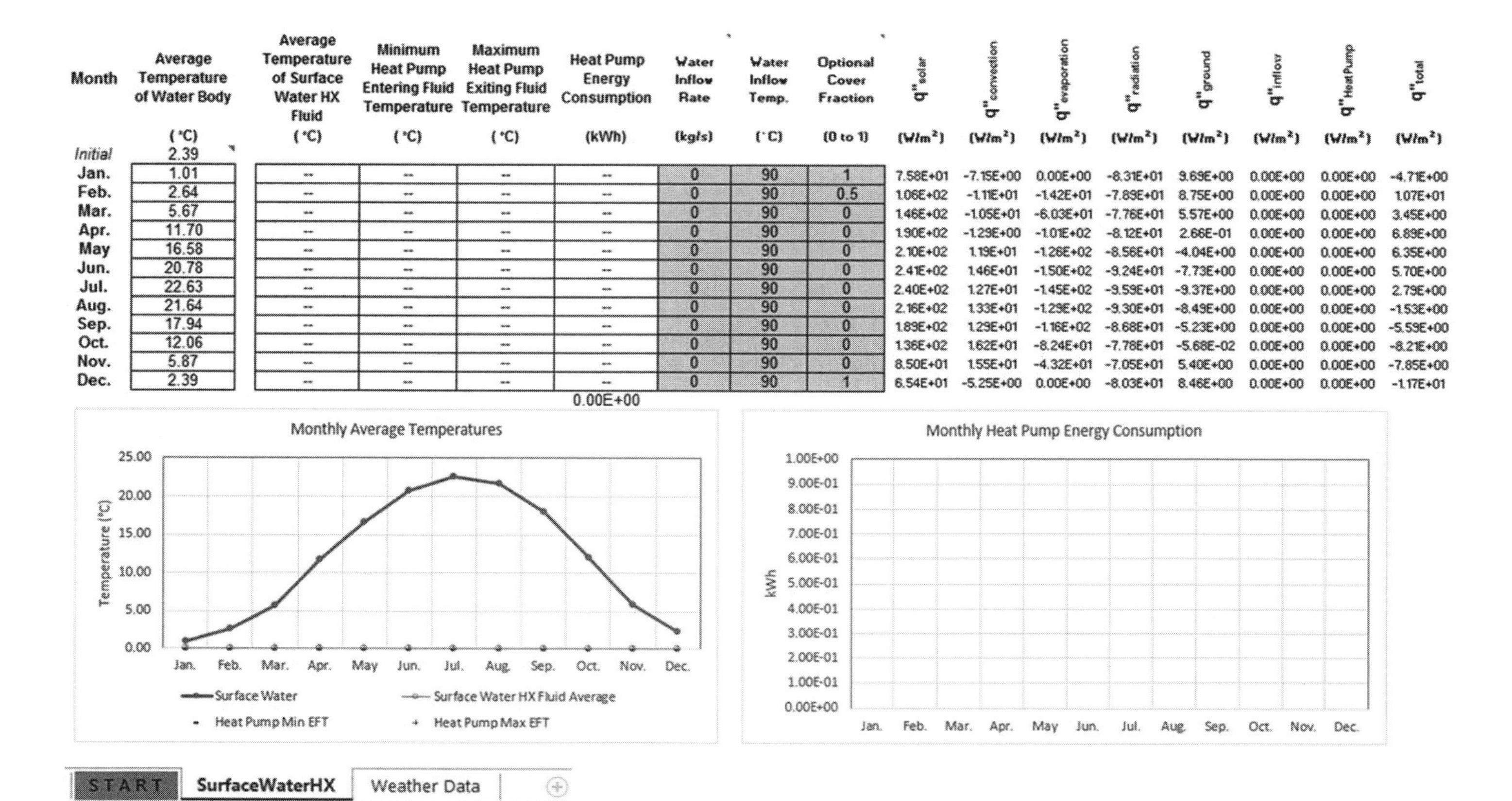

Month	Average Temperature of Water Body (°C)	Average Temperature of Surface Water HX Fluid (°C)	Minimum Heat Pump Entering Fluid Temperature (°C)	Maximum Heat Pump Exiting Fluid Temperature (°C)	Heat Pump Energy Consumption (kWh)	Water Inflow Rate (kg/s)	Water Inflow Temp. (°C)	Optional Cover Fraction (0 to 1)
Initial	2.39							
Jan.	1.01	--	--	--	--	0	90	1
Feb.	2.64	--	--	--	--	0	90	0.5
Mar.	5.67	--	--	--	--	0	90	0
Apr.	11.70	--	--	--	--	0	90	0
May	16.58	--	--	--	--	0	90	0
Jun.	20.78	--	--	--	--	0	90	0
Jul.	22.63	--	--	--	--	0	90	0
Aug.	21.64	--	--	--	--	0	90	0
Sep.	17.94	--	--	--	--	0	90	0
Oct.	12.06	--	--	--	--	0	90	0
Nov.	5.87	--	--	--	--	0	90	0
Dec.	2.39	--	--	--	--	0	90	1
					0.00E+00			

Month	q''_{solar} (W/m²)	$q''_{convection}$ (W/m²)	$q''_{evaporation}$ (W/m²)	$q''_{radiation}$ (W/m²)	q''_{ground} (W/m²)	q''_{inflow} (W/m²)	$q''_{HeatPump}$ (W/m²)	q''_{total} (W/m²)
Jan.	7.58E+01	-7.15E+00	0.00E+00	-8.31E+01	9.69E+00	0.00E+00	0.00E+00	-4.71E+00
Feb.	1.06E+02	-1.11E+01	-1.42E+01	-7.89E+01	8.75E+00	0.00E+00	0.00E+00	1.07E+01
Mar.	1.46E+02	-1.05E+01	-6.03E+01	-7.76E+01	5.57E+00	0.00E+00	0.00E+00	3.45E+00
Apr.	1.90E+02	-1.29E+00	-1.01E+02	-8.12E+01	2.66E-01	0.00E+00	0.00E+00	6.89E+00
May	2.10E+02	1.19E+01	-1.26E+02	-8.56E+01	-4.04E+00	0.00E+00	0.00E+00	6.35E+00
Jun.	2.41E+02	1.46E+01	-1.50E+02	-9.24E+01	-7.73E+00	0.00E+00	0.00E+00	5.70E+00
Jul.	2.40E+02	1.27E+01	-1.45E+02	-9.59E+01	-9.37E+00	0.00E+00	0.00E+00	2.79E+00
Aug.	2.16E+02	1.33E+01	-1.29E+02	-9.30E+01	-8.49E+00	0.00E+00	0.00E+00	-1.53E+00
Sep.	1.89E+02	1.29E+01	-1.16E+02	-8.68E+01	-5.23E+00	0.00E+00	0.00E+00	-5.59E+00
Oct.	1.36E+02	1.62E+01	-8.24E+01	-7.78E+01	-5.68E-02	0.00E+00	0.00E+00	-8.21E+00
Nov.	8.50E+01	1.55E+01	-4.32E+01	-7.05E+01	5.40E+00	0.00E+00	0.00E+00	-7.85E+00
Dec.	6.54E+01	-5.25E+00	0.00E+00	-8.03E+01	8.46E+00	0.00E+00	0.00E+00	-1.17E+01

Figure 8.9 Partial screen capture of the *SurfaceWaterHX* worksheet of the companion GHX tools, showing model results with no building loads

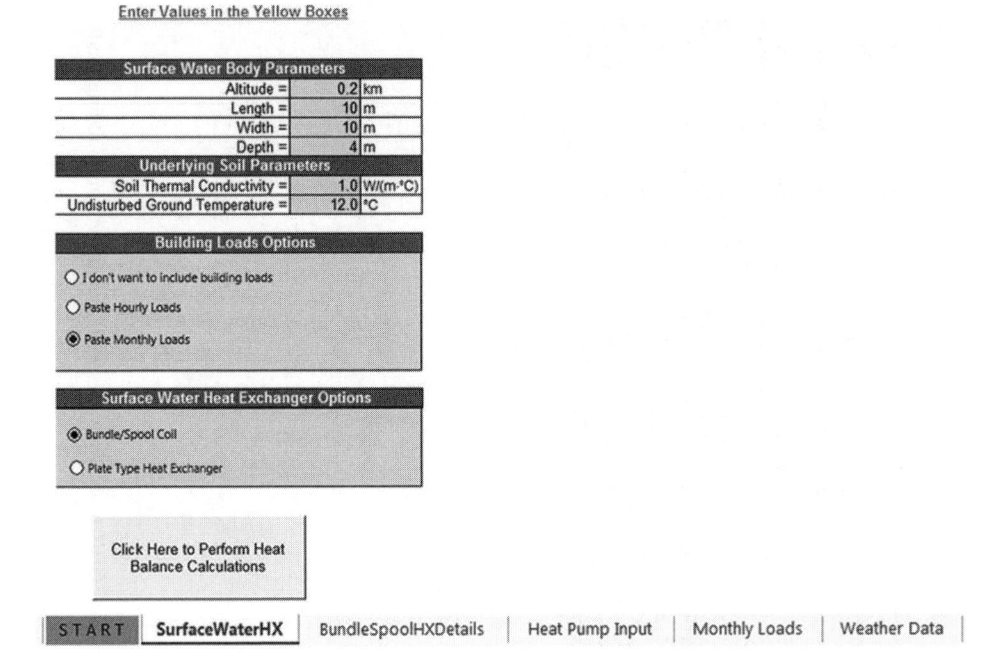

Figure 8.10 Partial screen capture of the *SurfaceWaterHX* worksheet of the companion GHX tools with loads and surface heat exchanger options

described in Chapter 6 for simulation of closed-loop vertical borehole GHXs (see Figures 6.13 and 6.14).

Screen captures of the *BundleSpoolHXDetails* and the *PlateHXDetails* worksheets are shown in Figures 8.11 and 8.12 respectively.

Performing the Calculations. The calculation of the fluid temperature exiting the surface water heat exchanger is iterative; it depends on its own thermal properties, and on the surface water temperature, which are unknown.

Mouse-clicking on the command button *Click Here to Perform Heat Balance Equations* executes a Visual Basic for Applications (VBA) code that performs the same tasks as described previously in Section 8.4.2.1. In addition, an iterative loop internal to the surface water iterative loop is included to determine the fluid temperatures in the surface water heat exchanger. When converged, based on the current iterate of the surface water temperature, the heat transfer rate due to the heat exchanger is added to the heat balance equation for the surface water body.

A partial screen capture of the *SurfaceWaterHX* worksheet displaying the model results is shown in Figure 8.13, based on the input data shown in Figure 8.10 (Surface Water HX input), Figure 8.11 (Bundle Spool HX details), Figure 4.22 (Heat Pump Input), Figure 6.13 (Monthly Loads), and weather data representative of Dayton, Ohio. Monthly surface water and heat pump fluid temperatures are plotted, monthly heat pump energy consumption is plotted, and monthly heat fluxes are tabulated.

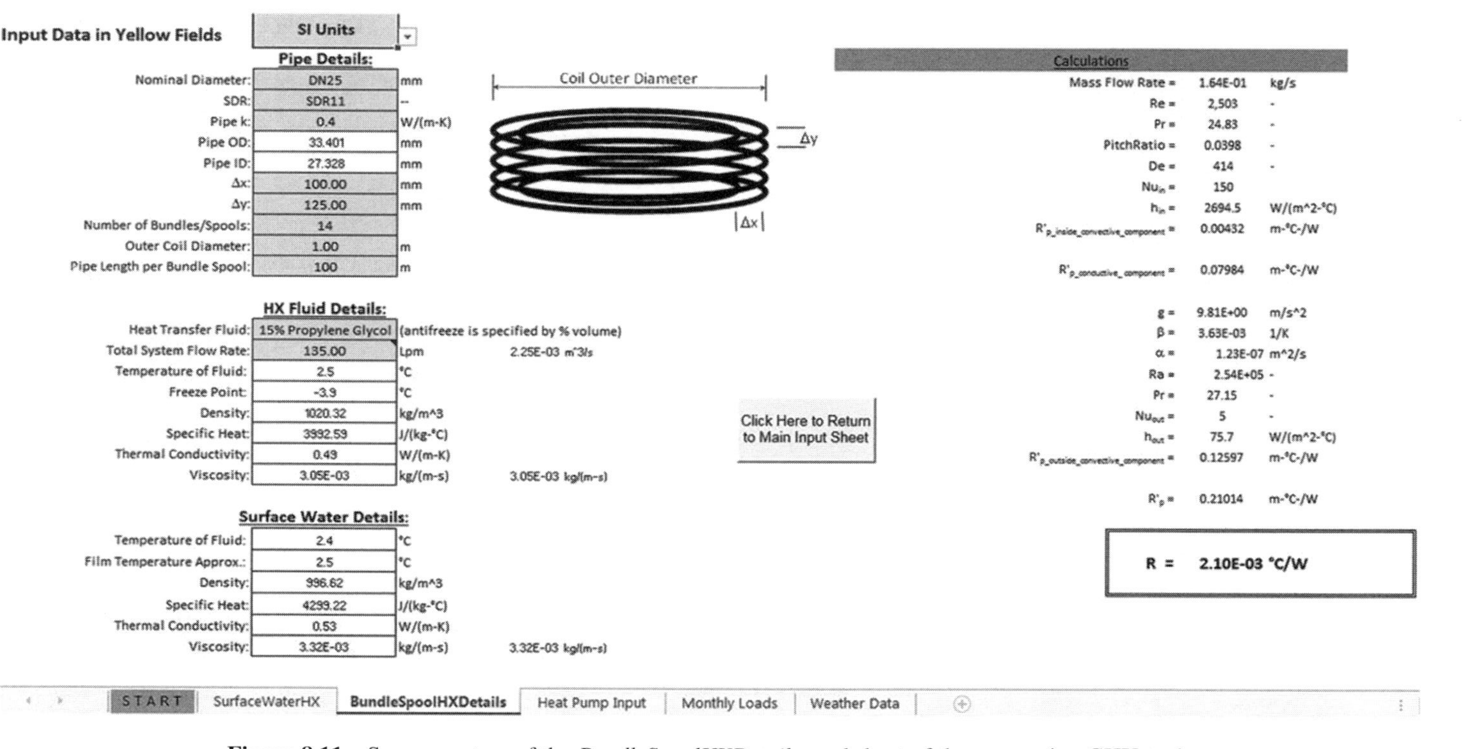

Figure 8.11 Screen capture of the *BundleSpoolHXDetails* worksheet of the companion GHX tools

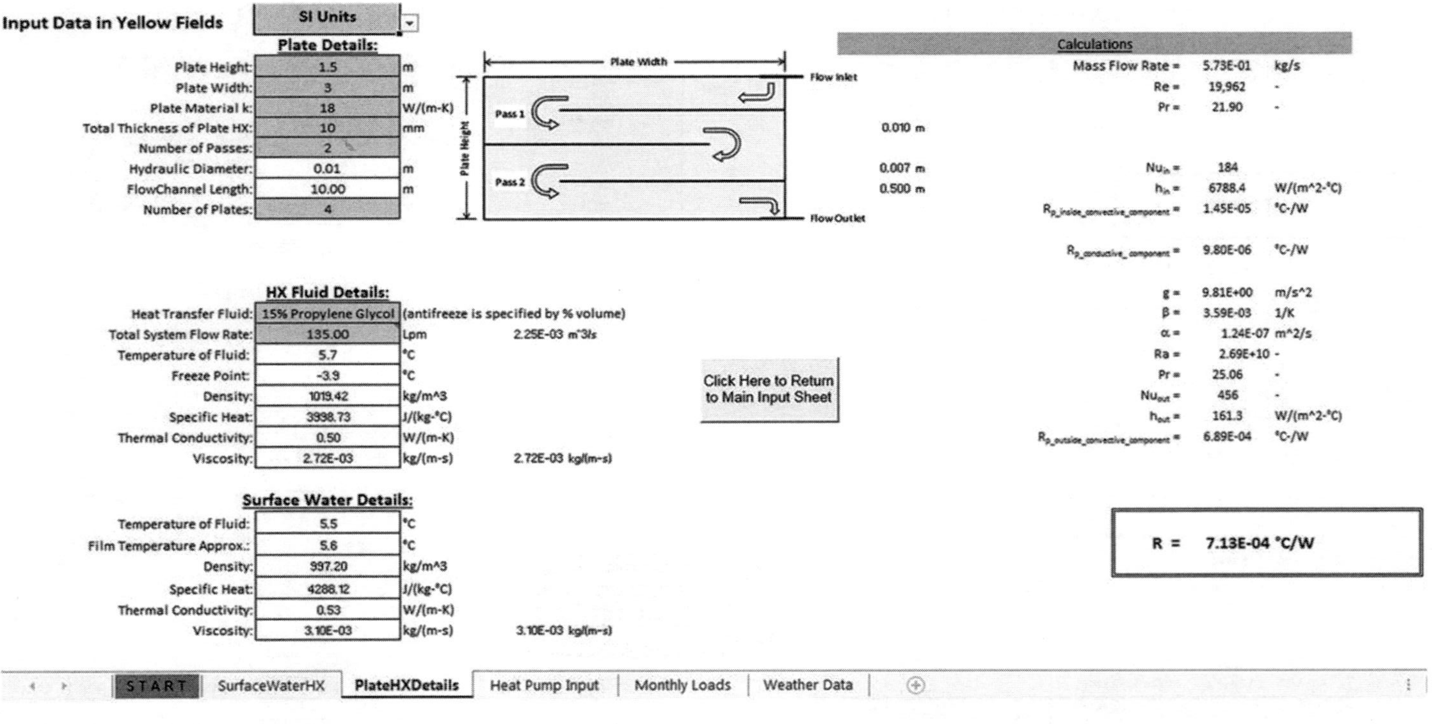

Figure 8.12 Screen capture of the *PlateHXDetails* worksheet of the companion GHX tools

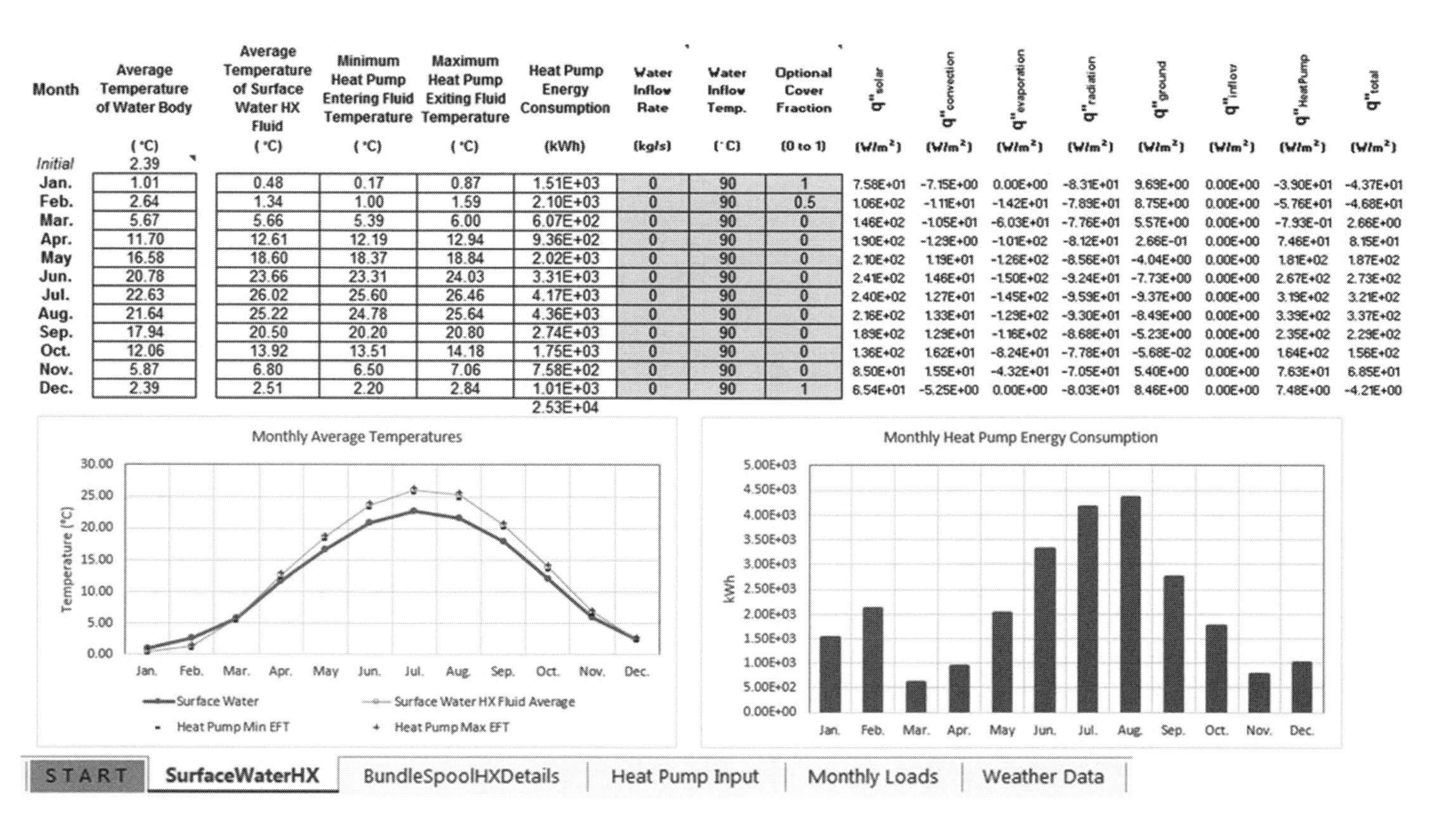

Month	Average Temperature of Water Body (°C)	Average Temperature of Surface Water HX Fluid (°C)	Minimum Heat Pump Entering Fluid Temperature (°C)	Maximum Heat Pump Exiting Fluid Temperature (°C)	Heat Pump Energy Consumption (kWh)	Water Inflow Rate (kg/s)	Water Inflow Temp. (°C)	Optional Cover Fraction (0 to 1)
Initial	2.39							
Jan.	1.01	0.48	0.17	0.87	1.51E+03	0	90	1
Feb.	2.64	1.34	1.00	1.59	2.10E+03	0	90	0.5
Mar.	5.67	5.66	5.39	6.00	6.07E+02	0	90	0
Apr.	11.70	12.61	12.19	12.94	9.36E+02	0	90	0
May	16.58	18.60	18.37	18.84	2.02E+03	0	90	0
Jun.	20.78	23.66	23.31	24.03	3.31E+03	0	90	0
Jul.	22.63	26.02	25.60	26.46	4.17E+03	0	90	0
Aug.	21.64	25.22	24.78	25.64	4.36E+03	0	90	0
Sep.	17.94	20.50	20.20	20.80	2.74E+03	0	90	0
Oct.	12.06	13.92	13.51	14.18	1.75E+03	0	90	0
Nov.	5.87	6.80	6.50	7.06	7.58E+02	0	90	0
Dec.	2.39	2.51	2.20	2.84	1.01E+03	0	90	1
					2.53E+04			

Month	q''_{solar} (W/m²)	$q''_{convection}$ (W/m²)	$q''_{evaporation}$ (W/m²)	$q''_{radiation}$ (W/m²)	q''_{ground} (W/m²)	q''_{inflow} (W/m²)	$q''_{HeatPump}$ (W/m²)	q''_{total} (W/m²)
Jan.	7.58E+01	-7.15E+00	0.00E+00	-8.31E+01	9.69E+00	0.00E+00	-3.90E+01	-4.37E+01
Feb.	1.06E+02	-1.11E+01	-1.42E+01	-7.89E+01	8.75E+00	0.00E+00	-5.76E+01	-4.68E+01
Mar.	1.46E+02	-1.05E+01	-6.03E+01	-7.76E+01	5.57E+00	0.00E+00	-7.93E-01	2.66E+00
Apr.	1.90E+02	-1.29E+00	-1.01E+02	-8.12E+01	2.66E-01	0.00E+00	7.46E+01	8.15E+01
May	2.10E+02	1.19E+01	-1.26E+02	-8.56E+01	-4.04E+00	0.00E+00	1.81E+02	1.87E+02
Jun.	2.41E+02	1.46E+01	-1.50E+02	-9.24E+01	-7.73E+00	0.00E+00	2.67E+02	2.73E+02
Jul.	2.40E+02	1.27E+01	-1.45E+02	-9.59E+01	-9.37E+00	0.00E+00	3.19E+02	3.21E+02
Aug.	2.16E+02	1.33E+01	-1.29E+02	-9.30E+01	-8.49E+00	0.00E+00	3.39E+02	3.37E+02
Sep.	1.89E+02	1.29E+01	-1.16E+02	-8.68E+01	-5.23E+00	0.00E+00	2.35E+02	2.29E+02
Oct.	1.36E+02	1.62E+01	-8.24E+01	-7.78E+01	-5.68E-02	0.00E+00	1.64E+02	1.56E+02
Nov.	8.50E+01	1.55E+01	-4.32E+01	-7.05E+01	5.40E+00	0.00E+00	7.63E+01	6.85E+01
Dec.	6.54E+01	-5.25E+00	0.00E+00	-8.03E+01	8.46E+00	0.00E+00	7.48E+00	-4.21E+00

Figure 8.13 Partial screen capture of the *SurfaceWaterHX* worksheet of the companion GHX tools, showing model results with building loads

8.5 Chapter Summary

This chapter extended our knowledge of ground heat exchangers to that of surface water. Several new environmental heat transfers that impact the temperature distribution in a surface water body were considered. The seasonal dynamics of water movement within a surface water body was discussed, which is mainly attributed to the unique property of water, where its maximum density is ~4 °C.

The two general configurations of surface water heat exchange systems were described: (i) open-loop systems and (ii) closed-loop systems. In open-loop systems, surface water is used directly for heating or cooling purposes (usually cooling). Closed-loop systems are generally integrated with a heat pump, and are sometimes known as surface water heat pump (SWHP) systems.

A further breakdown of open-loop surface water heat exchange systems was discussed: (i) direct surface water cooling (DSWC) systems, (ii) surface water heat pump (SWHP) systems, and (iii) hybrid surface water heat pump (HSWHP) systems.

Mathematical models of closed-loop surface water heat exchange systems were described, namely the bundle/spool type and the surface water plate type. The software tool provided on the book companion website was illustrated for use in simulation of closed-loop surface water heat exchanger problems.

Discussion Questions and Exercise Problems

8.1 The peak cooling load for a newly planned five-floor classroom building in Dayton, Ohio, is 155 tons or 540 kW. On an annual basis, the cooling demand for the building exceeds the heating demand by a factor of about 6.5. Hourly loads can be found on the loads spreadsheet on the book companion website.

A large pond is being planned for surface run-off control, as well as for landscaping and aesthetics. The pond has been proposed as a heat source/sink for a distributed water-source heat pump system, and you have been consulted to address the questions below. The site survey indicates that the local soil is clay. There is a maximum land area of 1½ acres or 0.6 hectares in which to construct the pond. The maximum pond depth that can be easily dug is 15 ft or 4.5 m.

(a) Design a surface water heat exchanger using HDPE coils. How many coils do you need and what diameter pipe? What is the minimum pond size (surface area and depth) that is needed to meet the cooling loads? (By 'meeting the cooling loads', the entering heat pump fluid temperature should not exceed 90 °F or 32 °C.)

(b) Suggest engineering options (at least two) that you could implement to reduce the pond size that you determined in (a).

Hint: As a starting point, try one 100 m coil per refrigeration ton.

8.2 Complete problem 8.1 with surface water plate-type heat exchangers.

9

Opportunistic Heat Sources and Sinks

9.1 Overview

This chapter describes opportunistic heat sources and sinks for incorporation into geothermal energy systems. In a 'design with the environment' approach, opportunistic considerations were mentioned in Chapter 3 with regard to site characterization. Here, we take a more detailed, but high-level look.

First, what is meant by 'opportunistic heat sources and sinks'? By the inherent nature of geothermal energy systems, they make use of energy stored in the Earth, and transfer that energy to loads via a liquid heat transfer medium. Thus, any heat source or sink may be incorporated into a geothermal energy system if the following desirable attributes are available: (i) infrastructure exists or is planned that can advantageously make use of Earth heat exchange, (ii) synergies exist among loads, (iii) the system is hydronic, and/or (iv) thermal energy storage is advantageous.

Learning objectives and goals:

1. Appreciate the numerous opportunities for incorporation of synergistic hydronic systems into geothermal energy systems.
2. Develop an open mind about coupling various heat/source sink opportunities to Earth heat exchange, limited by your imagination.

9.2 Use of Existing Water Wells

As discussed in Chapter 4 (Section 4.5.1), groundwater heat pumps make practical sense in residential buildings on well water if the groundwater is of good quality. In these applications, the heat pump can be integrated into the household water system and considered as just another water-using appliance. In either new construction or retrofit application, design care must be

Geothermal Heat Pump and Heat Engine Systems: Theory and Practice, First Edition. Andrew D. Chiasson.

Companion website: www.wiley.com/go/chiasson/geoHPSTP

taken to ensure that the well has adequate yield, and that the pressure tank has adequate capacity to handle the additional flow demand of the heat pump.

9.3 Heat Exchange With Building Foundations

Building foundations represent a potential opportunity for incorporation of a geothermal heat pump system into a building, as the infrastructure for a building foundation requires excavation and/or drilling for its construction. By incorporating a GHX into the foundation excavation or boreholes, the first cost of the GHX may be significantly reduced when compared with a conventional GHX.

9.3.1 Shallow Foundations and Basements

A shallow foundation heat exchanger (FHX) is essentially a closed-loop horizontal GHX installed in the excavation made for a basement or foundation, or along with other excavations used for utility trenching. However, owing to the limitations of available space in the shallow foundation of basement excavations, these FHXs are generally suited to low-energy or passive residential buildings. The heat transfer analysis is complicated by the proximity of the GHX to a basement and/or foundation structure. For further details, readers are referred to Shonder and Spitler (2009), Spitler *et al.* (2010), Xing *et al.* (2012), Cullin *et al.* (2012), and Cullin *et al.* (2014).

A variant on this type of opportunity is with a building that incorporates an underground parking garage or parkade. In those situations, a vertical GHX can be installed underneath the parking garage prior to building construction. Several examples of this type of GHX exist in multistory buildings in Western Canada.

9.3.2 Deep Foundations

A deep foundation heat exchanger (FHX) is essentially a closed-loop vertical GHX installed in the structure of a deep foundation of a building. A deep foundation is a type of foundation that transfers the structural loads of a building to a sound subsurface layer or a range of depths within the Earth. Some common reasons for a geotechnical engineer to specify a deep foundation over a shallow foundation are: (i) very large structural design loads (i.e., tall buildings), (ii) inadequate bearing capacity of soil at shallow depth, or (iii) small site footprint. There are different terms used to describe different types of deep foundation, including the pile, the pier, drilled shafts, and caissons. Piles are generally driven into the ground, while other deep foundations are typically constructed using excavation and drilling. Deep foundations can be made out of timber, steel, reinforced concrete, or prestressed concrete.

9.4 Utilization of Infrastructure from Other Energy Sectors

9.4.1 Underground Coal Fires

A significant amount of combustible material is present in underground coal beds, and thousands of underground coal fires are estimated to exist worldwide. In underground coal fires, the initial ignition of the coal bed is typically forced through natural causes due to lightning, brush and forest fires, or other uncontrolled natural and anthropogenic sources of heat. A spontaneous combustion is also possible owing to exothermic reactions in the underground coal bed. Although many underground coal fires have anthropogenic origins through mining, where partially mined seams may combust because of increased availability of oxygen, a significant number of underground coal fires are activated by natural causes.

Ignition of coal deposits is dependent on many factors, such as the temperature of the heat source, the extent of previous oxidation, the concentration of oxygen in contact with coal, the type of coal, the area of exposed coal surface, and the attitude of the deposit. Once the conditions are conducive for ignition, the fire develops by oxygen that reaches the point of ignition through existing cracks, fissures, and shafts in the overburden, the outcrop, and other entries. Most underground coal bed fires exhibit smoldering combustion that may involve only small amounts of coal capable of burning with as little as 2% oxygen content in the air, and may burn for extended periods of time – decades to centuries.

In an underground coal fire, heat transfer occurs through convection within the coal bed and radiation and conduction between the fire and the overburden. The overburden (surface soil above the coal seam), owing to its relatively low thermal conductivity, acts somewhat as an insulator that impedes the transfer of heat away from the burn zone. Considering that 1 tonne of medium-volatility bituminous coal, when combusted, releases about 31 650 MJ of energy that is then absorbed by the surrounding rock formation, temperatures in the overburden above large coal deposits may be as high as 500 °C (at some locations in Northern Wyoming, USA, temperatures as high as 815 °C have been recorded). Under such conditions, significant amounts of heat continuously transferred to the overburden create a large thermal energy reservoir in the overburden and surrounding rock.

Employing a vertical GHX array, Chiasson and Yavuzturk (2005) presented a concept for thermal energy extraction from geologic formations above burning coal seams, as shown in Figure 9.1. The GHX fluid (pure water) would be used to boil a secondary organic working fluid in an organic Rankine cycle power plant (see Chapter 14). The proposed BHE construction consisted of a vertical concentric steel arrangement, cemented in place. Horizontally bored heat exchangers are also possible.

Chiasson and Yavuzturk (2005) rather conservatively concluded that, with an average underground temperature of 150 °C (low to moderate for underground coal fires), a borehole heat exchanger of 120 m in depth could generate 1 kW of electrical energy with a 10% thermal-to-electrical energy conversion efficiency. Other uses of the extracted heat are possible, such as those typical of direct geothermal uses.

9.4.2 Abandoned Oil and Gas Wells

Oil and gas wells are drilled to typical depths of 1000–3000 m, depending on the occurrence of hydrocarbons in the target basin. Thus, fluids in these hydrocarbon reservoirs are naturally warmed to temperatures according to the local geothermal gradient. These reservoirs tend to become 'water flooded' with development age, and when production wells are no longer economical, they must be decommissioned, at significant cost. Use of these wells presents an opportunity to extract geothermal energy and defer costs for well abandonment and surface reclamation.

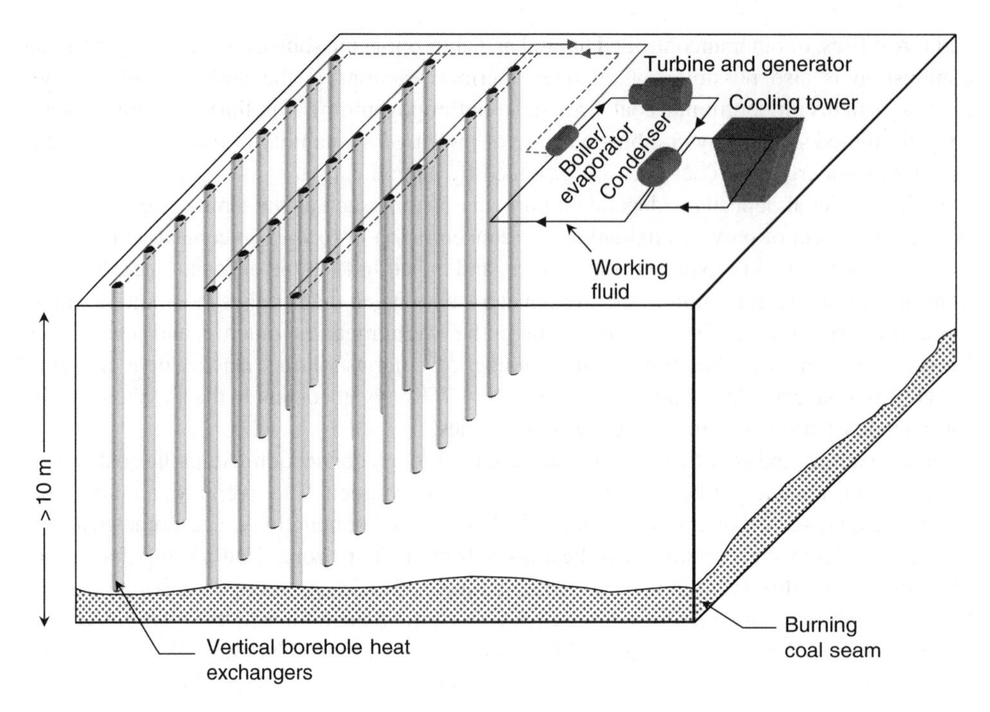

Figure 9.1 Conceptual model of a system for thermal energy extraction from an underground coal fire for electricity generation

There are a few different conceivable open- and closed-loop thermal energy harnessing methods from abandoned oil and gas wells, but conversion of the well to a closed-loop concentric borehole heat exchanger is simple and reliable, as shown in Figure 9.2 and described in what follows.

Deep wells are generally 'telescoped' to allow tooling to drill to successively greater depths. The production well casing is commonly a nominal diameter of 4.5 in or 115 mm, and would be left in place to convert the well to a concentric BHE. Production hardware such as pumps, tubing, and rods would be removed so that the well casing is fully clear and unobstructed along its full length to the top of the perforated interval. The perforated section of the well would be plugged with cement (isolating fluids in the well casing from formation fluids).

To outfit the well as a concentric tube heat exchanger, the existing casing would be used to form the outside tube of the concentric pair and an inner 'dip-tube' would be centralized within the casing to form the inner tube. The new borehole heat exchanger would be flushed with pure water (perhaps with added corrosion inhibitors) to displace the brine in the old well.

9.4.3 Water-Filled Abandoned Mines

Flooded mine workings provide an opportunity for the economic development of low-grade geothermal energy for heating and cooling of buildings and other applications. Examples exist in several countries around the world, such as Canada, the United States, Germany, and the United Kingdom. Mines are typically dewatered during mining activities, and then become

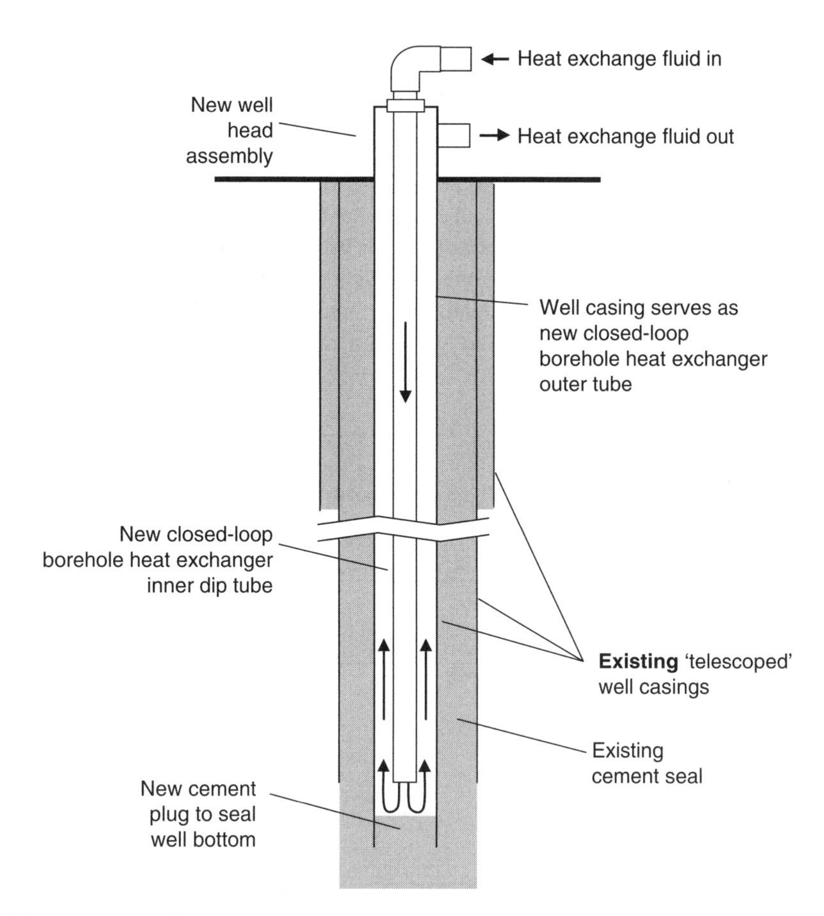

Figure 9.2 Schematic of a conceptual concentric borehole heat exchanger for thermal energy extraction from an end-of-life petroleum well (arrows show fluid flow direction)

water filled again after the mine closes. Alternatively, mines can be artificially flooded to take advantage of this relatively large thermal storage opportunity.

Mine development for geothermal purposes depends on the type of mineral deposit, the distribution of mineralization, depth to water, proximity to a thermal use, and economic factors. The size of the mine workings also depends on the extent and grade of the mineralization and the type of equipment utilized during mining. Thermal energy extraction may be accomplished via either open- or closed-loop methods.

9.5 Cascaded Loads and Combined Heat and Power (CHP)

'Cascading loads' with geothermal energy represent efficient use of the geothermal resource by sequentially extracting heat from a geothermal fluid for successively lower temperature applications, thereby improving the economics of the overall geothermal system. Thus, cascading is most applicable to geothermal resources with temperatures above ambient.

Figure 9.3 depicts an idealized cascaded geothermal energy concept, where the cascaded use, after being used for electrical power generation, can include space or district heating, greenhouse heating, and aquaculture pond or swimming pool heating. Downstream of direct geothermal uses, heat pump applications are conceivable. High-temperature power generation with geothermal energy is usually economic as standalone plants, but low-temperature power generation is often not economic, with net plant efficiencies normally below 10% owing to the low source temperature and relatively high parasitic loads from pumps. Thus, cascading loads often make most economic sense with low-temperature geothermal power plants.

The concept shown in Figure 9.3 is idealized, as there would need to be a suitable business case for each application. Further, fluid geochemistry must also be considered because

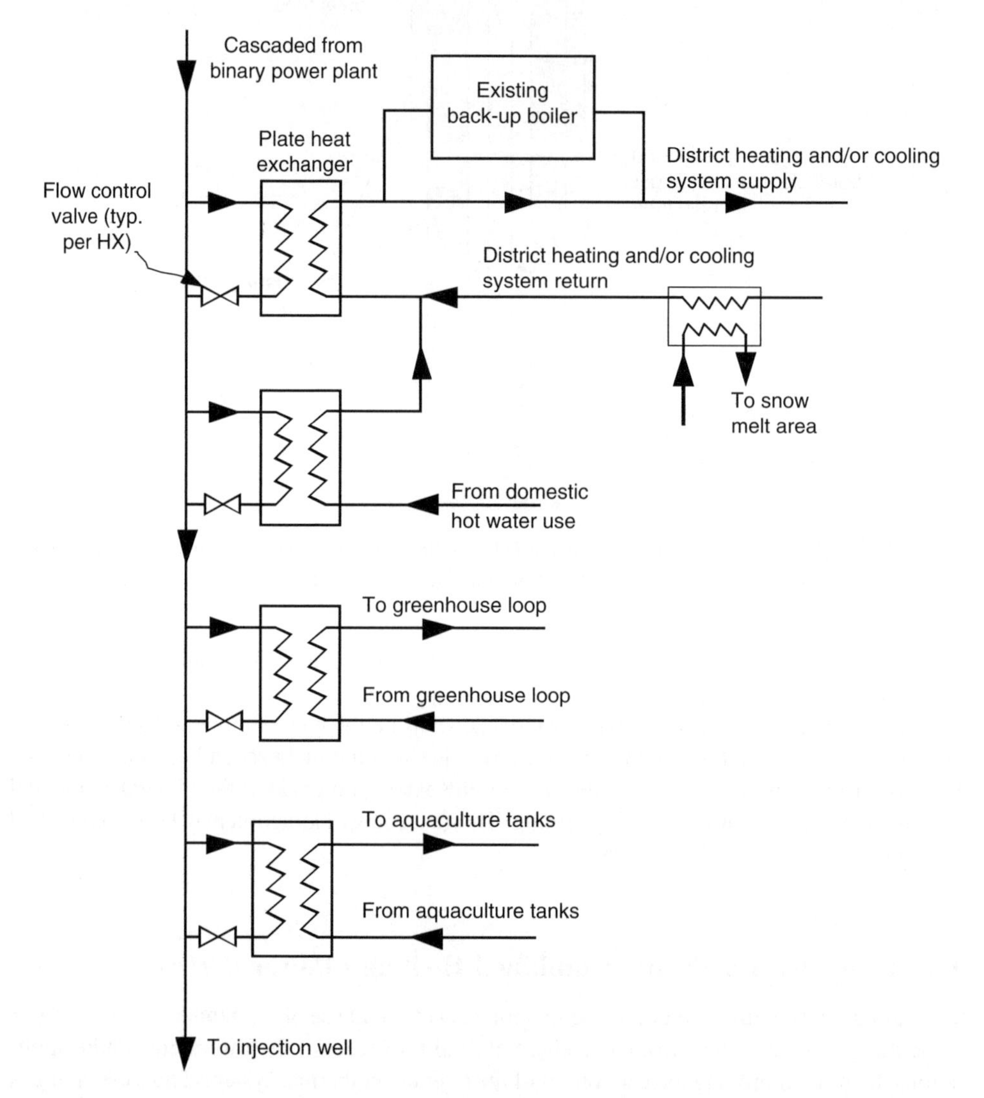

Figure 9.3 Idealized concept of cascading geothermal energy loads

excessive cooling of the geothermal fluid can result in undesirable mineral precipitation and clogging of the piping network.

Lund and Chiasson (2007) describe examples of geothermal power plants with cascaded loads. Examples of high-temperature CHP installations include the Sudurnes Regional Heating Corporation plant at Svatsengi, Iceland, and the Nesjavellir Geothermal plant near Reykjavik, Iceland. Low-temperature installations described include ones at Bad Blumau and Altheim, Austria, Neustadt-Glewe, Germany, and Egat, Thailand. From a simple economic analysis, Lund and Chiasson (2007) showed that, as would be expected, the greater the load factor for the direct-use application, the more thermally efficient will be the operation and the shorter will be the payback period of the combined heat and power project. A higher load factor essentially spreads the capital cost over a greater quantity of heat over the annual cycle, and thus results in a lower cost of delivered heat. Higher load factors are usually achieved with industrial operations using high-temperature resources that are not entirely dependent on weather factors, and thus operate more frequently during the year. Another benefit of some direct-use applications is job creation.

Another good example of a geothermal CHP system is at the Oregon Tech Campus in Klamath Falls, Oregon. Two geothermal power plants (rated at 280 kW_e and 1750 kW_e) utilize geothermal fluid at approximately 90 °C from production wells. These power plants have first-law thermodynamic efficiencies of less than 7% when parasitic loads are included. However, the overall efficiency is dramatically improved with the CHP arrangement. After heat is provided to the geothermal power plants, the geothermal fluid is cascaded to heat the campus in winter, and provide year-round domestic hot water. The campus consists of approximately 93 000 m^2 of floor space in total. Prior to injection, the geothermal fluid (in winter) is used in a hydronic snow-melting system to keep about 5000 m^2 of outdoor sidewalks, stairways, and pavement areas snow and ice free.

9.6 Integrated Loads and Load Sharing with Heat Pumps

Integrating loads and load sharing are inherent beneficial attributes of water-source heat pumps. Buildings with diverse floor plans (i.e., schools and large office buildings) make good candidates for a geothermal heat pump system where interzonal loads can be shared via a common liquid heat transfer loop. In other words, zones in cooling mode will reject heat to the hydronic loop, which can be used beneficially by zones in heating mode. In this subsection, situations with obvious load-sharing potential are described.

9.6.1 Swimming Pool Heating

Incorporation of swimming pool heating into a geothermal heat pump system is an opportunity for load sharing in certain climates. Particularly in cooling-dominated buildings, heat rejected from the building can be used to heat the pool. Further, the seasonal swimming pool heating loads can serve to balance the annual load on the GHX in cooling-dominated climates, resulting in a smaller, lower-cost GHX.

Chiasson (2005a) examined the economics of swimming pool heating in residential buildings in cold and warm climates, where the pool heating system was incorporated into a closed-loop GHX. It was found that GHX lengths may be reduced by up to about 20% in southern US

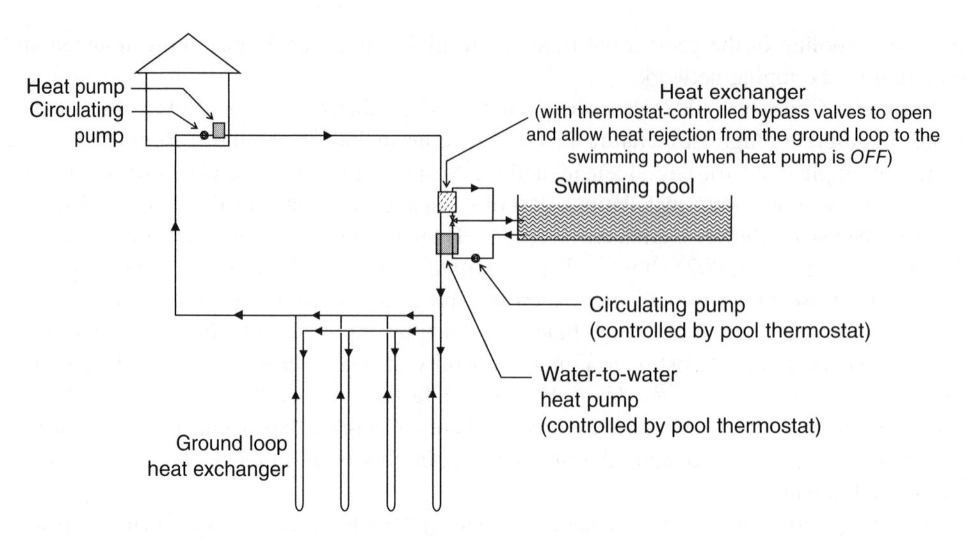

Figure 9.4 Conceptual diagram of incorporating swimming pool heating into a cooling-dominated building with a GHX

climates with the addition of a swimming pool, but may be as much as double in northern US climates. A simple economic analysis demonstrated that it would not be economically justifiable to heat a swimming pool with a GHP system in northern US climates owing to the extra GHX required to meet additional heating demands. Of course, this depends on the conventional cost of heating. In contrast, immediate savings could be realized in southern US climates, as the pool can accept heat from the heat pump system that would be otherwise rejected to the ground.

A concept for such a scenario is shown in Figure 9.4. Fluid exiting the GHP in the building can be used to heat the pool under two scenarios: (i) directly through a plate heat exchanger if the fluid is warm enough, OR (ii) as source fluid to a water-to-water heat pump. When the pool does not call for heating, the fluid exiting the GHP in the building would bypass the swimming pool and enter the GHX.

9.6.2 Simultaneous Need for Hot and Chilled Liquids

The need for simultaneous hot and chilled liquids arises in several hydronic applications, such as:

- multizone buildings with central four-pipe HVAC systems, and
- applications needing simultaneous refrigeration, space heating, and hot water, such as:
  - ice arenas,
  - supermarkets,
  - convenience stores, and
  - gas stations with convenience stores and car washes.

An excellent example of this type of opportunistic GHP application is described by Lohrenz (2005), where a GHP system is used to generate heating, cooling, refrigeration, and hot water for a building with a hockey and curling rink.

A concept for such a scenario is shown in Figure 9.5 for a bank of n-staged liquid-to-liquid heat pumps meeting any number of simultaneous heating and/or cooling loads. The concept is essentially a simple one, but requires careful control. The basic system control is summarized in Figure 9.5. In summary, the liquid-to-liquid heat pumps are staged on/off to maintain liquid temperature set-points in the respective storage tanks that feed the loads. If the hot tank liquid becomes too hot, or the cold tank liquid becomes too cold, valves are opened to allow the hot or cold fluid to flow to the GHX, thereby rejecting or extracting heat to/from the Earth.

9.6.3 Sewer Heat Recovery

A sewer is an underground conduit used for conveying effluent from buildings to a point of discharge or treatment. Flow in sewers is accomplished either by gravity or with pumps. Material in sewers is effluent from buildings, and thus is at a relatively uniform temperature year-round. Flow rates in sewers depend on the population served; smaller-diameter sewers feed into larger-diameter ones as the effluent makes its way to the point of discharge or treatment. It is these larger sewer trunks that make excellent opportunities for use with water-source heat pumps. One example of a large-scale, district sewer heat recovery project is in the False Creek neighborhood of Vancouver, British Columbia, Canada.

There are a number of practical considerations for successful sewer heat recovery projects: (i) available flow rates in the sewer, (ii) year-round temperature of effluent, (iii) allowable temperature drop of effluent (to maintain acceptable biological decay parameters), (iv) solids in the effluent, (v) chemical aggressiveness of effluent, and (vi) heat exchanger design (internal or external to sewer).

9.6.4 District Energy Systems

District energy (DE) may be defined as the heating and/or cooling of two or more structures from a central source. There are many district energy systems throughout the world using conventional energy sources; heat may be provided in the form of either steam or hot water and may be utilized to meet process, space, or hot water requirements; chilled water or an icy slurry may be distributed to meet the needs for space cooling.

A district energy system has three major components: (i) a thermal-energy-generating plant, (ii) a distribution system (piping), and (iii) building interconnections (e.g., meters, valves, pumps), often referred to as energy transfer stations. Design guidebooks for district heating and cooling systems have been developed by ASHRAE (2013a, 2013b).

In the United States, the first direct-use geothermal district heating system was built in Boise, Idaho, in 1892, utilizing groundwater at about 72 °C. This system, originally known as the Artesian Hot and Cold Water Co. and later as the Boise Warm Springs Water District, still serves the Warm Springs district of Boise. Throughout the western United States, numerous geothermal district heating systems were developed through the 1980s, and growth continues today.

For direct-use geothermal district energy systems, district load factor is a major determinant of system feasibility. The load factor is the ratio of total annual energy use to the total possible annual consumption if the peak were supplied continuously for a year. A high load

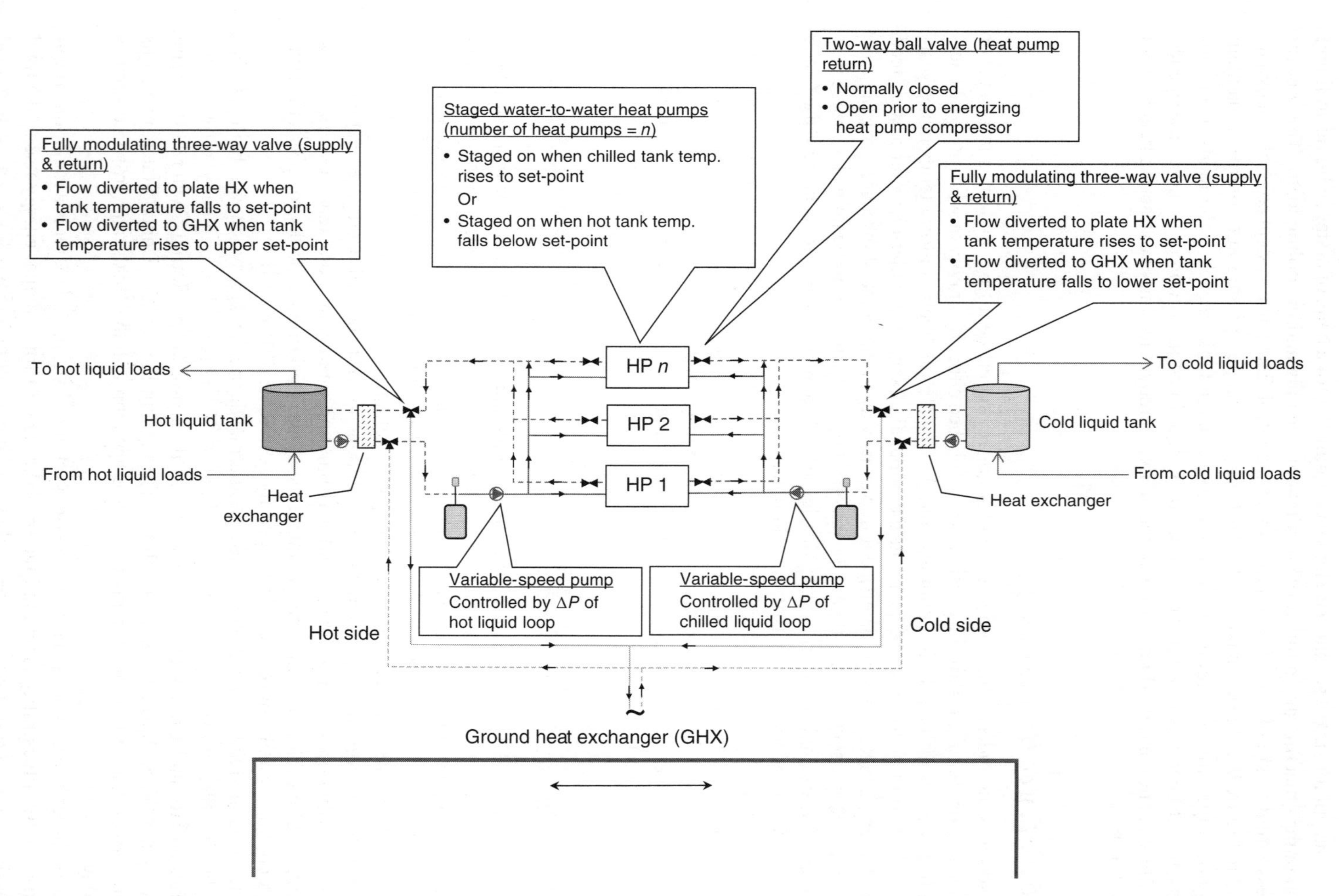

Figure 9.5 Conceptual diagram of a GHP system to produce simultaneous hot and cold liquids to loads (HP = heat pump; ΔP = differential pressure)

factor is critical, and is attainable through a pattern of mixed land uses. For example, an area of both commercial and residential units will have a higher load factor than separately. Obtaining a high load factor through mixed load uses is the key to success in many district energy systems.

Another metric of district energy favorability is the thermal load density. The thermal load density of a community is determined by the type and arrangement of its land uses in its neighborhoods, the space and water heating and cooling loads, the total amount of floor space per neighborhood, and local industrial process heat and refrigeration loads.

For geothermal heat pump system feasibility, load diversity and load synergy among buildings is key; economies of scale are not as prevalent in GHP district systems as in conventional district energy systems. Open-loop systems offer a higher energy density than closed-loop systems, but either option is viable, depending upon the circumstances. An example of an open-loop district GHP system is given by Bloomquist (2005), and an example of a closed-loop district GHP system is given by Lohrenz (2011).

There are numerous community benefits to district geothermal energy systems:

- use of indigenous resources translates directly into a reduction in the reliance on foreign imports and related strategic vulnerability,
- impacts can range from economic revitalization projects to a major industrial growth incentive,
- reduction in carbon emissions on larger scales than with individual buildings,
- increased employment opportunities that result in improving the local tax base,
- local retention of money spent on energy,
- intangible benefit of community pride, and
- various ownership models (public, private, or public-private) can result in revenue streams for local companies and/or government entities.

District energy systems employing a low-temperature GHX allow the incorporation of many of the opportunistic choices described above, and as shown in Figure 9.6. The basic design concept shown in Figure 9.6 takes advantage of a modular 'plug-and-play' structure such that heat sources or sinks can be added as practical. The concept is centered around a common low-temperature supply pipeline that serves to distribute energy in the form of an aqueous antifreeze solution to the sources and sinks. A low-temperature distribution loop was conceived in this design so that lower-grade heat sources could be rejected to or extracted from the loop. A lower-temperature fluid distribution loop typically requires larger-diameter pipe relative to that used in a high-temperature loop, but the added advantage of larger pipe diameter means more fluid volume in the loop and correspondingly more thermal mass (or thermal inertia) of fluid in the pipe, which helps to damp large fluid temperature excursions during peak load times. Amplification of the low-temperature source loop to useful temperatures for space heating is accomplished with water-to-air or water-to-water heat pumps distributed throughout the district in the buildings they serve. Heat rejection from heat pumps in cooling occurs to the same loop.

An integral component of the district energy system is the ground heat exchanger (GHX), which could consist of one central array or multiple decentralized arrays, and could be any combination of closed- and open-loop GHX designs. The GHX acts to provide a baseload heat

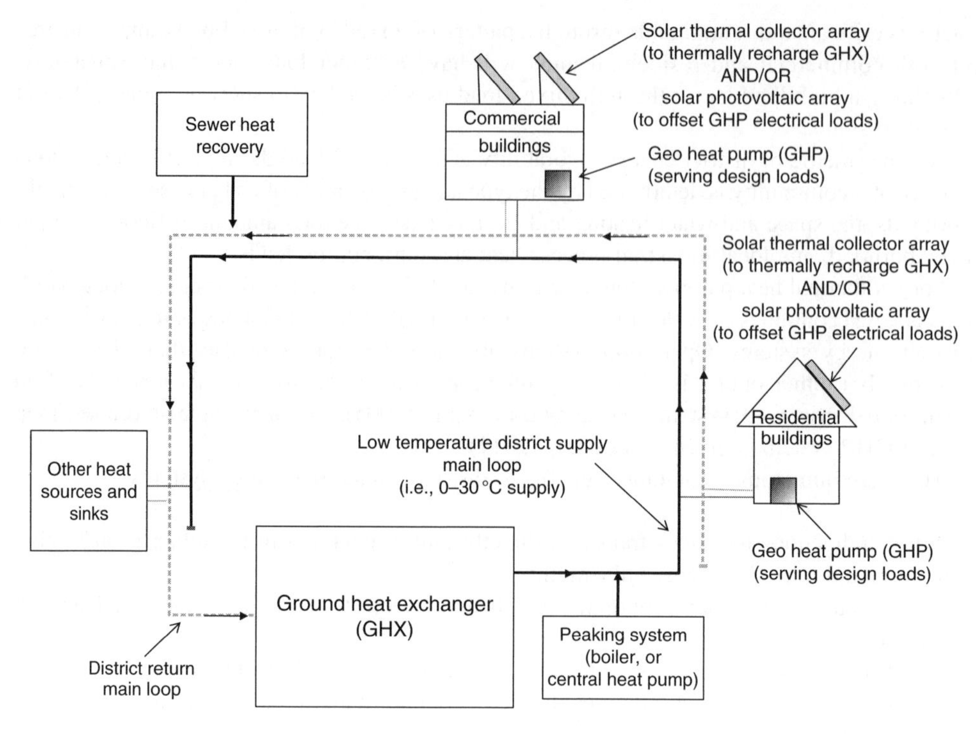

Figure 9.6 Conceptual diagram of a multisource, low-temperature district GHP system

source for heat pumps, supplemented by a peaking boiler during extreme cold periods. In addition, the GHX acts as a short-term and long-term (i.e., seasonal) thermal storage medium for various waste and other available heat sources (solar thermal or sewer heat), which helps to improve the GHX thermal performance during times when heat is needed.

As noted in Figure 9.6, general options exist for 'other heat sources and sinks', which could conceivably include waste heat addition from refrigeration systems (i.e., ice rinks), heat rejection from a fluid cooler, or any other source/sink deemed practical. This box could also represent another modular GHX as the district system expands and/or additional GHXs are incorporated at decentralized locations.

9.7 Chapter Summary

This chapter presented a general overview of opportunistic heat sources and sinks for incorporation into geothermal energy systems. In general, any heat source or sink may be incorporated into a geothermal energy system if infrastructure exists or is planned that can advantageously make use of Earth heat exchange, synergies exist among loads, the system is hydronic, and/or thermal energy storage is advantageous.

Discussion Questions and Exercise Problems

9.1 Research and write a short case study of use of underground mine water for heating and cooling.

9.2 Research the meaning of an Energy Pile® and describe how you would evaluate its heat transfer performance.

9.3 Research and write a short case study of use of sewer heat recovery for heating and cooling of buildings.

9.4 Research and write a short case study of a district geothermal energy system.

10

Piping and Pumping Systems

10.1 Overview

The foregoing chapters have focused mainly on the thermal considerations of geothermal energy systems. The focus of this chapter is hydraulic considerations of moving the heat transfer fluid from the geothermal resource to the point of use.

Circulating fluids provide heat transfer in geothermal systems. This sounds obvious, but engineering of fluid handling systems is often treated as secondary, or is overlooked as being unimportant. Poorly designed pumping and piping systems result in excessive pump energy consumption of the system. In addition, heat transfer may be inadequate, which can result in heat pump underperformance, which in turn can lead to occupant discomfort. Thus, this chapter focuses on the art and science behind the design of heat exchange fluid transfer piping, with particular attention to the GHX piping and mechanical room components. The chapter presents methods to calculate pressure drop in piping systems, methods to aid in properly sizing the transfer pipe, and a discussion of pumps and pump sizing.

Learning objectives and goals:

1. Review of the fluid power equation.
2. Calculation of the major and minor losses associated with pipe flow in piping networks and determination of the pumping power requirements.
3. Introduction to GHX piping system design and layout.
4. Key principles for designing and operating low-energy fluid flow systems.

Geothermal Heat Pump and Heat Engine Systems: Theory and Practice, First Edition. Andrew D. Chiasson.

Companion website: www.wiley.com/go/chiasson/geoHPSTP

10.2 The Fluid Mechanics of Internal Flows

Fluid flow in pipes and ducts (internal flows) provides the main heat transfer mechanism in geothermal energy systems. The fluid in such applications is usually forced to flow by a fan or pump through a flow section. We pay particular attention to friction, which is directly related to the pressure drop and head loss (or resistance to the flow) during flow through pipes and ducts. The pressure drop is then used to determine the pumping power or fan power requirement.

10.2.1 The Fluid Power Equation

The power required to move a fluid through a pipe or duct can be derived from an energy balance on the system. Referring to Figure 10.1, an energy balance on the control volume gives

$$\frac{dE_{CV}}{dt} = \dot{W}_f - \dot{q}_{out} + \dot{m}_{in}\left(h + \frac{\dot{V}^2}{2} + gz\right)_{in} - \dot{m}_{out}\left(h + \frac{\dot{V}^2}{2} + gz\right)_{out} \tag{10.1}$$

where dE_{CV}/dt is the rate of change in energy in the control volume, $\dot{W}_f$ is the rate of work transmitted to the fluid, $\dot{m}$ is the mass flow rate, h is the specific enthalpy, $\dot{V}$ is the fluid velocity, g is the acceleration due to gravity, z is the height above a fixed reference, and $\dot{q}_{out}$ is the rate of heat loss from the system.

Assuming steady-state conditions, and from conservation of mass and from the definition of enthalpy (i.e., $\dot{m}_{in} = \dot{m}_{out}$ and $h = u + Pv$, where u is the specific internal energy, P is the pressure, and v is the specific volume), Equation (10.1) is expressed as

$$\dot{W}_f = \dot{q}_{out} + \dot{m}\left(u + Pv + \frac{\dot{V}^2}{2} + gz\right)_{out} - \dot{m}\left(u + Pv + \frac{\dot{V}^2}{2} + gz\right)_{in} \tag{10.2}$$

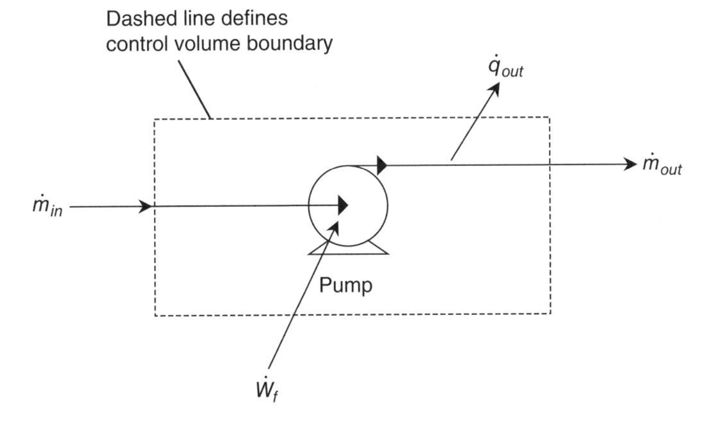

Figure 10.1 Control volume diagram for an energy rate balance on a pumping (or fan) system

Combining terms gives

$$\dot{W}_f = \dot{m}\left(\Delta P v + \frac{1}{2}\Delta\dot{V}^2 + g\Delta z\right) + (\dot{q}_{out} + \Delta U) \tag{10.3}$$

For incompressible flow, it is more common to use volumetric flow rate ($\dot{Q}$) rather than mass flow rate. Thus, assuming a constant fluid density, and substituting, $\dot{m} = \dot{V}\rho$ and $v = 1/\rho$, gives

$$\dot{W}_f = \dot{Q}\left(\Delta P + \frac{\rho}{2}\Delta\dot{V}^2 + \rho g\Delta z\right) + (\dot{q}_{out} + \Delta U) \tag{10.4}$$

where $\dot{Q}$ is the volumetric flow rate of the fluid.

The term ($\dot{q}_{out} + \Delta U$) represents the net energy added to the fluid from friction with the pipe/duct walls; fluid flow through pipes and ducts experiences a hydraulic resistance caused by shear stresses at the pipe wall. To be consistent with the other terms, it is useful to write ($\dot{q}_{out} + \Delta U$) in terms of pressure drop. Thus, we have

$$\dot{W}_f = \dot{Q}\left(\Delta P + \frac{\rho}{2}\Delta\dot{V}^2 + \rho g\Delta z + \rho h_l\right) \tag{10.5}$$

where h_l is the *head loss* in units of specific energy ($J \cdot kg^{-1}$ or Btu/lbm) due to friction between the fluid and pipes, ducts, and fittings.

A number of interesting observations can be made about Equation (10.5). First, each of the terms has units of pressure. Thus, the pump power necessary to propel the fluid can be written in terms *of* $\dot{W} = \dot{Q}\Delta P$. The first to fourth terms on the right-hand side of Equation (10.5) represent: (i) the 'static pressure difference' between the inlet and outlet, (ii) the 'velocity pressure difference' between the inlet and outlet, (iii) the 'elevation pressure difference' between the inlet and outlet, and (iv) the 'friction pressure drop' as the fluid flows through the pipes or ducts. Thus, the equation for the energy required to move an incompressible fluid through pipes or ducts, $\dot{W}_f$, can be written as

$$\dot{W}_f = \dot{Q}\left(\Delta P_{static} + \Delta P_{velocity} + \Delta P_{elevation} + \Delta P_{friction}\right) = \dot{Q}\Delta P_{total} \tag{10.6}$$

The first three components of the total pressure loss (ΔP_{static}, $\Delta P_{velocity}$, $\Delta P_{elevation}$) refer to differences between the inlet and outlet of the system. The fourth component of the total pressure loss, $\Delta P_{friction}$, refers to irreversible friction losses in the pipes and ducts and is always present (non-zero) in all real pump (and fan) applications. Thus, the total pressure drop can also be written as

$$\Delta P_{total} = \left(\Delta P_{static} + \Delta P_{velocity} + \Delta P_{elevation}\right)_{inlet-outlet} + \Delta P_{friction} \tag{10.7}$$

and the equation for $\dot{W}_f$ can be written as

$$\dot{W}_f = Q\left(\left(\Delta P_{static} + \Delta P_{velocity} + \Delta P_{elevation}\right)_{inlet-outlet} + \Delta P_{friction}\right) \tag{10.8}$$

10.2.2 Pressure and Head

A classic measure of pressure in internal flows is often accomplished using a piezometer or manometer, and the pressure difference between a fluid and the atmosphere is expressed in terms of the difference in height between levels of liquid in the manometer:

$$\Delta P_{total} = \rho g \Delta h \tag{10.9}$$

where g is the acceleration due to gravity, ρ is the density of the fluid in the manometer, and Δh is the height of the fluid column. When a pressure difference is characterized in terms of Δh, it is frequently called head. Thus, when pressure loss due to friction in pipes or ducts is measured in terms of Δh, it is often called friction head or head loss. Similarly, when the pressure required to lift a fluid against the force of gravity is measured in terms of Δh, it is often called elevation head. When the fluid in the manometer is water, the relationship between pressure and head is

$$\Delta h = \frac{\Delta P}{\rho_{H_2O}\, g} \tag{10.10}$$

In pumped systems, head is often expressed as the difference in height, Δh, between levels of a water-filled manometer. In IP units, this is expressed in feet of water (ft-H_2O) or, equivalently, ft-wg (arising from feet of water at gauge pressure). In fan systems, Δh is typically measured in inches of water (in-H_2O) or, equivalently, in-wg. In SI units, m-H_2O and Pa are common.

10.2.3 Dimensional Equations for Fluid Power

To summarize, we have the power necessary to propel a fluid through internal flow, given by $\dot{W}_f = \dot{Q}\Delta P = \dot{Q}\,(\rho g \Delta h)$.

In IP units, a useful dimensional equation to calculate the fluid power (in horsepower) to move water at standard conditions ($P = 1$ atm, $T = 60\,°F$) through pipes is

$$\dot{W}_f = \frac{\dot{Q} \times \Delta h_{total}}{3960} \tag{10.11}$$

where $\dot{W}_f$ is the fluid power in horsepower (hp), $\dot{Q}$ is the volumetric flow rate in gallons per minute (gpm), Δh_{total} is the total head in feet (ft), and 3960 is a conversion factor in units of gal-ft-H_2O/min-hp.

In SI units, a useful dimensional equation to calculate the fluid power (in W) to move water at standard conditions through pipes is

$$\dot{W}_f = \dot{Q} \times \Delta h_{total} \times \rho g \tag{10.12}$$

where $\dot{W}_f$ is in units of W, $\dot{Q}$ is in units of $m^3 \cdot s^{-1}$, and Δh_{total} is in units of m.

10.2.4 Inlet–Outlet Pressure Changes

The total pressure rise that a pump/fan must generate to move a fluid through a pipe/duct system is the sum of the pressure rise required to meet inlet and outlet conditions and the pressure rise to overcome friction in the pipe system. The pump/fan must generate a pressure rise to meet inlet and outlet conditions whenever the pressures, fluid velocities, or elevations are different between the inlet and outlet of the pipe system. The total pressure rise required to compensate for different inlet and outlet conditions is the sum of ΔP_{static}, $\Delta P_{velocity}$, and $\Delta P_{elevation}$. If the inlet and outlet pressures, velocities, and/or elevations are the same, the corresponding term will reduce to zero. If the inlet and outlet fluid pressures, velocities, and/or elevations are different, the corresponding terms must be evaluated.

The following pointers are important to keep in mind when evaluating pressure terms for internal flows:

- In *closed-loop systems*, fluid is pumped through a continuous loop, and therefore the inlet and outlet of the system are at the same location. Thus the pressure, velocity, and elevation of the inlet and outlet are identical, and the changes in static, velocity, and elevation pressures are zero.
- In *open-loop systems*, fluid is pumped from one location to a different location. In open systems the change between static, elevation, and velocity pressures between the inlet and outlet to the system must be considered; however, careful definition of the inlet and outlet locations can minimize the complexity of the calculations. For example, it is frequently possible to define the inlet and outlet locations so that both the inlet and outlet pressures are equal, resulting in the change in static pressure equal to zero.
- For *incompressible flow*, such as the flow of water through a pipe, fluid velocity is inversely proportional to the square of the pipe diameter. Thus, if the pipe diameter remains constant, the inlet and outlet velocities are equal, and the change in velocity pressure, $\Delta P_{velocity}$, is zero.

10.2.5 Pressure Loss Due to Friction

Total pressure loss due to friction, $\Delta P_{friction}$, is the sum of the total pressure loss from friction with the pipes and the total pressure loss from friction through the fittings (i.e., valves, elbows, etc.). Similarly, the total friction loss, $\Delta h_{friction}$, as fluid flows through pipes is the sum of the head loss from friction with the pipes, Δh_p, and the head loss from friction through the fittings, Δh_f. Head loss from friction through the fittings is also referred to as *minor losses*.

The suite of GHX tools (found on the book companion website) for use in GHX design problems contains algorithms for calculating pressure loss due to friction through pipes and fittings. Figure 10.2 is a screen capture of this highlighted calculation option. The next two subsections review the procedures for these calculations.

10.2.5.1 Pressure Loss Due to Friction in Straight Pipes

The Darcy–Weisbach equation relates the head loss (pressure drop, ΔP_p) due to friction along a given length of pipe to the average velocity of the fluid flow:

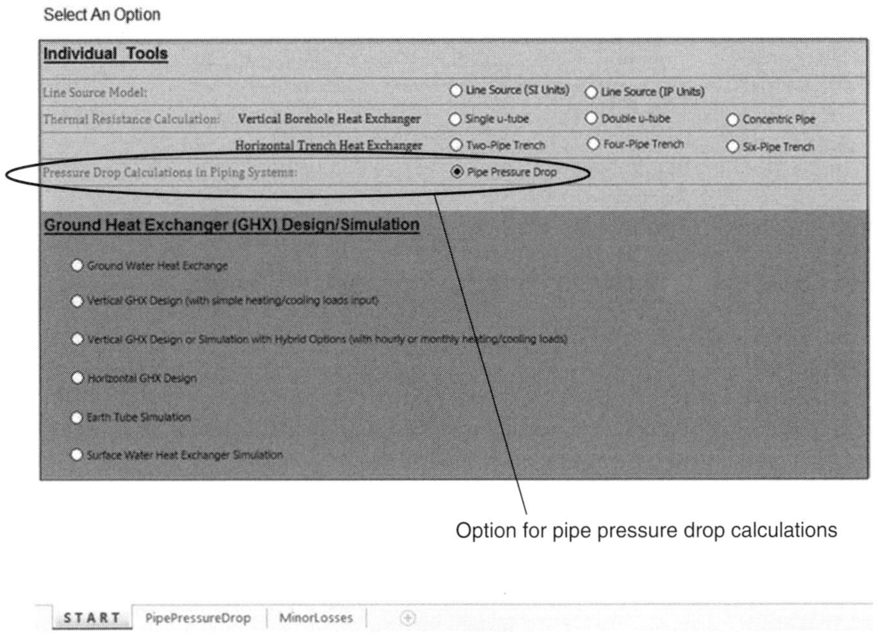

Figure 10.2 Screen capture of the START worksheet of the companion GHX tools, highlighting the pressure drop calculations in piping systems option

$$\Delta P_p = \frac{fL\rho_{fluid}\dot{V}^2}{2D} \tag{10.13}$$

or in terms of head loss:

$$\Delta h_p = f\frac{L}{D}\frac{\dot{V}^2}{2g} \tag{10.14}$$

where Δh_p is the head loss due to friction in straight pipe (m or ft), f is the dimensionless Moody (or Darcy) friction factor, L is the length of pipe (m or ft), D is the pipe inside diameter (m or ft), $\dot{V}$ is the fluid velocity ($\mathrm{m \cdot s^{-1}}$ or ft/s), and g is the acceleration due to gravity. Recall Equation (5.31), which described the calculation of f for smooth pipes after Petukhov for a large range of Reynolds numbers ($3000 < \mathrm{Re} < 5 \times 10^6$):

$$f = (0.790 \cdot \ln Re - 1.64)^{-2} \tag{10.15}$$

10.2.5.2 Pressure Loss Due to Friction Through Fittings (Minor Losses)

The total pressure loss from friction through the fittings, ΔP_f, is proportional to the velocity pressure. The constant of proportionality depends on the fitting. Thus, total pressure loss from friction through a fitting is calculated as

Table 10.1 Typical Pipe Fitting Loss Coefficients

Fitting	Loss Coefficient (—)
180° U-Bend	1.00
90° Elbow	0.90
Ball Valve	0.10
Reducer	0.20
Tee-Piece (branch flow)	1.00
Tee-Piece (in-line flow)	0.35

$$\Delta P_f = k_f \frac{\rho_{fluid} \dot{V}^2}{2} \quad (10.16)$$

where k_f is measured empirically and reported by fitting manufacturers. The head loss from friction through the fittings, Δh_f, can be calculated from

$$\Delta h_f = k_f \frac{SG \dot{V}^2}{2g} \quad (10.17)$$

where SG is the specific gravity of the fluid.

Table 10.1 lists some commonly used values for fitting loss coefficients. This table is included in the *Minor Losses* worksheet of the pipe pressure drop calculator. Users can enter different values as necessary.

10.3 Pipe System Design

When designing a piping or ducting system, flow requirements and piping/duct distances are typically known. Based on that information, the engineer must then select the pipe/duct diameter, select the fittings, determine a piping/ducting configuration that results in sufficient flow to the end uses, and determine the total pressure drop caused by the piping/ducting system. Thus, the task at hand involves determining the pressure drop (or head loss) when the pipe length and diameter are known for a specified flow rate (or velocity).

Pipe system design contains many analogies to electric circuit design. Current is analogous to fluid flow, and voltage drop/rise is analogous to pressure drop/rise. Thus, some pointers to keep in mind in piping system design are summarized in Table 10.2.

10.3.1 Initial Selection of Pipe/Duct Diameter

The selection of pipe/duct diameter generally involves a trade-off between the first cost of the pipe/duct and pump/fan energy costs of the lifetime of the system, both of which are highly dependent on pipe/duct diameter. Large-diameter pipes/ducts have a higher initial material and installation cost, but result in reduced friction losses and pump/fan costs. A rule of thumb that is often used as a starting place for selecting pipe diameters is to select the pipe diameter such that:

Table 10.2 Electric Circuit/Hydraulic Piping Analogies

Scenario	Electric Circuits	Hydraulic Piping
Series Wiring/Piping	• ΔVoltage is additive. • Current is constant.	• ΔPressure is additive. • Flow is constant.
Parallel Wiring/Piping	• Total current equals the sum of current in all branches. • Total ΔVoltage is that across path with greatest ΔVoltage.	• Total flow equals the sum of flow in all branches. • Total ΔPressure is that across path with greatest ΔPressure.

Table 10.3 Maximum Recommended Water Flow Rates in Pipes[a]

Nominal Diameter (in, DN)	SDR 11 HDPE (gpm; Lpm)	Sch. 40 Steel (gpm; Lpm)
¾ in, DN20	4.5; 17	4.0; 15
1 in, DN25	8; 30	7; 27
1¼ in, DN32	15; 57	15; 58
1½ in, DN40	22; 83	23; 87
2 in, DN50	40; 151	45; 170
3 in, DN75	110; 416	130; 490
4 in, DN100	220; 833	260; 985
6 in, DN150	600; 2270	800; 3030
8 in, DN200	1200; 4540	1600; 6050
10 in, DN250	2200; 8330	3000; 11350
12 in, DN300	3500; 13250	4600; 17400

[a] Based on approximate head loss gradient of 4 ft (m) per 100 ft (m) of pipe.
Multipliers for antifreeze mixtures:
- 20% propylene glycol: 0.85
- 20% methanol: 0.90

$$\Delta h_{friction} \cong 2.5-4.0\,\text{ft-H}_2\text{O per 100 ft pipe}$$

OR

$$\Delta h_{friction} \cong 2.5-4.0\,\text{m-H}_2\text{O per 100 m pipe}$$

Thus, the head loss gradient per unit length of pipe is 0.025–0.040. These design guidelines ensure that the fluid velocity is low enough to avoid pipe erosion and excess noise, and provide a reasonable balance between the cost of the pipes/ducts and pump/fan energy costs. Using this as a starting place, subsequent design iterations can identify economically optimum pipe/duct diameters. In many cases, the economically optimum pipe/duct diameter will be larger than that suggested by the design guideline. Table 10.3 shows maximum recommended flow rates in nominal pipe sizes (i.e., flow rates to achieve a head loss gradient of approximately 0.040).

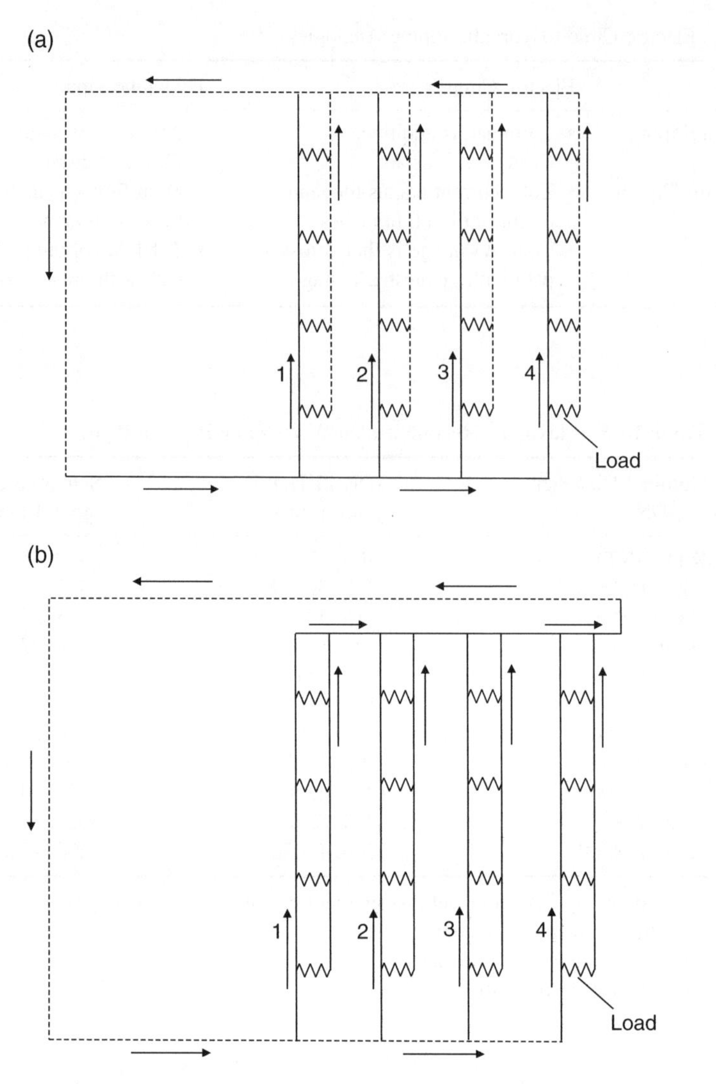

Figure 10.3 Parallel flow piping systems: (a) direct return; (b) indirect or reverse return

10.3.2 Parallel Flow Piping Arrangements

Many piping designs employ parallel flow. In parallel flow designs, the total pressure drop for sizing the pump and calculating pump energy costs is the total pressure drop for the path with the highest pressure drop. Figure 10.3 shows two common piping configurations that employ parallel flow.

The configuration shown in Figure 10.3a is called direct return. In this configuration, the total pressure drop for flow through branch 1 is less than the total pressure drop for flow through branch 4. Thus, if no balancing valves are installed, more fluid will flow through branch 1 than through branch 4, and the total pressure drop across the pump will be set by the pressure drop through branch 4.

The configuration shown in Figure 10.3b is called indirect or reverse return. In this configuration, the pressure drop and flow through all branches are equal. Thus, indirect return guarantees equal flow through all branches in the absence of balancing or flow control valves.

10.4 Configuring a Closed-Loop Ground Heat Exchanger

10.4.1 Laying Out the Pipe Network

The following are some good pointers to keep in mind in laying out a closed-loop GHX:

- It's an art as well as a science.
- It's desirable to minimize ground disturbance (trenching and landscape restoration).
- The GHX must be designed so it can be adequately flushed and purged of air and dirt during commissioning and throughout its lifetime. This means flow circuits or boreholes should be piped together in groups that are manageable to flush and purge.
- Provisions are necessary to isolate circuits for flushing/purging, and in case sections of the GHX need isolation to diagnose any future leaks.
- Isolation valves and pressure/temperature gauges (or P/T ports) should be placed in strategic locations (however, this can be overdone, which adds to the system cost).
- As a general rule, it's desirable to keep the overall pressure drop across the GHX to <15–20 ft-H_2O or <5–7 m-H_2O.
- Plumbing each GHX circuit or borehole in parallel with reverse return piping is the typical design approach. However, circuits of short pipe length (i.e., vertical boreholes of shorter depth) can be connected in series to keep flow rates turbulent.

Figure 10.4 is a photograph of a row of vertical boreholes being connected together in a vertical borehole field installed under an existing asphalt parking lot. Note the connections in the

Figure 10.4 A vertical GHX undergoing connections to horizontal transfer piping (photo taken by the author)

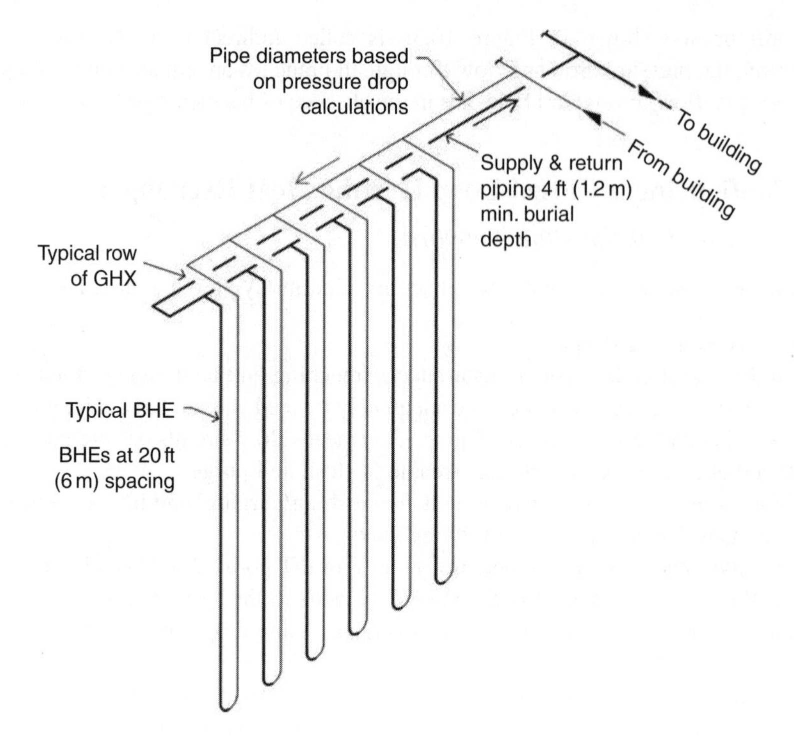

Figure 10.5 Conceptual diagram of a row of vertical GHX BHEs connected in a reverse-return piping configuration

horizontal piping lying on the ground being prepared for fusion to the vertical U-pipes. Note also the varying diameters of the horizontal transfer piping; these diameters are determined to conform with the flow rates shown in Table 10.3, which helps to balance the flows. Sand bedding is used on the floor of the trench to provide a suitable placement material for the horizontal transfer piping. U-pipes sticking out of the ground, remaining to be connected together, are evident on the left-hand side of the photograph. A conceptual diagram of a connected row of BHEs is shown in Figure 10.5.

Let us now illustrate the foregoing concepts with an example of laying out the horizontal transfer piping for a group of vertical BHEs.

Example 10.1 Configuring the Piping Network for a Closed-Loop Vertical GHX – IP Units

A closed-loop vertical GHX consists of 32 boreholes in a 4 × 8 grid pattern. The GHX is to be divided into four circuits, each consisting of a 2 × 4 grid. Design the horizontal transfer piping for one of the 2 × 4 groups of eight boreholes, as shown in Figure E.10.1a. Each borehole is planned to be 250 ft deep, and the total flow rate through the entire 32-borehole GHX is 160 gpm. The borehole-to-borehole spacing is 20 ft, and the first pair of boreholes are 50 ft from the building.

Figure E.10.1a A 2×4 vertical U-tube GHX borehole grid for Example 10.1

Solution

- As the total flow rate through the GHX is 160 gpm, the flow rate through each circuit is 40 gpm. Thus, 5 gpm is required per borehole.
- Choose HDPE, SDR11 pipe.
- Using Table 10.3 as a guide to size the pipe, the following reverse-return layout option is obtained (Figure E.10.1b). For clarity, the supply and return piping are shown in separate diagrams.

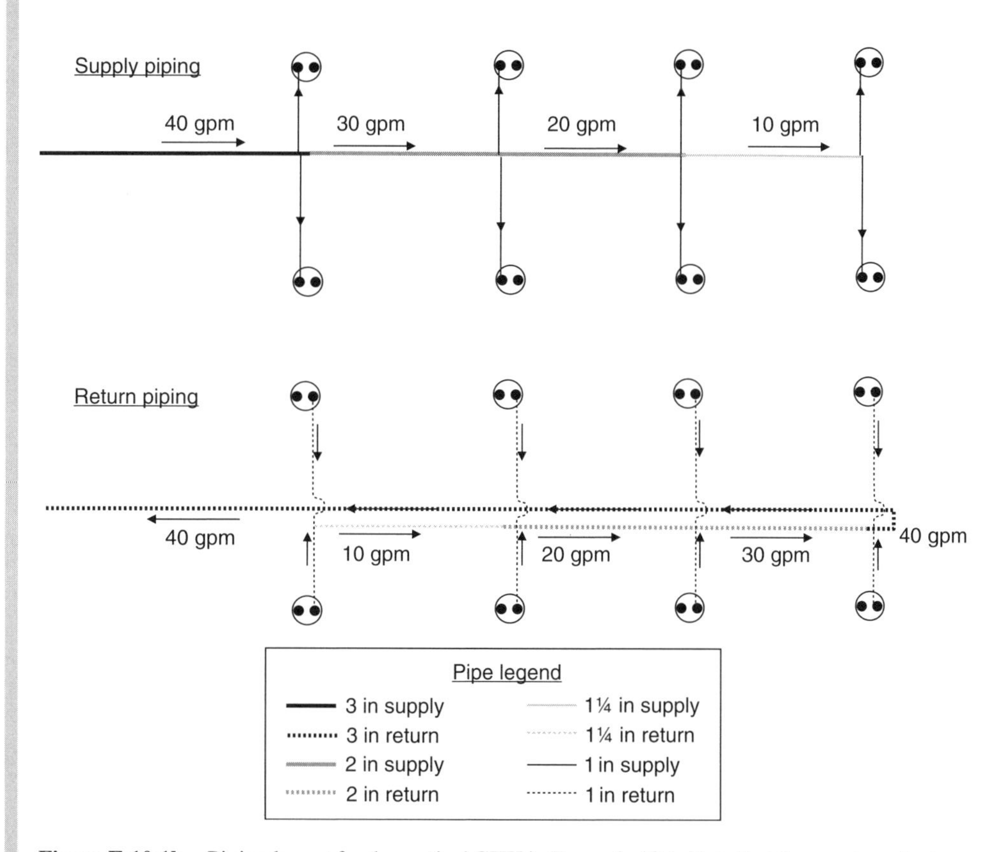

Figure E.10.1b Piping layout for the vertical GHX in Example 10.1. Note that the supply and return piping are shown in separate diagrams for clarity

Discussion: As should be evident, there are numerous ways of connecting these boreholes. Another possible design is to trench around the perimeter of the rectangle formed by the eight boreholes, and tee off 1 in pipes to each BHE, as shown in Figure E.10.1c. Yet another design involves trenching to the 2×4 grid of boreholes, and teeing off a close header, as shown in Figure E.10.1d. Again, this is an art as well as a science.

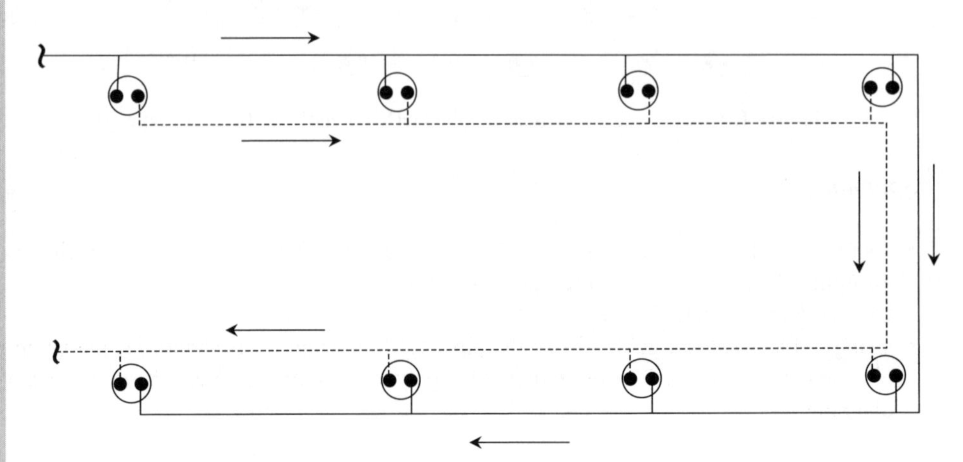

Figure E.10.1c Alternative piping layout for the vertical GHX in Example 10.1. Note that the supply and return piping are not sized

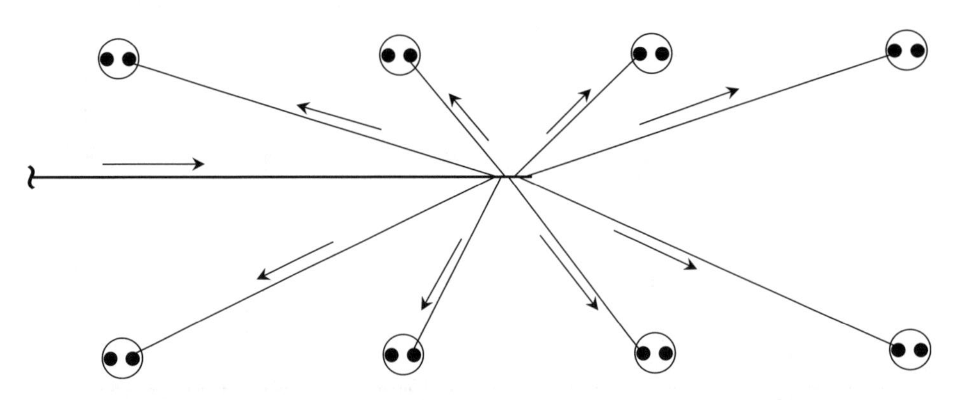

Figure E.10.1d Another alternative piping layout for the vertical GHX in Example 10.1. Note that only the supply piping is shown for clarity

Example 10.2 Calculating Pressure Drop and Fluid Power Requirements in a Closed-Loop Vertical GHX – IP Units

Calculate the pressure drop through the GHX (outside the building) for Example 10.1. Also, determine the fluid power requirement. The heat transfer fluid is pure water.

Solution

- As noted above, it is desirable to keep the pressure drop through the GHX to <15–20 ft-H_2O.

- Let's assume that the four circuits of eight boreholes are each grouped in a parallel, reverse-return arrangement, so we only need to work with one group of eight BHEs.
- As the design of the 2 × 4 BHE grid is parallel reverse return, we only need to calculate the pressure drop due to one BHE flow circuit. Let's choose the furthest one.
- Let's assume an average fluid temperature of 50 °F for *Re* calculations.
- We'll start with the flow in the branch supplying the GHX (leaving the building), and calculate the ΔP in each branch where there's a change in the flow rate or pipe diameter. A labelled sketch of branch number is shown in Figure E.10.2a. Note that branch 5 includes all 1 in piping from the header to the BHE, down and back up the BHE, and then returning to the header.
- Pressure drop calculation results are shown in Figures E.10.2b and E.10.2c above, which were determined using the pressure drop calculator with the GHX design tools found on the book companion website.

The total pressure drop through the GHX outside the building is 11.16 ft-H_2O.

Therefore, the power required to move the fluid through this GHX is given by Equation (10.11):

$$\dot{W}_f = \frac{\dot{Q} \times \Delta h_{total}}{3960} = \frac{160\,\text{gpm} \times 11.16\,\text{ft-H}_2\text{O}}{3960} = 0.45\,\text{hp}$$

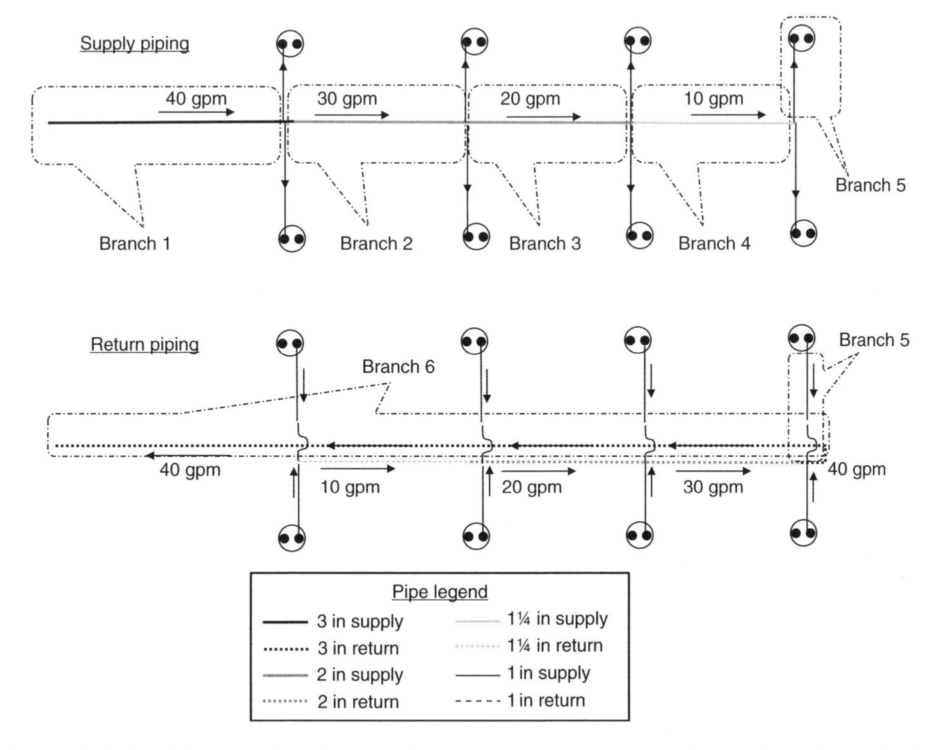

Figure E.10.2a Pipe branches of concern for pressure drop calculations for the piping layout for the vertical GHX in Example 10.1

IP Units | Input Data in Yellow Fields

Heat Transfer

Water

Pipe System Branch	Nominal Diameter	SDR	Flow Rate	Length	Fluid Temperature	Pipe OD	Pipe ID	Fluid Freeze Point	Fluid Density	Fluid Viscosity	Fluid Velocity	Reynolds Number (Re)	Friction Factor (f)	Pressure Drop in Pipe		Minor Losses (fittings, valves)		Totals	
	(in)	(-)	(gpm)	(ft)	(°F)	(in)	(in)	(°F)	(lb/ft$_3$)	(lbm/(ft-s))	(ft/s)	(-)	(-)	(ft)	(psi)	(ft)	(psi)	(ft)	(psi)
Branch 1	3	SDR11	40	50.0	50.0	3.50	2.86	32	62.4	8.77E-04	1.99	3.38E+04	2.30E-02	0.30	0.13	0.06	0.02	0.35	0.15
Branch 2	2	SDR11	30.00	20.00	50.0	2.38	1.94	32	62.4	8.77E-04	3.25	3.74E+04	2.24E-02	0.45	0.20	0.11	0.05	0.57	0.25
Branch 3	2	SDR11	20.00	20.00	50.0	2.38	1.94	32	62.4	8.77E-04	2.16	2.49E+04	2.47E-02	0.22	0.10	0.07	0.03	0.29	0.12
Branch 4	1.25	SDR11	10.00	20.00	50.0	1.66	1.36	32	62.4	8.77E-04	2.21	1.78E+04	2.69E-02	0.36	0.16	0.08	0.03	0.44	0.19
Branch 5	1	SDR11	5.00	520.00	50.0	1.32	1.08	32	62.4	8.77E-04	1.76	1.13E+04	3.05E-02	8.55	3.71	0.14	0.06	8.68	3.76
Branch 6	3	SDR11	40.00	110.00	50.0	3.50	2.86	32	62.4	8.77E-04	1.99	3.38E+04	2.30E-02	0.65	0.28	0.17	0.07	0.83	0.36
TOTALS:														10.54	4.57	0.62	0.27	11.16	4.84

Figure E.10.2b Results of pressure drop calculations for the piping layout for the vertical GHX in Example 10.1

Pipe System Branch	Fitting	Qty.	Pressure Drop		Fitting	Qty.	Pressure Drop		Fitting	Qty.	Pressure Drop		Total Pressure Drop in Fittings	
			(ft)	(psi)			(ft)	(psi)			(ft)	(psi)	(ft)	(psi)
Branch 1	Reducer	1	0.01	0.01	Tee-Piece (in-line flow)	2	0.04	0.02					0.06	0.02
Branch 2	Tee-Piece (in-line flow)	2	0.11	0.05									0.11	0.05
Branch 3	Reducer	1	0.01	0.01	Tee-Piece (in-line flow)	2	0.05	0.02					0.07	0.03
Branch 4	Tee-Piece (branch flow)	1	0.08	0.03									0.08	0.03
Branch 5	90° Elbow	2	0.09	0.04	180° U-Bend	1	0.05	0.02					0.14	0.06
Branch 6	Tee-Piece (branch flow)	1	0.06	0.03	90° Elbow	2	0.11	0.05					0.17	0.07

Figure E.10.2c Results of pressure drop calculations through fittings for the piping layout for the vertical GHX in Example 10.1

Discussion: The total pressure drop through this GHX is a desirable 11.16 ft-H_2O. Note that the flow through the U-tube branch is the most significant source of pressure drop. The pressure drop calculator also includes provisions for calculation of minor losses through fittings. Users can enter typical values for fitting loss coefficients.

10.4.2 Pipe Materials and Joining Methods

Most GHX pipe in the geothermal industry is HDPE, which is a thermoplastic pipe. As such it can be repeatedly melted and reformed, thus allowing HDPE pipe to be joined using heat fusion welding. Heat fusion welding is done in three ways: (i) socket fusion welding, (ii) butt fusion welding, and (iii) electrofusion.

In general, pipes of less than 2 in diameter or DN50 are joined with socket fittings. Pipes of 2 in diameter or DN50 and greater are butt fused. With socket fusion, the inside surface of a socket fitting and pipe are heated to ~260 °C using a specialized heating tool. After a predetermined time, the pipe and socket are removed from the heating tool and immediately pushed together. The semi-molten plastic surfaces bond as the joint cools. Butt fusion welding requires more specialized tools and training. The end faces of the pipes to be joined are shaved, heated to ~260 °C for a specific time, precisely aligned, and then pressed together such that the ends of the pipes 'butt' against each other to form a weld.

Electrofusion is the process of joining HDPE pipe with specialized socket fittings that are manufactured with internal resistance heating wires surrounding the inner face of their sockets. These wires are connect to electrodes projecting from the sides of fitting, which are connected to a specialized power supply that heats the fitting to the specified temperature for a specified time.

Thermal fusion of HDPE pipe requires trained personnel and careful attention to detail. Pipes are often fused in the field in muddy conditions with adverse weather. Care must be taken to ensure that the fusion surfaces of the pipe are clean and dry. Data logging devices are available for tracking key details of the fusion process, which are useful for quality assurance/quality control of the GHX installation.

Crosslinked polyethylene (PEX) is gaining in popularity in geothermal systems because it is well known in the plumbing trades. PEX is a thermoset polymer and thus cannot be thermally welded; joints are made mechanically with special fittings. Thus, any buried joints must be made using mechanical couplings approved for burial by the pipe manufacturer. In many applications where PEX is used for a GHX, the pipe network is designed so that continuous lengths of pipe, free of any joints, are used for the buried portion of the GHX; the ends of the pipe are joined to a manifold station, mounted in an accessible location, either inside or outside the building. PEX pipe of small nominal diameters (<1¼ in or DN30) is most commonly used in GHX applications.

10.4.3 Manifolds in GHXs

Example 10.1 illustrated a common GHX design where multiple parallel piping paths are joined to a common header system. The header is typically fabricated using methods described above in Section 10.4.2. The size of the header piping is typically varied in diameter along the length of the header in order to approximate a head loss gradient of 0.040, which helps to balance the flow through individual circuits connected in a reverse-return arrangement.

An alternative method of constructing parallel GHX loops involves bringing all parallel circuits to a specially designed manifold station, similar to the configuration shown in Figure E.10.1d. A specially designed manifold station consists of attachment assemblies for each flow circuit, isolating and balancing valves, purge valves, and pressure and temperature gauges. This type of manifolding is familiar to the plumbing trades, particularly in potable water system, radiant floor, and radiant panel designs. Manifold stations for GHX applications can be located inside the building served by the geothermal system, or they can be located outside the building in an accessible vault or service chamber.

There are several advantages of using a manifold-based GHX, including: (i) handling of smaller-diameter pipe, resulting in less labor, (ii) avoidance of buried joints, (iii) avoidance of thermal fusion in adverse weather conditions, (iv) balancing valves avoid the need for reverse-return piping and allow adjustment of flow rates in each circuit, (v) isolation valves allow independent flushing, purging, and abandonment of individual GHX loops, and (vi) addition of future loops can be accomplished with modular manifolds.

10.4.4 Air and Dirt Management

Elimination of air, dirt, sediment, and debris in piping systems is essential to proper operation of any hydronic system. There are obviously numerous ways for these to get into a geothermal system. Pipes sit outdoors at the job site and accumulate dirt and dust. Sediment easily gets into pipes when placed in dug trenches. Plastic pipe shavings accumulate in pipe when the pipe is cut or faced.

Not only must air, dirt, and debris be removed from the piping before system commissioning, it must also be managed throughout the life of the system. This involves explicit, proper construction specifications to ensure an air- and dirt-free system at start-up, along with a proper engineering design to include air and dirt elimination devices and expansion tanks. Left unmanaged, possible air-related problems in hydronic heating and cooling systems include:

- noise in piping,
- inadequate flows due to 'air locks' in pipes,
- circulating pump failure due to cavitation (as air pockets move through pumps),
- inadequate heat transfer in equipment (i.e., heat exchangers and heat pumps) owing to heat transfer surfaces not being fully wetted,
- improper performance of balancing valves,
- accelerated corrosion due to oxygen coming in contact with ferrous metals on metallic piping, and
- formation of iron precipitation solids with the appearance of sludge or mud.

10.4.4.1 Flushing and Purging and Pressure-Testing of the GHX

After the GHX construction is completed, it must be filled with water and flushed and purged of air and debris. Designers must design the piping system with readily accessible purge valves and provisions to allow for proper circuit isolation (i.e., properly placed shut-off valves) to achieve adequate purging. Purging should not be done with the system pump(s), as this could damage the pump(s) or distribute air and debris throughout the pipe system. Rather, an external device is used, consisting of a portable circulating pump with a tank open to atmosphere to allow air to escape, and to allow solids and debris to settle. Many contractors add antifreeze to the purge tank if the system design calls for antifreeze, and mix this in with water.

Purging should be done by qualified personnel. Construction documents should reference standards and be specific about flushing and purging activities, who is responsible, and that the results be documented. For example, IGSHPA recommends purge velocities in pipes of 2 ft/s (0.6 m/s) for at least 15 min, and until fluid is clear.

Pressure-testing of pipe is also essential in hydronic systems to certify that there are no leaks. Construction documents should reference standards and be specific about pressure-testing activities, who is responsible, and that the results be documented. Any GHX pipe should be pressure-tested relative to some standard prior to burial. At system commissioning, the entire system is pressure-tested again. Pipe pressure-testing standards vary. For example, IGSHPA recommends testing at a pressure at some percentage above operating pressure for a specified period of time. A more detailed standard is ASTM F2164, *Standard Practice for Field Leak Testing of Polyethylene (PE) and Crosslinked Polyethylene (PEX) Pressure Piping Systems Using Hydrostatic Pressure.*

10.4.4.2 Designing for Air and Dirt Elimination

Hydronic heating and cooling systems should be designed to handle air and dirt in the system during its operation. Even the best purging activities cannot remove entrained air and fine sediment in water, or 'microbubbles' adsorbed onto pipe walls. As the GHX fluid undergoes extreme temperature changes during the year, dissolved air will come in and out of solution as the fluid expands and contracts. If the building loop contains metallic piping or other components, residues from metal corrosion will be carried by the heat transfer fluid. Further, pipes or valves may need disassembling at some time for routine maintenance, causing air and debris to be reintroduced into the system. Thus, prudent designs of closed-loop GHX systems include: air separators, dirt/sediment separators, and expansion tanks.

Figures 10.6 and 10.7 show schematics of closed-loop geothermal systems with air and dirt/sediment management devices. Note the placement of the expansion tank on the suction side of

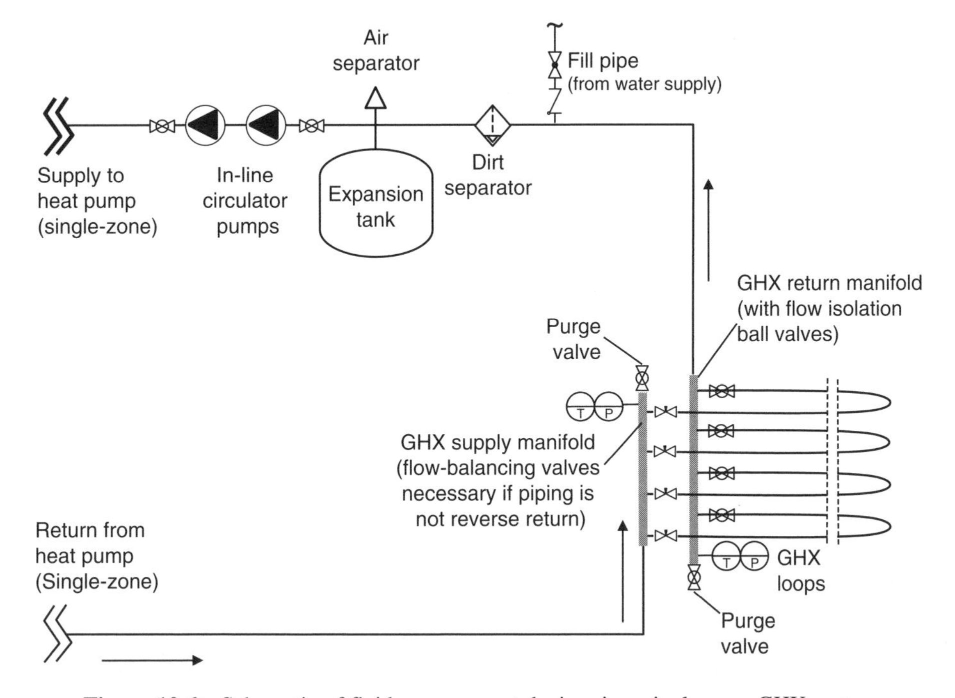

Figure 10.6 Schematic of fluid management devices in a single-zone GHX system

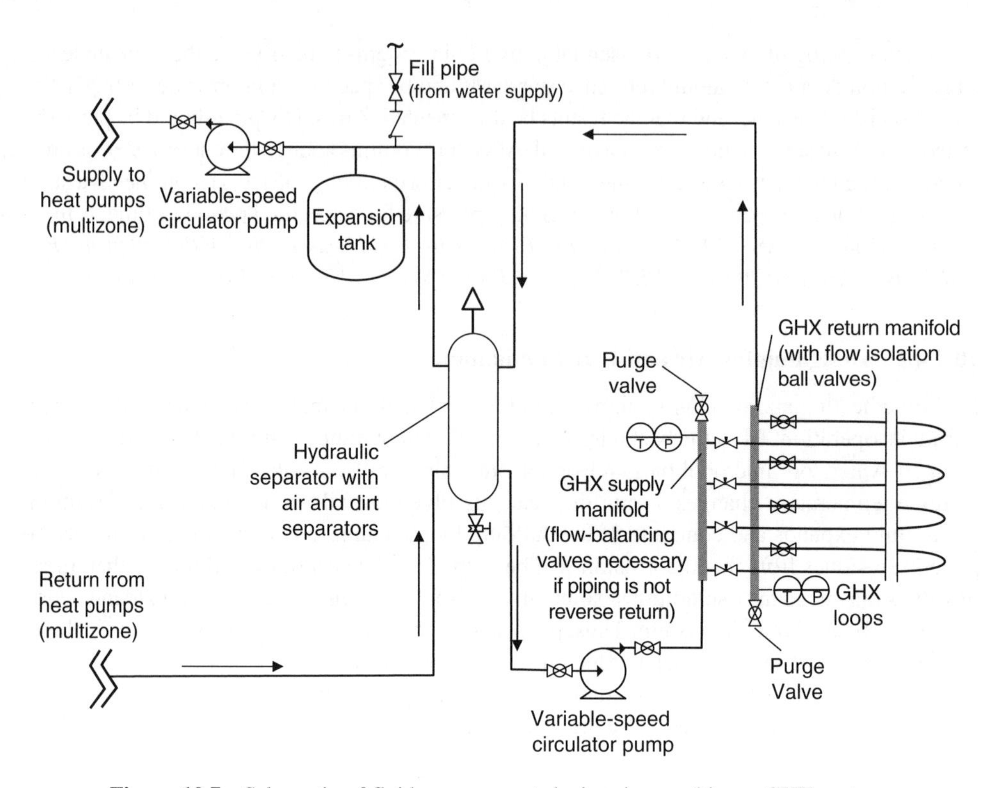

Figure 10.7 Schematic of fluid management devices in a multizone GHX system

the pump to avoid pump cavitation. Closed-loop systems in which the fluid may undergo large temperature changes should employ an expansion tank to handle increased fluid volume with increased temperature. The simplest type of expansion tank consists of a diaphragm with compressed air above.

A hydraulic separator is shown in Figure 10.7 as an energy-efficient pumping scheme in multizone systems. It also combines air and dirt/sediment separation. A hydraulic separator acts as an interface between the GHX and building loops, thereby creating a primary/secondary pumping scenario. Such an arrangement allows the flow rate in the GHX to be different from the flow rate through the building loop, and each pump can be operated at variable speeds, depending on the heating/cooling load demands in the building.

10.5 Circulating Pumps

The purpose of a pump is to add energy to a fluid (liquid), resulting in an increase in fluid pressure, not necessarily an increase in fluid speed across the pump. Pumping applications can generally be divided into two categories: (i) low flow at high pressure and (ii) high flow at low pressure. 'Low flow at high pressure' applications include hydraulic power systems and typically employ positive-displacement pumps. The majority of fluid-flow applications are 'high flow at low pressure' and use centrifugal pumps. In centrifugal pumps, the fluid enters along the centerline of the pump, is pushed outward by the rotation of the impeller blades, and exits along the outside of the pump.

10.5.1 Pump Curves

Pumps can generate high volume flow rates when pumping against low pressure or low volume flow rates when pumping against high pressure. The possible combinations of total pressure and volume flow rate for a specific pump can be plotted to create a pump curve. The curve defines the range of possible operating conditions for the pump. If a pump is offered with multiple impellers with different diameters, manufacturers typically plot a separate pump curve for each size of impeller on the same pump performance chart. Smaller impellers produce less pressure at lower flow rates. A typical pump performance chart with multiple pump curves is shown in Figure 10.8.

Recall from the foregoing that the power required to push the fluid through the pipe, $\dot{W}_f$, is the product of the volumetric flow rate and system pressure drop, $\dot{Q}\Delta P$. Graphically, fluid work is represented by the area under the rectangle defined by the operating point on a pump performance chart. However, the efficiency of the pump (η_{pump}) at converting the electrical power supplied to the pump into kinetic energy of the fluid is also plotted on the pump performance chart (shown as dashed lines in Figure 10.8). Pump efficiencies typically range from about 50% to 80%. Power that is not converted into kinetic energy is lost as heat. The power required by the pump is called the 'shaft work' or 'brake horsepower'. Pump efficiency is the ratio of fluid work to shaft work. Pump power can be calculated from the flow rate, total pressure, and efficiency values from the pump curve, using the following simple relation:

$$\dot{W}_{pump} = \frac{\dot{W}_{fluid}}{\eta_{pump}} = \frac{\dot{Q}\Delta P}{\eta_{pump}} \tag{10.18}$$

From Equation (10.11), a useful dimensional form of this equation for pumping water at standard conditions (in IP units) is

$$\dot{W}_{pump} = \frac{\dot{Q} \times \Delta h_{total}}{3960 \times \eta_{pump}} \tag{10.19}$$

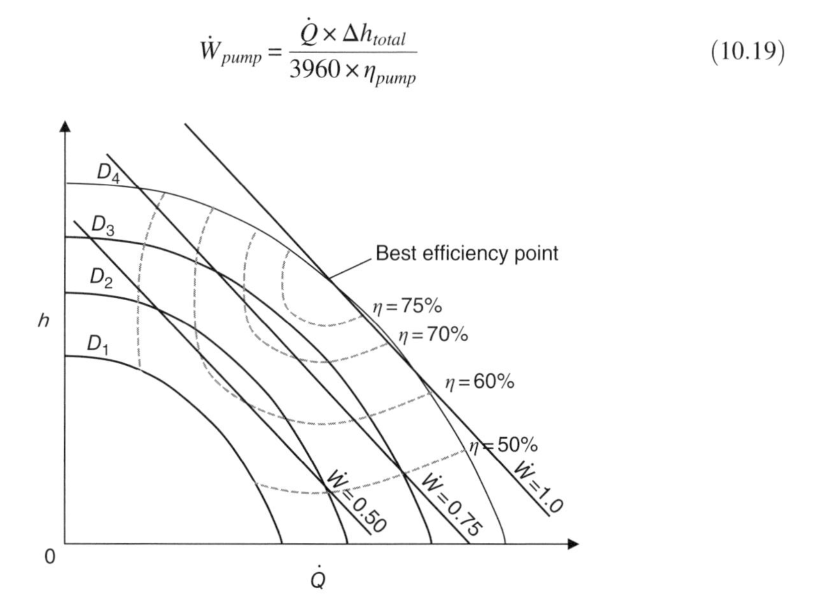

Figure 10.8 Generic pump curves at various impeller diameters (D_1–D_4)

where $\dot{W}_{pump}$ is the pump brakepower in horsepower (hp), $\dot{Q}$ is the volumetric flow rate in gallons per minute (gpm), Δh_{total} is the total head in feet (ft), and 3960 is a conversion factor in units of gal-ft-H_2O/min-hp.

From Equation (10.12), a useful dimensional form of this equation for pumping water at standard conditions (in SI units) is

$$\dot{W}_{pump} = \frac{\dot{Q} \times \Delta h_{total} \times \rho g}{\eta_{pump}} \tag{10.20}$$

where $\dot{W}_{pump}$ is the pump shaft power in units of W, $\dot{Q}$ is in units of $m^3 \cdot s^{-1}$, and Δh_{total} is in units of m.

Many pump performance graphs also plot curves showing the power required by the pump to produce a specific flow and pressure. These are shown as thin black lines in Figure 10.8. Note that these curves show the power required by the pump, including the efficiency of the pump. Calculating the work supplied to the pump using the preceding equations and comparing it with the value indicated on an actual pump performance graph is a useful exercise.

10.5.2 System Curves

The total pressure that a pump must produce to move the fluid is determined by the piping system. This total pressure of the piping system is the sum of the pressure due to inlet and outlet conditions and the pressure loss due to friction, as described above. In a piping system, pressure loss due to friction increases with increasing fluid flow; thus, system curves have positive slopes on pump performance charts. The operating point of a pump is determined by the intersection of the pump and system curves.

To determine the form of a system curve, recall Equation (10.7) for total pressure in a piping system. The total pressure caused by a piping system is the sum of the pressure due to inlet and outlet conditions and the pressure required to overcome friction through the pipes and fittings.

The inlet/outlet pressure that the pump must overcome is the sum of the static, velocity, and elevation pressures between the inlet and outlet of the piping system. For closed-loop piping systems, the inlet and outlet are at the same location; hence, the static, velocity, and elevation pressure differences are all zero. For open systems, the differences in static, velocity, and elevation pressures must be calculated. In many pumping applications, the velocity pressure difference between the inlet and outlet is zero or negligible, and inlet/outlet pressure is simply the sum of the static and elevation heads. In these cases, the inlet/outlet pressure is independent of flow and is represented on a pump performance chart as the pressure at zero flow.

Recall Equations (10.13) and (10.16) for calculating pressure loss from friction through pipes and through fittings. These equations clearly show that, for a given pipe system, the pressure drop is proportional to the square of the velocity, and hence the square of the volumetric flow rate. Thus, for a given pipe system, the pressure drop due to friction can be described as

$$\Delta h_{friction} = C\dot{Q}^2 \tag{10.21}$$

where C is a proportionality constant of the system.

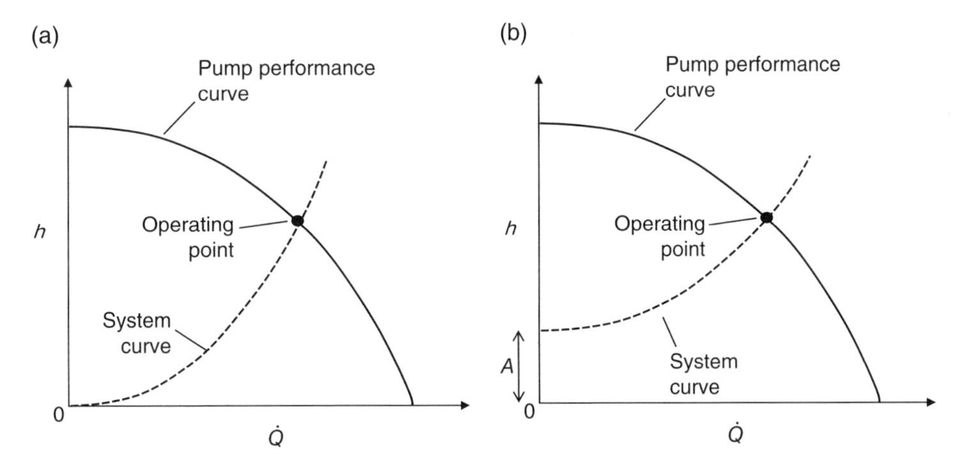

Figure 10.9 A system curve plotted on a pump curve for (a) a closed-loop piping system and (b) an open-loop piping system

The foregoing discussion shows that system curves have a flow-independent component (of inlet/outlet pressure) and a flow-dependent component that varies with the square of flow rate (of friction pressure). A system curve for a closed-loop piping system with no inlet/outlet pressure difference is shown Figure 10.9a. The curve is parabolic, of the form described by Equation (10.21). The curve passes through the origin because the inlet/outlet pressure difference, sometimes called the static head, is zero. The coefficient C can be determined if the operating point is known by substituting the known pressure drop and flow rate into Equation (10.21) and solving for C. The fluid work required to push the fluid through the pipe is the product of the volume flow rate and system pressure drop and is represented graphically by the area under the rectangle defined by the operating point.

A system curve for an open-loop piping system with a 'static' or 'inlet/outlet' pressure greater than zero is shown in Figure 10.9b. This system curve is of the form $\Delta h_{friction} = A + C\dot{Q}^2$, where A is the 'static' or 'inlet/outlet' pressure drop. As described above, the coefficient C can be determined if the operating point and inlet/outlet pressure are known by substituting the known values into Equation (10.21) and solving for C.

Example 10.3 Calculating a System Curve for a Closed-Loop Piping System – IP Units
Calculate and plot the system curve for the piping system of Example 10.2.

Solution
The total pressure drop determined from Example 10.2 through the GHX outside the building was 11.16 ft-H_2O at 160 gpm at the known operating point.

From Equation (10.21), we have

$$\Delta h_{friction} = C\dot{Q}^2$$

$$\therefore C = \frac{11.16\,\text{ft-H}_2\text{O}}{(160\,\text{gpm})^2} = 4.359 \times 10^{-4}\,\frac{\text{ft-H}_2\text{O}}{\text{gpm}^2}$$

Substituting values of $\dot{Q}$ into Equation (10.21), with $C = 4.359\text{E-4}$ ft-H_2O/gpm^2, we have the following results:

$\dot{Q}$ (gpm)	Δh (ft-H_2O)
0	0.00
10	0.04
20	0.17
40	0.70
60	1.57
100	4.36
160	11.16

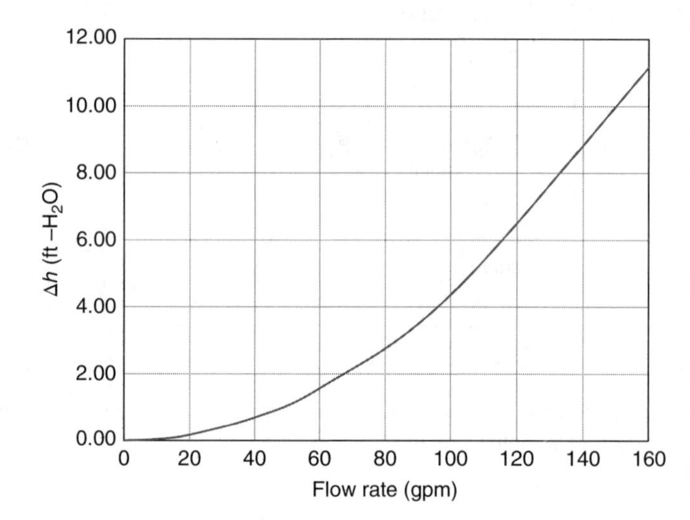

10.5.3 Pump Motor Work

Pumps and fans are typically driven by electrical motors. The power required by the motor is greater than the fluid work because the pump/fan, power transmission, and motor all incur irreversibilities. Thus, fluid power must be divided by the product of the efficiencies of all components of the pump power delivery system to determine the electricity required by the motor:

$$\dot{W}_{motor} = \frac{\dot{W}_{fluid}}{\eta_{motor} \times \eta_{drive} \times \eta_{pump}} = \frac{\dot{W}_{pump}}{\eta_{motor} \times \eta_{drive}} \tag{10.22}$$

For example, if the efficiency of the motor at converting electrical power to motor shaft power is 90%, the efficiency of belt drives (or other drives) at transferring motor shaft power to a pump is 92%, and the efficiency of a pump at converting pump shaft power to fluid power is 70%, the electrical power use required by the motor would be 73% greater than the required fluid power. In other words, $\dot{W}_{motor} = \dot{W}_{fluid} \div (0.90 \times 0.92 \times 0.70) = 1.73\ \dot{W}_{fluid}$.

10.5.4 Pump/Fan Affinity Laws

The pump and fan affinity laws describe useful relationships between the design parameters of rotational speed (i.e., the peripheral velocity of the impeller or RPM), friction head loss, and fluid power:

1. Volumetric flow rate varies proportionately with impeller rotating speed.
2. Friction head loss varies as the square of the volumetric flow rate.
3. Fluid power varies as the cube of the volumetric flow rate.

The affinity relations show that, in systems where all fluid power added by the pump/fan is to overcome pipe/duct friction, a small reduction in the volume flow rate results in a large reduction in the fluid power. For example, reducing the volume flow rate by one-half reduces fluid work by 88%!

10.5.5 Energy-Efficient Pumping

Kavanaugh and Rafferty (1997) and ASHRAE (2015) give benchmarks for pumping power in closed-loop GHX systems. The best pumping systems have an installed pump power less than 0.05 hp per ton of thermal load, or 0.01 kW_e per kW_t.

The fluid power equation shows that power required by fan/pump systems is a function of the volumetric flow rate, inlet/outlet conditions, and system friction. Thus, addressing these parameters is a good starting point for energy-efficient pumping opportunities.

In open-loop systems, elevation head is often a primary contributor to pump energy consumption. Thus, as discussed in Chapter 4, avoidance of deep aquifers and pumping as little groundwater as necessary are prudent measures.

The primary methods to reduce friction in a pipe/duct system are: (i) increase pipe diameter, (ii) use smooth pipes, and (iii) use fewer fittings or fittings with low pressure drop.

There are many pump control options in geothermal systems. Individual circulators can be placed at individual heat pumps and only be operated when needed (i.e., simple on/off control). This is possible in small and large systems; in larger systems, central circulators are needed to overcome head losses. In larger systems, primary/secondary pump systems are an option, as shown in Figure 10.7. Pumps of all sizes are available for use with variable-speed drives (VSDs). This common flow control option will be discussed in more detail in what follows.

Pump motor speed can be varied by varying the voltage or the frequency of the motor. With voltage variation, the pump motor is a DC motor equipped with an rectifier for AC–DC rectification. Electronic variable-frequency drives (VFDs) control the speed of AC motors by converting the frequency and voltage of the AC line supply from fixed to variable values. VSDs work best with premium-efficiency motors.

VSD pump applications typically require the following considerations relative to a constant-flow pumping system:

1. Install a VSD on the power supply to the pump motor. In parallel pumping configurations, one VSD is generally needed for each operational pump.
2. Eliminate valves on bypass pipes.
3. Install a differential pressure sensor between the supply and return headers at the process load located the farthest distance from the pump or the process load with the greatest required head to generate flow through the load. Determine the pressure drop needed to guarantee sufficient flow through the farthest process load at this point. Control the speed of the VSD to maintain this differential pressure. Alternatively, VSDs, such as on a dedicated GHX pump or well pump, could be controlled on the basis of GHX fluid temperatures.

To illustrate the energy savings opportunities with VSDs, consider the idealized system curve shown in Figure 10.10. The operating point is shown as ‘Operating Point A’, operating

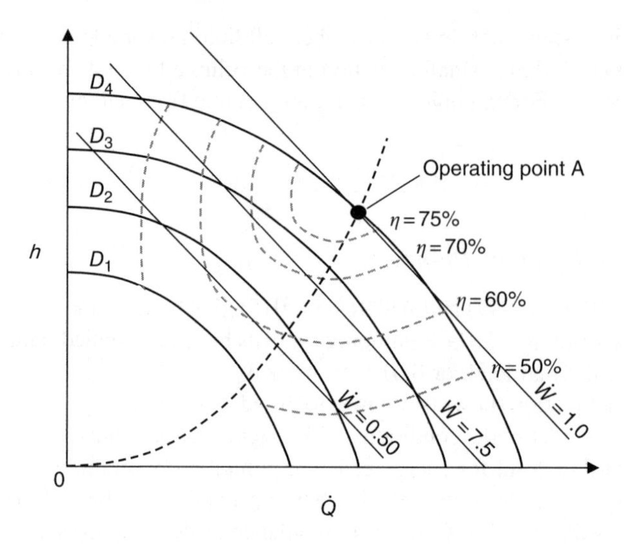

Figure 10.10 A system curve plotted on a family of pump curves, showing operating point A

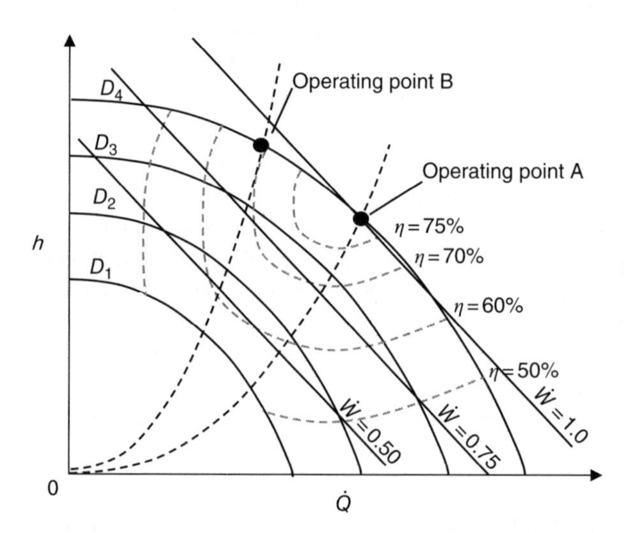

Figure 10.11 A system curve plotted on a family of pump curves, showing initial operating point A and operating point B with two-thirds the flow rate

near the best efficiency point on a pump curve with a pump impeller diameter D_4. The pumping power is 1.0 (dimensionless units).

Let's say that this pump is serving a building with several distributed geothermal heat pumps, and during a hypothetical part load condition, the flow decreases by one-third of the value at point A. This would occur as automatic valves at each heat pump shut off, causing the pressure in the pipe system to rise. The new operating point must follow the pump curve to operating point B at two-thirds the value of $\dot{Q}$ for operating point A. As the position of operating point B is known, it can be plotted as shown in Figure 10.11, and using methods described in Example

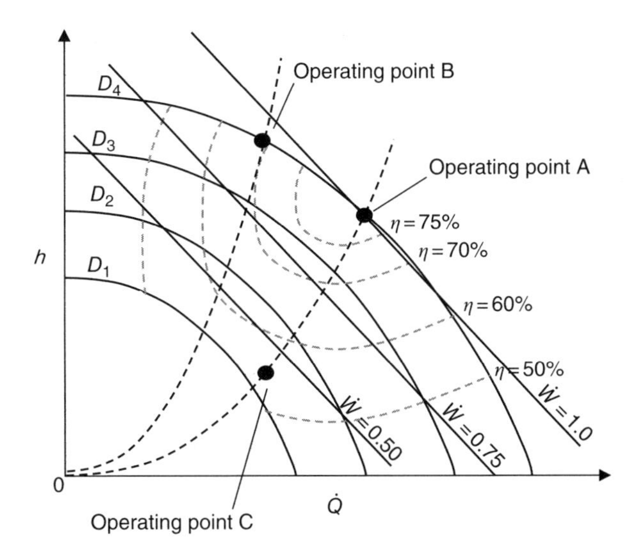

Figure 10.12 A system curve plotted on a family of pump curves, showing initial operating point A, operating point B with two-thirds the flow rate (no VSD), and operating point C with two-thirds the flow rate (with VSD)

10.3, a new system curve can be readily developed. A review of the curves in Figure 10.11 shows that the new pump energy requirement at point B would decrease to about 0.92 (—), or about 8% less than for point A.

With a VSD installed on this system to maintain a minimum required differential pressure (i.e., a pressure set-point that is large enough to push the fluid through end uses), the VSD would slow the pump motor speed, thus decreasing $\dot{Q}$ by one-third to the same $\dot{Q}$ value as that for operating point B. However, to get that flow rate, the new operating point (point C) would follow the *system curve* for operating point A, not the pump curve. Thus, this new operating point with a VSD would be as shown in Figure 10.12. A review of the curves in Figure 10.12 shows that the new pump energy requirement at point C would decrease to less than half that for point A.

10.6 Chapter Summary

This chapter presented an overview of fundamental fluid mechanics of internal flows, with the focus on determining pumping energy requirements for GHX systems. Methods were presented to size and configure piping for a closed-loop GHX, to calculate pressure drop, and to select a pump. A practical discussion of air, dirt, sediment, and debris management in hydronic systems was also presented.

Exercise Problems

10.1 A vertical, closed-loop GHX consisting of 4 × 5 vertical boreholes requires a piping design. Each BHE consists of a 1 in diameter or DN25 U-tube. Each borehole is drilled to a depth of 250 ft or 75 m, and boreholes are placed 20 ft or 6 m apart. Complete the design by doing the following:

(a) Provide a layout of the supply and return piping, and show all pipe diameters and associated branch flow rates. First, decide on how many flow circuits you want in your design. Next, decide on whether to bring your flow circuits to a manifold in a vault or a manifold in the building.
(b) Provide a detail sketch of your piping manifold (in the vault or building).
(c) Calculate the total pressure drop through the GHX. Assume 50 ft or 16 m to the mechanical room.
(d) Recommend a circulating pump capacity for this GHX (in hp or kW) and show your calculations.

10.2 A GHX has been designed in a primary/secondary pumping arrangement. The total flow and head loss at peak design conditions through the GHX are 100 gpm at 15 ft-H_2O, or 20 $m^3 \cdot h^{-1}$ at 5 m-H_2O.

(a) Conduct a web-based search to select a suitable pump for this application. You may follow the steps below to facilitate your choice:
- Locate the design flow and head point on the pump performance graph.
- Select a pump whose curve is on or just above the design point with an efficiency near the best efficiency point (BEP). If the point is not near the BEP, proceed to another set of pump curves.
- Select a pump power and rpm based on the pump curve (not the design point).
- If the calculated design point is below the pump curve, the actual operating point will be the intersection of the system and pump curves.

(b) Calculate and plot the system curve, and show the operating point.
(c) Determine the electrical power requirement at the operating point.

Part III

Geothermal Energy Conversion

In Part III, we discuss the conversion of geothermal energy to useful energy. **Chapter 11** leads readers through the parallels of heat pumps and heat engines by first reviewing the Carnot cycle, and then by looking at practical heat pump and heat engine cycles. **Chapter 12** discusses mechanical vapor compression heat pumps, **Chapter 13** discusses thermally driven heat pumps, and **Chapter 14** discusses organic Rankine (binary) geothermal power plants. Each of Chapters 12 to 14 discusses general principles of the cycle, with an emphasis on geothermal applications. Detailed analysis of the thermodynamic cycles are addressed through computer simulation.

Geothermal Heat Pump and Heat Engine Systems: Theory and Practice, First Edition. Andrew D. Chiasson.

Companion website: www.wiley.com/go/chiasson/geoHPSTP

11

Heat Pumps and Heat Engines: A Thermodynamic Overview

11.1 Overview

This chapter begins our discussion of methods and equipment to convert stored thermal energy into useful energy. A general thermodynamic overview and comparison of heat pump and heat engine cycles is presented. The Carnot cycle is discussed and compared with real and practical cycles.

Learning objectives and goals:

1. Review and strengthen thermodynamic concepts as they pertain to heat pumps and heat engines.
2. Define what a heat pump and heat engine is.
3. Calculate Carnot, first-law, and second-law efficiencies of heat pump and heat engine cycles.

11.2 Fundamental Theory of Operation of Heat Pumps and Heat Engines

The operating principles of heat engines and heat pumps are integral to the study of *Engineering Thermodynamics*, where *Engineering Thermodynamics* is generally described as the study of the interactions of heat and work. For many thousands of years, humans knew how to do work in order to produce heat (i.e., rubbing two sticks together to make a fire), but when we figured out how to make the opposite happen, that is, generate heat to produce work, the industrial revolution began and the entire world changed forever.

Geothermal Heat Pump and Heat Engine Systems: Theory and Practice, First Edition. Andrew D. Chiasson.

Companion website: www.wiley.com/go/chiasson/geoHPSTP

The realizations that heat and work are interchangeable forms of energy are ideas that did not happen overnight. It took centuries for philosophers and scientists to organize thoughts and scientific observations into what are known today as scientific laws. In this book, we will rely heavily on the laws of thermodynamics. The first law of thermodynamics states that, in any process, energy is conserved; it is simply the 'law of conservation'. The second law of thermodynamics is not so easy to define, but classic statements of this law offer guiding principles of operation of heat engines and heat pumps:

1. The Clausius Statement: *It is impossible for any system to operate in such a way that the sole result would be an energy transfer by heat from a cooler to a hotter body.*
2. The Kelvin–Planck Statement: *It is impossible for any system to operate in a thermodynamic cycle and deliver a net amount of energy by work to its surroundings while receiving energy by heat transfer from a single thermal reservoir.*
3. The Entropy Statement: *It is impossible for any system to operate in a way such that entropy is destroyed.*

In the study of thermodynamics, a *heat engine* is defined as a system operating in a cycle that converts thermal energy to mechanical energy by bringing a working substance from a higher-temperature state (T_H) to a lower-temperature state (T_C). This mechanical energy can then be used to do work, such as drive a turbine to generate electricity. A rendering of this concept is simplified schematically in Figure 11.1a. The first law of thermodynamics tells us that for a simple heat engine

$$\dot{q}_{in} = \dot{q}_{out} + \dot{w}_{out} \tag{11.1}$$

On the other hand, a heat pump is defined as a device that moves (or 'pumps') thermal energy opposite to the direction of spontaneous heat flow by absorbing heat from a cold space (T_C) and releasing it to a warmer one (T_H). A rendering of this concept is simplified schematically in

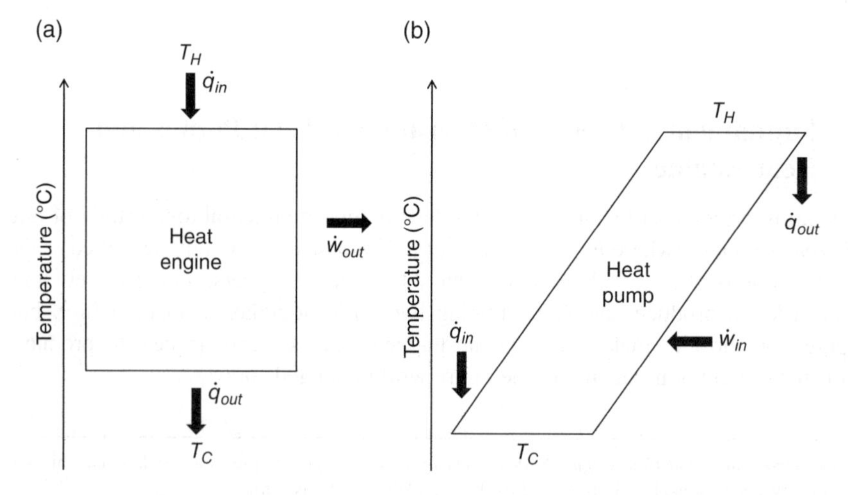

Figure 11.1 Simplified schematic diagrams of energy flows in (a) a heat engine and (b) a heat pump

Figure 11.1b. Note that, in contrast to the heat engine, heat rejection in a heat pump occurs at the high-temperature reservoir (T_H) and heat extraction occurs at the lower-temperature reservoir (T_C). Thus, heat pump cycles must operate at temperatures greater than the sink temperature (at $\dot{q}_{out}$) and lower than the source temperature (at $\dot{q}_{in}$).

The first law of thermodynamics tells us that for a simple heat pump

$$\dot{q}_{in} + \dot{W}_{in} = \dot{q}_{out} \tag{11.2}$$

The diagrams in Figure 11.1 are meant to be simplistic and quite general at this point. This is to illustrate the similarities between heat engines and heat pumps: you supply two energy flows and you'll get the third. How you choose to accomplish this is, in part, the subject of this book. We will first explore more of the details of the thermodynamic cycles before delving into the practical aspects. There are numerous cycles and methods of achieving the goals depicted in Figure 11.1. Some actual power cycles that exist today include: Stirling, Ericsson, Rankine, Brayton, and solid-state technologies (thermoelectric). Similarly, actual heat pump cycles that exist today include: mechanical vapor compression, thermally driven technologies (e.g., absorption cycles, adsorption cycles), electromechanical technologies (e.g., evaporative cooling, thermoacoustic devices, piezoelectric devices), and solid-state technologies (thermoelectric).

11.3 The Carnot Cycle

The Carnot cycle, described by Sadi Carnot in the 1820s, is a theoretical, ideal cycle for heat engines and heat pumps. With respect to Figure 11.1, a Carnot cycle dictates that there are four ideal, thermodynamically reversible processes. The heat transfers at the high- and low-temperature reservoirs (T_H and T_C) occur isothermally, and the processes to and from the high- and low-temperature reservoirs occur adiabatically. Thus, in conjunction with the Kelvin temperature scale (an absolute thermodynamic temperature scale), the ratio of the heat transfers $\dot{q}_C/\dot{q}_H$ in a fully reversible cycle depends only on the temperatures of the reservoirs:

$$\left(\frac{\dot{q}_C}{\dot{q}_H}\right)_{reversible} = \frac{T_C}{T_H} \tag{11.3}$$

Here, the word *reversible* is used in the pure thermodynamic sense, and should not be confused with its meaning when used in the context of a reversible heat pump. Recall that in the study of thermodynamics, a process that is *reversible* means that there are no irreversibilities, or *unrecoverabilities*. Thermodynamic irreversibilities common in engineering practice include friction, vibration, undesirable heat losses or gains, chemical reactions, deformation, etc.

A practical use of the Carnot cycle is that it determines best theoretical performance of heat engine and heat pump cycles as a function of reservoir temperature. This performance is defined in general as the ratio of useful energy out of a cycle to the energy input.

For a heat engine, performance is described by its efficiency (η), where the maximum efficiency (η_{max}), or Carnot efficiency, is defined as

$$\eta_{max} = 1 - \frac{T_C}{T_H} \tag{11.4}$$

By simple mathematical inspection of Equation (11.4), it is obvious that η_{max} must be less than 1 because the ratio T_C/T_H is less than unity. Remember that T_C and T_H are expressed in absolute units (K or °R). Mathematical inspection of Equation (11.4) also reveals that the greater T_H is relative to T_C, the greater is the maximum possible efficiency.

For a heat pump, performance is described by its *coefficient of performance* (COP), where the maximum efficiency, or Carnot efficiency, is defined differently for heating and cooling modes. The maximum possible COP for a heat pump in heating mode is given by

$$\text{COP}_{max,h} = \frac{T_H}{T_H - T_C} \tag{11.5}$$

and the maximum possible COP for a heat pump in cooling mode is given by

$$\text{COP}_{max,c} = \frac{T_C}{T_H - T_C} \tag{11.6}$$

By simple mathematical inspection of Equations (11.5) and (11.6), it is obvious that the maximum COP values of heat pumps must be greater than unity. Mathematical inspection of Equations (11.5) and (11.6) also reveals that the smaller the difference between T_H and T_C, the greater is the maximum possible COP. Further, any reservoir with a temperature above 0 K could be used in a theoretical heat pump application.

11.4 Real-World Considerations: Entropy and Exergy

Carnot cycles are theoretical because they do not consider the fact that heat transfer does not occur perfectly in real-world situations. Thus, we can combine our knowledge of the first law of thermodynamics and the maximum efficiencies from Carnot cycles to describe the efficiency of real-world heat engines and heat pumps; the actual performance of heat engines and heat pumps is related to the heat transfers actually achieved at the high- and low-temperature reservoirs. Thus, Equations (11.4) to (11.6) are modified by replacing T with $\dot{q}$ as described in what follows.

The performance of an actual heat engine is given by

$$\eta = 1 - \frac{\dot{q}_C}{\dot{q}_H} = \frac{\dot{W}_{out}}{\dot{q}_{in}} \tag{11.7}$$

The performance of an actual heat pump in heating mode is given by

$$\text{COP}_h = \frac{\dot{q}_H}{\dot{q}_H - \dot{q}_C} = \frac{\dot{q}_{out}}{\dot{W}_{in}} \tag{11.8}$$

while the performance of an actual heat pump in cooling mode is given by

$$\text{COP}_c = \frac{\dot{q}_C}{\dot{q}_H - \dot{q}_C} = \frac{\dot{q}_{in}}{\dot{W}_{in}} \tag{11.9}$$

How closely the actual performance of heat engines and heat pumps approaches their maximum possible performance can be readily seen by the so-called Clausius inequality, which forms the basis of the concept of entropy. The concept was proposed by Rudolf Clausius in the 1850s. In rearranging Equation (11.3), we can show that for simple heat engine and heat pump cycles

$$\frac{\dot{q}_C}{T_C} - \frac{\dot{q}_H}{T_H} = \Delta S \tag{11.10}$$

where S is entropy, and $\Delta S = 0$ for ideal Carnot cycles (i.e., those that are fully reversible), $\Delta S > 0$ for real cycles (i.e., those that have thermodynamic irreversibilities), and $\Delta S < 0$ for cycles that are not possible. Another thermodynamic definition of entropy for a process is $\dot{q} = \int T dS$. This is a handy definition of entropy because it is analogous to the familiar thermodynamic definition of work done by or on a compressible substance that undergoes a volume change in response to a pressure change (i.e., $\dot{W} = \int P dV$). Thus, these definitions illustrate that pressure is to work what temperature is to heat, and that a process undergoing a volume has an associated work transfer, just as a process undergoing an entropy change has an associated heat transfer. Keep in mind that the temperature of a compressible substance can be changed adiabatically if there is a pressure–volume change.

Heat engine and heat pump cycles are frequently depicted on temperature–entropy (T–s) diagrams. Figure 11.2 shows a Carnot cycle on such a diagram. The heat pump cycle is the reverse of the heat engine cycle. Note that Carnot cycles on a T–s are rectangular, as there is no change in entropy between points 1 and 2, or between points 3 and 4, and heat is added and rejected at constant temperature. In real cycles, the second law tells us that entropy increases owing to thermodynamic irreversibilities. Also, in practice, there are finite limits to perfect heat transfer at the high- and low-temperature reservoirs, and thus real cycles have significant departures from Carnot cycles. This will be the subject of the next section of this chapter.

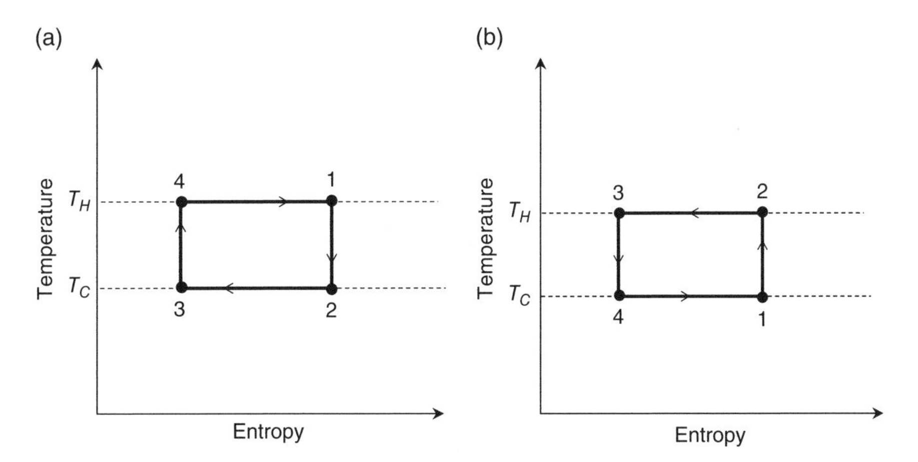

Figure 11.2 Temperature–entropy (T–s) diagrams for (a) a Carnot heat engine and (b) a Carnot heat pump

Example 11.1 Real, Ideal, or Impossible Heat Engine Cycle

Consider a geothermal power plant where groundwater wells are supplying heat according to the energy balance equation $\dot{q}=\dot{m}c_p\Delta T$, where $\dot{m}$ is the mass flow rate of the geothermal water, c_p is the specific heat of the water, and ΔT is the temperature change of the water. We will be discussing this energy balance equation later, but notice that it gives us a heat transfer rate regardless of the actual groundwater temperature. The plant is producing 1000 kW net work with a geothermal heat input of 5000 kW. Heat is rejected to the environment at 10 °C. Determine whether the cycle is real, ideal, or impossible if the geothermal reservoir temperature is (a) 400 K, (b) 353.75 K, and (c) 325 K. Assume the geothermal fluid is kept under a pressure such that it remains liquid.

Solution

The information given is shown schematically in Figure E.11.1.

By Equation (11.1)

$$\dot{q}_{in} = \dot{q}_{out} + \dot{W}_{out}$$

$$5000kW = \dot{q}_{out} + 1000\,\text{kW}$$

$$\therefore \dot{q}_{out} = 4000\,\text{kW}$$

By Equation 11.10

$$\frac{\dot{q}_C}{T_C} - \frac{\dot{q}_H}{T_H} = \Delta S$$

(a) For $T_H = 400$ K, we have

$$\frac{4000\,\text{kW}}{283\,\text{K}} - \frac{5000\,\text{kW}}{400\,\text{K}} = 1.634\frac{\text{kJ}}{\text{K}}\cdot\text{s}^{-1}$$

As $\Delta S > 1$, this cycle is a real cycle with irreversibilities.

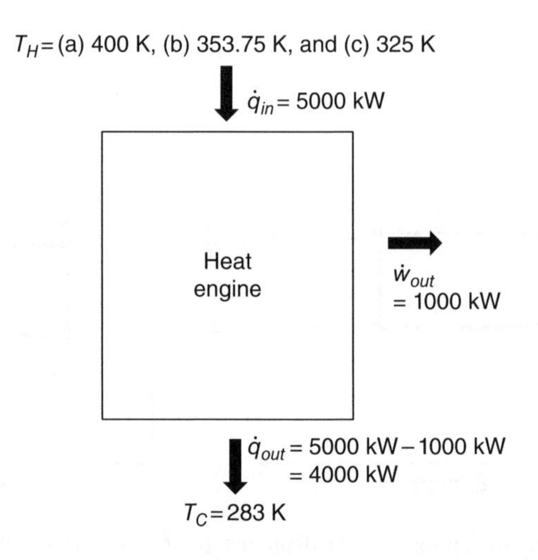

Figure E.11.1 Schematic diagram for Example 11.1

(b) For $T_H = 353.75$ K, we have

$$\frac{4000\,\text{kW}}{283\,\text{K}} - \frac{5000\,\text{kW}}{353.75\,\text{K}} = 0\frac{\text{kJ}}{\text{K}}\cdot\text{s}^{-1}$$

As $\Delta S = 0$, this cycle is an ideal cycle (i.e., a Carnot cycle), fully reversible.

(c) For $T_H = 325$ K, we have:

$$\frac{4000\,\text{kW}}{283\,\text{K}} - \frac{5000\,\text{kW}}{325\,\text{K}} = -1.25\frac{\text{kJ}}{\text{K}}\cdot\text{s}^{-1}$$

As $\Delta S < 0$, this cycle is not possible.

Example 11.2 Real, Ideal, or Impossible Heat Engine Cycle Revisited

Redo Example 11.1 by comparing the actual thermal performance of the geothermal power plant with the maximum possible for each geothermal reservoir temperature case.

Solution

The actual thermal performance of the plant is given by Equation (11.7):

$$\eta = 1 - \frac{\dot{q}_C}{\dot{q}_H} = \frac{\dot{W}_{out}}{\dot{q}_{in}} = \frac{1000\,\text{kW}}{5000\,\text{kW}} = 0.20\text{ or }20\%$$

(a) For $T_H = 400$ K, we have by Equation (11.4)

$$\eta_{max} = 1 - \frac{T_H}{T_C} = \frac{400\,\text{K}}{283\,\text{K}} = 0.2925\text{ or }29\%$$

As $\eta < \eta_{max}$, this cycle is a real cycle with irreversibilities.

(b) For $T_H = 353.75$ K, we have by Equation (11.4)

$$\eta_{max} = 1 - \frac{T_H}{T_C} = \frac{353.75\,\text{K}}{283\,\text{K}} = 0.20\text{ or }20\%$$

As $\eta = \eta_{max}$, this cycle is an ideal cycle (i.e., a Carnot cycle), fully reversible.

(c) For $T_H = 325$ K, we have by Equation (11.4)

$$\eta_{max} = 1 - \frac{T_H}{T_C} = \frac{325\,\text{K}}{283\,\text{K}} = 0.1293\text{ or }12.9\%$$

As $\eta > \eta_{max}$, this cycle is not possible.

The foregoing example illustrates that the thermal performance may not be the best means of expressing efficiency; if you tell someone that a geothermal power plant is 20% efficient,

Table 11.1 Summary of Efficiency Equations for Heat Engines and Heat Pumps

Cycle	Maximum Possible Efficiency (*Carnot efficiency*)	Actual or Thermal Efficiency (*first-law efficiency*)	Exergy Efficiency (*second-law efficiency*)
Heat Engine	$\eta_{max}=1-\frac{T_C}{T_H}$ Equation (11.4)	$\eta=1-\frac{\dot{q}_C}{\dot{q}_H}=\frac{\dot{W}_{out}}{\dot{q}_{in}}$ Equation (11.7)	$\eta_E=\frac{\eta}{\eta_{max}}$ Equation (11.11)
Heat Pump (heating)	$COP_{max,h}=\frac{T_H}{T_H-T_C}$ Equation (11.5)	$COP_h=\frac{\dot{q}_H}{\dot{q}_H-\dot{q}_C}=\frac{\dot{q}_{out}}{\dot{W}_{in}}$ Equation (11.8)	$\eta_E=\frac{COP_h}{COP_{max,h}}$ Equation (11.12)
Heat Pump (cooling)	$COP_{max,c}=\frac{T_C}{T_H-T_C}$ Equation (11.6)	$COP_c=\frac{\dot{q}_C}{\dot{q}_H-\dot{q}_C}=\frac{\dot{q}_{in}}{\dot{W}_{in}}$ Equation (11.9)	$\eta_E=\frac{COP_c}{COP_{max,c}}$ Equation (11.13)

they may not be impressed. However, if you qualify your statement by saying that the maximum plant efficiency is 29%, as in Example 11.2(a), then their opinion may be quite different. Therefore, energy must have both quantity and quality. Herein lies the concept of *exergy*, where we seek a means of expressing the potential work or available work of a process.

As we know from the first law of thermodynamics, energy is conserved in every process or device. Thus, Equations (11.7), (11.8), and (11.9) may be regarded as first-law efficiencies. On the other hand, *exergy* is not conserved in processes where entropy increases; exergy is destroyed by thermodynamic irreversibilities that occur in every real process. Readers are encouraged to consult a thermodynamics textbook for a more detailed treatment of exergy, but here we will use it in a very simple form to express so-called second-law efficiency, utilization efficiency, or exergy efficiency of heat engines and heat pumps as

$$\eta_E=\frac{\eta}{\eta_{max}} \tag{11.11}$$

Inspection of Equation (11.11) reveals that the upper theoretical limit of η_E is 1. Similar equations can be expressed for heat pump cycles as summarized in Table 11.1.

Example 11.3 Exergy Efficiency of a Geothermal Power Plant

Calculate the exergy efficiency for the geothermal power plant described in Example 11.1 for each of the geothermal reservoir temperature cases. Comment on the result of each.

Solution

(a) For $T_H = 400$ K, we have by Equation (11.11)

$$\eta_E=\frac{\eta}{\eta_{max}}=\frac{0.20}{0.2925}=0.684 \text{ or } 68.4\%$$

As $\eta_E < 1$, this cycle is a real cycle with irreversibilities.

(b) For $T_H = 353.75$ K, we have by Equation (11.11)

$$\eta_E = \frac{\eta}{\eta_{max}} = \frac{0.20}{0.20} = 1 \text{ or } 100\%$$

As $\eta_E = 1$, this cycle is an ideal cycle (i.e., a Carnot cycle), fully reversible.

(c) For $T_H = 325$ K, we have by Equation (11.11)

$$\eta_E = \frac{\eta}{\eta_{max}} = \frac{0.20}{0.1293} = 1.548 \text{ or } 154.8\%$$

As $\eta_E > 1$, this cycle is not possible.

Example 11.4 Exergy Efficiency of a Heat Pump

Consider a heat pump that is heating a residence. The building is being kept at an indoor temperature of 20 °C while it is 0 °C outside. The heat pump power usage is 5 kW and its COP is 2.5. Calculate the second-law efficiency of the heat pump at these conditions.

Solution

From Equation (11.5)

$$\text{COP}_{max,h} = \frac{T_H}{T_H - T_C} = \frac{293\text{ K}}{20\text{ K}} = 14.65$$

From Equation (11.12)

$$\eta_E = \frac{\text{COP}_h}{\text{COP}_{max,h}} = \frac{2.5}{14.65} = 17.1\%$$

Discussion: The ideal version of this heat pump could deliver 14.65 times more heating energy than work input energy! Unfortunately, actual heat pumps have lots of irreversibilities. Therefore, there is tremendous potential for improving the performance of heat pumps by reducing friction, turbulence, temperature differences, heat exchanger inefficiencies, and motor inefficiencies.

11.5 Practical Heat Engine and Heat Pump Cycles

The Carnot cycle is theoretical. It does not include actual methods of accomplishing anything. All of the Carnot processes must be reversible; heat transfers at T_H and T_C must occur without a temperature gradient. Even to come close to the efficiency of a Carnot cycle is met with practical challenges; significant thermodynamic irreversibilities exist in real processes; heat addition and rejection to cycles can only be done with finite-sized heat exchangers. In addition, there are environmental impact considerations in many of the heat engine and heat pump technologies employed today.

As engineers, our main goal involves balancing physics within practical constraints, and that is certainly the case with Carnot cycles vs. practical heat engine and heat pump cycles. There are numerous ways to achieve the goals presented in Figure 11.1, as we will discuss in what follows, and each method has its own strengths and weaknesses. Our discussion of practical heat engine and heat pump cycles is not meant to be an exhaustive discussion of heat engine and heat pump technologies, but rather an open-minded approach to technologies that possess current or potential geothermal application.

11.5.1 Practical Heat Engine Cycles

Here, we limit our discussion to heat engine cycles that are used to generate large, utility-scale electrical power. The thermodynamic cycles when it comes to large, utility-scale electrical power generation are the Rankine and Brayton cycles. These cycles dominate, not necessarily because they are more efficient, but because of their scalability: they can be built modularly to produce electricity on very large scales.

The Rankine cycle is a vapor power cycle where a heat source external to the cycle is used to drive a closed-loop fluid undergoing cyclic evaporation, expansion, condensation, and compression. Work is produced as the fluid expands through a turbine. In 'standard Rankine cycles', water/steam is used as the working fluid. In organic or binary Rankine cycles (abbreviated ORC), the working fluid is typically a low-boiling-point hydrocarbon fluid, thus allowing application with heat sources at temperatures below 100 °C. The Kalina cycle, a variant of the Rankine cycle, uses an ammonia–water mixture as the working fluid. As we shall see in a subsequent section of this chapter, more energy can be extracted when ammonia–water is used as the working fluid.

The Rankine cycle is sometimes referred to as an external (or non-internal) combustion cycle, based on the configuration of the heat source. Heat can be supplied to the working fluid in several different configurations: fossil fuel (coal or natural gas), nuclear reactor, solar concentrating, or geothermal.

The Brayton cycle, another popular turbine cycle for large, utility-scale electrical power generation, is a gas power cycle that can be configured in an open- or closed-loop configuration. The open configuration, considered to be an internal combustion engine cycle, consists of air undergoing an open cycle of compression, mixing and combustion with natural gas, and expansion through a turbine. In the closed configuration, the cycle is considered to be an external combustion engine cycle, where the working fluid (typically air) undergoes a closed cycle of compression, heating, expansion through a turbine, and cooling.

As energy engineers, we are continually seeking more efficient and sustainable means of generating electrical energy. Power cycles are undergoing continuous improvement, with more efficient turbines, heat exchangers, and research into alternative working fluids. Significant research is also being conducted with regard to non-turbine technologies. In reference to Figure 11.2a, Table 11.2 summarizes heat engine cycles that have potential with geothermal applications.

Table 11.2 Heat Engine Cycles with Possible Geothermal Application

Cycle	Process 1–2	Process 2–3	Process 3–4	Process 4–1	Comments
			Gas Power Technologies		
Closed Brayton	Adiabatic expansion through a turbine	Isobaric heat rejection	Adiabatic compression using a compressor	Isobaric heat input	• The working fluid is a gas (typically air) that undergoes cyclic compression and expansion to produce mechanical work.
Ericsson and Stirling	Regeneration	Isothermal heat rejection, gas compression	Regeneration	Isothermal heat input, gas expansion	• Ericsson and Stirling cycles are variants of the Carnot cycle. In Ericsson {Stirling} cycles, regeneration is done at constant pressure {volume}.
			Vapor Power Technologies		
Kalina	Adiabatic expansion through a turbine	Varying pressure and temperature heat rejection	Chemical absorption	Varying pressure and temperature heat rejection	• The working fluid is alternately vaporized and condensed. • The working fluid is a mixture of two substances, typically ammonia and water, causing heat to be absorbed and rejected neither isothermally nor isobarically.
Organic Rankine	Adiabatic expansion through a turbine	Isobaric heat rejection	Adiabatic compression using a pump	Isobaric heat input	• The working fluid is alternately vaporized and condensed or cooled, depending on whether the cycle is subcritical or supercritical. • The working fluid is an organic fluid (typically a hydrocarbon) with a boiling point lower than water. 'Standard' cycles use water/steam. • Analogous to the mechanical vapor compression cycle for heat pumps.
Triangular or trilateral cycle	Adiabatic expansion through a two-phase screw expander	Isobaric heat rejection	Adiabatic compression using a pump	Isobaric heat input	• The working fluid is cycled between a two-phase fluid and a liquid. • Similar to an organic Rankine cycle, except that the fluid enters the expansion process as a saturated liquid. • A practical difficulty with this cycle is the limitation with developing an efficient two-phase expander.
			Solid-State Technologies		
Thermo-electric	Seebeck effect	Peltier/Thompson effects	Not applicable	Peltier/Thompson effects	• Direct conversion of temperature differences to electrical energy via the Seebeck effect. • Heat is added and rejected via Peltier/Thompson effects, where heat is generated or removed in the presence of an electric current flowing through dissimilar conductors.

11.5.2 Practical Heat Pump Cycles

Mechanical vapor compression systems are by far the dominant heat pump technology owing to their scalability, reliability, familiarity, and relatively compact size. Vapor compression systems have been serving HVAC needs since the 1920s after the invention of Freon®, and replaced earlier absorption and other cooling systems, where refrigerants were toxic and/or flammable.

Mechanical vapor compression systems transfer heat through a closed-loop cycle by compressing, condensing, expanding, and evaporating a refrigerant fluid. There is a long list of working fluids that can serve as the refrigerant, but most systems use only a few synthetic refrigerants designed specifically to operate between the temperatures suitable to HVAC applications. However, these refrigerants have been found to have detrimental effects on the global environment when released into the atmosphere.

As energy engineers, we are continually seeking more efficient and sustainable means of providing heating and cooling technologies. Mechanical vapor compression cycles are undergoing continuous improvement, with more efficient compressors, heat exchangers, and research into alternative refrigerants. Significant research is also being conducted with regard to non-vapor compression technologies in view of the environmental impact of currently used refrigerants. In reference to Figure 11.2b, Table 11.3 summarizes heat pump cycles that have potential with geothermal applications.

11.6 The Working Fluids: Refrigerants

All refrigerants are denoted by a code that begins with the letter 'R' (denoting Refrigerant) and followed by a number. R000–R399 refrigerants are chemical refrigerants where individual digits code the number of C, H, and F atoms. Other information is used to code the number of halons. R400-series refrigerants are zeotrope mixtures of refrigerants that do not have an evaporation point, but rather an evaporation range. R500-series refrigerants are azeotrope mixtures with a fixed evaporation point. R600-series refrigerants are all other organic refrigerants, and R700-series refrigerants are inorganic refrigerants.

We will focus our discussion here on working fluids in ORC heat engines and geothermal heat pumps. In general, however, the ultimate working fluid in all ORC heat engines and heat pumps has been elusive. The selection of the working fluids (refrigerants) in these machines is based on several factors, namely thermal performance, safety, environmental impact, and cost, and there is no universal refrigerant for all cases. The desirable fluid would possess good thermal properties such as high latent heat of vaporization (large refrigerating effect), high thermal conductivity, low viscosity, and boiling and freezing points suitable for the refrigerant's application. Safety considerations are also important, such as flammability and toxicity, and the refrigerant should also have easy leak detection characteristics. The environmental impact of refrigerants has received worldwide attention and concern over the past decades, particularly with respect to ozone depletion and climate change. A worldwide metric to compare deleterious effects of refrigerants is the so-called *global-warming potential* (GWP), which is a relative measure of how much heat a greenhouse gas traps in the atmosphere. Calculation of GWP is a debated topic, but it is widely used, and is expressed as a factor relative to that of carbon dioxide, the GWP of which is normalized to a value of 1.

Cycle	Process 1–2	Process 2–3	Process 3–4	Process 4–1	Comments
			Gas Compression Technologies		
Reverse Brayton (Bell Coleman)	Adiabatic compression using a compressor	Isobaric heat rejection	Adiabatic expansion through a turbine	Isobaric heat input	• Mechanical work is applied to a gas undergoing cyclic compression and expansion to produce a heating or cooling effect.
Ericsson and Stirling	Regeneration	Isothermal heat rejection, gas compression	Regeneration	Isothermal heat input, gas expansion	• Ericsson and Stirling cycles are variants of the Carnot cycle. In Ericsson {Stirling} cycles, regeneration is done at constant pressure {volume}.
			Vapor Compression Technologies		
Mechanical vapor compression	Adiabatic compression using a compressor	Isobaric heat rejection	Adiabatic expansion through an expansion valve	Isobaric heat input	• The working fluid is alternately vaporized and condensed or cooled, depending on whether the cycle is subcritical or transcritical. • The working fluid is most commonly a synthetic refrigerant. CO_2 is used in transcritical cycles. • Analogous to the Rankine power cycle.
			Thermally Driven Technologies		
Absorption/ adsorption	Heat input plus adiabatic compression using a pump	Isobaric heat rejection	Adiabatic expansion	Isobaric heat input	• The refrigerant fluid is alternately vaporized and condensed. • A thermal energy source acts similarly to a compressor in vapor compression systems. • In absorption cycles, the working fluid is an absorbent–refrigerant mixture, commonly water–ammonia or lithium bromide–water. • In adsorption cycles, the refrigerant is typically water, working in conjunction with a desiccant adsorber.
			Solid-State Technologies		
Thermo- electric	Seebeck effect	Peltier/Thompson effects	Not applicable	Peltier/ Thompson effects	• Direct conversion of electrical energy to temperature differences via the Seebeck effect. • Heat is added and rejected via Peltier/Thompson effects, where heat is generated or removed in the presence of an electric current flowing through dissimilar conductors.
Magneto-caloric	Alternating magnetic field	Magnetocaloric effect	Not applicable	Magnetocaloric effect	• The magnetocaloric effect is observed in some magnetic alloys, where they heat up when they are placed in a magnetic field and cool down when they are removed from a magnetic field.
Thermoelastic	Alternating stress field	Thermoelastic effect	Not applicable	Thermoelastic effect	• The thermoelastic effect is observed in shape memory alloys that absorb and release heat as they undergo application and release of stress.

Table 11.4 Selected Working Fluids Common in Vapor Compression and/or Organic Rankine Cycles

Refrigerant	Family	Chemical Name	Approx. ODP	Approx. GWP	Application
R-11	CFC	CCl_3F	1.00	4750	Heat pump
R-12	CFC	CCl_2F_2	0.82	10900	Heat pump
R-114	CFC	$CClF_2CClF_2$	0.58	10000	Heat pump, ORC
R-22	HCFC	$CHClF_2$	0.04	1800	Heat pump
R-134a	HFC	CH_2FCF_3	0	1370	Heat pump, ORC
R-1234yf	HFC	$CH_2=CFCF_3$	0	<5	Heat pump
R-245fa	HFC	$CHF_2CH_2CF_3$	0	1050	Heat pump, ORC
R-32	HFC	CH_2F_2	0	675	Heat pump
R-404a	HFC Blend	R-125, R-143a, R-134a	0	3700	Heat pump
R-407C	HFC Blend	R-32, R-125, R-134a	0	1700	Heat pump
R-410a	HFC Blend	R-32, R-125	0	2100	Heat pump
R-290 (propane)	Natural	C_3H_8	0	20	ORC
R-600 (butane)	Natural	C_4H_{10}	0	20	ORC
R-717 (ammonia)	Natural	NH_3	0	<1	Heat pump, Kalina
R-718 (water)	Natural	H_2O	0	<1	Heat pump
R-729 (air)	Natural	N_2, O_2, Ar	0	0	Brayton
R-744 (carbon dioxide)	Natural	CO_2	0	1	Heat pump

Prior to the 1990s, heat pump equipment predominantly operated with a working fluid from the refrigerant classes known as chlorofluorocarbons (CFCs) and hydrochlorofluorocarbons (HCFCs). When scientists discovered that chemicals contributed significantly to the depletion of the ozone layer, industry adopted an alternative class of synthetic refrigerants known as hydrofluorocarbons (HFCs), which are still widely used at the time of this publication. Various international agreements regulate the use of CFCs and HCFCs.

Although HFCs have zero ozone depletion potential, and have contributed to the phase-out of CFCs and HCFCs, they contribute significantly to the global greenhouse gas (GHG) inventory when released to the atmosphere. For example, the common HFC refrigerants (HFC-134a and HFC- 410A) have global-warming potentials (GWP) of 1370 and 2100 times that of carbon dioxide (CO_2), respectively (UNEP 2011). Table 11.4 summarizes working fluids in vapor compression and organic Rankine cycles, along with their ozone depletion potential (ODP) and their GWP values after (UNEP 2011).

The trend to reduce HFC use will likely impose constraints on standard vapor compression equipment, including increased cost, reduced efficiency, and potential trade-offs of safety. Potential phase-out of HFCs has stimulated intense interest in alternative refrigerants with low GWP, and ironically there is currently a renewed interest to revisit some of the earliest, natural refrigerants: ammonia and carbon dioxide. Use of these refrigerants will be discussed in more detail in subsequent sections of this chapter.

11.7 Chapter Summary

This chapter has provided an overview of the first and second laws of thermodynamics as applied to simple heat pumps and heat engines. Practical applications have been summarized with an open mind that there are numerous possibilities with geothermal potential. Finally, current and past working fluids have been discussed, highlighting the trend toward seeking refrigerants with as little environmental impact as possible.

Discussion Questions and Exercise Problems

11.1 Describe the operation of a thermoelectric generator.

11.2 An inventor claims to have developed a new cold-climate, air-source heat pump that boasts a coefficient of performance of 5.5 while keeping a house at 20 °C when the outdoor air temperature is −40 °C. Is this claim reasonable and/or acceptable?

11.3 A company claims to have developed a power cycle that receives 1000 kW by heat transfer from a thermal reservoir at a temperature of 500 K while discharging 800 kW by heat transfer to a thermal reservoir at 200 K. Is this claim reasonable and/or acceptable?

11.4 A power plant is using biomass to generate steam for a vapor power cycle. 4 MW of useful power is being produced while dissipating 14 MW of waste heat to the atmosphere.

(a) How much heat is being released from combustion of the biomass?
(b) What is the power plant efficiency?

11.5 Consider an air-source heat pump heating a home. The useful heat produced is 5 kW, with a heat pump coefficient of performance (COP) of 2.5.

(a) How much electricity (in kW) is this heat pump using?
(b) How much heat (in kW) is being extracted from the air?

11.6 A heat pump heating a residence is maintaining an indoor temperature of 20 °C while it is −10 °C outside. The heat pump power usage is 5 kW and its COP is 2.0. Calculate the second-law efficiency of the heat pump at these conditions.

12

Mechanical Vapor Compression Heat Pumps

12.1 Overview

Mechanical vapor compression systems operating in a subcritical cycle represent the vast majority of heat pump, air-conditioning, and refrigeration applications today. The cycle operates by transfer of heat through a closed-loop cycle by compressing, condensing, expanding, and evaporating a refrigerant fluid. The term 'subcritical' means that the cycle operates well below the critical point of the refrigerant. Heat pump manufacturers install controls to monitor refrigerant temperature and pressure, and heat pumps will shut off when the refrigerant pressure approaches the critical pressure.

We will begin our discussion of vapor compression heat pumps by examining the ideal cycle and an associated conceptual engineering model. We will then extend these concepts to non-ideal cycles and associated representation on a pressure–enthalpy diagram. Next, we will look at classification of vapor compression heat pumps according to their possible source-sink configuration, followed by examination of the operating mechanics. We will then put everything together into a computer simulation model to examine heat pump performance. The section will be concluded with examination of transcritical cycles.

Learning objectives and goals:

1. Review the fundamentals of vapor compression heat pump operation.
2. Draw analogies between the thermodynamic cycles of the Carnot heat pump cycle, and connect these to real applications of the mechanical vapor compression refrigeration cycle.
3. Develop, analyze, and sketch heat pump thermodynamic cycles on pressure–enthalpy (P–h) diagrams, and calculate energy flows in all processes of the cycle.

Geothermal Heat Pump and Heat Engine Systems: Theory and Practice, First Edition. Andrew D. Chiasson.

Companion website: www.wiley.com/go/chiasson/geoHPSTP

4. Select a heat pump from manufacturer's catalog data, given a set of heating and cooling loads.
5. Appreciate the role of international standards for geothermal heat pumps.

12.2 The Ideal Vapor Compression Cycle

Recall the schematic of a general Carnot heat pump cycle shown in Figure 11.2b and its operation: there are four ideal, thermodynamically reversible processes. The heat transfers at the high- and low-temperature reservoirs (T_H and T_C) occur isothermally, and the processes to and from the high- and low-temperature reservoirs occur adiabatically. The *ideal vapor compression cycle* essentially represents an 'intermediate' step from the Carnot heat pump cycle toward a real cycle. It lays the groundwork for a practical heat pump cycle, but retains perfect heat transfers and isentropic processes of the Carnot cycle. Thus, the ideal vapor compression cycle is not achievable.

The concept of the vapor compression cycle arose out of the practical difficulties of implementing a Carnot cycle. The Carnot cycle requires isothermal expansion and compression involving simultaneous heat and work which are difficult to achieve in practice. It was recognized that one easy way to achieve the isothermal processes at T_H and T_C is to implement these processes in a phase-change condition. Thus, we move the Carnot cycle under the liquid–vapor dome to compare with the ideal vapor compression cycle as shown in Figure 12.1 on a scaled *T–s* diagram.

As shown in Figure 12.1, both heat pump cycles are operating between T_H and T_C of 25 °C and 0 °C, respectively, in the vapor dome with perfect heat transfer to/from the surroundings. However, this is achieved in the ideal vapor compression cycle by actually operating between two pressures, denoted in Figure 12.1 by P_H and P_C, corresponding to T_H and T_C in the vapor dome. Further, two-phase compression is not easily achievable in the vapor compression cycle, and thus process 1–2 is moved out of the vapor dome to represent 'dry' compression of gas only. This results in point 2 being well into the superheated region as its pressure is raised to P_H. We also define an intermediate point here, state 2′, where the process 2–2′ represents desuperheating of the refrigerant vapor. This concept will come in handy in our discussion of geothermal heat pumps. The liquid leaving the condenser is either saturated or slightly subcooled. Finally, the vapor compression cycle replaces isentropic expansion in process 3–4 with an isenthalpic throttling process.

Our conceptual engineering model of subcritical ideal vapor compression cycles consists of the four principal control volumes as shown schematically in Figure 12.2. All energy transfers by work and heat are taken as positive in the directions of the arrows on the schematic, and energy balances are written accordingly.

In our engineering model we may make a number of simplifying assumptions, where each component is analyzed as a control volume at steady state. Dry compression is presumed, and the refrigerant is a vapor with a quality (x) of 1. The refrigerant exits the condenser as a saturated liquid with a quality (x) of 0. The compressor operates adiabatically, and the refrigerant expanding through the valve undergoes a constant-enthalpy throttling process. Kinetic and potential energy changes may be ignored.

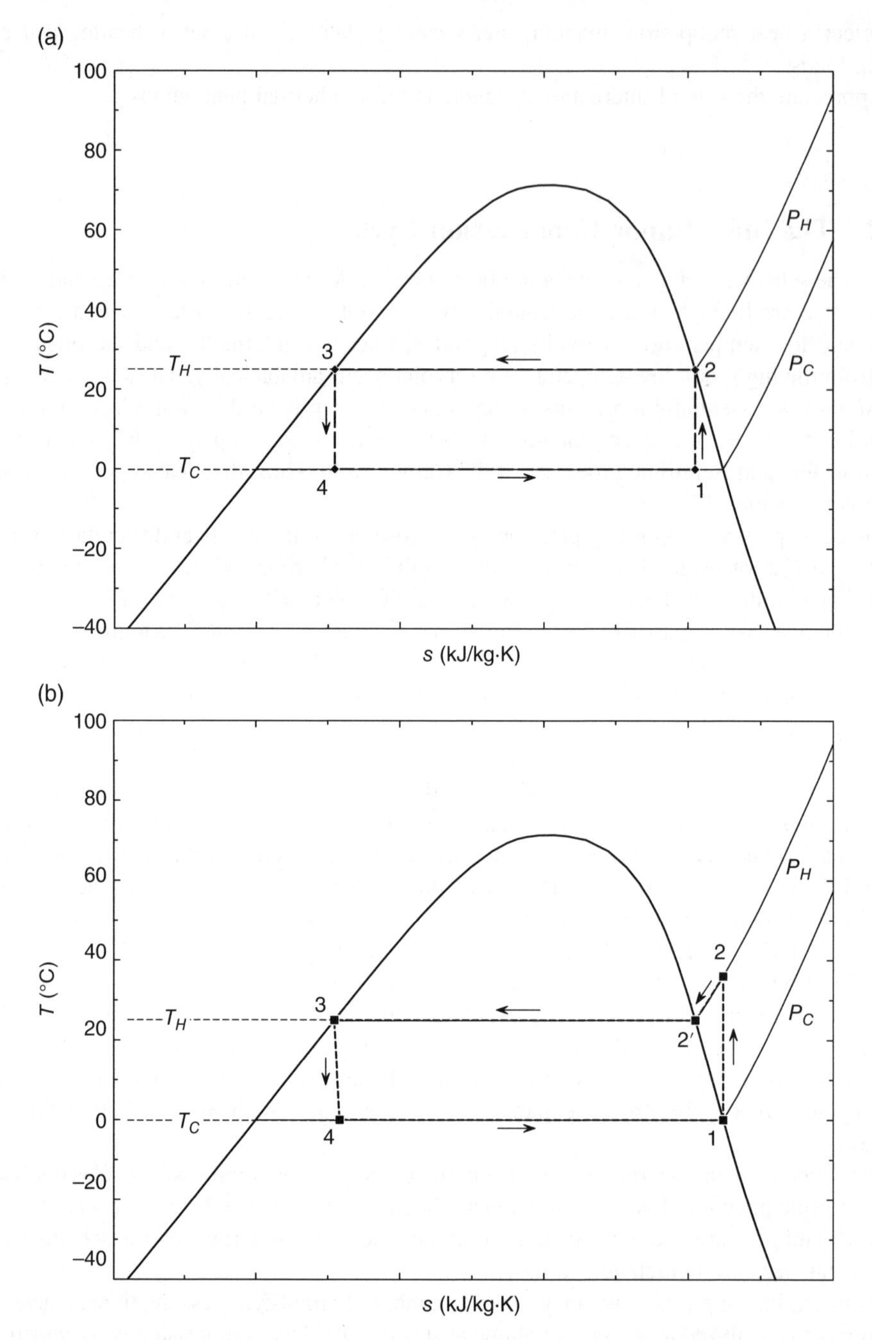

Figure 12.1 Analogous temperature–entropy (*T–s*) diagrams for (a) a Carnot heat pump cycle and (b) an ideal vapor compression heat pump cycle

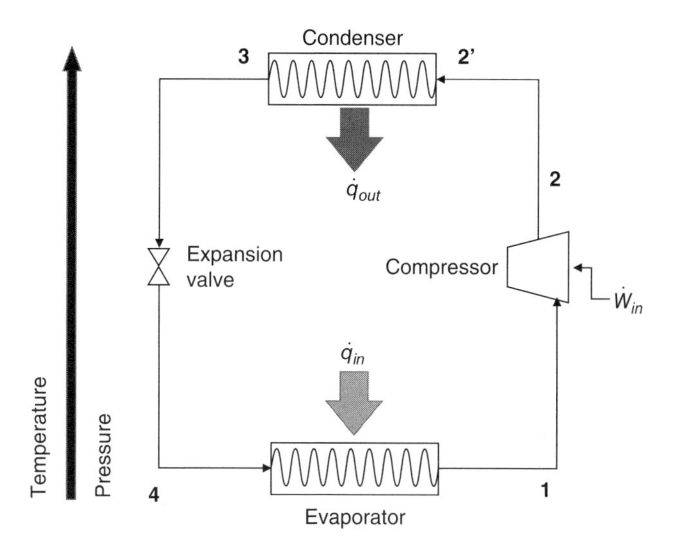

Figure 12.2 Conceptual model of refrigerant cycle components in a subcritical vapor compression heat pump

In process 1–2, vapor refrigerant is compressed to a relatively high temperature and pressure, requiring work input. The energy balance on the compressor, assuming an adiabatic process, is given by

$$\dot{W}_{in} = \dot{m}_R(h_2 - h_1) \tag{12.1}$$

where $\dot{m}_R$ is the mass flow rate of the refrigerant.

In process 2–3, vapor refrigerant condenses to liquid through heat transfer to the cooler surroundings. The energy balance on the condenser is given by

$$\dot{q}_{out} = \dot{m}_R(h_2 - h_3) \tag{12.2}$$

The maximum amount of superheat available is given by

$$\dot{q}_{superheat} = \dot{m}_R(h_{2'} - h_2) \tag{12.3}$$

In process 3–4, liquid refrigerant expands to the evaporator pressure. Assuming a throttling process, the energy balance on the expansion valve is

$$h_4 = h_3 \tag{12.4}$$

Finally, in process 4–1, two-phase liquid–vapor mixture of refrigerant is evaporated through heat transfer from the warmer surroundings. The energy balance on the evaporator is given by

$$\dot{q}_{in} = \dot{m}_R(h_1 - h_4) \tag{12.5}$$

The heat pump coefficient of performance for heating and cooling modes can therefore be expressed, respectively, as

$$\mathrm{COP}_h = \frac{h_2 - h_3}{h_2 - h_1} \tag{12.6}$$

$$\mathrm{COP}_c = \frac{h_1 - h_4}{h_2 - h_1} \tag{12.7}$$

12.3 The Non-Ideal Vapor Compression Cycle

12.3.1 Principal Irreversibilities and Isentropic Efficiency

The non-ideal vapor compression cycle has some significant departures from the ideal mechanical vapor compression cycle, as shown in Figure 12.3. There are irreversibilities in all real components: entropy is generated in the compressor (owing to friction and vibration, for example), and heat transfers at the condenser and evaporator must be accomplished under a temperature gradient with the surroundings.

Note that, as shown in Figure 12.3, the increase in entropy through the compressor drives point 2 further into the superheat region. This gives rise to the definition of the compressor isentropic efficiency (η_c) as

$$\eta_c = \frac{h_{2s} - h_1}{h_2 - h_1} \tag{12.8}$$

where $h_{2s} - h_1$ represents the specific work of the isentropic compression process, and $h_2 - h_1$ represents the actual specific work of the compression process. Thus, h_{2s} is evaluated at the pressure of point 2 and entropy of point 1.

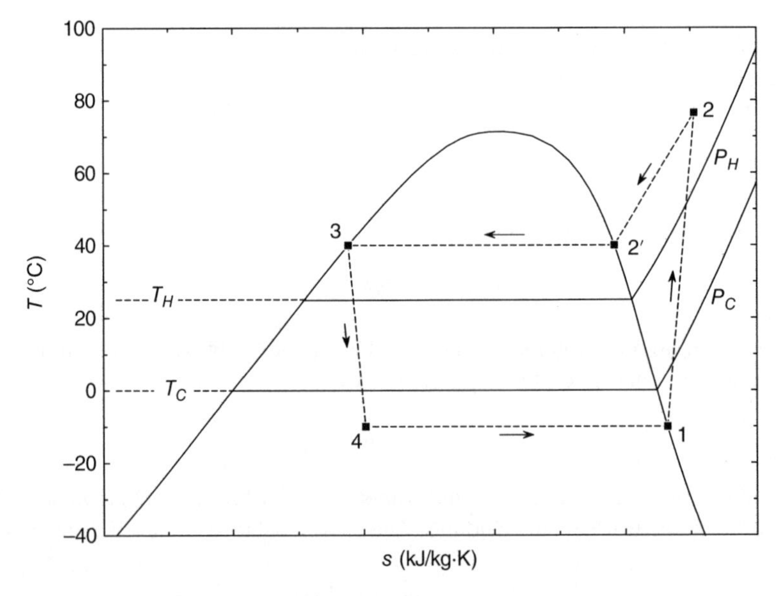

Figure 12.3 *T–s* diagram of a non-ideal vapor compression heat pump cycle

Note that, also as shown in Figure 12.3, the refrigerant's boiling and condensing temperatures are significantly higher than T_H and significantly lower than T_C. These temperature excursions are due to practical limitations of heat exchanger size and cost, and perfect heat transfer is not possible. The difference between T_H and T_C is sometimes referred to as the *temperature lift*. This expression comes from the analogy of the compressor having to do physical work of *lifting* the temperature from the lower-temperature source to the higher-temperature sink. Thus, as previously mentioned, it is evident that the greater difference between T_H and T_C, the greater is the work input required by the compressor.

12.3.2 Engineering Model of Geothermal Heat Pumps

With respect to geothermal heat pumps, we can define more specific conceptual engineering models. Here, in heating mode, $\dot{q}_{out}$ becomes the building heating load (or heat pump heating capacity) and $\dot{q}_{in}$ becomes the ground thermal load. Conversely, in cooling mode, $\dot{q}_{out}$ becomes the ground thermal load and $\dot{q}_{in}$ becomes the building cooling load (or heat pump cooling capacity). Thus, we can now add the terms $\dot{q}_{building}$ and $\dot{q}_{ground}$ to our analysis, as shown in Figure 12.4.

By rearranging Equations (11.8) and (11.9), and substituting $\dot{q}_{building}$ and $\dot{q}_{ground}$ where appropriate, we arrive at the following expressions for the ground thermal loads: in heating mode

$$\dot{q}_{ground} = \dot{q}_{building} \times \frac{\text{COP}_h - 1}{\text{COP}_h} \quad (12.9)$$

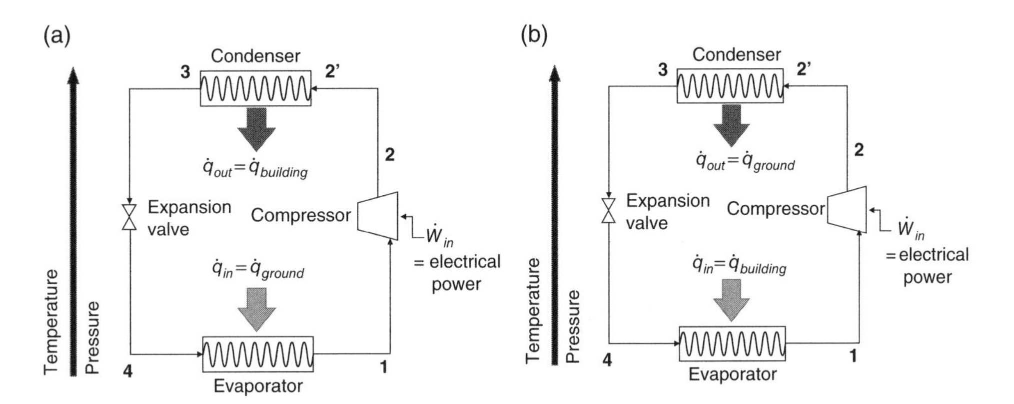

Figure 12.4 Conceptual model of refrigerant cycle components in a subcritical vapor compression geothermal heat pump for (a) heating mode and (b) cooling mode

and in cooling mode

$$\dot{q}_{ground} = \dot{q}_{building} \times \frac{COP_c + 1}{COP_c} \qquad (12.10)$$

*12.3.3 Graphical Representation on Pressure–Enthalpy (*P–h*) Diagrams*

Given that vapor compression cycles essentially work between two pressures (i.e., the condenser pressure and the evaporator pressure), and that energy flows are quantified by changes in enthalpy, it is very convenient to diagram these cycles on pressure–enthalpy (*P–h*) plots. A generalized *P–h* diagram is shown in Figure 12.5 and briefly described in what follows. The dark, heavy line on the *P–h* diagram of Figure 12.5 marks the boundary of the two-phase, liquid–vapor region shaded in gray. The portion of the dark, heavy line to the left (right) of the critical point is the saturated liquid (saturated vapor) line. Thinner solid lines represent lines of constant temperature (isotherms), and dashed lines represent lines of constant entropy (isentropes).

The subcritical vapor compression cycle, with the engineering approximations discussed above, takes the form of a quadrilateral, as shown in Figure 12.6. The relative differences between the condenser heat rejection and the evaporator heat extraction are evident, which shows the compressor work. The ideal or isentropic compressor process (process 1–2 s) is shown in Figure 12.6 as a dashed line, and the departure of the ideal and non-ideal compression process is evident.

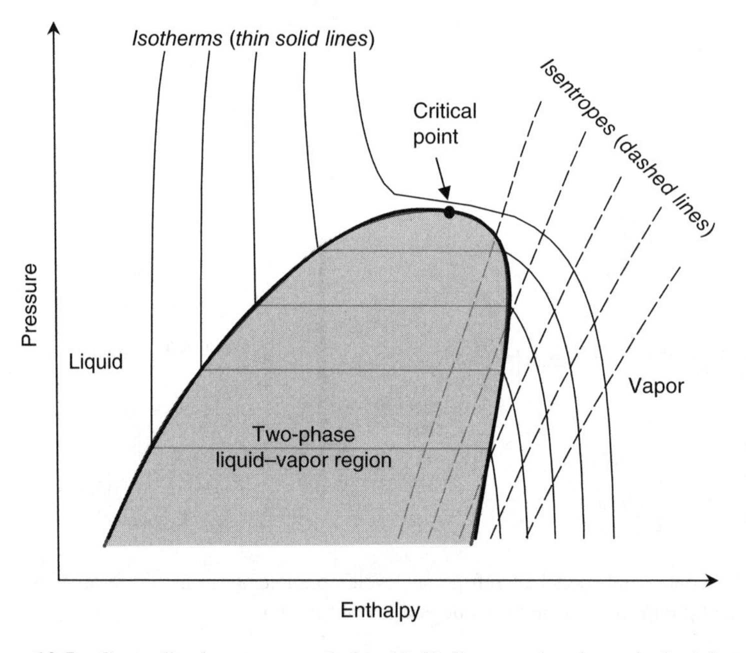

Figure 12.5 Generalized pressure–enthalpy (*P–h*) diagram showing principal features

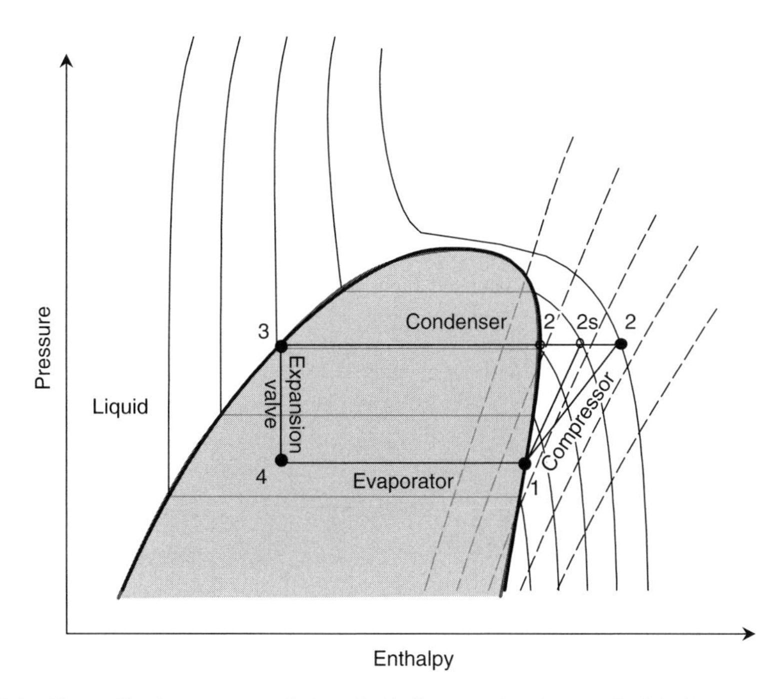

Figure 12.6 Generalized pressure–enthalpy (*P–h*) diagram showing a subcritical vapor compression cycle

Example 12.1 Simple Subcritical Heat Pump Cycle Analysis on a *P–h* Diagram

Consider a heat pump that is heating a residence while operating at a condenser pressure of 3000 kPa and an evaporator pressure of 700 kPa. The refrigerant is R410a. Assume that the isentropic efficiency of the compressor is 80%. For a refrigerant mass flow rate of $0.1\ \text{kg} \cdot \text{s}^{-1}$, determine/show the following:

(a) a sketch of the cycle on a P–h diagram and show the ideal (isentropic case),
(b) the heat rejected at the condenser,
(c) the heat extracted at the evaporator,
(d) the refrigerant temperature leaving the compressor,
(e) the condensing temperature of the refrigerant,
(f) the evaporating temperature of the refrigerant,
(g) the compressor power,
(h) the heating COP, and
(i) the maximum superheat available for use in a domestic hot water tank.

Solution

We could solve this manually or with the use of an equation solver. Here, we will use Engineering Equation Solver (EES). The information is entered in EES as follows:

```
"GIVEN:"
R$ = 'R410a'; $REFERENCE R410a ASH "String for refrigerant, with ASHRAE reference state"
P1 = 700 [kPa]; P2 = 3000 [kPa]
n_compressor = 0.80
mdot_refrigerant = 0.1 [kg/s]

"SOLUTION:"
"Simplifying Assumptions:"
x1 = 1 [-]; x3 = 0 [-]  "assume a saturated vapor at Point 1, and a saturated liquid at Point 3"

"Point 1"
h1 = Enthalpy(R$, P = P1, x = x1)
T1 = temperature (R$, P = P1, x = x1)
s1 = entropy(R$, P = P1, x = x1)

"Point 2, 2s, and 2' "
s2s = s1
h2s = enthalpy(R$, P = P2, s = s2s)
n_compressor = (h2s – h1) / (h2 - h1)
h2_prime = enthalpy(R$, P = P2, x = 1)
T2 = temperature (R$, P = P2, h = h2)

"Point 3"
P3 = P2
h3 = Enthalpy(R$, P = P3, x = x3)
T3 = Temperature (R$, P = P3, x = x3)

"Point 4"
P4 = P1
h4 = h3
x4 = quality(R$, P = P4, h = h4)

"Energy Balances:"
COP_h = (h2 – h3) / (h2 – h1)                        "heating COP"
q_evap = mdot_refrigerant * (h1 – h4)                 "evaporator heat transfer"
q_cond = mdot_refrigerant * (h2 – h3)                 "condenser heat transfer"
W_comp = mdot_refrigerant * (h2 – h1)                 "compressor work"
q_superheat = mdot_refrigerant * (h2 – h2_prime)      "maximum superheat"
```

The following results are obtained:

(a) a sketch of the cycle on a *P–h* diagram, showing the ideal (isentropic case): see Figure E.12.1
(b) the heat rejected at the condenser: $q_{cond} = 18.68$ [kW]
(c) the heat extracted at the evaporator: $q_{evap} = 13.64$ [kW]
(d) the refrigerant temperature leaving the compressor: $T_2 = 79.33$ [C]
(e) the condensing temperature of the refrigerant: $T_3 = 48.98$ [C]
(f) the evaporating temperature of the refrigerant: $T_1 = -4.048$ [C]
(g) the compressor power: $W_{comp} = 5.039$ [kW]
(h) the heating COP: $COP_h = 3.706$
(i) the maximum superheat available for use in a domestic hot water tank: $q_{superheat} =$ 4.808 [kW]

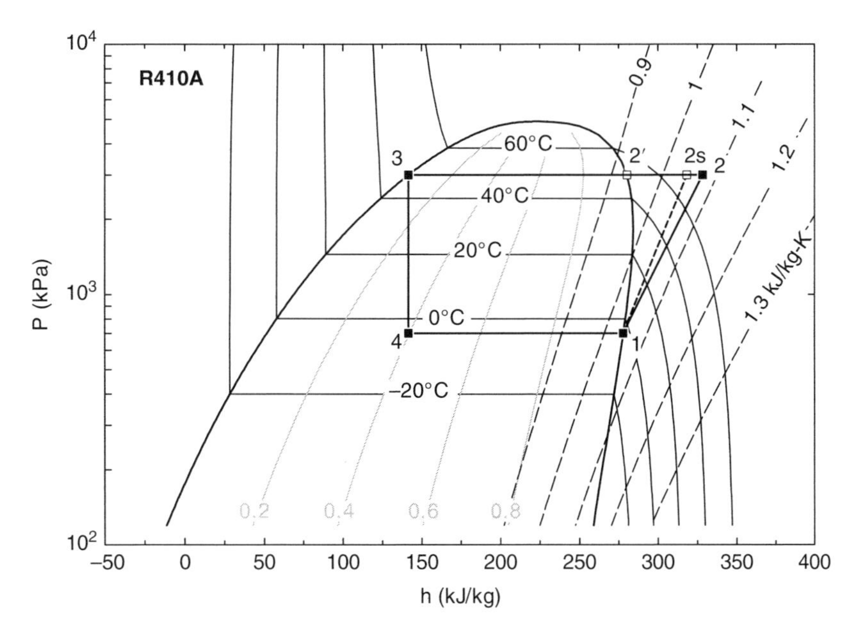

Figure E.12.1 *P–h* plot for Example 12.1

Discussion: This example demonstrates that we can make considerable progress by knowing the condensing and evaporating pressures, and by making some engineering approximations. Note that knowledge of the refrigerant mass flow rate is not necessary to analyze the cycle. When the refrigerant mass flow rate is known or assumed, calculation of all heat transfer rates is possible. Note also the temperatures in the evaporator and condenser; the condensing temperature appears high enough to provide heating to the residence. The source temperature would have to be significantly above –4.05 °C for this heat pump to provide heating.

12.3.4 Heat Exchanger Analysis

Example 12.1 illustrates an engineering model of a subcritical vapor compression cycle by knowing the condensing and evaporating pressure of the refrigerant. This simple model results in very useful outcomes, such as the thermodynamic state points, heat transfer rates (when the refrigerant mass flow rate is known or assumed), and the heat pump COP. However, it's the irreversibilities of the heat transfer to/from the sink/source that dictates the operating pressures of the refrigerant, and we are often interested in the temperatures and flow rates of these fluids. In other words, in a water-to-air geothermal heat pump, the entering air conditions and flow rate and the geothermal fluid temperature and flow rate control the pressures in the heat pump evaporator and condenser at any given time. Thus, in this subsection, we consider the heat transfer characteristics between the refrigerant and the fluids on the source and load sides of the heat pump.

Geothermal heat pumps may have one or two types of heat exchanger: (i) water-to-refrigerant heat exchangers of the counterflow type, or (ii) air-to-refrigerant heat exchangers of the cross-flow type. The effectiveness of heat transfer is governed by the *overall heat transfer coefficient*, which is a function of the total thermal resistance (conductive plus convective) to heat transfer between the two fluids. Here, it is useful to employ the *heat exchanger effectiveness* (ε) concept, where

$$\varepsilon = \frac{\dot{q}}{\dot{q}_{max}} \tag{12.11}$$

where $\dot{q}$ is the actual heat transfer rate, and $\dot{q}_{max}$ is the maximum possible heat transfer rate, and ε can range from 0 to 1.

In the study of heat exchangers in classic heat transfer theory, you may recall that ε is a function of NTU (the number of transfer units) and C, the heat capacity rate ($\dot{m}c_p$) of the fluids on both sides of the heat exchanger. Furthermore, $\dot{q}_{max}$ is actually experienced by the fluid with the minimum heat capacity rate (C_{min}) such that

$$\dot{q}_{max} = C_{min}(T_{h,i} - T_{c,i}) \tag{12.12}$$

where $T_{h,i}$ and $T_{c,i}$ are the hot inlet and cold inlet fluid temperatures respectively.

Fortunately, *for heat exchangers involving a fluid undergoing a phase change*, such as the case with vapor compression heat pumps, there are special considerations that simplify our analysis. That is, $C_{refrigerant} \to \infty$ (i.e., given the definition of c_p), and C_{min} always refers to the fluid (liquid or air) on the building or geothermal side of the heat exchanger (i.e., the fluid external to the refrigerant circuit). Temperatures in the evaporator and condenser as a function of percent heat transfer rate (or relative position in the heat exchanger) are shown schematically in Figure 12.7. These plots are sometimes referred to as *T–q* diagrams. The closer the values

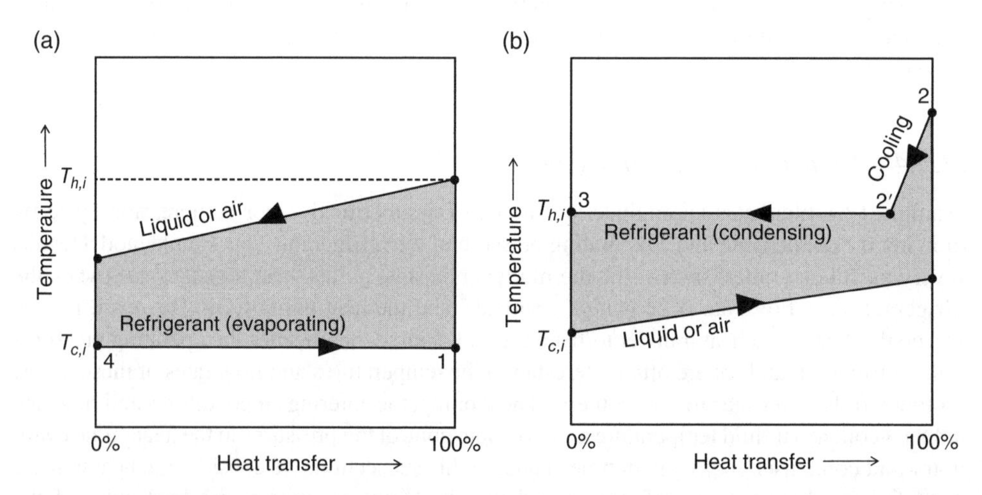

Figure 12.7 Schematic of temperature change vs. percent heat transfer (or *T–q* diagram) in (a) the evaporator and (b) the condenser. Note that labeled points correspond to points on the previous *P–h* diagrams, and arrows indicate flow direction

of $T_{h,i}$ and $T_{c,i}$, the more *effective* is the heat exchanger, which is typically exponentially proportional to the heat exchanger size and cost. Thus, as previously mentioned, heat pump manufacturers must strike a balance between the heat exchanger size and cost and acceptable temperature differences between the refrigerant and the source/sink fluid.

At this point, you may realize a possible difficulty in applying a single heat exchanger effectiveness equation at the condenser because the refrigerant actually enters as a hot gas and desuperheats prior to condensing (Figure 12.7b). Fortunately, Stoecker and Jones (1982) have found that reasonable approximations can be made if $T_{h,i}$ is taken as the refrigerant condensing temperature and $T_{c,i}$ is taken as the building return fluid temperature.

The heat exchanger effectiveness, when one fluid is undergoing a phase change, is further defined as

$$\varepsilon = 1 - e^{-UA/C} \tag{12.13}$$

where U is the heat transfer coefficient, including conductive and convective components, A is the area of the heat exchanger, and C is the heat capacity rate of the *external* fluid (i.e., not the refrigerant). The heat transfer coefficient (U) is a function of the heat exchanger material and configuration and the thermal properties of the heat transfer fluids and their mass flow rates. Determination of U is an essential, but often the most uncertain, part of heat exchanger analysis. Thus, the use of the heat exchanger effectiveness concept allows heat exchanger analysis with approximations of ε, and without exact knowledge of U. Therefore, the general heat exchanger equation we will use in our subsequent analysis is

$$\dot{q} = \varepsilon C(T_{h,i} - T_{c,i}) \tag{12.14}$$

There are additional psychrometric considerations with the use of Equation (12.14) for heat pumps with air as the heat distribution medium when modeling the heat pump in cooling mode. Thus, it is convenient to introduce an effectiveness–heat capacity rate product α and β, where

$$\dot{q}_E = \alpha(T_{h,i} - T_{c,i}) \tag{12.15a}$$

and

$$\dot{q}_C = \beta(T_{h,i} - T_{c,i}) \tag{12.15b}$$

where subscripts E and C refer to the evaporator and condenser respectively. We now have separate equations describing the heat transfer rate at the evaporator ($\dot{q}_E$) and condenser ($\dot{q}_C$), and we will apply these differently in the heating and cooling modes, as discussed in what follows.

12.3.4.1 Heating Mode

In heating mode, the condenser rejects heat to the load, and the evaporator extracts heat from the source. With respect to geothermal heat pumps, the heat distribution medium on the load side could either be air or liquid, while the source side fluid is liquid.

Table 12.1 Summary of Heat Exchanger Effectiveness Variables for a Geothermal Heat Pump in Heating Mode

Variable	Evaporator	Condenser
$\dot{q}$	$\dot{q}_{evaporator} = (\dot{m}(h_1 - h_4))_{refrigerant}$ $= \dot{q}_{ground} = (\dot{m}c_p\Delta T)_{geofluid}$ $= \alpha(T_{h,i} - T_{c,i})$	$\dot{q}_{condenser} = (\dot{m}(h_2 - h_3))_{refrigerant}$ $= \dot{q}_{load} = (\dot{m}c_p\Delta T)_{load\ fluid}$ $= \beta(T_{h,i} - T_{c,i})$
ε	$\varepsilon = \varepsilon_{evaporator}$	$\varepsilon = \varepsilon_{condenser}$
	Typical ε Values: Refrigerant-to-Air: $\varepsilon \approx 0.6$ Refrigerant-to-Liquid: $\varepsilon \approx 0.8$	
C	$C_{evaporator} = (\dot{m}c_p)_{geofluid}$	$C_{condenser} = (\dot{m}c_p)_{load\ fluid}$
	Typical Volumetric Flow Rate ($\dot{V}$) Values: Air: $\dot{V} \approx 400$ cfm/ton or 0.054 m^3/s/kW Liquid: $\dot{V} \approx 2$–3 gpm/ton or 2–3 Lpm/kW	
$T_{h,i}$	fluid temperature leaving geothermal heat exchanger	refrigerant condensing temperature
$T_{c,i}$	refrigerant evaporating temperature	load fluid return temperature
α, β	$\alpha = \varepsilon_{evaporator} \times C_{evaporator}$	$\beta = \varepsilon_{condenser} \times C_{condenser}$

Application of the heat exchanger effectiveness equations to a heat pump in heating mode is relatively straightforward. A summary of the equations and variables is given in Table 12.1. Finally, the energy balance on the geothermal fluid and the heat distribution fluid is given by

$$\dot{q} = \dot{m}c_p(T_i - T_o) \tag{12.16}$$

where T_i and T_o refer to the inlet and outlet fluid temperature respectively.

Example 12.2 Refrigerant-to-Air Heat Exchanger Providing Heating

Consider the heat pump in Example 12.1, where we calculated the temperature of the condensing refrigerant $T_3 = 48.98$ °C, and a heat rejection rate at the condenser of 18.68 kW. If the heat distribution medium is air flowing at 0.054 m^3/s per kW of heating output, calculate the temperature of the air entering and leaving the heat pump.

Solution

Let's again solve this problem using Engineering Equation Solver (EES). The information is entered in EES as follows:

```
"INPUT DATA:"
"Air Conditions"
P = 101.1 [kPa]                              "The local atmospheric air pressure"
Q_dot_air_per_kW = 0.054 [m^3/s]             "Air volumetric flow rate per peak kW heating"
T3 = 48.98 [C]                               "The refrigerant condensing temperature"

"SOLUTION:"
"Assumptions:"
epsilon_cond = 0.6 [-]                       "HX effectiveness estimate of condenser"

"Calculate fluid properties and fluid mass flow rates per kW"
"T_air_db is the dry bulb temperature of the air returning from the building, entering the heat
pump"
rho_air = Density(Air,T = T_air_db, P = P)
cp_air = Cp(Air,T = T_air_db)
m_dot_air_perkW = Q_dot_air_per_kW * rho_air

"Heat Transfer Relationships:"
1 [kW] = Beta*(T3 – T_air_db)                "condenser heat exchanger relationship
                                             (per kW)"
Beta = epsilon_cond * m_dot_air_perkW        "effectiveness-heat capacity rate product"
* cp_air
"calculate the air temperature leaving the heat pump, supplying the space"
1 [kW] = m_dot_air_perkW * cp_air * (T_air_supply – T_air_db )
```

The following results are obtained:

The building return air temperature (entering the heat pump): $T_{air,db} = 23.13$ [C]
The building supply air temperature (leaving the heat pump): $T_{air,supply} = 38.64$ [C]

12.3.4.2 Cooling Mode

In cooling mode, the condenser rejects heat to the ground heat exchanger, and the evaporator extracts heat from the load. With respect to geothermal heat pumps, the heat distribution medium on the load side could either be air or liquid, while the source side fluid is liquid.

Evaporator cooling coil analysis is complicated by the fact that both heat and mass transfer occur when air is the distribution medium. As moist air passes across a cooling coil, sensible heat is transferred as the air changes in temperature, and latent heat is transferred as water is condensed out of the air stream.

First, let's consider the simple case of a water-to-water heat pump in cooling mode, where application of the heat exchanger effectiveness equations is relatively straightforward, as described above for a heat pump in heating mode; only sensible heat transfer occurs. A summary of the equations and the variables is given in Table 12.2. The energy balance on the geothermal fluid and the heat distribution fluid is given by Equation (12.16).

Now, let's consider the case of a water-to-air heat pump in cooling mode, where moist air is the heat distribution medium. The term *moist air* refers to a mixture of dry air and water vapor in which the dry air is treated as a pure component. The study of systems involving moist air is known as psychrometrics.

Table 12.2 Summary of Heat Exchanger Effectiveness Variables for a Geothermal Water-to-Water Heat Pump in Cooling Mode

Variable	Evaporator	Condenser
$\dot{q}$	$\dot{q}_{evaporator} = (\dot{m}(h_1 - h_4))_{refrigerant}$ $= \dot{q}_{load} = (\dot{m}c_p\Delta T)_{load\ fluid}$ $= \alpha(T_{h,i} - T_{c,i})$	$\dot{q}_{condenser} = (\dot{m}(h_2 - h_3))_{refrigerant}$ $= \dot{q}_{ground} = (\dot{m}c_p\Delta T)_{geofluid}$ $= \beta(T_{h,i} - T_{c,i})$
ε	$\varepsilon = \varepsilon_{evaporator}$	$\varepsilon = \varepsilon_{condenser}$
	Typical ε Values: Refrigerant-to-Air: $\varepsilon \approx 0.6$ Refrigerant-to-Liquid: $\varepsilon \approx 0.8$	
C	$C_{evaporator} = (\dot{m}c_p)_{load\ fluid}$	$C_{condenser} = (\dot{m}c_p)_{geofluid}$
	Typical Volumetric Flow Rate ($\dot{V}$) Values: Air: $\dot{V} \approx$ 400 cfm/ton or 0.054 m^3/s/kW Liquid: $\dot{V} \approx$ 2–3 gpm/ton or 2–3 Lpm/kW	
$T_{h,i}$	load fluid return temperature	refrigerant condensing temperature
$T_{c,i}$	refrigerant evaporating temperature	fluid temperature leaving geothermal heat exchanger
α, β	$\alpha = \varepsilon_{evaporator} \times C_{evaporator}$	$\beta = \varepsilon_{condenser} \times C_{condenser}$

A generalized psychrometric chart is shown in Figure 12.8 and briefly described in what follows. Points on the chart represent thermodynamic state points of moist air. Five properties are displayed: (i) dry bulb temperature on the x-axis, (ii) humidity ratio (kg water per kg dry air) on the y-axis, (iii) relative humidity as curved lines, (iv) wet-bulb temperature as long-dashed lines, and (v) specific volume as short-dashed lines. Not shown on the chart in Figure 12.8 are lines of constant enthalpy, which are approximately coincident with wet-bulb temperature lines. A value of relative humidity of 1 represents the saturation line of air.

Classic moist air-conditioning processes are handy to diagram on the psychrometric chart, as shown in Figure 12.9. Processes that move either to the left or to the right involve sensible heat transfer, as the dry-bulb temperature either decreases or increases. Processes that move either up or down involve latent heat transfer, as the humidity ratio either increases or decreases.

With respect to cooling coils, we are interested in the process of simultaneous cooling and dehumidifying. The sensible heat ratio (SHR) is used to describe the relative amount of sensible cooling to total cooling in such a process:

$$\mathrm{SHR} = \frac{\dot{q}_{sensible}}{\dot{q}_{sensible} + \dot{q}_{latent}} \tag{12.17}$$

Thus, with no latent cooling, SHR = 1, and conversely SHR = 0 in a complete dehumidification process.

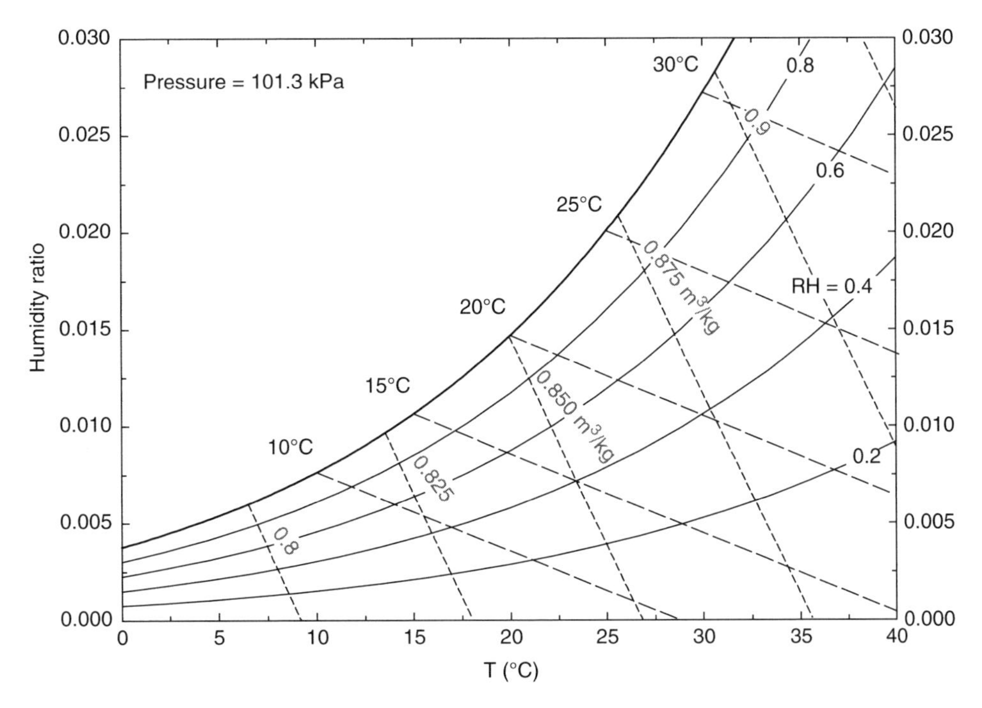

Figure 12.8 Generalized psychrometric chart showing principal features

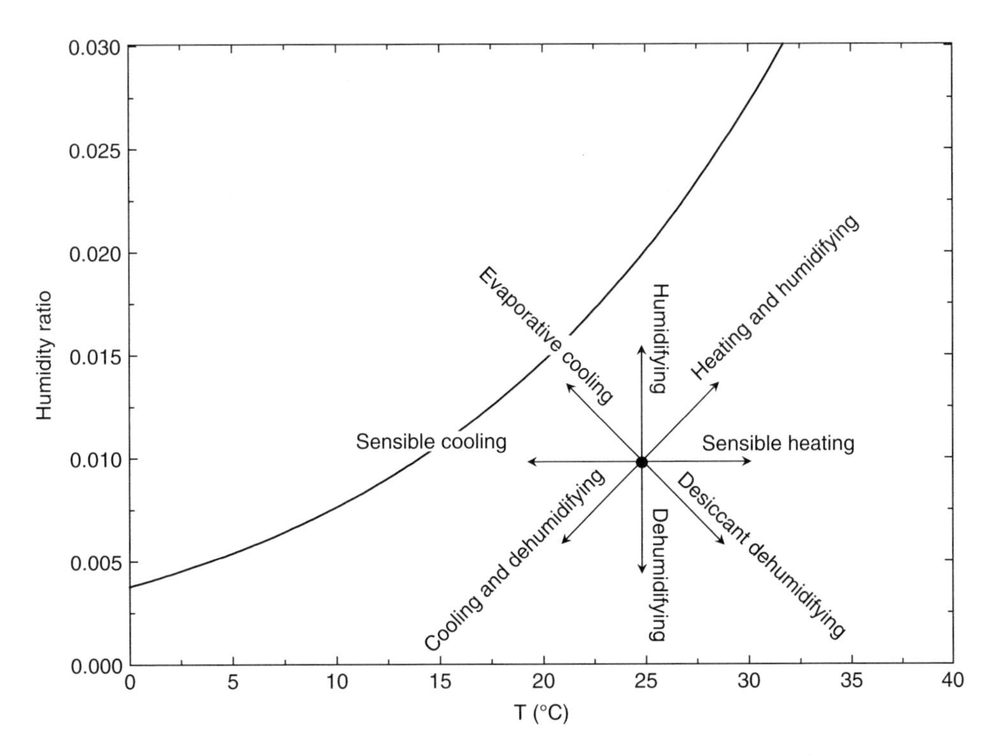

Figure 12.9 Moist air-conditioning processes on a psychrometric chart

Table 12.3 Summary of Heat Exchanger Effectiveness Variables for a Geothermal Water-to-Air Heat Pump in Cooling Mode

Variable	Evaporator	Condenser
$\dot{q}$	$\dot{q}_{evaporator} = (\dot{m}(h_1 - h_4))_{refrigerant}$ $= \dot{q}_{load} = (\dot{m}\Delta h)_{air}$ $= \alpha(T_{h,i} - T_{c,i})$	$q_{condenser} = (\dot{m}(h_2 - h_3))_{refrigerant}$ $= q_{ground} = (\dot{m}c_p\Delta T)_{geofluid}$ $= \beta(T_{h,i} - T_{c,i})$
ε	$\varepsilon = \varepsilon_{evaporator}$	$\varepsilon = \varepsilon_{condenser}$
	Typical ε Values: Refrigerant-to-Air: $\varepsilon \approx 0.6$ Refrigerant-to-Liquid: $\varepsilon \approx 0.8$	
C	$C_{evaporator} = (\dot{m}c_p)_{load\,fluid}$	$C_{condenser} = (\dot{m}c_p)_{geofluid}$
	Typical Volumetric Flow Rate ($\dot{V}$) Values: Air: $\dot{V} \approx$ 400 cfm/ton or 0.054 m^3/s/kW Liquid: $\dot{V} \approx$ 2–3 gpm/ton or 2–3 Lpm/kW	
$T_{h,i}$	load fluid return temperature	refrigerant condensing temperature
$T_{c,i}$	refrigerant evaporating temperature	fluid temperature leaving geothermal heat exchanger
α, β	$\alpha = \text{SHR} \times \varepsilon_{evaporator} \times C_{evaporator}$	$\beta = \varepsilon_{condenser} \times C_{condenser}$

We can now make use of the SHR and modify our heat exchanger equations in a simple way to model moist air processes as summarized in Table 12.3. Note that we have only added the SHR to modify our definition of α such that we can use the same heat exchanger effectiveness equations described previously. Also, the energy balance on the moist air stream is given by

$$\dot{q} = \dot{m}_{air}(h_i - h_o) \tag{12.18}$$

where h_i and h_o refer to the heat pump inlet and outlet enthalpies of the moist air.

We identify some further evaporator coil relationships here that will help us subsequently to model heat pump performance. The evaporator heat exchanger effectiveness may also be expressed as

$$\varepsilon_{evaporator} = \frac{T_{air,return} - T_{air,supply}}{T_{air,return} - T_{coil\,surface}} \tag{12.19}$$

where $T_{air,return}$ refers to the building return air temperature (entering the heat pump), $T_{air,supply}$ refers to the building supply air temperature (leaving the heat pump), and $T_{coil\,surface}$ refers to the surface temperature of the evaporator coil. The sensible heat ratio (SHR) may be further defined as

$$\text{SHR} = \frac{c_{p,\,air}\left(T_{air,return} - T_{air,supply}\right)}{h_{air,return} - h_{air,supply}} \tag{12.20}$$

Example 12.3 Refrigerant-to-Air Heat Exchanger Providing Cooling

Consider a water-to-air heat pump in cooling mode where the evaporating temperature of the refrigerant $T_1 = -10$ °C. The heat distribution medium is moist air flowing at 0.054 m^3/s per kW of cooling capacity. The air enters the heat pump at a dry-bulb temperature of 25 °C and relative humidity of 50%. Calculate/show:

(a) the dry-bulb temperature of the air leaving the heat pump,
(b) the relative humidity of the air leaving the heat pump,
(c) the sensible heat ratio of the cooling process,
(d) the surface temperature of the evaporator coil, and
(e) the cooling process on a psychrometric chart.

Solution

Let's again solve this problem using Engineering Equation Solver (EES). EES contains useful psychrometric features and capability to plot processes on a psychrometric chart. The information is entered in EES as follows:

```
"INPUT DATA:"
"Air Conditions"
P = 101.3 [kPa]      "The local atmospheric air pressure"
T_air_db = 25 [C]    "The building return (heat pump entering) air dry bulb temperature"
RH_air = 0.5         "The building return (heat pump entering) air relative humidity"
Q_dot_air_per_kW = 0.054 [m^3/s]   "Air volumetric flow rate per peak kW heating"
T1 = -10 [C]         "The refrigerant evaporating temperature"

"SOLUTION:"
"Assumptions:"
epsilon_evap = 0.6 [-] "HX effectiveness estimate of evaporator"

"Calculate fluid properties and fluid mass flow rates per kW"
"T_air_db is the dry bulb temperature of the air returning from the building, entering the heat pump"
rho_air = Density(AirH2O,T = T_air_db,R = RH_air,P = P)
cp_air = Cp(AirH2O,T = T_air_db,R = RH_air,P = P)
m_dot_air_perkW = Q_dot_air_per_kW * rho_air

"Heat Transfer Relationships:"
1 [kW] = Alpha*(T_air_db - T1)  "evaporator heat exchanger relationship (per kW)"
Alpha = SHR * epsilon_evap * m_dot_air_perkW * cp_air   "effectiveness-heat capacity rate product"
SHR * 1 [kW] = m_dot_air_perkW * cp_air * (T_air_db - T_air_supply ) "calculate the leaving air
temp."

"Coil Relationships"
epsilon_evap = ( T_air_db - T_air_supply) / (T_air_db - T_coil) "Approximation for T air supply"
h_air_return = enthalpy(AIRH2O,P = P,T = T_air_db, R = RH_air) "Enthalpy of building return
                                                                  air (entering evaporator coil)"
SHR = cp_air * ( T_air_db - T_air_supply) / (h_air_return - h_air_supply)
RH_supply = relhum(AIRH2O,P = P,T = T_air_supply, h = h_air_supply) "Relative humidity of
                                                                      supply air"

"For Psychrometric Plot"
T[1] = T_air_db; T[2] = T_air_supply                       "The x-axis dry-bulb temperatures"
W[1] = HumRat(AirH2O,T = T_air_db,R = RH_air,P = P)        "The humidity ratio of the rerurn air"
W[2] = HumRat(AirH2O,T = T_air_supply,h = h_air_supply,P = P)"The humidity ratio of the supply air"
```

The following results are obtained:

(a) the dry-bulb temperature of the air leaving the heat pump: $T_{air,supply} = 13.55$ [C]
(b) the relative humidity of the air leaving the heat pump: $RH_{supply} = 0.8526$
(c) the sensible heat ratio of the cooling process: SHR = 0.7383
(d) the surface temperature of the evaporator coil: $T_{coil} = 5.924$ [C]
(e) the cooling process on a psychrometric chart: see Figure E.12.3

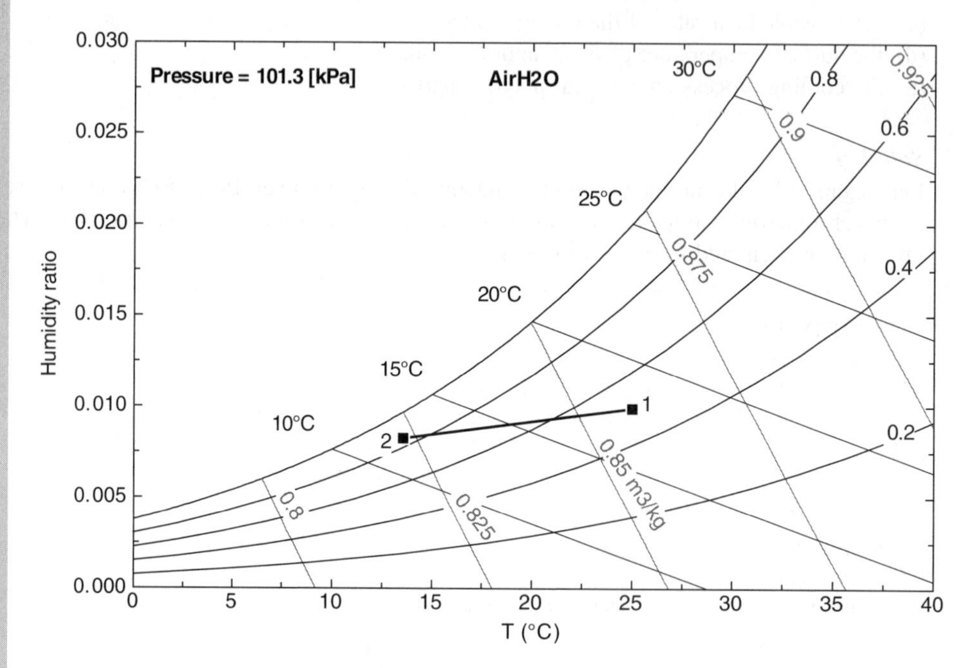

Figure E.12.3 Psychrometric plot for Example 12.3

12.4 General Source-Sink Configurations

Mechanical vapor compression heat pumps can be subdivided according to source-sink configuration, where the source can be: (i) air, (ii) water (or other liquid), or (iii) ground (Earth). Thus, such heat pumps are referred to as *air-source*, *water-source*, or *ground-source*. The term *ground-source* has many other synonyms, such as geothermal, Earth-coupled, ground-coupled, or Geoexchange®.

Air-source heat pumps are also sometimes referred to as DX (direct-expansion or direct-exchange) heat pumps because the refrigerant piping coil is in direct contact with the source/sink media. Here, we should keep in mind that there is a slight difference between the thermodynamic meaning of 'heat pump' and that used in HVAC jargon when referring to air-source heat pumps. The HVAC industry typically refers to a heat pump as a machine that provides both heating and cooling, while those that provide cooling only are referred to as *air-conditioners*. These machines are typically configured as a *packaged* or *unitary* machine or as a *split* system, and are available in fractions of a refrigeration ton (or fractions of a kW) up

Figure 12.10 Photos of various air-to-air heat pumps: (a) through-the-window unitary air-conditioner, (b) through-the-wall room heat pump, and (c) outdoor section of a residential split heat pump (photos taken by the author)

to tens of tons or kW. Small systems where the source and sink are air include the familiar through-the-window, room air-conditioner or the through-the-wall unit (i.e., wall-shaker unit) found in hotel rooms, and split systems common in residential applications. Larger systems include packaged rooftop units that serve single- or multizone applications, such as office buildings, retail buildings, schools, and other commercial and institutional buildings.

Figure 12.10 shows types of commonly existing air-source units in residential and other single-zone applications. These provide a cost-effective option for space air-conditioning when heating, cooling, and dehumidifying are needed at various times of the day or year. Although common and cost effective to install, air-to-air heat pumps do have limitations. Their COP and heating and cooling capacity are strongly impacted by the outdoor temperature. In heating mode, with decreasing outdoor air temperature, the heating capacity of the heat pump decreases to a point where the refrigerant temperature needs to be too low to extract heat from the air from a practical standpoint. Thus, these heat pumps are typically equipped with an electric resistance heating element; some air-source heat pumps are packaged as a 'dual-fuel' heat pump, where the unit switches to a natural gas furnace below an outdoor set-point temperature (e.g., 0 °C). Another phenomenon that may occur with air-to-air heat pumps is icing of the outdoor unit, caused by condensing and freezing of water vapor in the air. This may occur when the temperature of refrigerant temperature and the outdoor air drop significantly below the freeze point of water. Thus, air-to-air heat pumps typically come with defrosting provisions for the outdoor portion.

Although the air-to-air heat pump technology is improving, many air-to-air heat pumps are still not suitable for severe winter climates where temperatures frequently drop below −10 °C and large accumulations of snow and ice are possible around the outdoor unit. Theoretically, cold-climate applications can simply be dealt with by employing larger-sized equipment. Practically, however, larger-sized equipment comes with larger fans and more airflow that preclude human occupant comfort and can be cost prohibitive to operate without part-load controls. As we shall see subsequently, packaged heat pumps must have air flow rates that can properly transfer heat, while not being too high to cause objectionable noise or uncomfortable draftiness for occupants. New technologies are solving these issues with variable refrigerant flow technologies, fully variable-speed fans, and variable-speed compressors that match airflows and compressor power to the transient heating or cooling load.

Air-source heat pumps can also be configured to applications in which the source is air and the sink is water. Such air-to-water applications include swimming pool heating, hot tub heating, or domestic hot water heating.

Water-source heat pumps have application in essentially all other heat pump situations where the source medium is not air. The fluid conveying heat to/from the heat pump can be either water or an aqueous antifreeze solution, and it can come from a variety of sources. While one of these sources can be geothermal, we will exclude geothermal sources for now, and consider the 'source' as any other liquid.

As with air-source heat pumps, water-source heat pumps are typically configured as a *packaged* or *unitary* machine or as a *split* system, and are available in fractions of a refrigeration ton (or fractions of a kW) up to tens of tons or kW. Water-source heat pumps are available in many shapes and sizes to fit different needs: vertical, horizontal, console, and split.

The most common type of unitary water-source heat pump is the water-to-air heat pump. Commonly, this type of heat pump sees application in a closed-loop configuration in a multi-zone building where the fluid loop temperature is kept in the temperature range 10–30 °C (or about 50–90 °F). Thus, these types of building loop are heated typically by boilers and cooled typically by cooling towers. In other words, a typical control sequence might involve allowing the building fluid loop temperature (i.e., the aqueous fluid available to the heat pumps) to float between 10 and 30 °C as heat pumps extract or reject heat to the loop. Under that condition, the thermal mass of the fluid circulating in the building loop is enough to satisfy the requirements of the heat pumps. When the loop temperature falls below the lower set-point (10 °C), a boiler is energized, and conversely, when the loop temperature rises above the upper set-point (30 °C), the cooling tower is operated. In North America, these types of system have been termed a 'California loop'.

Water-to-air heat pumps may also be configured in a split-type arrangement, where the heat pump unit is remote (or split) from the load. Figure 12.11 shows one such unit that serves the cooling load of a computer server room. The water source is well water, and the refrigerant lines are directed to an air-handler at the load. The solenoid valve on the outlet water connection modulates the water flow based on the room thermostat. Note its position on the outlet water connection, allowing the water-to-refrigerant coil to remain under positive pressure.

Another type of water-source heat pump is the water-to-water type, where a water or aqueous fluid loop acts as both the heat source and heat sink. Again, we have to be careful with the HVAC jargon here: a water-to-water heat pump that provides cooling only (i.e., chilled water)

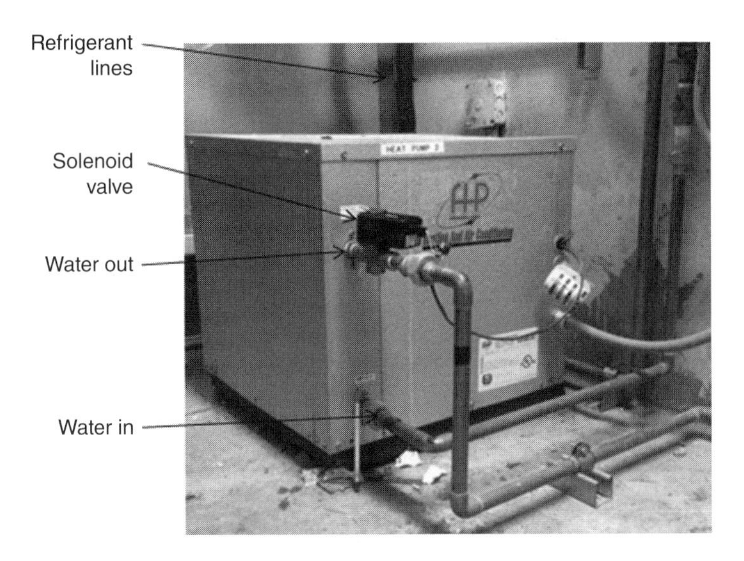

Figure 12.11 Photo of a split-type 5 ton (17.5 kW) water-source heat pump cooling a computer server room (note the pencil for scale under the water inlet connection) (photo taken by the author)

is called a *chiller*. In some instances, a water-to-water heat pump that provides both heating and cooling (i.e., chilled and hot water) is called a *reversible chiller*. *Chillers* connected to a cooling tower or a fluid cooler are perhaps one of the most common types of central cooling system in large buildings. Large water-to-water heat pumps have seen application in district systems. For example, the City of Vancouver, British Columbia, employs large heat pumps in a sewer heat recovery application, where the 'water' source is municipal sewage.

Smaller, unitary or packaged water-to-water heat pumps can be used in a wide variety of applications for building heating and cooling, as well as applications such as domestic water heating, pool heating, or processes where either heated water, chilled water, or both are required.

Ground-source or *geothermal* heat pumps are of two general types based on the source: (i) water-source and (ii) DX geothermal. They offer unique advantages over their air-source counterparts, but are more complex in their design owing to the Earth coupling and associated thermal storage effects, which is the remaining subject of this book.

Geothermal heat pumps are generally more efficient than their air-source counterparts because Earth temperatures are more stable and closer to room conditions than outdoor air, which results in lower compressor work and therefore greater COPs. Further, the geothermal heat pump itself is entirely indoors, whereas air-source heat pumps require outdoor equipment that is exposed to the elements. This outdoor exposure of air-source equipment lowers its operating efficiency and life expectancy relative to geothermal heat pumps. Further, part of the outdoor section of an air-source heat pump includes an outdoor fan to enhance heat transfer, which consumes significantly more energy than circulating pumps required in geothermal systems.

Geothermal heat pumps of the water-source type are the more common type, and essentially have all the features of water-source heat pumps described above. The main difference is that geothermal heat pumps are referred to as 'extended range', meaning that they accept a wide range of operating fluid temperatures (i.e., 25 to 110 °F or −4 to 43 °C). A photo of a 6 ton

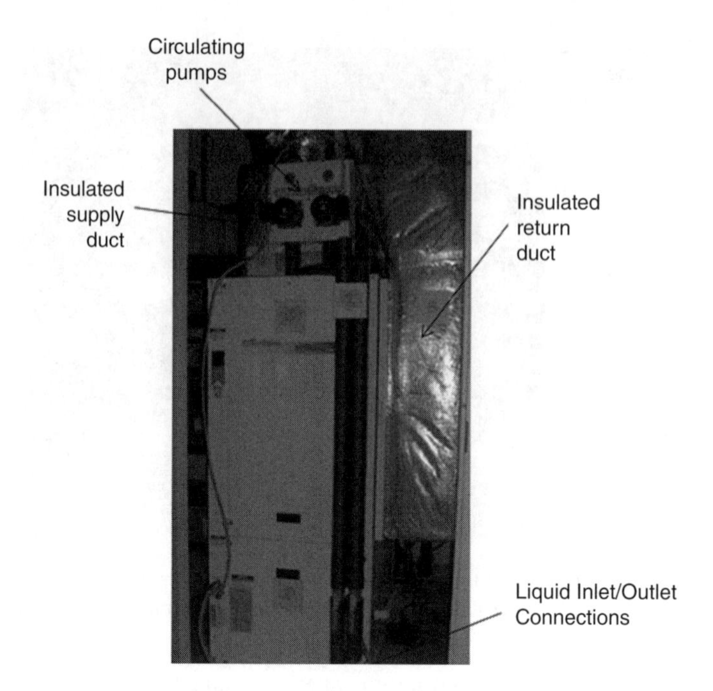

Figure 12.12 Photo of a 6 ton (21 kW) geothermal heat pump (vertical configuration) (photo taken by the author)

(21 kW) water-source geothermal heat pump is shown in Figure 12.12. This type of unit is referred to as a *vertical configuration*, which describes the position of the supply air discharge duct. It could be ordered with a right- or left-hand return. As described above for water-source heat pumps, horizontal configurations are also available for ceiling mounting. Note the dual circulation pumps configured in a so-called 'push-pull' arrangement, where one pump *pushes* fluid into the heat pump, and the other *pulls* fluid from the heat pump. These pumps are wired to the zone thermostat, and are energized when the thermostat calls for heating or cooling.

The second type of geothermal heat pump is the DX geothermal heat pump. These are far less common than geothermal heat pumps of the water-source type. They are so named because of their Earth coupling, where copper refrigerant lines are buried directly in the Earth in a closed-loop configuration; the ground coupling is essentially an extension of the refrigerant circuit. Because of this configuration, multiple heat pumps are usually not connected together on a common loop like water-source heat pumps. Thus, DX geothermal heat pumps are typically limited to single-zone applications, such as residential and light commercial.

In certain acidic soil conditions, DX geothermal heat pump systems require corrosion protection of the buried copper refrigerant lines. In cooling mode, some systems have provisions for soil wetting via irrigation drip lines in order to enhance soil thermal conductivity by keeping the soil moist.

In summary, we have identified eight possible configurations of mechanical vapor compression heat pumps, as shown in Figure 12.13: (1) air-to-air, (2) air-to-water, (3) water-to-air, (4) water-to-water, (5) geothermal water-to-air, (6) geothermal water-to-water, (7) DX geothermal-to-air, and (8) DX geothermal-to-water.

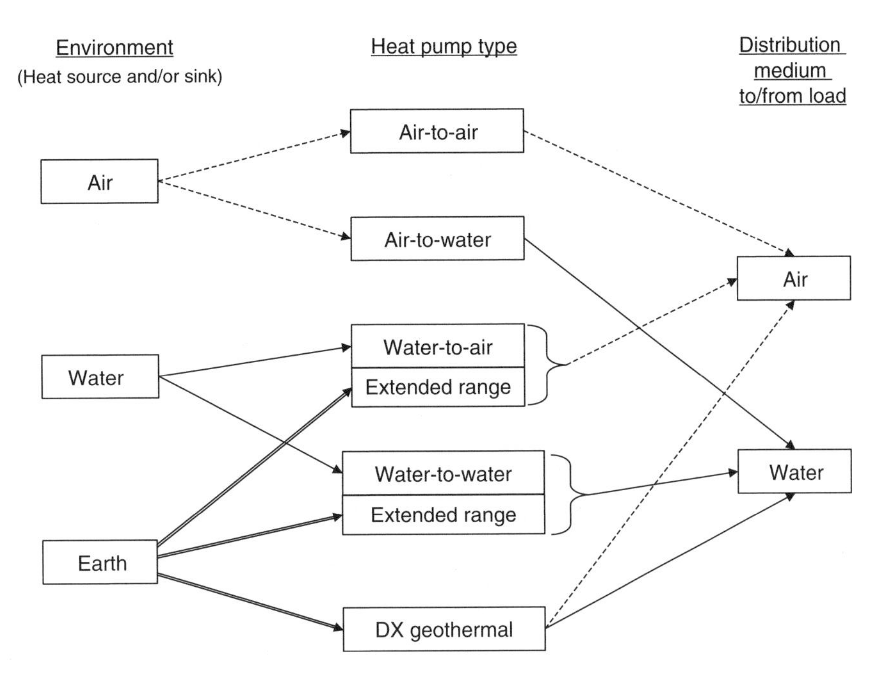

Figure 12.13 Summary of possible configurations of vapor compression heat pumps

12.5 Mechanics of Operation

Let us now examine the operating principles of the subcritical vapor compression cycle by following the refrigerant flow in air-source and water-source heat pumps in heating mode and cooling mode. We will base our discussion around the processes that occur in the four principal components of these cycles: (i) compressor, (ii) condenser, (iii) expansion valve, and (iv) evaporator. The discussion that follows refers to Figures 12.15 to 12.20.

First, actual components of a vapor compression heat pump are shown in Figure 12.14, which is a photo of a laboratory vapor compression heat pump apparatus where the source and sink are air. Fans behind the evaporator and condenser are used to enhance heat transfer. A reversing valve is used to redirect the refrigerant flow to either the top or bottom heat exchanger, which acts either as a condenser or evaporator heat exchanger, depending on whether it is on the discharge or suction side of the compressor. The tank on the suction line between the evaporator and the compressor is a suction accumulator, and the tank on the liquid line between the condenser and expansion valve is a liquid receiver. The primary function of the suction accumulator is to catch and hold any liquid refrigerant that didn't evaporate in the evaporator; liquid refrigerant in the compressor can damage internal parts of the compressor. The primary function of the receiver is the opposite; it retains refrigerant vapor not condensed in the condenser. Many heat pump manufacturers utilize refrigerant accumulators as standard equipment.

12.5.1 The Compressor

Process 1–2 represents the vapor compression process, and hence is common to all vapor compression heat pump cycles. Here, work input to a compressor raises the vapor refrigerant to a relatively

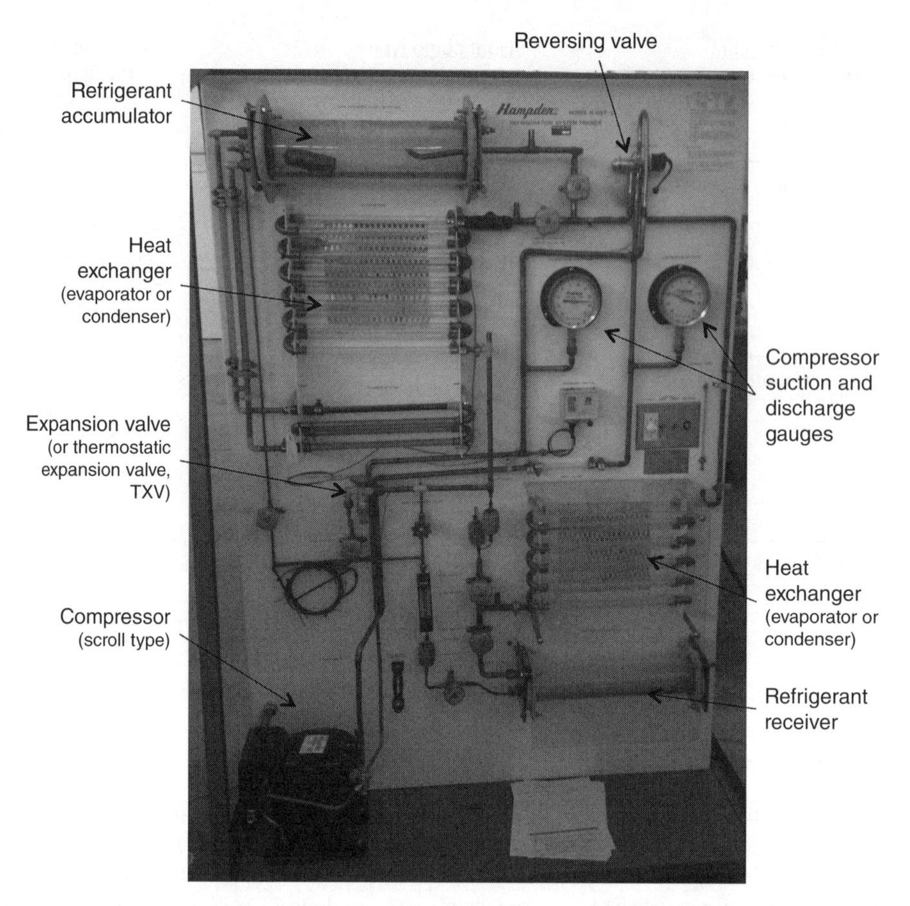

Figure 12.14 Photo of a laboratory heat pump apparatus, showing major components (photo taken by the author)

high temperature and pressure state; the temperature of the refrigerant must be raised high enough such that heat can be rejected to the environment in the subsequent condensing process (*2–3*). Note that in air-source heat pumps (Figures 12.15 and 12.18), the compressor is located outdoors.

Compressors, typically the scroll type in unitary heat pumps, are the main energy user in heat pumps, and thus electrical energy to run the compressor represents the main cost of operating a heat pump. Some unitary heat pumps are equipped with two compressors which are operated in stages. Others have a refrigerant bypass port on the compressor that allows it to operate at a lower capacity when the bypass port is open, and at a higher capacity when the bypass port is closed. These have common application where one peak load (usually heating) significantly exceeds the other (usually cooling). Another application is in climates that have high cooling and dehumidifying loads. In such *two-stage* heat pumps, the second stage is only utilized during peak load conditions. However, variable-speed compressors are now in the marketplace, rendering two-stage machines unnecessary. In heat pumps with variable-speed compressors, the motor speed is varied in one of two ways: (i) varying the frequency of the motor with variable-frequency drives, or (ii) rectifying alternating to direct current, and then varying the voltage input.

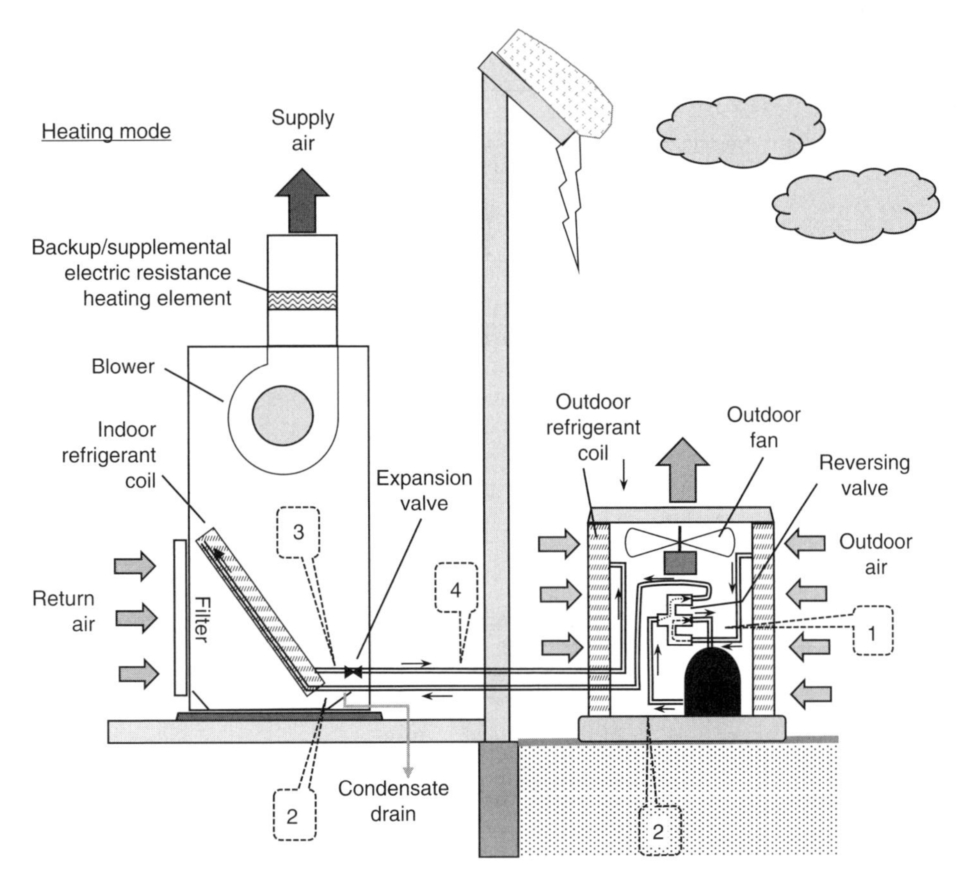

Figure 12.15 Basic operation of an air-to-air vapor compression heat pump cycle *in heating mode*

12.5.2 The Condenser

Process 2–3 in general represents condensing of the vapor refrigerant to liquid through heat transfer to cooler surroundings. Here, there are significant differences between what happens in the heating and cooling modes.

Optional *process 2–2′* represents desuperheating of the refrigerant vapor, which is a separate option found typically in residential geothermal heat pumps (see Figures 12.16 and 12.19). Sometimes referred to as a *heat recovery option*, the common application in residential systems involves heat rejection from the superheated refrigerant vapor to a domestic hot water storage tank. Here, a small refrigerant-to-water coaxial heat exchanger is installed in addition to the condenser, along with a small circulating pump. Typical control involves energizing the circulating pump when the temperature of the refrigerant vapor is sufficiently hotter than the water in the storage tank.

In heating mode (Figures 12.15 to 12.18), a reversing valve directs the refrigerant vapor to the heat exchanger that is coupled to the heat sink on the load side of the heat pump. The refrigerant enters this heat exchanger (the condenser) at relatively high temperature and

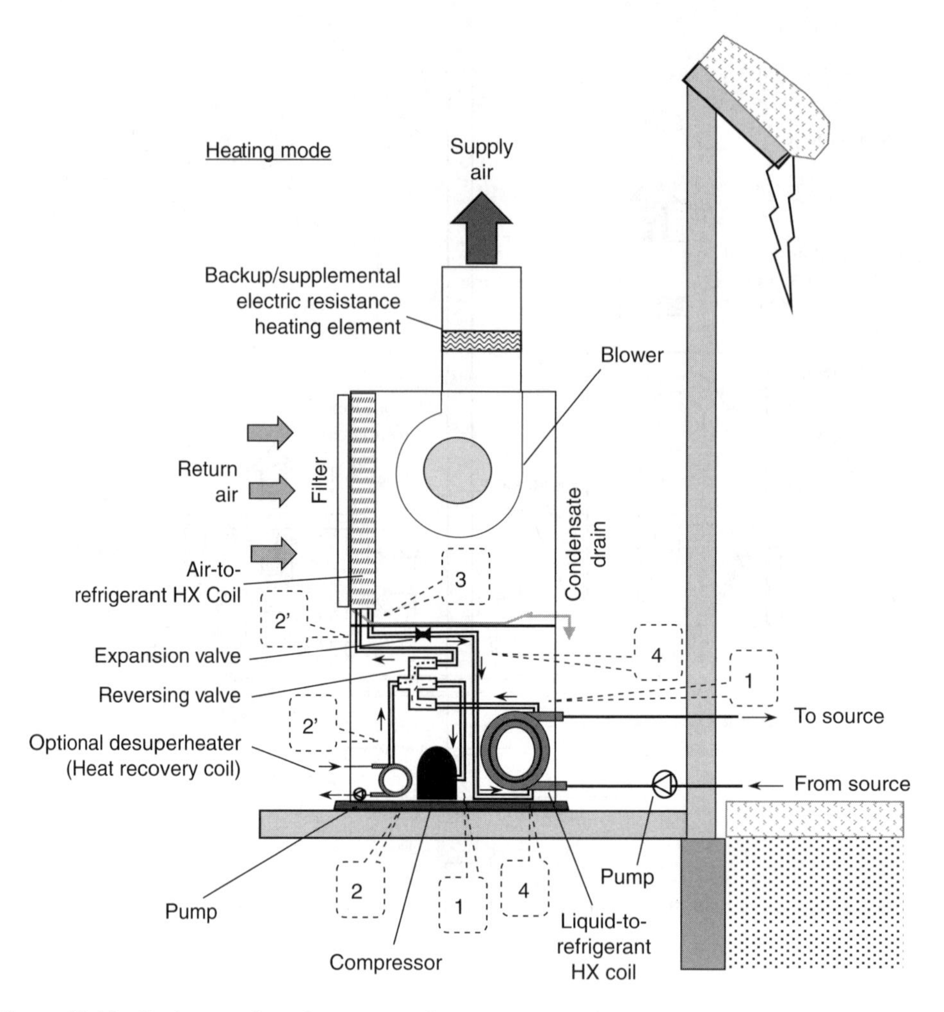

Figure 12.16 Basic operation of a water-to-air vapor compression heat pump cycle *in heating mode*

pressure, such that heat is rejected to the heat distribution medium in the building (or other heating application), thereby condensing the refrigerant. The refrigerant state at point 2 is either a superheated vapor or a saturated vapor, depending on whether or not the heat pump is equipped with a heat recovery desuperheater. As shown in Figure 12.3, the refrigerant temperature must be significantly above the sink temperature in order to transfer heat within practical limitations of the equipment. The pressure in the condenser is controlled by the downstream process (*3–4*) such that the refrigerant fully condenses to liquid at point 3.

When air is the distribution medium (Figures 12.15 and 12.16), a fan moves air to be heated across the air-to-refrigerant finned coil. Manufacturers' fan settings are typically 400 cfm/ton or 54 L/s/kW. Such airflow settings are a combination of practical considerations (heat pump size and duct size) in addition to human occupant comfort and perceptions of draftiness. Thus, these airflow considerations dictate that air be introduced to occupant spaces at a temperature of

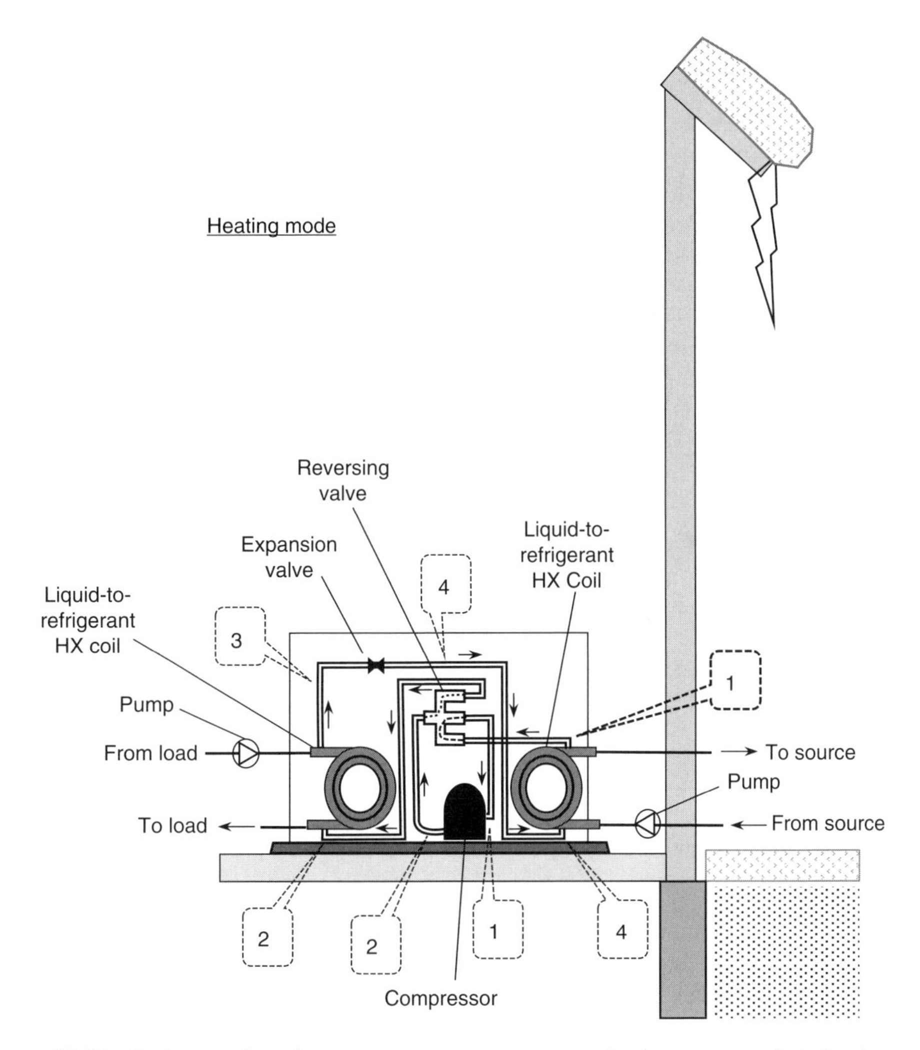

Figure 12.17 Basic operation of a water-to-water vapor compression heat pump cycle *in heating mode*

about 15 °C above the thermostat set-point. For example, in a room with a thermostat set-point of 20 °C, supply air is typically introduced at a temperature of at least 35 °C.

When water is the distribution medium (Figure 12.17), a pump circulates water to be heated across the water-to-refrigerant coaxial coil. These coaxial coils are manufactured such that they require 2–3 gpm/ton or Lpm/kW.

In cooling mode (Figures 12.18 to 12.20) a reversing valve directs the refrigerant vapor to the heat exchanger that is coupled to the heat sink on the 'source' side of the heat pump. The refrigerant enters this heat exchanger (the condenser) at relatively high temperature and pressure, such that heat is rejected to the environment, thereby condensing the refrigerant. The refrigerant state at point 2 is either a superheated vapor or a saturated vapor, depending on whether or not the heat pump is equipped with a heat recovery desuperheater. As shown in

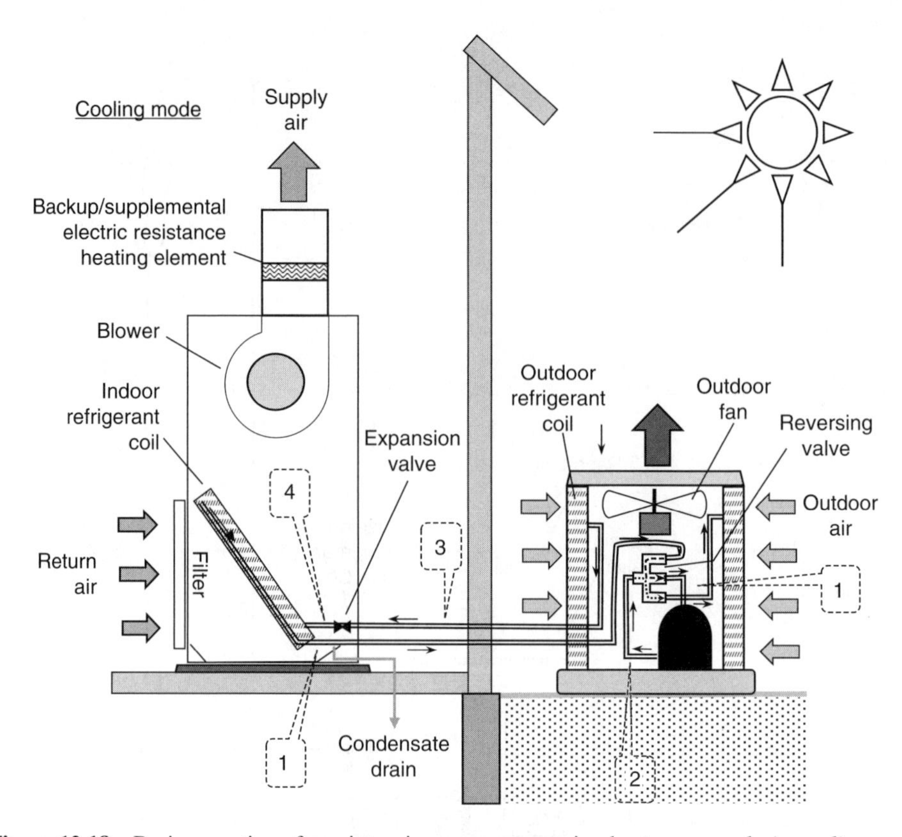

Figure 12.18 Basic operation of an air-to-air vapor compression heat pump cycle *in cooling mode*

Figure 12.3, the refrigerant temperature must be significantly above the environment sink temperature in order to transfer heat within practical limitations of the equipment. The pressure in the condenser is controlled by the downstream process (*3–4*) such that the refrigerant fully condenses to liquid at point 3.

In air-source heat pumps (Figure 12.18), an outdoor fan moves air across the outdoor air-to-refrigerant finned coil. Manufacturers' fan settings are typically 1000 cfm/ton or 135 L/s/kW. Such airflow settings are required for adequate transfer of heat through the outdoor unit at extreme temperature conditions. When water is the distribution medium (Figures 12.19 and 12.20), a pump circulates water through a water-to-refrigerant coaxial coil. These coaxial coils are manufactured such that they require 2–3 gpm/ton or Lpm/kW.

12.5.3 The Expansion Valve

Process 3–4 represents the process of expansion of the refrigerant as it flows through a thermostatically controlled expansion valve, or TXV. This valve is also sometimes referred to as the *metering device* because it controls (or meters) the flow of refrigerant. Upstream of the TXV the refrigerant is at the condensing pressure, and downstream of the TXV the refrigerant

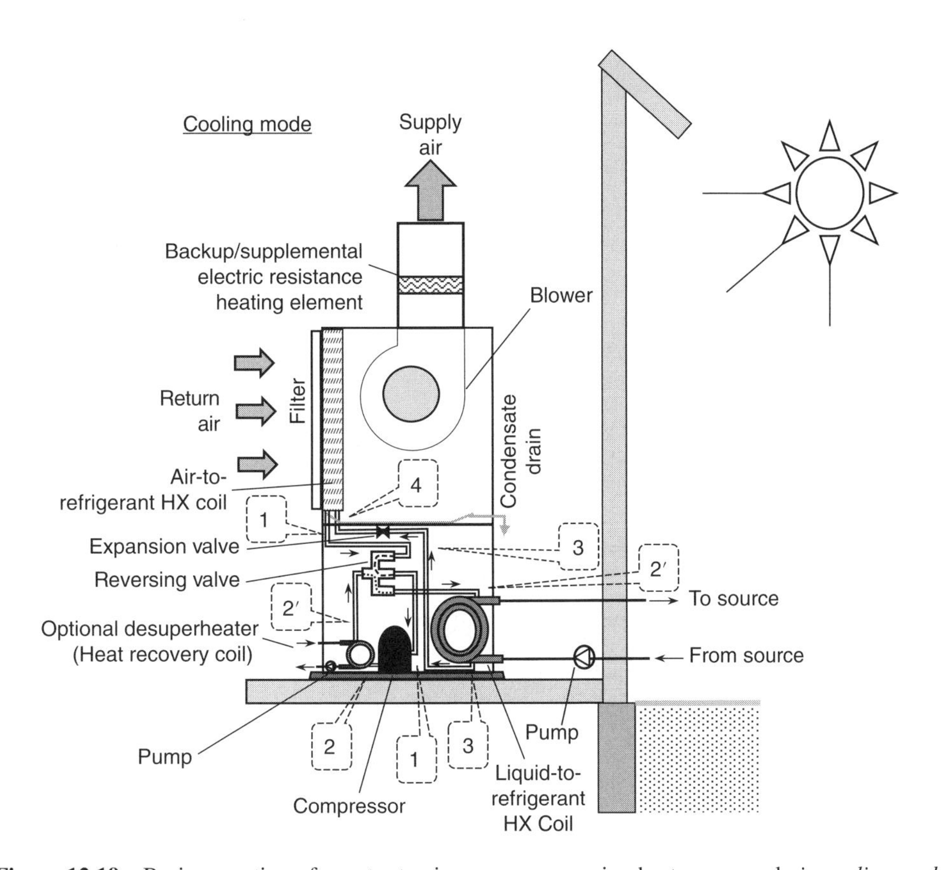

Figure 12.19 Basic operation of a water-to-air vapor compression heat pump cycle *in cooling mode*

is at the evaporating pressure. The TXV responds to conditions leaving the evaporator (*process 4–1*); when the refrigerant is at higher temperature and pressure, the TXV will open wider, allowing more refrigerant to flow. According to the kinetic theory of gases, the relatively large pressure drop occurring in process 3–4 results in a large corresponding temperature drop. As a result, some of the refrigerant begins to evaporate, and the refrigerant exists as a two-phase liquid–vapor mixture at state 4.

12.5.4 The Evaporator

Process 4–1 in general represents the process of evaporation of the two-phase liquid–vapor mixture through heat transfer from warmer surroundings. After leaving the evaporator, the refrigerant flows through the reversing valve, where it is directed back to the compressor, and the cycle is repeated. With process 4–1 there are significant differences between what happens in the heating and cooling modes.

In heating mode (Figures 12.15 and 12.16), the refrigerant enters the heat exchanger that is coupled to the source side of the heat pump. The refrigerant enters this heat exchanger (the evaporator in this case) at relatively low temperature and pressure, such that heat is

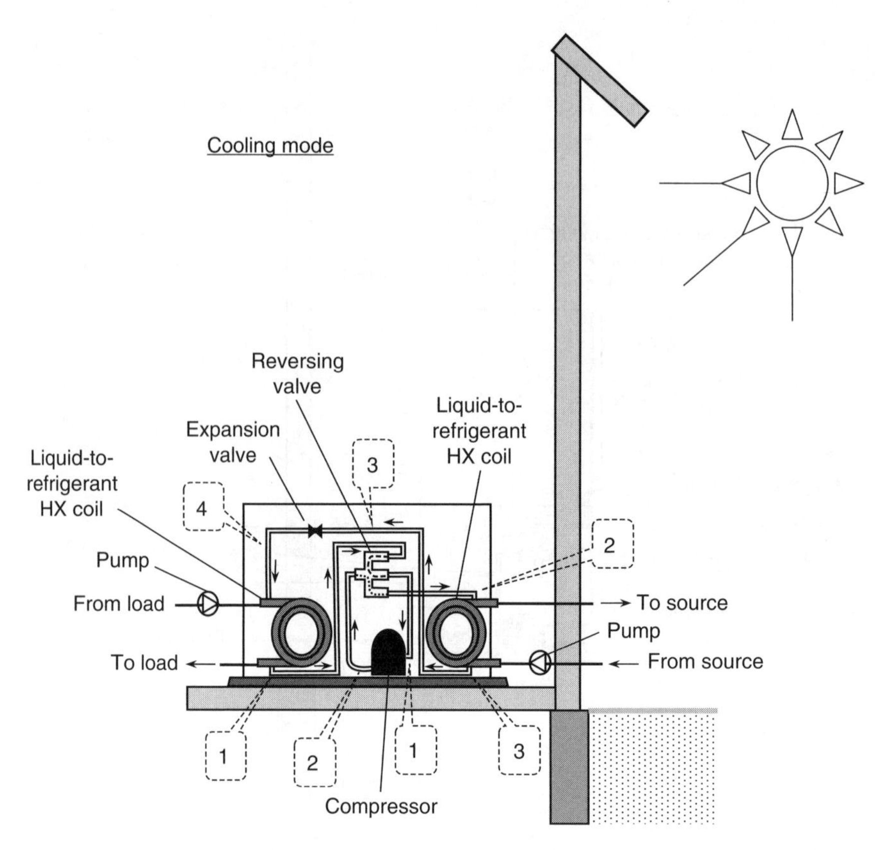

Figure 12.20 Basic operation of a water-to-water vapor compression heat pump cycle *in cooling mode*

extracted from the heat source medium, thereby evaporating the refrigerant. As shown in Figure 12.3, the refrigerant temperature must be significantly below the source temperature in order to extract heat within practical limitations of the equipment. The refrigerant at point 4 is a vapor.

When air is the heat source (Figure 12.15), an outdoor fan moves air across the outdoor air-to-refrigerant finned coil. Manufacturers' fan settings are typically 1000 cfm/ton or 135 L/s/kW. Such airflow settings are required for adequate transfer of heat through the outdoor unit at extreme temperature conditions. When water is the distribution medium (Figures 12.16 and 12.17), a pump circulates water through a water-to-refrigerant coaxial coil. These coaxial coils are manufactured such that they require 2–3 gpm/ton or Lpm/kW.

In cooling mode, the refrigerant enters the heat exchanger that is coupled to the load side of the heat pump. The refrigerant enters this heat exchanger (the evaporator in this case) at relatively low temperature and pressure, such that heat is extracted from the medium to be cooled, thereby evaporating the refrigerant. As shown in Figure 12.3, the refrigerant temperature must be significantly below the load fluid temperature in order to extract heat within practical limitations of the equipment. The refrigerant at point 4 is a vapor.

When air is the distribution medium (Figures 12.18 and 12.19), a fan moves air to be cooled across the air-to-refrigerant finned coil. Manufacturers' fan settings are typically

400 cfm/ton or 54 L/s/kW. Such airflow settings are a combination of practical considerations (heat pump size and duct size) in addition to human occupant comfort and perceptions of draftiness. As the refrigerant temperature in the evaporator is significantly below the dewpoint of the air, both sensible and latent cooling occur. Dehumidification of and moisture removal from the air stream necessitate installation of a condensate pan and drain. Typical supply air conditions to a room are of the order of 14 °C (dry bulb) and 12 °C (wet bulb).

When water is the distribution medium (Figure 12.20), a pump circulates water to be chilled across the water-to-refrigerant coaxial coil. These coaxial coils are manufactured such that they require 2–3 gpm/ton or Lpm/kW.

12.5.5 Other Components

The sketches in Figures 12.15 to 12.20 show the presence of a reversing valve, which is a necessary piece of equipment found in unitary heat pumps. A photo of an actual cut-away reversing valve in heating and cooling position is shown in Figures 12.21a and b respectively. The valve is actuated to the appropriate position when the heat pump thermostat calls for either heating or cooling. We note here that heating and cooling can be accomplished by systems without the use of a reversing valve. For example, rather than a refrigerant changeover with a reversing valve, the load and source fluids can experience a changeover. Alternatively, a load and source fluid changeover is unnecessary if the application requires simultaneous

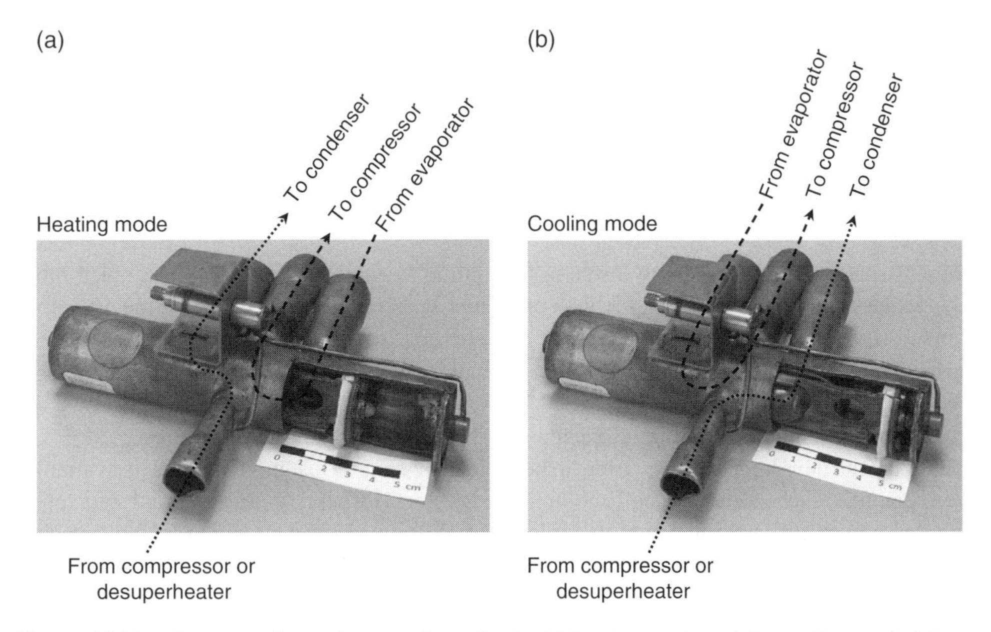

Figure 12.21 Cut-away photo of a reversing valve in (a) heating mode and (b) cooling mode (photos taken by the author)

Figure 12.22 Cut-away photo of a cupronickel, coaxial, water-refrigerant heat exchanger (photo taken by the author)

heating and cooling. Such systems were described in Chapter 9 with respect to large, central hydronic systems.

Figure 12.22 shows a cut-away photo of a typical cupronickel alloy water-to-refrigerant coaxial coil used in geothermal pumps. Note the rifling of the inner pipe to increase heat transfer area. Rifling also causes the fluids to flow in a 'corkscrew' pattern, which increases the convection heat transfer coefficient.

12.5.6 Heat Pump Power Requirements

Compressors are the main energy users in heat pump systems. Compressors use much more power than pumps to deliver an equivalent pressure rise, because the specific volume of a vapor is much larger than for a liquid. Gases are compressible substances, and recall that the compressor work is dictated by $\dot{W} = \int P dV$. In a nominal 3 ton (10 kW) heat pump, the compressor power requirement is of the order of 2.0–2.7 kW. As we shall see in subsequent sections, the actual energy requirement strongly depends on the source and sink temperatures.

Fans are the next-largest energy users in heat pump systems. Most modern units use variable-speed controls, such as electronically commutated motors (ECMs). At zero external static pressure, fan power requirements are of the order of 60 W/ton or 17 W/kW of heat pump capacity, but can be nearly double that value in ducted systems.

Circulating pumps in heat pump systems are typically the lowest energy users in the system. In properly designed systems, pumping power requirements can be less than

50 W/ton or 14 W/kW of heat pump capacity. Most modern designs use pumps with variable-speed motor controls.

12.5.7 Performance Modeling

We can now consolidate the foregoing to construct a fairly complete mathematical model of a subcritical vapor compression cycle heat pump. The objective is to create a model that is a useful representation of a real vapor compression cycle. Let's illustrate by modeling a heat pump via computer simulation in EES for both heating and cooling modes.

First, it is important to note that there are numerous combinations of possible input and output data for a heat pump simulation model, yielding many degrees of freedom. The choice of input and output data depends on what is known, and what is desired to be known. The inputs and outputs are chosen here to obtain as much relationship to real practice as possible, and thus we will construct steady-state models with the data as organized in Table 12.4. Other possible combinations of interest include the following as input data: COP, compressor power requirement, heating load, or cooling load.

In the examples that follow, the driving factors are the compressor power input and the heat pump entering fluid details. It is useful to model the compressor in more detail, as the isentropic efficiency varies as a function of compression ratio. The default function used in the examples that follow relates the compressor isentropic efficiency to the compression ratio

Table 12.4 Summary of Input and Output Data of Interest for Example Mechanical Vapor Compression Heat Pump Simulation Model

Fluid Cycle	Input Data	Output Data
Load	• thermal properties of fluid entering heat pump • moist air properties for cooling mode (where applicable) • fluid flow rate • heat exchanger effectiveness	• heat transfer rate • fluid state leaving heat pump
Refrigerant (Vapor Compression)	• refrigerant fluid • compressor rated power input • compressor efficiency at rated power input	• thermodynamic state of all points in the cycle • compressor power requirement • actual compressor efficiency • refrigerant mass flow rate
Environment	• thermal properties of fluid entering heat pump • fluid flow rate • heat exchanger effectiveness	• heat transfer rate • fluid state leaving heat pump
System	• none necessary	• coefficient of performance

as: $n_{compressor} = -0.08*(P2/P1) + 1.025$. This is a simple function based on generic compressor operating data; detailed compressor modeling is left to the reader.

Example 12.4 Simulation of a Water-to-Air Geothermal Heat Pump in Heating Mode
A water-to-air geothermal heat pump using R410A refrigerant is operating in heating mode at a residence with a thermostat set-point of 22 °C. The geothermal fluid is an aqueous mixture of water and 20% propylene glycol (by mass), and its temperature entering the heat pump is 5 °C. The rated power input to the compressor is 5 kW. The air flow rate is a typical $0.054\ m^3 \cdot s^{-1} \cdot kW^{-1}$, and the water flow rate is a typical $3.0\ L \cdot min^{-1} \cdot kW^{-1}$.

Assume that the heat exchanger effectiveness values are 0.6 and 0.8 on the air side and water side respectively, and that the compressor isentropic efficiency is 80% at the rated power input.

Determine/show the following:

(a) a sketch of the cycle on a *P–h* diagram and show the ideal (isentropic case),
(b) the heat rejected at the condenser,
(c) the heat extracted at the evaporator,
(d) the refrigerant temperature leaving the compressor,
(e) the condensing temperature of the refrigerant,
(f) the evaporating temperature of the refrigerant,
(g) the mass flow rate of the refrigerant,
(h) the compressor power,
(i) the heating COP,
(j) the supply air temperature to the space,
(k) the temperature of the geothermal fluid exiting the heat pump, and
(l) the maximum superheat available for use in a domestic hot water tank.

Solution
The EES input data are as follows:

```
"INPUT DATA:"
"Air Conditions"
P = 101.1 [kPa]                      "The local atmospheric air pressure"
T_air_db = 22 [C]                    "The building return (heat pump entering) air dry
                                      bulb temperature"
Q_dot_air_per_kW = 0.054 [m^3/s]     "Air volumetric flow rate per peak kW heating"
epsilon_cond = 0.6 [-]               "HX effectiveness estimate of condenser"

"Heat Pump"
R$ = 'R410a'; $REFERENCE R410a ASH "String for refrigerant, using ASHRAE reference state"
CompressorRatedInput = 5 [kW]        "The rated power input of the compressor"
CompressorEfficiency = 0.80 [-]      "isentropic efficiency at rated input"

"Geothermal Fluid Conditions"
T_geofluid = 5 [C]                   "The geofluid temperature entering the heat pump"
Q_dot_geo_per_kW = 3 [L/min]         "Geothermal volumetric flow rate per peak kW
                                      heating"
epsilon_evap = 0.6 [-]               "HX effectiveness estimate of evaporator coil"
```

```
"SOLUTION:"
"Simplifying Assumptions:"
x1 = 1 [-]; x3 = 0 [-]
n_compressor = -0.08 * (P2/P1) + 1.025 "assumed function of compression ratio"
W_comp = CompressorRatedInput * CompressorEfficiency/ n_compressor

"Calculate fluid properties and fluid mass flow rates per kW"
T_F = FreezingPt(PG,C = 20[%])                    "the freezing point of the geofluid"
rho_geofluid = Density(PG,T = T_geofluid,C = 20[%])
cp_geofluid = Cp(PG,T = T_geofluid,C = 20[%])
rho_air = Density(Air,T = T_air_db, P = P)
cp_air = Cp(Air,T = T_air_db)
m_dot_air_perkW = Q_dot_air_per_kW * rho_air
m_dot_geofluid_perkW = Q_dot_geo_per_kW * convert(L/min, m^3/s) * rho_geofluid

"Point 1"
h1 = Enthalpy(R$, T = T1, x = x1); P1 = pressure (R$, T = T1, x = x1); s1 = entropy(R$, T = T1, x = x1)

"Point 2, 2s, and 2' "
s2s = s1; P2 = P3; T2 = temperature (R$, P = P2, h = h2)
h2s = enthalpy(R$, P = P2, s = s2s)
n_compressor = (h2s - h1) / (h2 - h1)
h2_prime = enthalpy(R$, T = T3, x = 1)

"Point 3"
h3 = Enthalpy(R$, T = T3, x = x3); P3 = pressure (R$, T = T3, x = x3)

"Point 4"
P4 = P1; h4 = h3; x4 = quality(R$, P = P4, h = h4)

"Energy Balances:"
COP_h = (h2 - h3) / (h2 - h1)
q_evap = mdot_refrigerant * (h1 - h4)
q_cond = mdot_refrigerant * (h2 - h3)
W_comp = mdot_refrigerant * (h2 - h1)
q_superheat = mdot_refrigerant * (h2 - h2_prime)

"Heat Transfer Relationships:"
1 [kW] = Beta*(T3 - T_air_db)                "condenser heat exchanger relationship (per kW)"
1 [kW] = Alpha*(T_geofluid - T1)             "evaporator heat exchanger relationship (per kW)"
Beta = epsilon_cond * m_dot_air_perkW *      "effectiveness-heat capacity rate product"
cp_air
Alpha = epsilon_evap * m_dot_geofluid_-      "effectiveness-heat capacity rate product"
perkW * cp_geofluid
1 [kW] = m_dot_geofluid_perkW * cp_geofluid * (T_geofluid - T_GHXin) "find the leaving
geofluid temperature"
1 [kW] = m_dot_air_perkW * cp_air * (T_air_supply - T_air_db ) "find the leaving air temperature"
```

The following results are obtained:

(a) a sketch of the cycle on a P–h diagram, showing the ideal (isentropic case): see Figure E.12.4

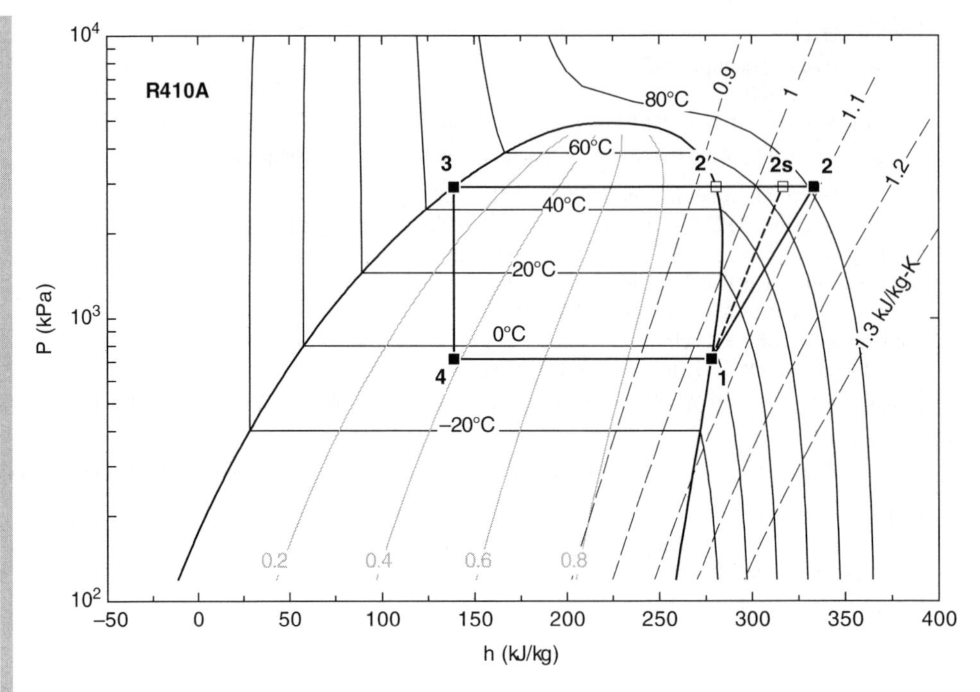

Figure E.12.4 *P–h* property plot for Example 12.4

(b) the heat rejected at the condenser: $q_{cond} = 20.08$ [kW]
(c) the heat extracted at the evaporator: $q_{evap} = 14.37$ [kW]
(d) the refrigerant temperature leaving the compressor: T2 = 82.38 [C]
(e) the condensing temperature of the refrigerant: T3 = 47.75 [C]
(f) the evaporating temperature of the refrigerant: T1 = −3.289 [C]
(g) the mass flow rate of the refrigerant: $mdot_{refrigerant} = 0.1033$ [kg/s]
(h) the compressor power: $W_{comp} = 5.714$ [kW]
(i) the heating COP: $COP_h = 3.515$
(j) the supply air temperature to the space: $T_{air,supply} = 37.45$ [C]
(k) the geothermal fluid temperature leaving the heat pump: $T_{GHXin} = 0.0264$ [C] _
(l) the maximum superheat available: $q_{superheat} = 5.437$ [kW]

Example 12.5 Effect of External Fluid Temperatures on the COP of a Geothermal Heat Pump in Heating Mode

With the model constructed in Example 12.4, develop an *x–y–z* plot of the heat pump COP as a function of the entering geothermal fluid and entering air temperatures.

Solution

Here, let's use the parametric feature in EES to vary the entering geothermal fluid temperature from 0 to 15 °C, and the entering air temperature from 15 to 30 °C.

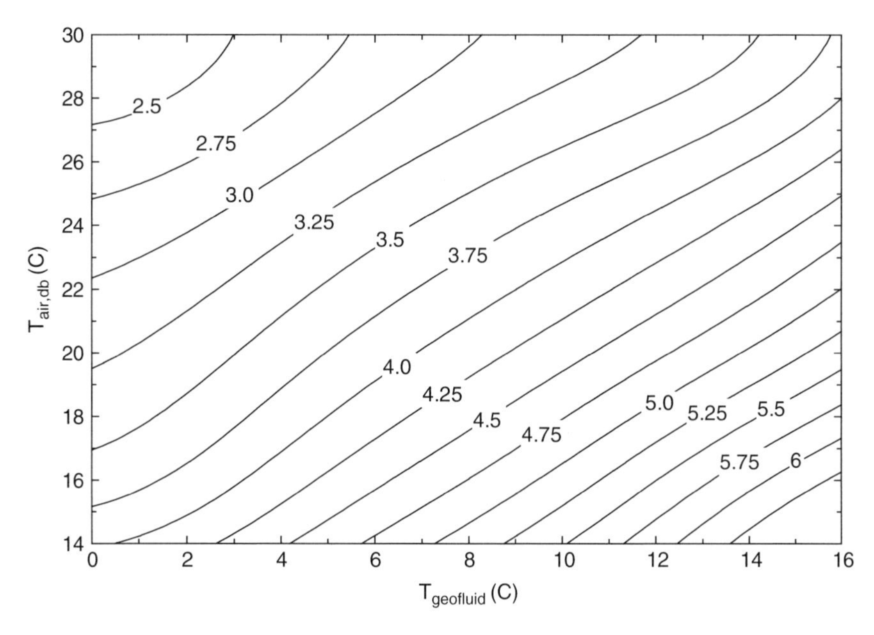

Figure E.12.5 Plot of heat pump COP as a function of heat pump entering air and geofluid temperature for Example 12.5

Discussion: A review of the plot shown in Figure E.12.5 shows, as one might expect, that the heat pump COP is quite sensitive to the load and environment fluid temperatures. For example, if the entering geothermal fluid temperature in the previous example were to drop from 5 to 0 °C, then the heat pump COP would decrease from 3.5 to approximately 3.0. Similarly, if the entering air temperature were to increase from 22 to 25 °C, more heat pump power would be required, and thus the COP would decrease to about 3.2.

Example 12.6 Simulation of a Water-to-Air Geothermal Heat Pump in Cooling Mode

A water-to-air geothermal heat pump using R410A refrigerant is operating in cooling mode. The heat pump entering air conditions are 25 °C dry-bulb temperature and 50% relative humidity. The supply air temperature to the space is 13 °C. The geothermal fluid is an aqueous mixture of water and 20% propylene glycol (by mass), and its temperature entering the heat pump is 27 °C. The rated power input to the compressor is 5 kW. The air flow rate is a typical $0.054\ \mathrm{m^3 \cdot s^{-1} \cdot kW^{-1}}$, and the water flow rate is a typical $3.0\ \mathrm{L \cdot min^{-1} \cdot kW^{-1}}$.

Assume that the heat exchanger effectiveness values are 0.6 and 0.8 on the air side and water side respectively, and that the compressor isentropic efficiency is 80% at the rated power input.

Determine/show the following:

(a) a sketch of the cycle on a P–h diagram and show the ideal (isentropic case),
(b) the heat rejected at the condenser,
(c) the heat extracted at the evaporator,

(d) the refrigerant temperature leaving the compressor,
(e) the condensing temperature of the refrigerant,
(f) the evaporating temperature of the refrigerant,
(g) the mass flow rate of the refrigerant,
(h) the compressor power,
(i) the cooling COP,
(j) the sensible heat ratio of the process,
(k) the temperature of the geothermal fluid exiting the heat pump,
(l) the maximum superheat available for use in a domestic hot water tank, and
(m) a sketch of the air process on a psychrometric chart.

Solution

The EES input data are as follows:

```
"INPUT DATA:"
"Air Conditions"
P = 101.1 [kPa]                          "The local atmospheric pressure"
T_air_db = 25 [C]                        "The building return (heat pump entering) air
                                         dry bulb temperature"
RH_air = 0.5                             "The building return (heat pump entering) air
                                         relative humidity"
Q_dot_air_per_kW = 0.054 [m^3/s]         "Air volumetric flow rate per peak kW cooling"
epsilon_evap = 0.6 [-]                   "HX effectiveness estimate of evaporator coil"
"Enter a value for only one of the following three inputs (T_air_supply. SHR, T_coil). The
remaining two will be outputs."
T_air_supply = 13 [C]                    "The building supply (heat pump exiting) air dry
                                         bulb temperature"
{SHR = ? [-] }                           "The sensible heat ratio of the heat pump"
{T_coil = ? [C] }                        "The coil surface temperature"

"Heat Pump"
R$ = 'R410a'; $REFERENCE R410a ASH "String for refrigerant, using ASHRAE reference state"
CompressorRatedInput = 5.0 [kW]          "The rated power input of the compressor"
CompressorEfficiency = 0.80 [-]          "isentropic efficiency at rated input"
"Geothermal Fluid Conditions"
T_geofluid = 27 [C]                      "The geofluid temperature entering the heat
                                         pump"
Q_dot_geo_per_kW = 3 [L/min]             "Geothermal volumetric flow rate per peak
                                         kW cooling"
epsilon_cond = 0.8 [-]                   "HX effectiveness estimate of condenser"

"SOLUTION:"
"Simplifying Assumptions:"
x1 = 1 [-]; x3 = 0 [-]
n_compressor = -0.08 * (P2/P1) + 1.025   "assumed function of compression ratio"
W_comp = CompressorRatedInput * CompressorEfficiency/ n_compressor
```

```
"Calculate fluid properties and fluid mass flow rates per kW"
T_F = FreezingPt(PG,C = 20[%])                    "the freezing point of the geofluid"
rho_geofluid = Density(PG,T = T_geofluid,C = 20[%])
cp_geofluid = Cp(PG,T = T_geofluid,C = 20[%])
rho_air = Density(AirH2O,T = T_air_db,R = RH_air,P = P)
cp_air = Cp(AirH2O,T = T_air_db,R = RH_air,P = P)
m_dot_air_perkW = Q_dot_air_per_kW * rho_air
m_dot_geofluid_perkW = Q_dot_geo_per_kW * convert(L/min, m^3/s) * rho_geofluid

"Point 1"
h1 = Enthalpy(R$, T = T1, x = x1); P1 = pressure (R$, T = T1, x = x1)
s1 = entropy(R$, T = T1, x = x1)

"Point 2, 2s, and 2′ "
T2 = temperature (R$, P = P2, h = h2); P2 = P3
s2s = s1; h2s = enthalpy(R$, P = P2, s = s2s)
n_compressor = (h2s - h1) / (h2 - h1)
h2_prime = enthalpy(R$, T = T3, x = 1)

"Point 3"
h3 = Enthalpy(R$, T = T3, x = x3); P3 = pressure (R$, T = T3, x = x3)

"Point 4"
P4 = P1; h4 = h3; x4 = quality(R$, P = P4, h = h4)

"Energy Balances:"
COP_c = (h1 - h4) / (h2 - h1)
q_evap = mdot_refrigerant * (h1 - h4)
q_cond = mdot_refrigerant * (h2 - h3)
W_comp = mdot_refrigerant * (h2 - h1)
q_superheat = mdot_refrigerant * (h2 - h2_prime)

"Heat Transfer Relationships:"
1 [kW] = Beta*(T3 - T_geofluid)              "condenser heat exchanger relationship
                                              (per kW)"
1 [kW] = Alpha*(T_air_db - T1)               "evaporator heat exchanger relationship
                                              (per kW)"
Beta = epsilon_cond * m_dot_geofluid_perkW * "effectiveness-heat capacity rate product"
cp_geofluid
Alpha = SHR * epsilon_evap * m_dot_air_-     "effectiveness-heat capacity rate product"
perkW * cp_air
1 [kW] = m_dot_geofluid_perkW * cp_geofluid  "find the leaving geofluid temperature"
* (T_GHXin - T_geofluid)
SHR * 1 [kW] = m_dot_air_perkW * cp_air *    "find the leaving air temperature"
(T_air_db - T_air_supply )
"Coil Relationships"
```

```
epsilon_evap = ( T_air_db - T_air_supply) /          "Approximation for T air supply"
(T_air_db - T_coil)
h_air_return = enthalpy(AIRH2O,P = P,T =             "Enthalpy of building return air (entering
T_air_db, R = RH_air)                                evaporator coil)"
SHR = cp_air * ( T_air_db - T_air_supply) / (h_air_return - h_air_supply)
RH_supply = relhum(AIRH2O,P = P,T = T_air_-          "Relative humidity of supply air"
supply, h = h_air_supply)
"For Psychrometric Plot"
T[1] = T_air_db; T[2] = T_air_supply                 "The x-axis dry-bulb temperatures"
W[1] = HumRat(AirH2O,T = T_air_db,R =                "The humidity ratio of the rerurn air"
RH_air,P = P)
W[2] = HumRat(AirH2O,T = T_air_supply, h =           "The humidity ratio of the supply air"
h_air_supply, P = P)
```

The following results are obtained:

(a) a sketch of the cycle on a *P–h* diagram, showing the ideal (isentropic case): see Figure E.12.6a

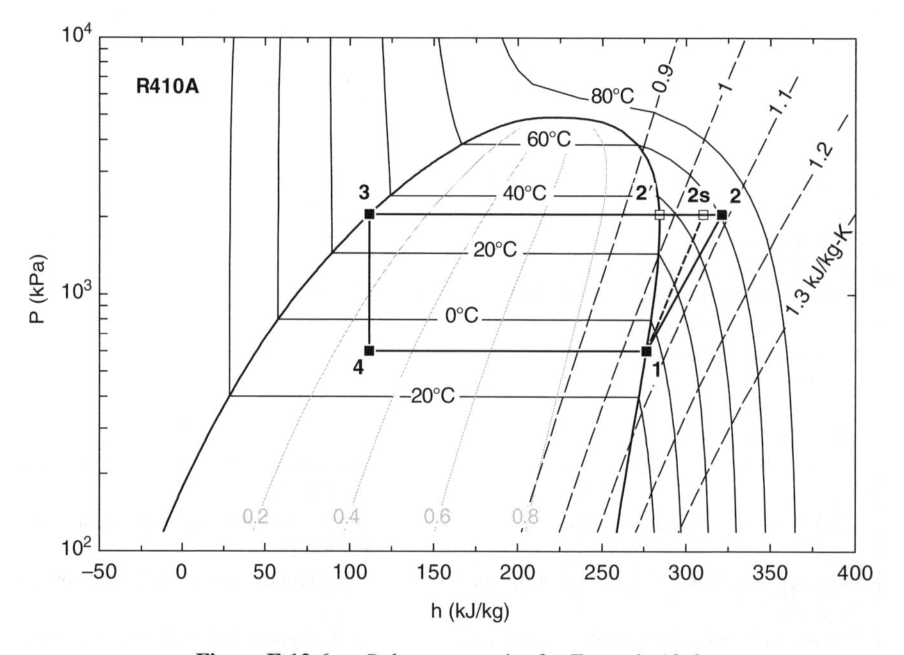

Figure E.12.6a *P–h* property plot for Example 12.6

(b) the heat rejected at the condenser: $q_{cond} = 24.8$ [kW]
(c) the heat extracted at the evaporator: $q_{evap} = 19.48$ [kW]
(d) the refrigerant temperature leaving the compressor: T2 = 61.24 [C]
(e) the condensing temperature of the refrigerant: T3 = 33.19 [C]
(f) the evaporating temperature of the refrigerant: T1 = −8.516 [C]

(g) the mass flow rate of the refrigerant: $mdot_{refrigerant} = 0.1183$ [kg/s]
(h) the compressor power: $W_{comp} = 5.312$ [kW]
(i) the cooling COP: $COP_c = 3.667$ [-]
(j) the sensible heat ratio of the process: $SHR = 0.7725$ [-]
(k) the temperature of the geothermal fluid exiting the heat pump: $T_{GHXin} = 31.95$ [C]
(l) the maximum superheat available: $q_{superheat} = 4.393$ [kW]
(m) a sketch of the air process on a psychrometric chart: see Figure E.12.6b

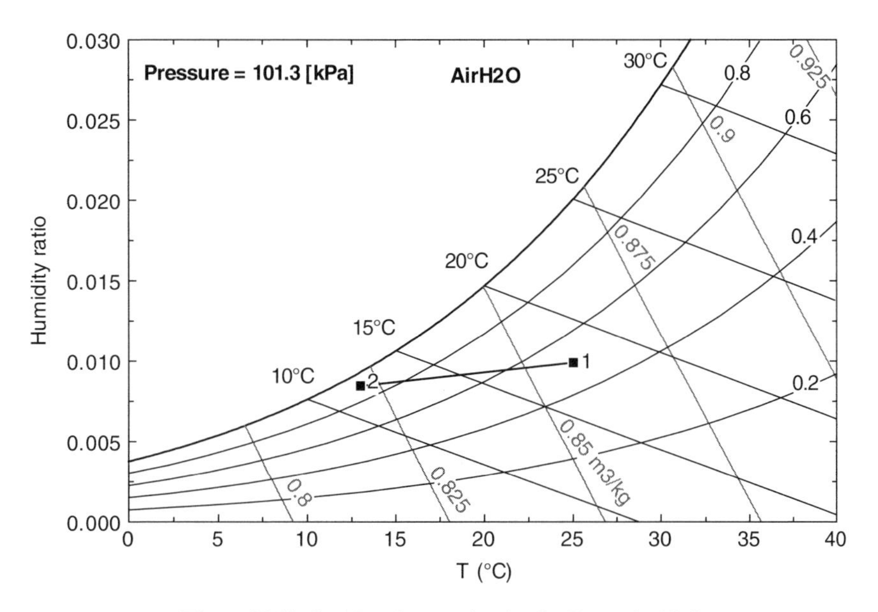

Figure E.12.6b Psychrometric plot for Example 12.6

Example 12.7 Effect of External Fluid Temperatures on the COP of a Geothermal Heat Pump in Cooling Mode

With the model constructed in Example 12.6, develop an *x–y* plot of the heat pump COP as a function of the entering geothermal fluid temperature.

Solution

Here, let's use the parametric feature in EES to vary the entering geothermal fluid temperature from 15 to 35 °C.

Discussion: A review of the plot shown in Figure E.12.7 shows, as one might expect, that the heat pump COP is quite sensitive to the geothermal fluid temperature. For example, if the entering geothermal fluid temperature in the previous example were to drop from 27 to 22 °C, then the heat pump COP would increase from 3.7 to approximately 4.5.

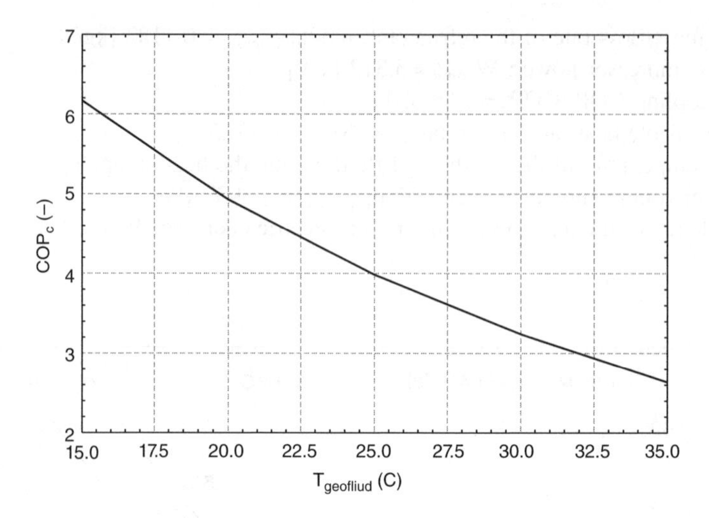

Figure E.12.7 Plot of heat pump COP as a function of heat pump entering air and geofluid temperature for Example 12.7

12.6 Transcritical Cycles

The transcritical cycle exhibits further departures from the Carnot cycle in that there is no distinct high-temperature reservoir; heat transfer occurs over a wide range of temperatures above the critical point. Thus, the high pressure level of the cycle is supercritical, and the low pressure level of the cycle is subcritical.

The use of transcritical cycles in heat pump applications is currently synonymous with CO_2 heat pumps. As mentioned previously, the quest for low-GWP refrigerants has sparked renewed interest in CO_2 (or R744) as a refrigerant, as it is considered a natural refrigerant, and its GWP is unity. CO_2 has a convenient critical temperature (30.98 °C), which sits in the middle of the range of temperatures that are frequently found in HVAC applications.

The critical pressure of CO_2 is relatively high (7.38 MPa), giving rise to its abandonment when CFCs entered the refrigerant scene in the 1920s–1930s. However, it is this relatively high pressure that is now seen as an advantage for CO_2 use because the associated high fluid density allows for application in small areas, such as automotive and marine applications. Heat pump components must be able to withstand these high operating pressures.

CO_2 as a refrigerant has great potential in geothermal heat pump applications, with both the water-source and DX types. CO_2 Earth heat pipes have also been employed in soil freezing applications, such as with the Alaska Oil Pipeline, and with railroad bed stabilization in permafrost regions. Recent trends are the use of CO_2 in supermarket applications where both high and low temperature needs exist (e.g., Hinde *et al.*, 2009).

Thermodynamically, there is not too much difference between the transcritical and subcritical mechanical vapor compression cycles. Figure 12.23 shows our engineering model of a transcritical cycle where two main differences are noted in comparison with the subcritical cycle: (1) the hot gas is not condensed and remains supercritical, and (2) an internal heat exchanger is employed to preheat the gas on its way to the compressor, thereby cooling the gas prior to expansion. Analogous to the desuperheater and condenser in the subcritical cycle, transcritical cycles may also have multiple gas coolers that reject heat to multiple loads such as hot-water

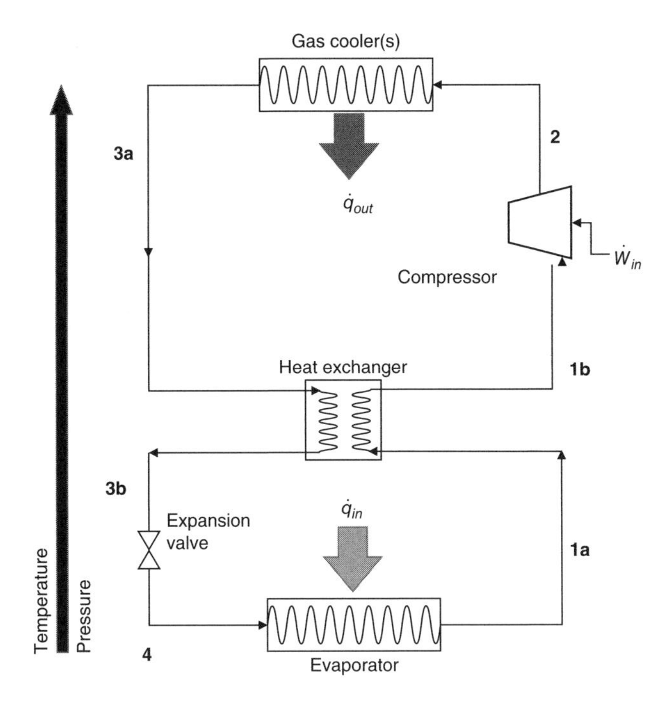

Figure 12.23 Conceptual model of refrigerant cycle components in a transcritical vapor compression heat pump

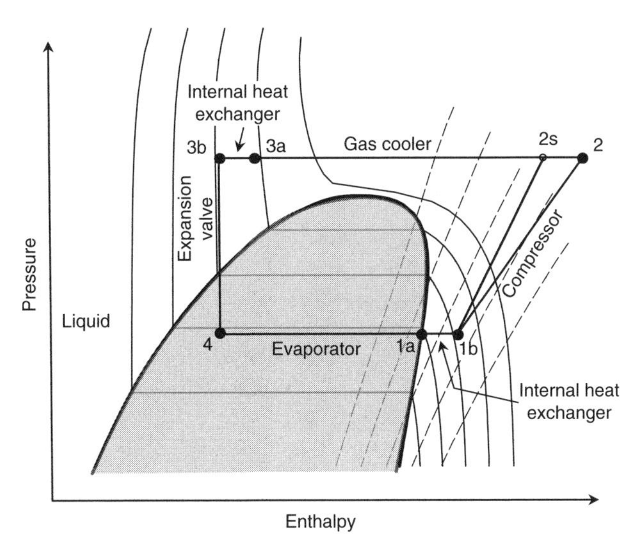

Figure 12.24 Generalized pressure–enthalpy (*P–h*) diagram showing a transcritical vapor compression cycle

heating and space heating. A generic transcritical cycle is shown on a general *P–h* diagram in Figure 12.24.

Let's use an example to illustrate the operation of a transcritical heat pump cycle with a simple comparison with a subcritical heat pump cycle.

Example 12.8 Basic Analysis of a Transcritical Heat Pump Cycle on a *P–h* Diagram

Reconsider the heat pump heating a residence in Example 12.1. Recall that that heat pump was operating with R410a at a condenser pressure of 3000 kPa and an evaporator pressure of 700 kPa, and we assumed an isentropic efficiency of the compressor of 80%. The refrigerant mass flow rate was $0.1\ kg \cdot s^{-1}$.

Now, to attempt a comparison with a heat pump using R744, let's use the same evaporator temperature we calculated in Example 12.1 ($T_1 = -4.048\ °C$) and use the same compression ratio (i.e., P2/P1 = 3000 kPa/700 kPa). Let's also assume the same isentropic efficiency of the compressor and refrigerant mass flow rate. Determine/show the following:

(a) a sketch of the cycle on a *P–h* diagram and show the ideal (isentropic case),
(b) the heat rejected at the gas cooler(s),
(c) the heat extracted at the evaporator,
(d) the refrigerant temperature leaving the compressor,
(e) the temperature of the refrigerant entering the expansion valve,
(f) the evaporating temperature of the refrigerant,
(g) the compressor work, and
(h) the heating COP.

Solution

The EES input data are as follows:

```
"GIVEN:"

R$ = 'carbondioxide'; $REFERENCE carbondioxide ash "String for refrigerant, with ASHRAE
reference state"
T1 = -4.05 [C]
P2 = P1 * 3000/700                          "the pressure ratio given in Example 12-1"
n_compressor = 0.80
mdot_refrigerant = 0.1 [kg/s]

"Simplifying Assumptions:"
x1a = 1 [-]; x4 = 0.3 [-]
epsilon_HX = 0.75 [-]                       "the heat exchanger effectiveness"

"SOLUTION:"
"Point 1a & 1b"
P1 = P_sat(R$,T = T1)                       "saturation pressure given the evaporating
                                            temperature"

h1a = Enthalpy(R$, P = P1, x = x1a)
T1a = temperature (R$, P = P1, x = x1a)
s1b = entropy(R$, P = P1, h = h1b)

"Point 2, and 2s "
s2s = s1b
h2s = enthalpy(R$, P = P2, s = s2s)
```

```
n_compressor = (h2s - h1b) / (h2 - h1b )
T2 = Temperature (R$, P = P2, h = h2)
"Point 3"
P3 = P2
h3b = h4
T3a = Temperature (R$, P = P3, h = h3a)
T3b = Temperature (R$, P = P3, h = h3b)
"Point 4"
P4 = P1
h4 = Enthalpy(R$, P = P4, x = x4)
"Energy Balances per Unit Mass:"
COP_h = (h2 - h3a) / (h2 - h1b)
q_extracted = mdot_refrigerant * (h1a - h4)      "evaporator heat transfer"
q_rejected = mdot_refrigerant * (h2 - h3a)       "heat rejected through gas cooler(s)"
W_comp = mdot_refrigerant * (h2 - h1b)           "compressor work"
q_HX = mdot_refrigerant * (h3a - h3b)            "heat exchanged (lost) by the hot gas"
q_HX = mdot_refrigerant * (h1b - h1a)            "heat exchanged (gained) by the cold gas"
"heat exchanger effectiveness"
q_HX = epsilon_HX * mdot_refrigerant             "HX effectiveness equation"
* cp_CO2 * (T3a - T1a)
T1b = Temperature(R$, P = P1, h = h1b)
cp_CO2 = cp(R$, T = (T1a + T1b)/2, P = P1)   "specific heat of CO2 at the average of T1 & T2"
```

The following results are obtained:

(a) a sketch of the cycle on a *P–h* diagram, showing the ideal (isentropic case): see Figure E.12.8

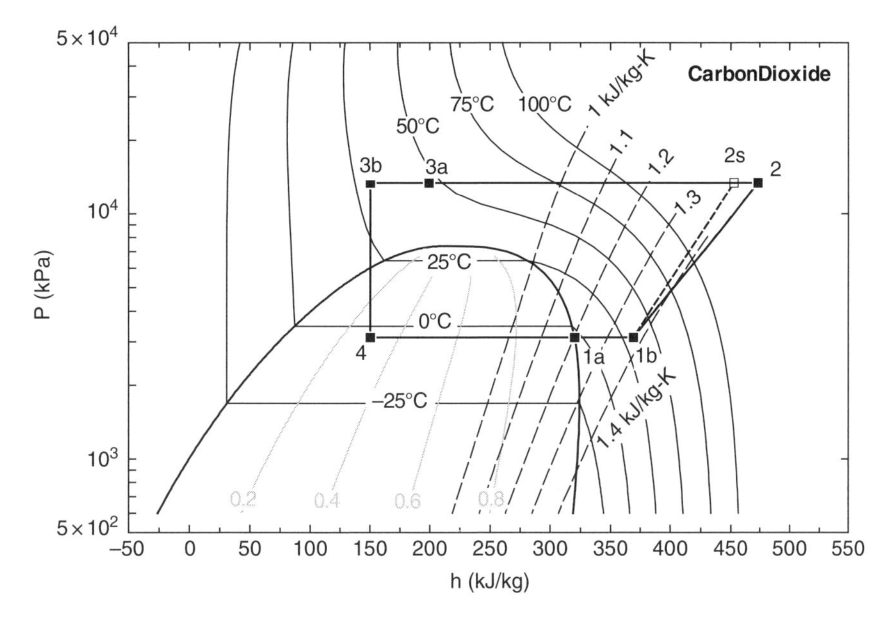

Figure E.12.8 *P–h* plot for Example 12.8

(b) the heat rejected at the gas cooler(s): $q_{rejected} = 27.41$ [kW]
(c) the heat extracted at the evaporator: $q_{extracted} = 16.99$ [kW]
(d) the refrigerant temperature leaving the compressor: T2 = 173.5 [C]
(e) the temperature of the refrigerant entering the expansion valve: T3b = 30.17 [C]
(f) the evaporating temperature of the refrigerant: T1a = -4.05 [C]
(g) the compressor work: $W_{comp} = 10.43$ [kW]
(h) the heating COP: $COP_h = 2.629$

Discussion: This example demonstrates some of the significant differences between a CO_2 transcritical cycle and a subcritical R410 cycle at the same evaporator temperature and gas compression ratio. Let's compare the results as follows:

Item	Transcritical CO_2 Cycle	Subcritical R410a Cycle
Useful heat rejected	27.41 kW	18.68 kW
Het extracted	16.99 kW	13.64 kW
Compressor leaving temp.	173.5 °C	79.33 °C
Compressor work	10.43 kW	5.04 kW
Heating COP	2.69	3.71

Perhaps the most notable difference is the much greater temperature exiting the compressor in the transcritical cycle (i.e., 173.5 °C compared with 79.33 °C). Thus, the transcritical cycle has a tremendous advantage over the subcritical cycle in meeting higher-temperature heating loads (and lower-temperature cooling loads as well!). The transcritical cycle also rejects about 47% more useful heat than the subcritical R410a cycle. The additional heating benefits of the transcritical cycle mean that the compressor work in the transcritical cycle is more than double that of the subcritical cycle. This greater work input results in a lower COP with the transcritical cycle, about 27% lower than the R410a cycle.

12.7 Vapor Compression Heat Pump Performance Standards and Manufacturer's Catalog Data

12.7.1 International Standards

The basic purpose of heat pump standards, aside from consumer protection, is to establish performance testing and rating criteria for factory-made units. Thus, comparison of one unit with another is possible. Performance standards are also used in government policy and energy codes. The main metric of comparison is the heat pump COP. In the United States, the cooling COP is reported as an energy efficiency ratio (EER), which is in units of Btu/h cooling output per W of electrical input (Btu/h/W).

Unfortunately, the ratings used for different types of equipment such as furnaces, air-source heat pumps, and geothermal heat pumps are not directly comparable because different methods are used. Air-source heat pump and furnace ratings use seasonal efficiency values, whereas geothermal heat pumps are rated under a specific set of laboratory conditions.

Air-source heat pumps are rated according to ANSI/AHRI Standard 210/240 and ISO Standard 13253. In the United States, the cooling performance metric is the seasonal energy efficiency ratio (SEER), calculated by dividing the seasonal cooling energy (in Btu) by the power input (in W), thus being reported in units of Btu/W. The heating performance metric is the heating seasonal performance (HSPF), and is similarly calculated. A number of assumptions are made regarding operation of the heat pump; the rating is based on a moderate climate and as a result is not representative of either very cold or very warm climates. The US Department of Energy sets out minimum SEER and HSPF standards to which equipment must be manufactured.

The annual fuel utilization efficiency (AFUE) is the most widely used measure of the heating efficiency of a boiler or furnace. It measures the ratio of the amount of heat delivered to the heating load to the amount of fuel supplied. Thus, a furnace that has an 80% AFUE rating converts 80% of the fuel supplied into useful heat. One drawback of the AFUE metric is that the electricity required to operate the furnace fan (and the combustion air fan if applicable) is not included in the rating. Consequently, the AFUE gives a somewhat less than complete picture of annual energy requirements of the equipment.

Water-source and brine-source heat pumps, which include the geothermal types, are rated under ISO Standard 13256. Part 1 (ISO 13256-1) deals with water-to-air and brine-to-air heat pumps, and Part 2 (ISO 13256-2) deals with water-to-water and brine-to-water heat pumps. The standard contains three subcategories, depending on the nature of the water source: water loop, groundwater, and ground loop. The main metric of performance is the heat pump COP, except in the United States, where the cooling COP is reported as an EER.

The 'water loop' subcategory assumes that the heat pump source/sink is a boiler-chiller-type loop, and thus ratings are standardized at an entering water temperature of 20 °C in heating and 30 °C in cooling. The 'groundwater' subcategory assumes that the heat pump source/sink is groundwater, and thus ratings are standardized at an entering water temperature of 10 °C in heating and 15 °C in cooling. The 'ground loop' subcategory assumes that the heat pump source/sink is a closed-loop ground heat exchanger where extended-range temperatures can be expected, and thus ratings are standardized at an entering fluid temperature of 0 °C in heating and 25 °C in cooling.

In the water-to-air and brine-to-air heat standard (ISO 13256-1), heat pumps are rated with entering air conditions fixed at 27 °C dry bulb and 19 °C wet bulb for cooling, and 20 °C dry bulb for heating. In the water-to-water and brine-to-water heat standard (ISO 13256-2), heat pumps are rated with entering load water temperatures fixed at 12 °C for cooling and 40 °C for heating.

Under the ISO 13256 performance standard, there is also a correction for pump and fan power consumption, which affects the overall COP/EER. Flow rates are set to result in a fluid temperature change of 10 °F or 5.55 °C, and measured pumping power is added to the overall power consumption, thus lowering the COP/EER. Blower power is corrected to zero external static pressure, and the nominal airflow is rated at a specific external static pressure. This effectively reduces the power consumption of the unit and increases cooling capacity, but decreases heating capacity.

DX geothermal heat pumps are rated under CSA Standard C748-13. Entering air conditions are similar to those described above. Entering refrigerant temperatures are standardized, and testing is performed with the DX refrigerant coil placed in a water bath.

12.7.2 Data from Manufacturer's Catalogs

Dozens of geothermal heat pump manufacturers exist worldwide. A good list of manufacturers, at least in the United States, can be found in the AHRI Directory: www.ahridirectory.org/ahridirectory

Manufacturers in the AHRI Directory exist under the following residential categories:

- Geothermal – Direct Geoexchange Heat Pumps
- Geothermal – Water-to-Air Heat Pumps
- Geothermal – Water-to-Water Heat Pumps

and under the following commercial categories:

- Geothermal – Water-to-Air Heat Pumps
- Geothermal – Water-to-Water Heat Pumps

Many geothermal heat pump manufacturers supply performance data on their websites. To achieve optimal performance of a heat pump, proper selection procedures are necessary. Some geothermal heat pump manufacturers provide computer software selection programs that can select the best heat pump given the loads and other input data.

The process of sizing and selecting a heat pump is different from that of sizing an air-conditioner or furnace. The size of conventional air-conditioners and furnaces is typically selected so that the equipment can deliver the required cooling or heating at the most extreme conditions expected in a given climate. However, heat pumps supply both heating and cooling, and peak heating and cooling loads are rarely identical. Thus, a heat pump could be sized to meet the peak cooling load or the peak heating load, but rarely can it be sized to meet both peak loads.

In most cases, the size of the heat pump is determined to meet the peak cooling load. This eliminates the problem of oversizing the unit and creating humidity control problems due to excessive on-off cycling, or undersizing the unit and not being able to deliver enough cooling. Sizing so that the actual sensible capacity of the equipment will satisfy the sensible capacity of the zone is typically recommended by manufacturers for best comfort and equipment life. If sizing for the peak cooling load results in extra heating capacity, there is no significant heating performance penalty. If sizing for the peak cooling load results in insufficient heating capacity, the additional heat is typically supplied by electric resistance heating elements. Electric resistance heating is a factory-supplied option with geothermal heat pumps.

In cold climates, a prudent design choice involves the use of water-to-water heat pumps serving radiant panels (floors or walls) or fan coil units. Such units are typically connected to a buffer tank, where the tank temperature is kept at a desired set-point. The tank temperature can be reset based on outdoor air temperature.

The following are general manual procedures for selecting a heat pump from manufacturer's catalog data:

1. Obtain peak heating and cooling loads (total and sensible), as calculated with industry-accepted procedures as described in Chapter 2.
2. Obtain the following design parameters:
   (a) entering water temperature and flow rates, and
   (b) entering air conditions and flow rates (if applicable).
3. For water-to-air heat pump applications, preliminarily select a unit that is closest to, but not larger than, the total and sensible cooling conditions. For water-to-water heat pump applications, preliminarily select a unit closest to the desired load (either heating or cooling).
4. Manufacturers organize catalog data differently.
   (a) *Water–Air Heat Pumps*. Enter the catalog tables at the design entering water temperature. Interpolation is acceptable. Next, for various entering water temperatures, manufacturers typically either fix the water flow rate and vary the entering air conditions, or fix the entering air conditions and vary the water flow rate. If the water flow rate is fixed, note the total and sensible cooling capacities for the design entering air temperatures. If the entering air temperatures are fixed, note the total and sensible cooling capacities for the design water flow rate.
   (b) *Water–Water Heat Pumps*. Enter the catalog tables at the *source* design entering water temperature and flow rate. Locate the correct *load* entering water temperature and flow rate, and note the heating or cooling capacity.
5. For water-to-air heat pump applications, note the heating capacity. If the heating capacity exceeds the design load, then this is acceptable. If not, supplemental heating may be required.
6. For water-to-air heat pump applications, determine the correction factors associated with the fixed factors in the tables (i.e., either entering air conditions or water flow rate). This will be the case if your design criteria have dry-bulb and wet-bulb or water flow rates different from those fixed in the tables. Correction factors are also needed if your design air flow differs from that fixed in the tables. The total cooling and sensible cooling noted in step 4a are then multiplied by these correction factors.
7. For both water-to-air and water-to-water applications, cooling and heating capacities need to be corrected for use of antifreeze.
8. Compare the corrected capacities to the design loads. If they are within tolerance (usually within 10%), then the equipment is acceptable. It is generally better to undersize than oversize, as undersizing improves humidity control, reduces sound levels, and extends the life of the equipment.
9. If the corrected capacities are not within tolerance limits of the design loads, then the next larger or smaller unit should be considered, and steps 1 to 8 repeated.

12.8 Chapter Summary

This chapter presented an overview of the vapor compression heat pump cycle by first comparing it with the Carnot heat pump cycle. Ideal and non-ideal vapor compression heat pump cycles were examined in some detail, and cycles were analyzed on pressure–enthalpy diagrams.

The mechanics of operation of geothermal heat pumps were discussed in some detail. Thermodynamic irreversibilities were considered, specifically a detailed heat exchanger analysis, allowing us to develop detailed computer simulation models of geothermal heat pumps in both heating and cooling modes. Transcritical cycles using CO_2 as a refrigerant were also examined. Finally, international heat pump standards were briefly discussed, along with a procedure for selecting a heat pump from manufacturer's catalog data.

Discussion Questions and Exercise Problems

12.1 Prove the following for a theoretical geothermal heat pump and show your work:

(a) In heating mode: the Ground Load = (COP − 1)/COP × (Building Heating Load).
(b) In cooling mode: the Ground Load = (COP + 1)/COP × (Building Cooling Load).
(c) For a ground-source heat pump operating with a COP of 4 in cooling mode, what is the thermal load to be rejected to the ground if the cooling load is 10 kW?
(d) For a ground-source heat pump operating with a COP of 3 in heating mode, what is the thermal load to be extracted from the ground if the heating load is 10 kW?

Hint: for (a) and (b), perform an energy balance on the heat pump.

12.2 Consider a heat pump that is heating a residence while operating at a condenser pressure of 3200 kPa and an evaporator pressure of 1000 kPa. The refrigerant is R410a. Assume that the isentropic efficiency of the compressor is 80%. For a refrigerant mass flow rate of $0.1\ kg \cdot s^{-1}$, determine/show the following:

(a) a sketch of the cycle on a *P–h* diagram and show the ideal (isentropic case),
(b) the heat rejected at the condenser,
(c) the heat extracted at the evaporator,
(d) the refrigerant temperature leaving the compressor,
(e) the condensing temperature of the refrigerant,
(f) the evaporating temperature of the refrigerant,
(g) the compressor power,
(h) the heating COP, and
(i) the maximum superheat available for use in a domestic hot water tank.

12.3 Repeat problem 12.2, but with (i) R134a and (ii) R22, and compare the results. To compare, use the same condensing and evaporating temperatures of each refrigerant and answer/show the following:

(a) a sketch of the cycle on a *P–h* diagram and show the ideal (isentropic case),
(b) the heat rejected at the condenser,
(c) the heat extracted at the evaporator,
(d) the refrigerant temperature leaving the compressor,
(e) the condensing pressure of the refrigerant,
(f) the evaporating pressure of the refrigerant,
(g) the compressor power,

(h) the heating COP, and
(i) the maximum superheat available for use in a domestic hot water tank.

12.4 Using the supplied EES file, redo Example 12.4 with refrigerant R134a and compare and discuss your results.

12.5 Using the supplied EES file, redo Example 12.6 with another refrigerant of your choice and compare and discuss your results.

12.6 Using the supplied EES file, redo Example 12.8 with the following natural refrigerants: (a) ammonia, (b) propane, and (c) isobutene.

12.7 The total heating and cooling loads of a residence have both been calculated to be 48 000 Btu/h. The sensible cooling load is 42 000 Btu/h. The indoor design temperatures are 75 °F for cooling and 70 °F for heating. The geothermal designer is designing a closed-loop system to return a maximum entering water temperature to the heat pump of 85 °F in cooling mode and 40 °F in heating mode.

Search the web for a water-to-air geothermal heat pump manufacturer's catalog dataset. You may start with the AHRI Directory to facilitate your search.

(a) Choose a suitable heat pump to meet the above loads and clearly state the model name, number, and manufacturer. List any assumptions you make.
   i. What is your heat pump's cooling capacity at the design condition?
   ii. What is your heat pump's SHR at the design condition?
(b) What is the 'ground load' at the design cooling condition?
(c) What is the power consumption at the design cooling condition?
(d)
   i. What is your heat pump's heating capacity at the design condition?
   ii. Is your heat pump's heating capacity adequate?
   iii. What size of a supplemental electric resistance heating element would you recommend if the heating capacity is inadequate?
(e) What is the 'ground load' at the design heating condition?
(f) What is the power consumption at the design heating condition?

12.8 A geothermal heat pump system is under consideration for a residence with a radiant floor heating system. The peak heating load is determined to be 15 kW. The geothermal consultant expects that a closed-loop ground heat exchanger is most plausible, which will supply a fluid temperature of 0 °C to the heat pump during peak heating. The radiant floor heating system is designed for a supply temperature of 40 °C.

Search the web for a water-to-water geothermal heat pump manufacturer's catalog dataset. You may start with the AHRI Directory to facilitate your search.

(a) Choose a suitable heat pump to meet the above load and clearly state the model name, number, and manufacturer. List any assumptions you make.
(b) What is your heat pump's heating capacity at the design condition?
(c) What is the 'ground load' at the design heating condition?
(d) What is the power consumption at the design heating condition?
(e) What is the heat pump COP at the design condition?

13

Thermally Driven Heat Pumps

13.1 Overview

Both vapor compression and thermally driven cycles accomplish the removal of heat from a substance through the evaporation of a refrigerant at a low pressure, and the rejection of heat to a substance through the condensation of the refrigerant at a higher pressure. The method of creating the pressure difference and circulating the refrigerant is the primary difference between the two cycles. The vapor compression cycle employs a mechanical compressor to create the pressure differences necessary to circulate the refrigerant. In a thermally driven cycle, two types are the most prevalent: (1) *absorption* cycles, where the refrigerant is transported through the cycle by being absorbed and dissolved into another fluid, and (2) *adsorption* cycles, where the refrigerant is transported through the cycle by being adsorbed onto a solid. Other thermally driven cycles are possible, but here we will refer to thermally driven heat pumps as those employing an absorption or adsorption cycle.

Learning objectives and goals:

1. Review the fundamentals of thermally driven heat pump operation.
2. Draw analogies between thermally driven and mechanical vapor compression refrigeration cycles.
3. Appreciate the role of geothermal energy exchanges in thermally driven heat pump applications.
4. Develop, analyze, and sketch absorption and adsorption heat pump thermodynamic cycles on $\log P$–$(-1/T)$ plots, and calculate energy flows in all processes of the cycle.

Geothermal Heat Pump and Heat Engine Systems: Theory and Practice, First Edition. Andrew D. Chiasson.

Companion website: www.wiley.com/go/chiasson/geoHPSTP

13.2 Cycle Basics

Thermally driven heat pumps exchange only thermal energy with their surroundings; no significant mechanical energy is exchanged, and no significant conversion of heat to work or work to heat occurs in the cycle. They are used in applications where one or more of the exchanges of heat with the surroundings is the useful product (e.g., chilled water or hot water). Thus, they have several advantages relative to mechanical vapor compression heat pumps: (i) they do not rely on mechanical work, (ii) they reduce summer peak electric demand, (iii) combined with solar energy as the heat source, cooling loads coincide with solar availability, (iv) refrigerants used have less global warming potential and are less ozone depleting than conventional refrigerants, and (iv) they utilize available heat sources such as solar energy, industrial waste heat, and geothermal energy.

All thermally driven heat pump cycles include at least three thermal energy exchanges with their surroundings (i.e., energy exchange at three different temperatures) (Figure 13.1). The highest- and lowest-temperature heat flows ($\dot{q}_h$ and $\dot{q}_c$ respectively) are in one direction, and the mid-temperature heat flow ($\dot{q}_{mid}$) is in the opposite direction. In the *forward cycle*, the extreme (hottest and coldest) heat flows are into the cycle, and the mid-temperature level heat is usually rejected to the environment. Such a cycle is also called the heat amplifier, heat pump, conventional cycle, or type I cycle. When the extreme-temperature heat flows are out of the cycle, it is called a *reverse cycle*, heat transformer, temperature amplifier, temperature booster, or type II cycle. Type I cycles are used for cooling and refrigeration, or mid-level heating; medium- and low-temperature-level heat can also be used simultaneously for heating and cooling purposes. Type II cycles are used for heating by supplying low-grade heat and amplifying it.

The efficiency of thermally driven heat pumps is given by the definition of COP. For cooling, the COP (COP_c) is given by

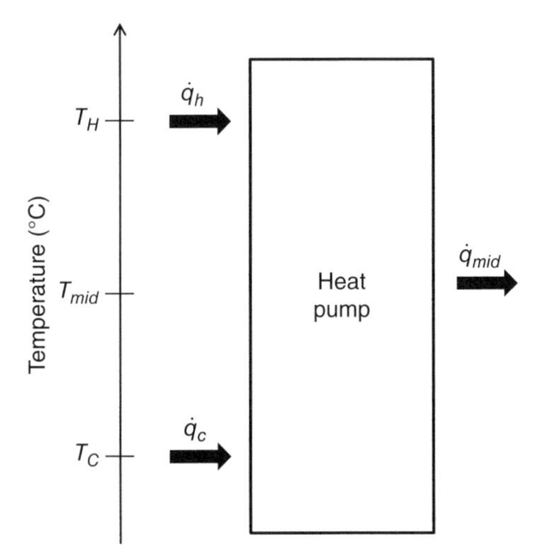

Figure 13.1 Schematic of energy exchanges in a thermally driven heat pump cycle

$$\mathrm{COP}_c = \frac{\dot{q}_c}{\dot{q}_h} \tag{13.1}$$

and for heating, the COP (COP_h) is given by

$$\mathrm{COP}_h = \frac{\dot{q}_{mid}}{\dot{q}_h} = 1 + \mathrm{COP}_c \tag{13.2}$$

13.3 Absorption Cycles

Herold *et al.* (1996) and Kühn (2013) present detailed discussions on, and mathematical analysis of, absorption heat pump cycles. Absorption refrigeration machines were developed and perfected between the 1920s and 1940s. Today, absorption cooling technology is considered to be a mature technology, but the major growth appears to be the Asia-Pacific region. Gas-fired absorption chillers have been used for economic reasons in areas with high electrical loads, and propane-fired refrigeration for use in the recreational and camping sectors has been a stable market for decades. Uses of waste-heat-driven and solar-driven trigeneration installations (i.e., combined heat and power plus cooling) are increasing worldwide. Particularly in Europe, so-called solar high-combi systems have been examined and employed for trigeneration. Further information on hi-combi systems can be found at http://www.highcombi.eu. Energy strategy policies of governments also play an increasing role in the growth of non-vapor compression cycles, but their popularity, at least in the United States, seems to coincide with high electricity rates.

13.3.1 Source-Sink Configurations and Refrigerant–Absorbent Pairs

Absorption machines are commercially available today in two basic refrigerant–absorbent pairs. For applications above 0 °C (primarily space cooling), the cycle uses lithium bromide as the absorbent and water as the refrigerant. For applications below 0 °C, an ammonia/water cycle is employed, with ammonia as the refrigerant and water as the absorbent. Use of other refrigerant–absorbent pairs is an area of ongoing research and development.

Prevalent machines are configured as packaged, reversible water-cooled chillers. In a potential geothermal application, any of the three temperature levels could be handled by a geothermal resource. Figure 13.2 shows possible configurations in cooling mode (Figure 13.2a) and heating mode (Figure 13.2b). For example, H_2O–LiBr absorption machines operate with T_h of the order of 70–95 °C, which could easily be supplied by a moderate-temperature geothermal resource. The low- and mid-temperature levels of H_2O–LiBr absorption machines typically operate at temperatures of the order of 10 and 30 °C respectively, which could easily be handled by a closed- or open-loop ground heat exchange system.

The main advantage of the H_2O–LiBr pair is that it is known to have the highest thermal efficiency of the thermally driven heat pump cycles. The machines are economical and relatively compact. Advantages of using water as a refrigerant are obvious: water has a high latent heat, and it is chemically stable and non-toxic. Aqueous LiBr solution has a low vapor pressure

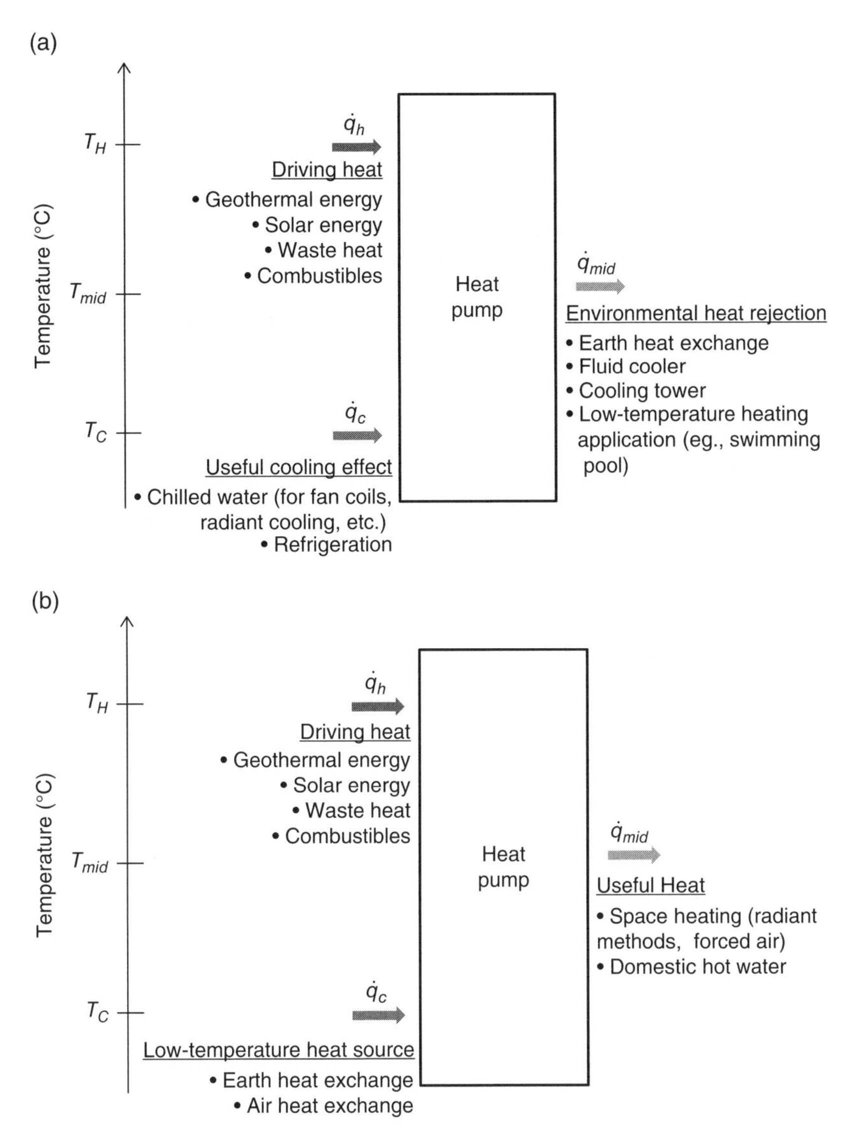

Figure 13.2 Schematic of possible source/sink energy exchanges in an absorption heat pump cycle in (a) cooling mode and (b) heating mode

and a low viscosity, and it is also non-toxic. The disadvantages of using water as a refrigerant are its vapor pressure and its freeze point; low vapor pressures necessitate a vacuum-tight construction of the heat exchange chambers, and cooling applications are limited to 0 °C. Another disadvantage of the H_2O–LiBr pair is that undesirable LiBr crystallization at high concentration may occur, thus limiting the potential temperature difference between the low and medium temperature levels.

The advantages and disadvantages of the ammonia–water pair are essentially opposite to those of the H_2O–LiBr pair. Perhaps the main advantage of the ammonia–water pair is that its use permits the generation of very low refrigeration temperatures down to −40 °C. Further, there are no concerns of crystallization with the ammonia–water pair, thus permitting use of very high temperatures. The main disadvantage of the ammonia–water pair is that ammonia is toxic and flammable, and it is explosive in air under certain conditions. The vapor pressure is high, similar to that of vapor compression heat pumps, and thus the use of rated pressure vessels is needed in construction of ammonia–water absorption machines. The higher operating vapor pressures also translate to a higher energy requirement for the solution pump. Because of the relative vapor pressures of ammonia and water, the two fluids don't fully separate at high temperature conditions, thus typically requiring the need for rectification (i.e., removal of traces of water from the ammonia refrigerant prior to entering the condenser). Such rectification complicates the cycle relative to the H_2O–LiBr pair, and also detracts from cycle efficiency.

13.3.2 Mechanics of Operation

Let us now examine the operating principles of a simple absorption cycle by following the refrigerant and absorbent flow. Analogous to the mechanical vapor compression cycle, we will base our discussion around the processes that occur in the four principal components of these cycles: (i) 'thermal compressor', (ii) condenser, (iii) expansion valve, and (iv) evaporator. The discussion that follows refers to Figure 13.3.

Processes 2–3, *3–4*, and *4–1* on the left-hand side of the schematic shown in Figure 13.3 include components familiar from our discussion of the vapor compression system, so we will not repeat that discussion here. These components are the evaporator, condenser, and expansion valve, and only refrigerant flows through these components.

Process 1–2 represents the *thermal compressor*, so named as an analog to the mechanical compressor in vapor compression heat pump cycles. The right-hand side of the schematic shown in Figure 13.3 includes components that replace the mechanical compressor of the vapor compression refrigeration system: absorber, pump, and generator (or desorber). These components involve interactions of refrigerant–absorbent liquid solutions. A principal advantage of the absorption system is that, for comparable useful thermal duty, the work input to the solution pump is almost negligible relative to that required for the compressor of a vapor compression system.

In the absorber, refrigerant vapor leaving the evaporator at state 1 is absorbed by the weak refrigerant–absorbent solution leaving the generator (point C). The absorption process is exothermic, and the amount of refrigerant that can be dissolved in the absorbent decreases with increasing fluid temperature. Thus, a cooling fluid is circulated through the absorber in order to maintain the absorber temperature as low as practical. The solution leaves the absorber as a *strong solution* (with respect to the refrigerant) at point a, where it is pumped up to the pressure of the generator at point b.

Not shown in our simple diagram in Figure 13.3 is a common modification that adds an internal heat exchanger between the absorber and generator. The function of this heat exchanger is to preheat the strong solution entering the generator with the weak solution leaving the generator, thereby reducing the heat requirement in the generator.

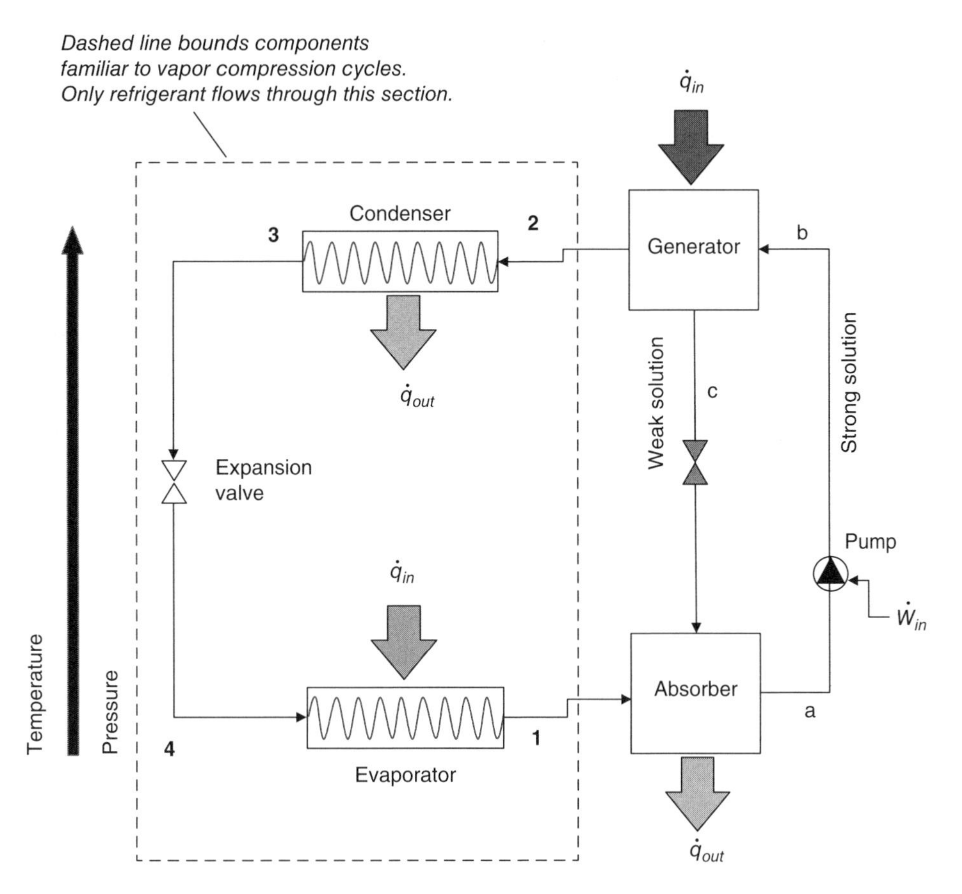

Figure 13.3 Basic operation and conceptual model of principal components in an absorption-cycle heat pump

In the generator (or desorber), heat transfer from a high-temperature heat source boils the refrigerant, separating it from the absorbent. This process is endothermic. Remaining in the generator is a weak solution (with respect to the refrigerant), which leaves at point c. Pure refrigerant vapor leaves at point 2, making its way to the condenser. Not shown in our simple diagram in Figure 13.3 is another common modification, specifically in ammonia–water absorption cycles, which adds a rectifier between the generator and condenser. As mentioned previously, the function of the rectifier is to remove traces of water from the ammonia refrigerant, mainly to avoid the formation of ice in the downstream components.

The schematic in Figure 13.3 shows four heat transfers, but there are typically only three fluid connections. The high-temperature and low-temperature fluid loops are obviously separate, but there is typically only one heat rejection loop connected to both the absorber and condenser. Such a loop can be plumbed in parallel or in series. When plumbed in series, the fluid enters the absorber first, as it is important to maintain low temperatures in the refrigerant absorption process.

The foregoing has addressed a so-called single-effect cycle. To increase overall cycle efficiency, double- or multi-effect cycles have been developed. The term 'effect' is used mainly in cooling applications to describe how often high-temperature heat is supplied to achieve a regeneration effect and thus a useful cooling effect. While multi-effect cycles offer higher COPs, they require much higher inlet heating medium temperatures to achieve the multiple cooling effects, and therefore the potential use of renewable energy sources is limited.

13.3.3 Thermodynamic Considerations

Our conceptual engineering model of an absorption cycle consists of the principal control volumes as shown schematically in Figure 13.3. All energy transfers by work and heat are taken as positive in the directions of the arrows on the schematic, and energy balances are written accordingly. Energy balances on the condenser, evaporator, and expansion valve are similar to those discussed in Section 12.2, and thus won't be repeated here. In the thermal compressor of absorption cycles, both mass and energy balances must be considered.

A mass balance on the solution in the thermal compressor is

$$\dot{m}_{strong} = \dot{m}_{weak} + \dot{m}_R \tag{13.3}$$

where the subscripts *strong*, *weak*, and R refer to the strong solution, weak solution, and refrigerant respectively. Another characteristic relationship of the fluid is given by the circulation ratio:

$$\frac{\dot{m}_{weak}}{\dot{m}_R} = \frac{x_{strong}}{\Delta x} \tag{13.4}$$

where x is the mass fraction of the refrigerant, and Δx is the concentration difference ($x_{strong} - x_{weak}$).

An energy balance on the absorber is given by

$$\dot{m}_{weak} \times h_c + \dot{m}_R \times h_1 = \dot{q}_{absorber} + \dot{m}_{strong} \times h_a \tag{13.5}$$

where $\dot{q}_{absorber}$ is the heat rejected to the cooling fluid and h is enthalpy, with subscripts referring to state points shown in Figure 13.3. An energy balance on the generator is given by

$$\dot{q}_{generator} + \dot{m}_{strong} \times h_a + \dot{W}_{pump} = \dot{m}_{weak} \times h_c + \dot{m}_R \times h_2 \tag{13.6}$$

where $\dot{q}_{generator}$ is the heat supplied by the driving heat source, $\dot{W}_{pump}$ is the pump energy added, and h is enthalpy, with subscripts referring to state points shown in Figure 13.3. Finally, an energy balance on the thermodynamic cycle is given by

$$\dot{q}_{generator} + \dot{q}_{evaporator} + \dot{W}_{pump} = \dot{q}_{absorber} + \dot{q}_{condenser} \tag{13.7}$$

Recall that graphical representation of mechanical vapor compression cycles on P–h diagrams was very useful in performing cycle analysis. This is not the case for absorption cycles

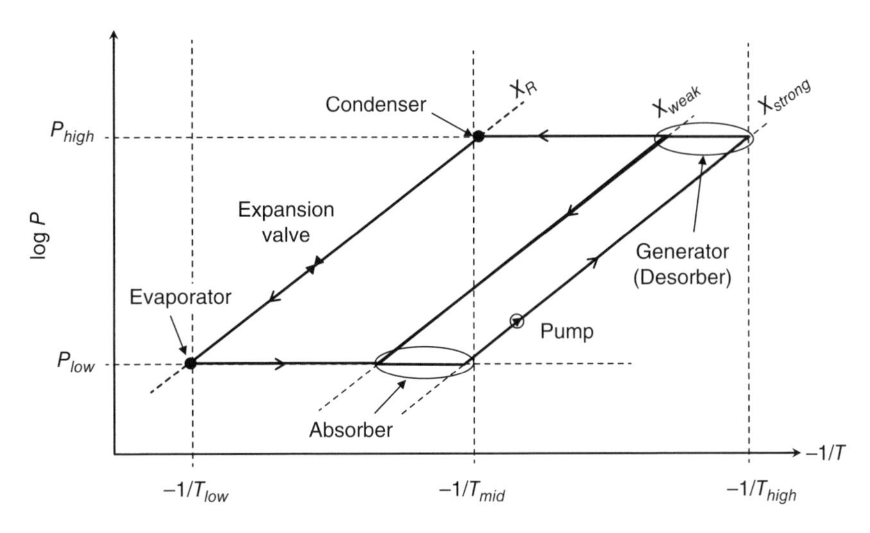

Figure 13.4 Generic log$P$–($-1/T$) plot showing a single-effect absorption heat pump cycle

because the working fluid is a mixture. Thus, absorption cycles are typically represented on a specific type of P–T–x plot, where pressure is plotted on a logarithmic scale versus $-1/T$. These are commonly referred to as *logP–(−1/T)* plots. As shown in Figure 13.4, such a plot results in saturation curves for various refrigerant mass fractions plotting as near-straight lines. On these diagrams, the high and low pressure levels and the three (low, medium, and high) temperature levels are evident. The refrigerant mass fraction in each of the processes is also easily distinguished.

Figure 13.5 shows actual log$P$–($-1/T$) plots for the most common refrigerant–absorbent pairs (i.e., water–lithium bromide and ammonia–water). The aforementioned attributes of each pair are evident for the hypothetical cycles shown on the plots. That is, near-vacuum pressures (of the order of 10 mbar) are needed in the evaporator of H_2O–LiBr absorption cycles in order to produce reasonable temperatures for chilled water applications (i.e., ~7 °C). Operating temperatures in the condenser, absorber, and generator are easily seen on the chart. The higher operating pressures in NH_3–H_2O absorption cycles are evident in Figure 13.5b. At an evaporator pressure of 1 bar, temperatures of the order of −30 °C can be achieved. The potentially higher generator temperatures in NH_3–H_2O cycles are also easily seen in Figure 13.5b.

Example 13.1 Basic Analysis of an H_2O–LiBr Absorption Heat Pump Cycle on a log$P$–($-1/T$) Plot

Consider an H_2O–LiBr absorption chiller providing chilled water to an air handing unit that is cooling a small office building. The chiller is operating at a high pressure condition of 80 mbar and low pressure condition of 8 mbar. The LiBr mass fraction on the strong and weak solution is 60 and 54% respectively, and the refrigerant is pure water (i.e., the LiBr mass fraction is zero).

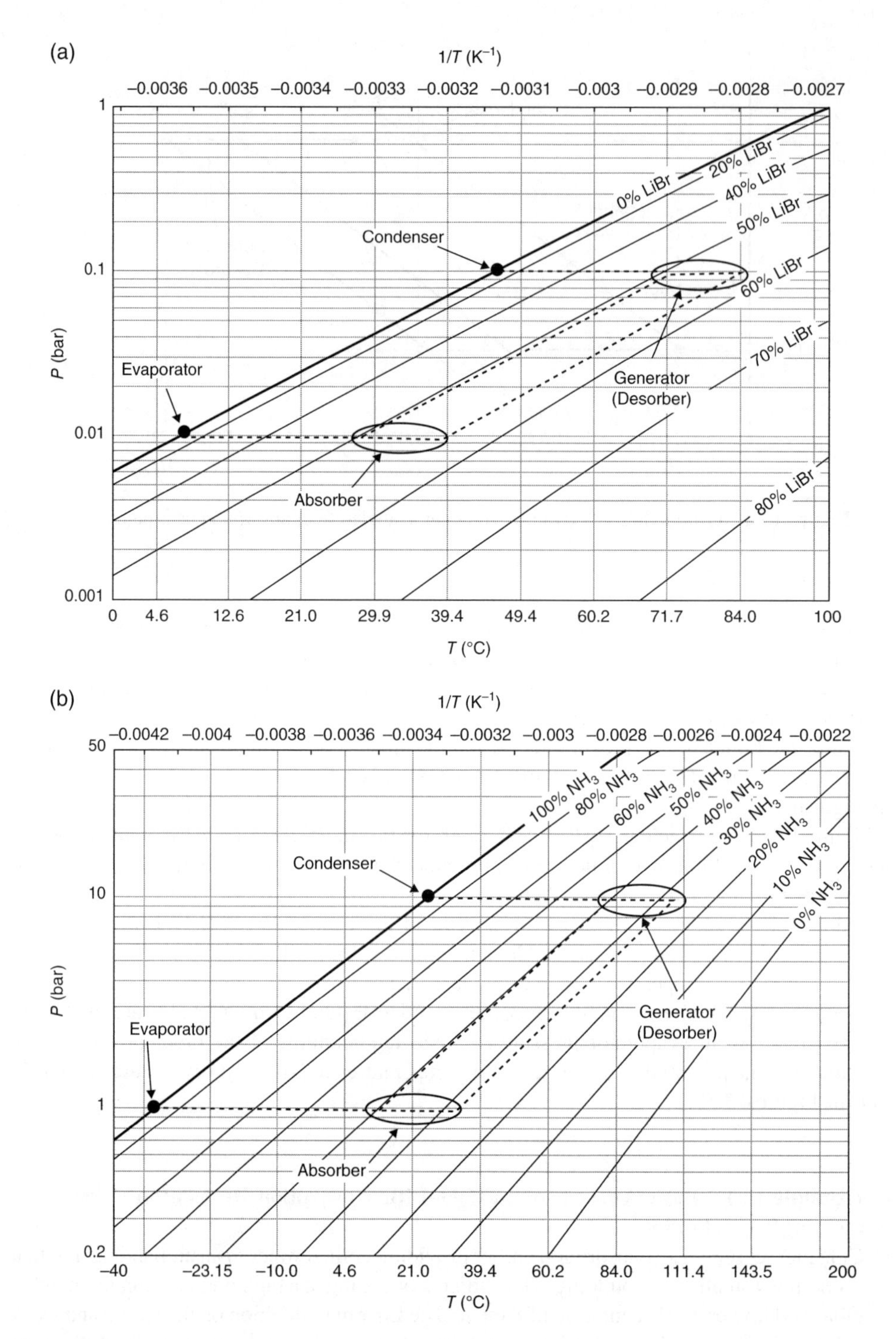

Figure 13.5 log$P$–(−1/T) plot showing a single-effect absorption heat pump cycle for (a) water–lithium bromide and (b) ammonia–water refrigerant–absorbent pairs

For a refrigerant (water) mass flow rate of $0.0075\ kg \cdot s^{-1}$, determine/show the following:

(a) a sketch of the cycle on a logP–(−1/*T*) plot,
(b) the heat rejected at the condenser,
(c) the heat rejected at the absorber,
(d) the heat input at the evaporator,
(e) the heat input at the generator,
(f) the condensing temperature of the refrigerant,
(g) the evaporating temperature of the refrigerant,
(h) the temperature in the absorber,
(i) the temperature in the generator,
(j) the solution pump power (assuming a pump efficiency of 0.6), and
(k) the cooling COP.

Solution

We will once again use EES to solve this problem because we need to perform thermodynamic calculations for the H_2O–LiBr mixture. These thermodynamic functions are available in EES through the EES Library Routines. The EES input data are as follows:

```
"GIVEN:"
P1 = 0.008 [bar]; P2 = 0.08 [bar]                 "The evaporator and condenser pressure"
x_s = 0.60                                        "LiBr mass fraction of the strong solution"
x_w = 0.54                                        "LiBr mass fraction of the weak solution"
m_dot_H2O = 0.0075 [kg/s]                         "mass flow rate of the refrigerant (water)"
"SOLUTION:"
"Determine the Mass Flow Rates"
D_x = x_s - x_w                                   "the concentration difference"
m_dot_weak/ m_dot_H2O = x_s / D_x                 "the fluid circulation ratio"
m_dot_strong = m_dot_weak + m_dot_H2O             "mass balance"
"Determine the Absorber Temperature"
T1a = T_LiBrH2O(P1,x_w); Ta = T_LiBrH2O(P1,x_s)
T_absorber = (T1a + Ta) / 2
"Determine the Generator Temperature"
Tb = T_LiBrH2O(P2,x_s); Tc = T_LiBrH2O(P2,x_w)
T_generator = (Tb + Tc) / 2
"Water (refrigerant) Cycle:"
"Entering Condenser"
T2 = T_generator
h2 = enthalpy('water', P = P2, T = T2)
"Leaving Condenser"
T3 = temperature('water', P = P2, x = 0)          "No sub-cooling is assumed, so x = 0"
h3 = enthalpy('water', P = P2, x = 0)
"Entering Evaporator"
T4 = temperature('water', P = P1, x = 0)
```

```
h4 = h3                                                    "throttling process"
"Leaving Evaporator"
T1 = Temperature('water', P = P1, x = 1)                   "No super-heating is assumed,
                                                           so x = 1"
h1 = enthalpy('water', P = P1, x = 1)
"LiBr (absorbent) Cycle:"
ha = h_LiBrH2O(T_absorber,x_s)                             "Leaving Absorber (Point a)"
hc = h_LiBrH2O(T_generator,x_w)                            "Leaving Generator (Point c)"
"Energy Balances:"
q_evaporator = m_dot_H2O * (h1 - h4)
q_condenser = m_dot_H2O * (h2 - h3)
n_pump = 0.6 [-]                                           "assumed pump efficiency"
W_pump = m_dot_strong * (P2-P1) * convert(bar,kPa)/(rho_-  "Pump work"
LiBrH2O(Ta,x_s) * n_pump)
q_generator + m_dot_strong * ha + W_pump = m_dot_weak * hc + m_dot_H2O * h2 "energy
balance on generator"
m_dot_weak * hc + m_dot_H2O * h1 = q_absorber + m_dot_strong * ha "energy balance on
absorber"
COP_c = q_evaporator / q_generator
Cycle_Balance = q_evaporator + q_generator + W_pump - (q_absorber + q_condenser )
```

The following results are obtained:

(a) a sketch of the cycle on a logP–(−1/T) plot: see Figure E.13.1

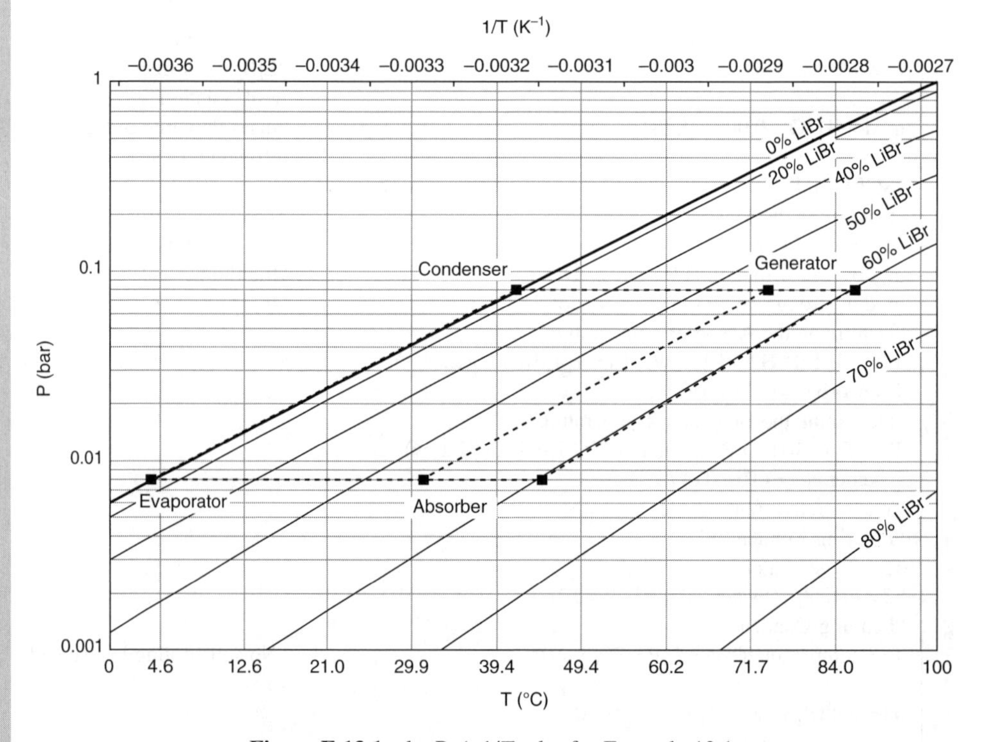

Figure E.13.1 logP–(−1/*T*) plot for Example 13.1

(b) the heat rejected at the condenser: $q_{condenser} = 18.57$ [kW]
(c) the heat rejected at the absorber: $q_{absorber} = 22.76$ [kW]
(d) the heat input at the evaporator: $q_{evaporator} = 17.5$ [kW]
(e) the heat input at the generator: $q_{generator} = 23.83$ [kW]
(f) the condensing temperature of the refrigerant: T3 = 41.52 [C]
(g) the evaporating temperature of the refrigerant: T1 = 3.761 [C]
(h) the temperature in the absorber: $T_{absorber} = 37.1$ [C]
(i) the temperature in the generator: $T_{generator} = 80.59$ [C]
(j) the solution pump power (assuming a pump efficiency of 0.6): $W_{pump} =$ 0.0005807 [kW]
(k) the cooling COP: $COP_c = 0.7345$

Discussion: This example demonstrates some of the operating features of an H_2O–LiBr absorption chiller. Note the negligibly small power requirement of the solution pump. The evaporator temperature (3.8 °C) is low enough to provide chilled water to an air handler at a typical temperature of 6–7 °C. The absorber and condenser temperatures are of similar magnitude, at 37.1 and 41.5 °C respectively. The generator temperature is 80.6 °C, which could be provided by a solar or geothermal energy source, for example. The COP is less than 1, which is the case for a single-effect H_2O–LiBr absorption chiller.

13.3.4 Heat Transfer Considerations

Example 13.1 illustrates an engineering model of an absorption cycle where the operating pressures and the mass fractions of the strong, weak, and refrigerant fluids were inputs. This simple model results in very useful outcomes, such as the thermodynamic state points, heat transfer rates (when the refrigerant mass flow rate is known or assumed), and the heat pump COP. However, just as was discussed with geothermal heat pumps of the vapor compression type, it is what's happening outside the heat pump that dictates the operating pressures of the absorber, generator, condenser, and evaporator, and we are often interested in the temperatures and flow rates of these fluids.

In Section 12.3.4, the use of the *heat exchanger effectiveness* (ε) concept in heat exchanger analysis was discussed. However, this concept is difficult to apply in the absorber and generator because there is more than one inlet and outlet. Therefore, we seek an alternative means of describing these heat exchangers in absorption heat pumps.

A very simple concept used to describe heat exchanger performance is the *approach temperature* concept. The word 'approach' is derived from how closely one fluid approaches the other in a heat exchanger. The closer the approach temperature, the more effective is the heat exchanger. Relatively small approach temperatures and relatively high heat effectiveness values are associated with larger and more costly heat exchangers.

Different fields of engineering may have different definitions of the approach temperature, and there is no universal definition. The use of the approach temperature can be ambiguous and confusing because there are at least two inlets and two outlets in a heat exchanger, so what temperature differences do we take? In general, this depends on the purpose of the heat exchanger. The approach temperature is more convenient to define, depending on whether the heat exchanger is used to: (i) cool a fluid stream or (ii) heat a fluid stream.

We will define the approach temperature as follows. For a heat exchanger used *to heat a fluid*:

$$T_{approach,\ heating} = T_{h,i} - T_{c,o} \tag{13.8}$$

and for a heat exchanger used *to cool a fluid*:

$$T_{approach,\ cooling} = T_{h,o} - T_{c,i} \tag{13.9}$$

where the subscripts h and c represent the hot and cold fluid respectively, and the subscripts i and o represent the inlet and outlet of the heat exchanger respectively.

It should also be noted that when one of the fluids in a heat exchanger is undergoing a phase change, such as in an evaporator or a condenser, we are usually interested in the *condenser approach* or the *evaporator approach*. Thus, to apply Equation (13.8) or (13.9) in those cases, the condensing or evaporating temperature is used.

Example 13.2 Determining the Approach Temperature of a Refrigerant-to-Air and Refrigerant-to-Water Heat Exchanger for a Geothermal Heat Pump in *Heating* Mode

Recall Example 12.4 where we simulated a heat pump in heating mode. Our analysis concluded that the heat pump was heating air from 22 to 37.45 °C by passing it through a refrigerant-to-air heat exchanger, where the condensing temperature of the refrigerant was 47.75 °C after entering the condenser at 82.38 °C. At the same time, the heat pump was extracting heat from the Earth via a fluid stream entering the heat pump at 5 °C and leaving at 0.026 °C and boiling the refrigerant at –3.29 °C. Calculate the condenser and evaporator approach temperatures.

Solution

The object of the condenser in this case is to heat an air stream in the building. Thus, the hot fluid is the refrigerant and the cold fluid is the air, and the approach temperature is calculated by Equation (13.8). The actual hot inlet temperature is 82.38 °C, but in cases such as this one, it's more common to use the condensing temperature:

$$T_{approach,\ heating} = T_{h,i} - T_{c,o}$$
$$T_{approach,\ heating} = 47.75\,^{\circ}C - 37.45\,^{\circ}C$$
$$T_{approach,\ heating} = 10.3\,^{\circ}C$$

The object of the evaporator in this case is to boil the refrigerant by extracting heat from the Earth. Thus, the hot fluid is the geothermal fluid and the cold fluid is the refrigerant, and the approach temperature is also calculated by Equation (13.8):

$$T_{approach,\ heating} = T_{h,i} - T_{c,o}$$
$$T_{approach,\ heating} = 5^{\circ}C - (-3.29\,^{\circ}C)$$
$$T_{approach,\ heating} = 8.3\,^{\circ}C$$

Example 13.3 Determining the Approach Temperature of a Refrigerant-to-Air and Refrigerant-to-Water Heat Exchanger for a Geothermal Heat Pump in *Cooling* Mode
Recall Example 12.6 where we simulated a heat pump in cooling mode. Our analysis concluded that the heat pump was cooling air from 23 to 13 °C by passing it through a refrigerant-to-air heat exchanger, where the evaporating temperature of the refrigerant was −8.5 °C. At the same time, the heat pump was rejecting heat to the Earth via a fluid stream entering the heat pump at 27 °C and leaving at 31.85 °C, and condensing the refrigerant to 33.19 °C. The refrigerant entered the condenser at 61.24 °C. Calculate the condenser and evaporator approach temperatures.

Solution
The object of the condenser in this case is for the geothermal fluid to cool and condense the refrigerant. Thus, the hot fluid is the refrigerant and the cold fluid is the geothermal fluid, and the approach temperature is calculated by Equation (13.9):

$$T_{approach,\ cooling} = T_{h,o} - T_{c,i}$$
$$T_{approach,\ cooling} = 33.19\,°C - 27\,°C$$
$$T_{approach,\ cooling} = 6.19\,°C$$

The object of the evaporator in this case is to cool the air stream. Thus, the hot fluid is the air and the cold fluid is the refrigerant, and the approach temperature is also calculated by Equation (13.9):

$$T_{approach,\ cooling} = T_{h,o} - T_{c,i}$$
$$T_{approach,\ cooling} = 13\,°C - (-8.5\,°C)$$
$$T_{approach,\ cooling} = 21.5\,°C$$

13.3.5 Performance Modeling

We can now consolidate the foregoing to construct a fairly complete mathematical model of an absorption cycle heat pump. The objective is to create a model that is a useful representation of a real absorption cycle. Let's illustrate by modeling an absorption chiller via computer simulation in EES.

Again, we note that there are numerous combinations of possible input and output data for a heat pump simulation model, yielding many degrees of freedom. The choice of input and output data depends on what is known, and what is desired to be known. The inputs and outputs are chosen here to obtain as much relationship to real practice as possible, and thus we'll construct a steady-state absorption chiller model with the data as organized in Table 13.1.

Table 13.1 Summary of Input and Output Data of Interest for Example Absorption Chiller Simulation Model

Fluid Cycle	Input Data	Output Data
Load (Chilled Fluid)	• thermal properties of fluid • chilled fluid set-point • evaporator approach temperature	• fluid state leaving heat pump • mass flow of fluid required rate to meet the load

(*continued overleaf*)

Table 13.1 (*continued*)

Fluid Cycle	Input Data	Output Data
	• allowable ΔT entering/leaving chiller	
Generator	• thermal properties of fluid • source temperature entering generator • generator approach temperature • allowable ΔT entering/leaving chiller	• required mass flow rate of the heat source fluid • heat transfer rate
Absorber and Condenser	• cooling fluid temperature entering absorber (and condenser) • absorber approach temperature • condenser approach temperature • allowable ΔT entering/leaving absorber and condenser	• required mass flow rates of the cooling fluid through the absorber and condenser • heat transfer rates
Refrigerant–Absorbent Pair	• refrigerant–absorbent fluid pair (e.g., H_2O–LiBr or NH_3–H_2O) • concentration difference between the strong and weak solution • solution pump efficiency	• thermodynamic state of all points in the cycle • solution pump power requirement • mass flow rates of solutions
System	• none necessary	• coefficient of performance

Example 13.4 Simulation of an H_2O–LiBr Absorption Chiller

A water–lithium bromide chiller is providing chilled water at 7 °C to an apartment complex with a fan coil unit in each apartment. The peak cooling load is 100 kW. The heat source is a solar thermal system heating a storage tank of pure water kept at 90 °C for supply to the generator. The cooling fluid is water from a geothermal heat exchanger designed for 30 °C supply to the absorber and condenser of the chiller at peak cooling conditions. Assuming a 5 °C approach temperature for each of the generator, absorber, and condenser, and a 3 °C approach temperature for the evaporator, show/calculate the following:

(a) a sketch of the cycle on a logP–($-1/T$) plot,
(b) the heat transfer rates at the condenser, absorber, and generator,
(c) the mass flow rates of the generator fluid, cooling water, and chilled water,
(d) the mass flow rates of the refrigerant, strong solution, and weak solution,
(e) temperatures in the generator, absorber, condenser, and evaporator,
(f) the solution pump power, and
(g) the cooling COP.

Solution

The EES model is constructed as follows:

```
"GIVEN:"
"Evaporator Conditions"
q_evaporator = 100 [kW] "The chilled water load"
T_Evaporator_Approach = 3 [C] "Approach temperature for chilled water and evaporator"
DT_chilled_fluid = 5 [C] "The temperature differential of the chilled water"
T_chilled_water_out = 7 [C] "Chilled water set point"
"Generator Conditions"
T_Generator_Approach = 5 [C] "Approach temperature for generator and heat source"
DT_Generator_fluid = 5 [C] "The temperature differential of the generator fluid"
T_heat_source_in = 90[C] "The heat source temperature inlet"
"Absorber Conditions"
T_Absorber_Approach = 5 [C] "Approach temperature for absorber and cooling water"
T_cooling_source_in = 30[C] "The absorber cooling temperature inlet"
"Condenser Conditions"
T_Condenser_Approach = 5 [C] "Approach temperature for condenser and cooling water"
"Cooling Fluid for Absorber + Condenser"
DT_Cooling_fluid = 5 [C] "The temperature differential of the cooling fluid"
"H2O-LiBr Conditions"
D_x = 0.06 [-] "Assumed concentration difference (between weak and strong solutions)"
"SOLUTION:"
"Water (refrigerant) Cycle:"
"Entering Condenser"
T2 = T_generator
h2 = enthalpy('water', P = P2, T = T2)
"Leaving Condenser"
P2 = Pressure(water, T = T3, x = 0)                 "No sub-cooling is assumed, so x =0"
h3 = enthalpy('water', P = P2, x = 0)
"Entering Evaporator"
T4 = temperature('water', P = P1, x = 0)
h4 = h3                                             "throttling process"
"Leaving Evaporator"
P1 = Pressure(water, T = T1, x = 1)                 "No super-heating is assumed, so x
                                                    =1"
h1 = enthalpy('water', P = P1, x = 1)
"LiBr (absorbent) Cycle:"
T1a = T_LiBrH2O(P1,x_w); Ta = T_LiBrH2O(P1,x_s)     "The Inlet and Outlet Absorber
                                                    Temperatures"
ha = h_LiBrH2O(T_absorber,x_s)                      "Leaving Absorber (Point a)"
Tb = T_LiBrH2O(P2,x_s); Tc = T_LiBrH2O(P2,x_w)      "The Inlet and Outlet Generator
                                                    Temperature"
hc = h_LiBrH2O(T_generator,x_w)                     "Leaving Generator (Point c)"
"Determine the Mass Flow Rates and Mass Fractions"
x_generator = x_LiBrH2O(T_generator ,P2)            "LiBr mass fractions in the gener-
                                                    ator"
x_s = x_generator + D_x/2; x_w = x_generator - D_x/2
```

```
                                                 "LiBr mass fractions of the strong
                                                 and weak solutions"
m_dot_weak/ m_dot_H2O = x_s / D_x                "the fluid circulation ratio"
m_dot_strong = m_dot_weak + m_dot_H2O            "mass balance"
"Energy Balances:"
q_evaporator = m_dot_H2O * (h1 - h4)
q_condenser = m_dot_H2O * (h2 - h3)
n_pump = 0.6 [-]                                 "assumed pump efficiency"
W_pump = m_dot_strong * (P2-P1) * convert(bar,kPa)/   "Pump work"
(rho_LiBrH2O(Ta,x_s) * n_pump)
q_generator + m_dot_strong * ha + W_pump = m_dot_weak * hc + m_dot_H2O * h2 "energy
balance on generator"
m_dot_weak * hc + m_dot_H2O * h1 = q_absorber + m_dot_strong * ha "energy balance on
absorber"
COP_c = q_evaporator / q_generator
Cycle_Balance = q_evaporator + q_generator + W_pump - (q_absorber + q_condenser )
"Calculate Temperature, Flow Requirements of the External Fluids"
T_cooling_source_in = T_absorber - T_Absorber_Approach "This determines the absorber tem-
perature"
T_heat_source_in = T_generator + T_Generator_Approach "This determines the generator tem-
perature"
T_chilled_water_out = T1 + T_Evaporator_Approach "This determines the refrigerant evaporat-
ing temperature T1"
T_cooling_source_in = T3 - T_condenser_approach "This determines the refrigerant condensing
temperature T3"
q_evaporator = m_dot_chilled_water * cp(water, T = T_chilled_water_out, P = 1 [bar]) *
DT_chilled_fluid
q_generator = m_dot_generator_water * cp(water, T = T_heat_source_in, P = 1 [bar]) * DT_Ge-
nerator_fluid
q_absorber + q_condenser = m_dot_cooling_water * cp(water, T = T_cooling_source_in, P = 1
[bar]) * DT_Cooling_fluid
```

The following results are obtained:

(a) a sketch of the cycle on a log$P$–($-1/T$) plot:

(b) the heat transfer rates at the condenser, absorber, and generator:

$$q_{condenser} = 106.4\ [kW],\ q_{absorber} = 124.3\ [kW],\ \text{and}\ q_{generator} = 130.7\ [kW]$$

(c) the mass flow rates of the generator fluid, cooling water, and chilled water:

$$m_{generator,water} = 6.219\ [kg/s],\ m_{cooling,water} = 11.03\ [kg/s],\ \text{and}\ m_{chilled,water} = 4.769\ [kg/s]$$

(d) the mass flow rates of the refrigerant, strong solution, and weak solution:

$$m_{H2O} = 0.04235\ [kg/s],\ m_{strong} = 0.5049\ [kg/s],\ \text{and}\ m_{weak} = 0.4626\ [kg/s]$$

(e) temperatures in the generator, absorber, condenser, and evaporator:

$$T_{generator} = 85\ [C],\ T_{absorber} = 35\ [C],\ T3 = 35\ [C],\ \text{and}\ T1 = 4\ [C]$$

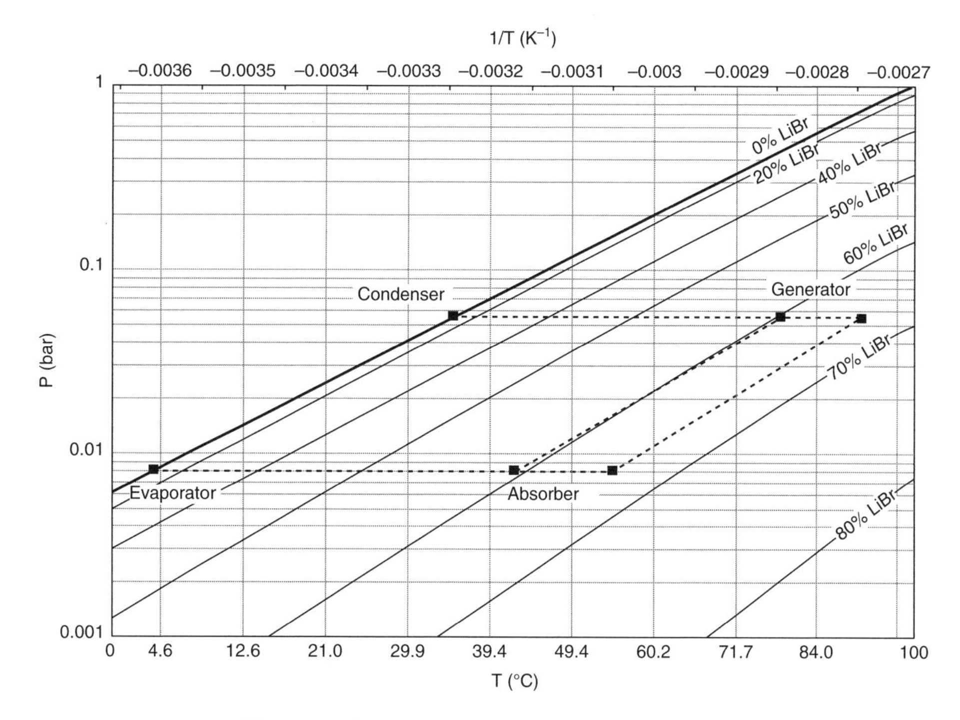

Figure E.13.4 logP–(−1/T) plot for Example 13.4

(f) the solution pump power: $W_{pump} = 0.002229$ [kW]
(g) the cooling COP: $COP_c = 0.7649$

Example 13.5 Simulation of an NH_3–H_2O Absorption Chiller

A town with a geothermal resource estimated at a temperature of 90 °C is in the preliminary planning stage of building a new ice rink. The idea has come up to consider using the geothermal resource as a heat source for an absorption chiller to provide refrigeration for making the ice surface, but the groundwater yield from a geothermal well may be a limiting factor. Upper estimates of potential well yield are placed at 1000 L/min.

The refrigeration load for the ice rink has been estimated at 300 kW. The ice surface will be kept at its design temperature using an underslab network of pipes carrying a secondary refrigerant at −15 °C. This secondary refrigerant is planned to be an aqueous solution of 20% calcium chloride. The cooling fluid will be water from a cooling tower designed for 30 °C supply to the absorber and condenser of the chiller at peak conditions. Assuming a 5 °C approach temperature for each of the generator, absorber, and condenser, and a 3 °C approach temperature for the evaporator, determine whether there will be enough water available from the geothermal well. Also, calculate the COP of the chiller and show the process on a log*P*–(−1/*T*) plot.

Solution

Given the low-temperature requirements, an absorption chiller for this application would have to be an ammonia–water type. Our EES model is constructed as follows:

```
"GIVEN:"
"Evaporator Conditions"
q_evaporator = 100 [kW] "The chilled water load"
T_Evaporator_Approach = 3 [C] "Approach temperature for chilled water and evaporator"
DT_chilled_fluid = 5 [C] "The temperature differential of the chilled water"
T_chilled_fluid_out = -15 [C]              "Chilled fluid set point"
$REFERENCE Ammonia ASH "String for refrigerant, using ASHRAE reference state"
"Generator Conditions"
T_Generator_Approach = 5 [C] "Approach temperature for generator and heat source"
DT_Generator_fluid = 5 [C]                 "The temperature differential of the generator
                                           fluid"
T_heat_source_in = 90[C] "The heat source temperature inlet"
"Absorber Conditions"
T_Absorber_Approach = 5 [C] "Approach temperature for absorber and cooling water"
T_cooling_source_in = 30[C]                "The absorber cooling temperature inlet"
"Condenser Conditions"
T_Condenser_Approach = 5 [C] "Approach temperature for condenser and cooling water"
"Cooling Fluid for Absorber + Condenser"
DT_Cooling_fluid = 5 [C] "The temperature differential of the cooling fluid"
"Ammonia- H2O Conditions"
D_x = 0.15 [-] "assumed concentration difference (between weak and strong solutions)"
"SOLUTION:"
"Ammonia (refrigerant) Cycle:"
"Entering Condenser"
T2 = T_generator
h2 = enthalpy(Ammonia, P = P2, T = T2)
"Leaving Condenser"
P2 = Pressure(Ammonia, T = T3, x = 0)      "No sub-cooling is assumed, so x =0"
h3 = enthalpy(Ammonia, P = P2, x = 0)
"Entering Evaporator"
T4 = temperature(Ammonia, P = P1, x = 0)
h4 = h3                                    "throttling process"
"Leaving Evaporator"
P1 = Pressure(Ammonia, T = T1, x = 1)      "No super-heating is assumed, so x =1"
h1 = enthalpy(Ammonia, P = P1, x = 1)
"Water (absorbent) Cycle:"
T1a = Temperature(NH3H2O,x = x_w,P = P1,   "The Inlet Absorber Temperature"
Q = 0)
Ta = Temperature(NH3H2O,x = x_s,P = P1,Q   "The Outlet Absorber Temperature"
= 0)
ha = Enthalpy(NH3H2O,x = x_s,T =           "Leaving Absorber (Point a)"
T_absorber,Q = 0)
Tb = Temperature(NH3H2O,x = x_s,P = P2,Q   "The Inlet Generator Temperature"
= 0)
Tc = Temperature(NH3H2O,x = x_w,P = P2,Q   "The Outlet Generator Temperature"
= 0)
hc = Enthalpy(NH3H2O,x = x_w,T = T_gener-  "Leaving Generator (Point c)"
ator,Q = 0)
"Determine the Mass Flow Rates and Mass Fractions"
```

```
x_generator = MassFraction(NH3H2O,T = T_generator,
P = P2,Q = 0) "NH3 mass fractions in the generator"
x_s = x_generator - D_x/2; x_w = x_generator      "NH3 mass fractions of the strong and weak
+ D_x/2                                            solutions"
m_dot_weak/ m_dot_NH3 = x_s / D_x                  "the fluid circulation ratio"
m_dot_strong = m_dot_weak + m_dot_NH3              "mass balance"
"Energy Balances:"
q_evaporator = m_dot_NH3 * (h1 - h4)
q_condenser = m_dot_NH3 * (h2 - h3)
n_pump = 0.6 [-]                                   "assumed pump efficiency"
W_pump = m_dot_strong * (P2-P1) * convert(bar,kPa)/(DENSITY(NH3H2O,T = Ta, x = x_s,Q
= 0) * n_pump) "Pump work"
q_generator + m_dot_strong * ha + W_pump = m_dot_weak * hc + m_dot_NH3 * h2 "energy
balance on generator"
m_dot_weak * hc + m_dot_NH3 * h1 = q_absorber + m_dot_strong * ha "energy balance on
absorber"
COP_c = q_evaporator / q_generator
Cycle_Balance = q_evaporator + q_generator + W_pump - (q_absorber + q_condenser )
"Calculate Temperature, Flow Requirements of the External Fluids"
T_cooling_source_in = T_absorber - T_Absorber_Approach "This determines the absorber tem-
perature"
T_heat_source_in = T_generator + T_Generator_Approach "This determines the generator tem-
perature"
T_chilled_fluid_out = T1 + T_Evaporator_Approach "This determines the refrigerant evaporating
temperature T1"
T_cooling_source_in = T3 - T_condenser_approach "This determines the refrigerant condensing
temperature T3"
q_evaporator = m_dot_chilled_fluid* cp(CaCl2,T = T_chilled_fluid_out, C = 20[%]) * DT_
chilled_fluid
q_generator = m_dot_generator_water * cp(water, T = T_heat_source_in, P = 1 [bar]) * DT_
Generator_fluid
q_absorber + q_condenser = m_dot_cooling_water * cp(water, T = T_cooling_source_in, P = 1
[bar]) * DT_Cooling_fluid
```

Solution

The process diagram on a log$P$–($-1/T$) plot is shown below on Figure E.13.5.

The EES model calculates the COP of the chiller at 0.49, and the required mass flow of water through the generator at 9.8 kg/s, or 609 L/min. This flow estimate is within the upper estimate of the geothermal well yield, so the design is potentially acceptable.

Discussion: Note the much lower COP values in this example relative to the previous example. Lower COPs of NH_3–H_2O absorption chillers relative to H_2O–LiBr absorption chillers are typical, owing to the differing thermal properties of the refrigerant (i.e., water relative to ammonia) and owing to the higher operating pressures. In order fully to evaluate the system COP, the geothermal well pumping power should be included. Thus, in reality, further analysis would be required.

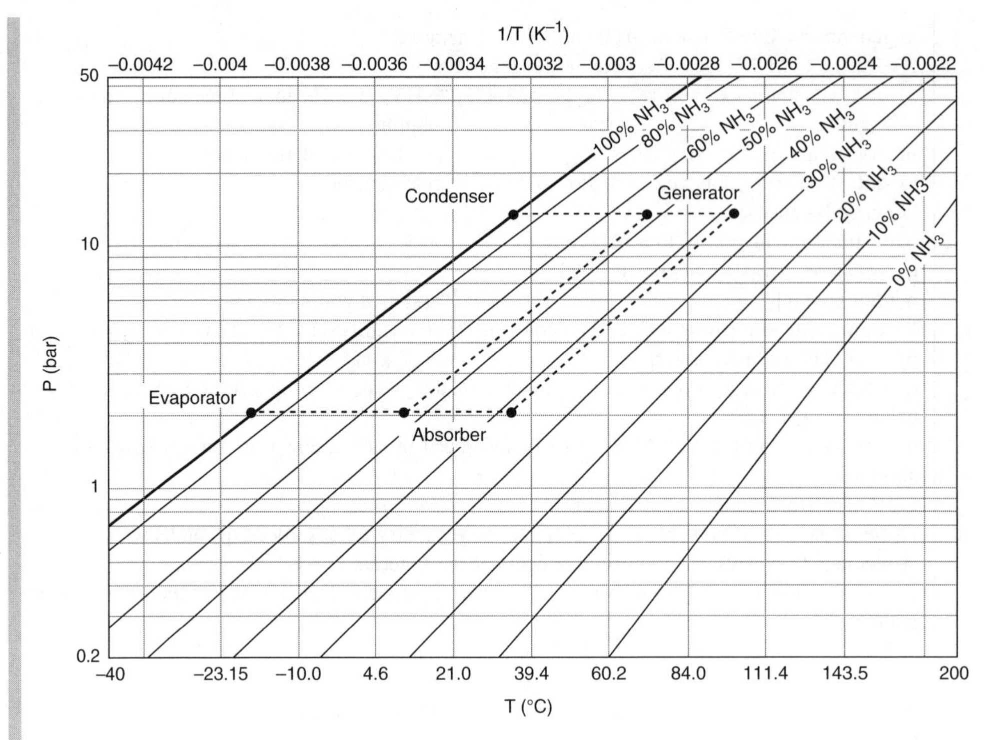

Figure E.13.5 log$P$–($-1/T$) plot for Example 13.5.

13.4 Adsorption Cycles

Adsorption cycles for refrigeration were first used in the early 1900s. Kühn (2013), citing Cristoph (2012) in a historical review, reports on a domestic activated carbon/methanol refrigerator called 'Eskimo' and sold in the 1930s. That article also reports on research activities in the United States in the 1980s, and the manufacturing of several solar adsorption refrigerators in the 1990s. Two Japanese companies produced the first adsorption chillers operating with the working pair water–silica gel in the late 1980s, and since the early 2000s, several small-scale adsorption chillers (8–50 kW cooling capacity) for solar cooling or use in trigeneration systems have been developed, mainly in Europe.

The principle of operation of adsorption heat pumps is thermodynamically similar to that of an absorption heat pump. The only difference is that an adsorption heat pump uses solid sorption instead of the liquid sorption used in absorption machines. Referring to Figure 13.3, adsorption heat pumps have the same components as absorption heat pumps, except that the generator and absorber are chambers filled with a desiccant material. The desiccant is a porous material possessing a high surface area-to-volume ratio, and material pairs typically applied in adsorption heat pumps are: (i) silica gel–H_2O, (ii) zeolite–H_2O, (iii) activated carbon–MeOH, and (iv) activated carbon/salt–NH_3.

Adsorption heat pumps have some additional operating concerns when compared with absorption heat pumps. Given the obvious, that solid absorbent cannot be moved from one chamber to the other like liquid absorbent can be pumped from the absorber to the generator in absorption heat pumps, some means is necessary to transfer heat from the absorber to the generator in adsorption heat pumps. This introduces the necessity of a cycling time, after which there is a changeover of the absorber and desorber chambers. Such a changeover is accomplished by circulating the refrigerant through the absorber and desorber chambers. This discontinuously working process may result in fluctuating outlet temperatures.

As with absorption heat pump cycles, it is common to represent the adsorption process on a $\log P$–$(-1/T)$ diagram. Material pairs using water as the refrigerant operate under similar pressure ranges as H_2O–LiBr absorption machines. Other thermodynamic and heat transfer considerations for adsorption heat pump cycles are similar to those of absorption heat pump cycles.

13.5 Thermally Driven Heat Pump Performance Standards and Manufacturer's Catalog Data

The basic purpose of heat pump standards, aside from consumer protection, is to establish performance testing and rating criteria for factory-made units. Thus, comparison of one unit with another is possible. Performance standards are also used in government policy and energy codes. One such standard for absorption heat pumps is ANSI/AHRI Standard 560-2000.

According to Kühn (2013), manufacturers of large absorption chillers are mainly concentrated in the countries with the strongest markets: China, Japan, and South Korea. Manufacturers include Broad, Shuangliang, Sanyo, Yazaki, Hitachi, Kawasaki, Mitsubishi, Ebara, LG, and Century. A few absorption chillers are manufactured in the United States by Carrier, Trane, and York/Johnson Controls, and in India by Thermax. Robur, an Italian company, has manufactured gas-fired ammonia/water absorption heat pumps for many decades.

Since the early 2000s, several small-scale absorption and adsorption chillers (8–50 kW cooling capacity) have been developed, mainly in Europe. According to Kühn (2013), companies in active production include Yazaki, EAW Energieanlagenbau, Thermax, Pink, and AGO (absorption) and Sortech and Invensor (adsorption). Mayekawa is the only known manufacturer of medium-scale adsorption chillers.

Selection of a water-fired chiller is similar to that described in Chapter 12 for vapor compression heat pumps. Cooling capacities and heating capacities are tabulated in catalog data as a function of heating medium temperature, cooling water temperature, and associated flow rates. Manufacturers present these data differently, with some in tables while other data could be in the form of performance curves.

13.6 Chapter Summary

This chapter presented an overview of absorption and adsorption thermally driven heat pump cycles. Analogies to the familiar vapor compression cycle were discussed. The mechanics of operation was described, along with potential geothermal application. Details of the thermodynamic cycle were analyzed on $\log P$–$(-1/T)$ plots, and the approach temperature was described as a simple means of heat exchanger analysis. Finally, standards and manufacturers of absorption and adsorption heat pumps were briefly discussed, along with brief mention of manufacturer's catalog data.

Discussion Questions and Exercise Problems

13.1 Conduct a parametric study of the heating medium supply temperature for the absorption chiller in Example 13.4. Using the supplied EES file, vary the generator supply temperature from 75 up to 90 °C, and develop an *x–y* plot of the COP as a function of the generator temperature.

13.2 Conduct a parametric study of the cooling water supply temperature for the absorption chiller in Example 13.4. Using the supplied EES file, vary the cooling water supply temperature from 15 up to 35 °C, and develop an *x–y* plot of the COP as a function of the cooling water supply temperature.

13.3 Regarding the geothermal resource in Example 13.5, there is some uncertainty in the actual resource temperature. Using the supplied EES file, conduct a sensitivity analysis of the heat pump COP to the geothermal temperature. Do this by developing an *x–y* plot of the COP as a function of the resource temperature, which could range from 80 up to 120 °C.

14

Organic Rankine Cycle (Binary) Geothermal Power Plants

14.1 Overview

We extend our discussion of heat pumps now to heat engines. A discussion of heat engines employing the Rankine Cycle is appropriate here because of the strong similarities to the mechanical vapor compression cycle. Further, there is a continuing interest and trend to employ power cycles in low-temperature geothermal applications similar to those previously discussed for absorption cooling (i.e., <100 °C). Even at geothermal resource temperatures up to ~175 °C, closed-loop Rankine cycles are typically employed in a binary-type configuration. Such cycles are referred to as organic Rankine cycles (ORCs) because a secondary fluid is employed as the working fluid. The 'organic' working fluid is typically a hydrocarbon with a boiling point lower than water.

For an in-depth treatment of geothermal power plants, readers are referred to DiPippo (2012). As mentioned in Chapter 1, many of the geothermal power plants operating worldwide are binary plants.

We will first begin our discussion of ORC heat engines by examining the ideal cycle and an associated conceptual engineering model. We will then extend these concepts to non-ideal cycles and associated representation on a pressure–enthalpy diagram. Next, we will briefly look at their operating mechanics, and then put everything together into a computer simulation model to examine ORC performance. The section will be concluded with some modifications to the basic ORC cycle to improve performance in geothermal applications.

Learning objectives and goals:

1. Review the fundamentals of Rankine cycle vapor power operation.
2. Draw analogies between the Rankine heat engine cycle and the mechanical vapor compression heat pump cycle.

Geothermal Heat Pump and Heat Engine Systems: Theory and Practice, First Edition. Andrew D. Chiasson.

Companion website: www.wiley.com/go/chiasson/geoHPSTP

3. Develop, analyze, and sketch Rankine cycles on pressure–enthalpy (P–h) diagrams, and calculate energy flows in all processes of the cycle.

14.2 The Ideal Rankine Cycle

Recall the schematic of a general Carnot heat engine cycle shown in Figure 11.2a and its operation: there are four ideal, thermodynamically reversible processes. The heat transfers at the high- and low-temperature reservoirs (T_H and T_C) occur isothermally, and the processes to and from the high- and low-temperature reservoirs occur adiabatically. The *ideal Rankine cycle* essentially represents an 'intermediate' step from the Carnot heat engine cycle toward a real cycle. It lays the groundwork for a practical heat engine cycle, but retains the perfect heat transfers and isentropic processes of the Carnot cycle. Thus, the ideal Rankine cycle is not achievable.

The concept of the Rankine cycle arose out of the practical difficulties of implementing a Carnot cycle. The Carnot cycle requires isothermal expansion and compression involving simultaneous heat and work, which are difficult to achieve in practice. It was recognized that one easy way to achieve the isothermal processes at T_H and T_C is to implement these processes in a phase-change condition. Thus, as with vapor compression cycles, we move the Carnot cycle under the liquid–vapor dome to compare with the ideal Rankine cycle, as shown in Figure 14.1.

Similar to the vapor compression cycle, the Rankine cycle actually operates between two pressures corresponding to the boiling and condensing temperature of the working fluid. Further, two-phase expansion is not easily achievable in the Rankine cycle, and thus process 1–2 is moved out of the vapor dome to represent 'dry' expansion. As shown in Figure 14.1b, the Rankine cycle has some variants, depending on the working fluid. The turbine inlet condition may either be a saturated vapor (point 1) or a superheated vapor (point 1b). The desired result is point 2 or 2b being a near-saturated vapor as its pressure is decreased to the condensing pressure of the working fluid. As we shall see with some working fluids used in geothermal ORC plants, owing to the skewed shape of the liquid–vapor dome, the working fluid in process 1–2 will remain entirely a vapor if point 1 is a saturated vapor. The fluid leaving the condenser at point 3 is either saturated or slightly subcooled liquid. The rejection of enough heat from the working fluid in the condenser to bring the fluid to a saturated liquid is done for practical reasons; pumping of two-phase fluid is difficult in practice. Finally, in the ideal Rankine cycle, the liquid at point 3 is compressed isentropically to high pressure at point 4 in process 3–4 via liquid pumping. In the basic cycle, point 4 represents the inlet to the boiler. Here, we define an intermediate point, point 4′, which will come in handy for geothermal applications, where an optional preheater is used to heat the liquid working fluid to saturation.

Our conceptual engineering model of ORCs consists of the four principal control volumes as shown schematically in Figure 14.2. All energy transfers by work and heat are taken as positive in the directions of the arrows on the schematic, and energy balances are written accordingly.

In process 1–2, the working fluid expands through a turbine or expander, developing work. The energy balance on the turbine or expander, assuming an adiabatic process, is given by

$$\dot{W}_{out} = \dot{m}_R(h_1 - h_2) \tag{14.1}$$

where $\dot{m}_R$ is the mass flow rate of the refrigerant.

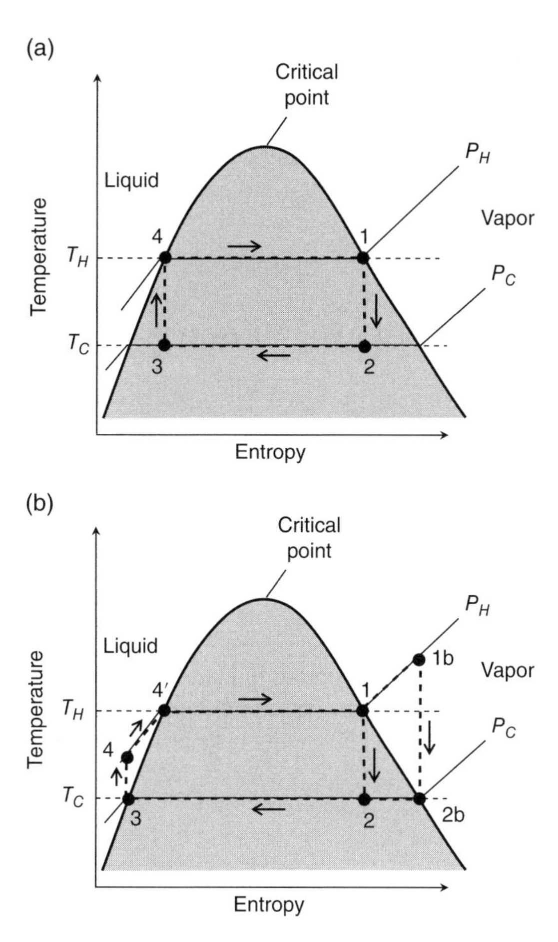

Figure 14.1 Analogous temperature–entropy (T–s) diagrams for (a) a Carnot heat engine cycle and (b) an ideal Rankine cycle

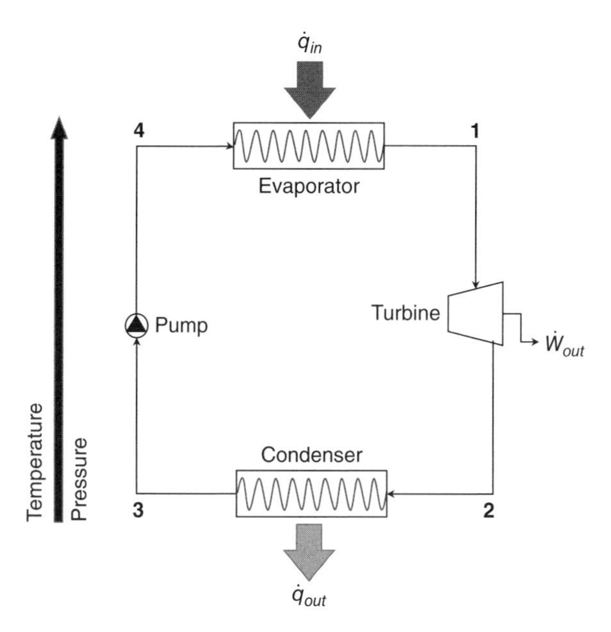

Figure 14.2 Conceptual model of working fluid cycle components in an organic Rankine cycle heat engine

In process 2–3, vapor is desuperheated (where applicable), and is then condensed to liquid through heat transfer to a cooling fluid, typically water or air. The energy balance on the condenser is given by

$$\dot{q}_{out} = \dot{m}_R(h_2 - h_3) \tag{14.2}$$

In process 3–4, liquid is pumped into the boiler (or evaporator), requiring work input. The pump power is given by

$$\dot{W}_{pump} = \dot{m}_R(h_4 - h_3) \tag{14.3}$$

The specific work required for the pump may also be evaluated as

$$\frac{\dot{W}_{pump}}{\dot{m}_R} \approx v_3(P_4 - P_3) \tag{14.4}$$

where v_3 is the specific volume of the working fluid at the pump inlet.

Finally, in process 4–1, liquid is heated to saturation and evaporated in the boiler through heat transfer from the energy source. The energy balance on the boiler is given by

$$\dot{q}_{in} = \dot{m}_R(h_1 - h_4) \tag{14.5}$$

The cycle efficiency is given as

$$\eta_{cycle} = \frac{\dot{W}_{out} - \dot{W}_{pump}}{\dot{q}_{in}} = \frac{(h_1 - h_2) - (h_4 - h_3)}{(h_1 - h_4)} \tag{14.6}$$

14.3 The Non-Ideal Rankine Cycle

14.3.1 Principal Irreversibilities and Isentropic Efficiency

The non-ideal Rankine Cycle has some significant departures from the ideal cycle. The components described above to implement the Rankine cycle are not perfect and exhibit thermodynamic irreversibilities. The turbine and pump have isentropic efficiencies less than 100%. Flow through the boiler and condenser is accomplished through pressure drops, and the approach temperatures between the boiler and condenser and their external heat sources/sinks will not be zero. In other words, the working fluid in the boiler will not be heated to the source temperature, and will not be cooled to the cooling fluid temperature in the condenser.

The turbine isentropic efficiency (η_t) is defined as

$$\eta_t = \frac{h_1 - h_2}{h_1 - h_{2s}} \tag{14.7}$$

where $h_1 - h_{2s}$ represents the specific work of the isentropic expansion process, and $h_1 - h_2$ represents the actual specific work of the expansion process. Thus, h_{2s} is evaluated at the pressure of point 2 and entropy of point 1.

Similarly, the pump isentropic efficiency (η_p) is defined as

$$\eta_p = \frac{h_{4s} - h_3}{h_4 - h_3} \tag{14.8}$$

where $h_{4s} - h_3$ represents the specific work of the isentropic compression process, and $h_4 - h_3$ represents the actual specific work of the compression process. Thus, h_{4s} is evaluated at the pressure of point 4 and entropy of point 3.

14.3.2 Engineering Model of ORC Geothermal Power Plants

With respect to ORC geothermal power plants, we can define more specific conceptual engineering models. The following features are distinct to low-temperature geothermal ORCs.

First, the choice of the working fluid in ORCs precludes the turbine inlet from operating in the superheat region. Thus, we can make the engineering approximation that the quality at point 1 (x_1) is 1. Further, we can make the engineering approximation that point 3 is a saturated liquid, and thus $x_3 = 0$.

Second, geothermal ORCs may include the addition of an optional, separate preheater heat exchanger. As seen in Figure 14.1b, a significant amount of heating of the working fluid is required to bring it to its boiling point. In systems with a single boiler, this is typically accomplished in a counterflow, shell-and-tube-type heat exchanger that handles both stages of heating of the subcooled liquid, followed by complete boiling. Systems with a separate preheat heat exchanger, as shown in Figure 14.3, allow for a combination of larger heat exchange equipment, closer boiler approach temperature, and overall improved thermal efficiency. We shall analyze these heat exchange details in a subsequent section, and consider the preheating and boiling of the working fluid in distinct processes. This approach is similar to our consideration of desuperheating in vapor compression heat pumps.

Third, as shown in Figure 14.3, there are various methods of heat rejection to the environment as a consequence of the second law of thermodynamics. ORC geothermal power plants may either be water cooled or air cooled, typically depending on the availability of water. An overview of waste-heat rejection methods in geothermal power plants is given by Chiasson (2015). To obtain good thermal efficiencies of geothermal power generation, the turbine exhaust pressure must be as low as can be economically achieved with the available cooling water or air temperature. Heat rejection methods from geothermal power plants seek to minimize this turbine exhaust pressure by efficiently rejecting heat to the environment. The thermodynamic efficiency of geothermal power generation is strongly influenced by the quantity of heat that can be rejected and at what temperature. With increasing global concern for water shortages, future trends in rejecting heat from geothermal power plants appear to be toward optimizing hybrid methods. In fully air-cooled condensers, the condensing temperature is a function of the ambient air dry-bulb temperature, and consequently, air-cooled plants perform poorly in hot summer weather. Thus, evaporative condensers with minimal water consumption appear to be the main development trend in the near term.

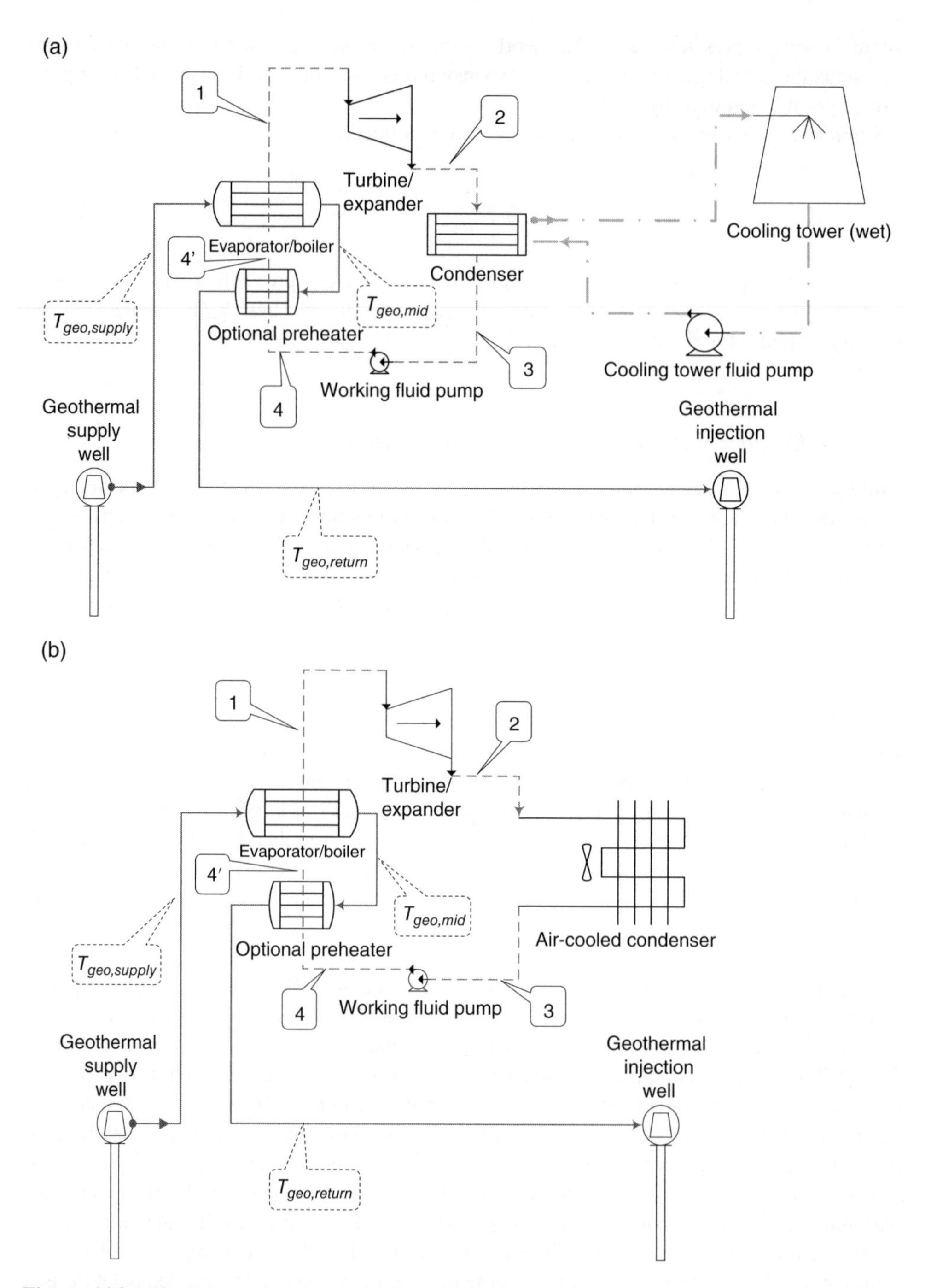

Figure 14.3 Binary Rankine cycle example schematic for (a) a water-cooled plant and (b) an air-cooled plant

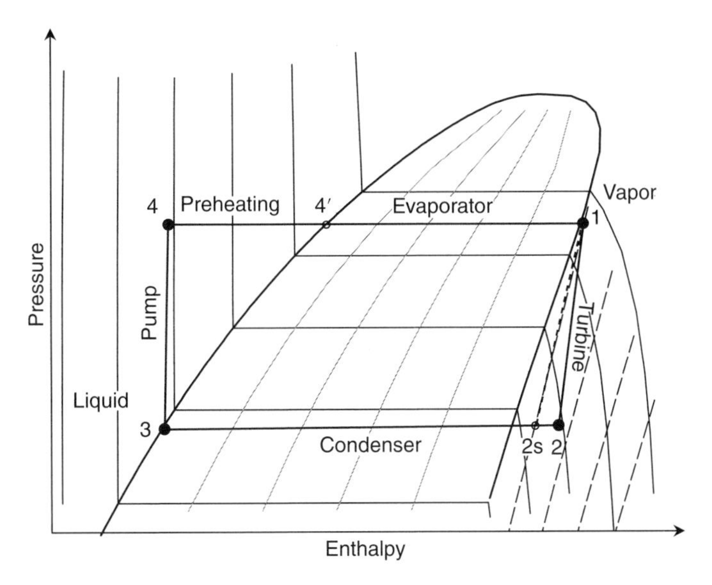

Figure 14.4 Pressure–enthalpy (*P–h*) diagram for R245fa, showing a Rankine cycle

14.3.3 Graphical Representation on Pressure–Enthalpy (P–h) *Diagrams*

Given the similarities between ORC and vapor compression cycles, the *P–h* diagram provides a powerful analysis tool for ORCs. For some reason, most thermodynamics textbooks prefer *T–s* diagrams for Rankine cycle presentations.

A generalized *P–h* diagram was shown in Figure 12.5 and briefly described in Section 12.3.3. With the engineering approximations discussed above, the shape of a Rankine cycle on a *P–h* diagram (Figure 14.4) takes the same form as the subcritical vapor compression cycle. The only difference between the two is that the process flows are reversed. The relative differences between the condenser heat rejection and the evaporator (boiler) plus preheating are evident. The ideal or isentropic turbine process (process 1–2s) is shown in Figure 14.4 as a dashed line, and the departure of the ideal and non-ideal expansion process is evident.

Regarding the overall heating and boiling process of the working fluid (process 4–1), approximately at least 33% of the heat added is attributed to preheating of the subcooled liquid to a saturated liquid. In contrast, regarding the overall desuperheating and condensing process (process 2–3), less than 10% of the heat rejected is attributed to desuperheating.

As mentioned above, the choice of the working fluid in ORCs precludes the turbine inlet from operating in the superheat region. This is evident by the process 1–2 line and the slope of the isentropes for R245fa shown in Figure 14.4. Further examination of Figure 14.4 reveals that heating point 1 beyond the saturated vapor state doesn't make much practical sense, as it is the change in enthalpy through the turbine that produces work, not the actual enthalpy value itself. In other words, for a fixed turbine isentropic efficiency, little to no benefit exists in superheating point 1.

Example 14.1 Simple Organic Rankine Cycle Analysis on a *P–h* Diagram

Consider a geothermal binary power plant with R245fa as the working fluid, operating with a fluid boiling and condensing pressure of 1000 kPa and 200 kPa respectively. Assume an isentropic efficiency of 85 and 75% for the turbine and pump respectively. For a turbine power output of 1000 kW, determine/show the following:

(a) a sketch of the cycle on a *P–h* diagram and show the ideal (isentropic case),
(b) the heat rejected at the condenser,
(c) the heat required to preheat the working fluid,
(d) the heat required to boil the working fluid,
(e) the working fluid temperature leaving the turbine,
(f) the condensing temperature of the working fluid,
(g) the boiling temperature of the working fluid,
(h) the required mass flow rate of the working fluid, and
(i) the thermal efficiency of the cycle.

Solution

This example is readily solved with EES. The input data are as follows:

```
//GIVEN:

//Cycle Input Data
W_turbine = 1000 [kW]                "the turbine desired output"
n_turbine = 0.8; n_pump = 0.75       "the turbine and pump isentropic effi-
                                     ciency"
R$ = 'R245fa'                        "string describing the working fluid"
//Boiler Input Data
P_sat_boiler = 1000 [kPa]            "the fluid boiling pressure"
//Condenser Input Data
P_sat_condenser = 200 [kPa]          "the fluid condensing pressure"

//SOLUTION:
//Assumptions:
x1=1; x3 = 0
//Point 1:
P[1] = P_sat_boiler
h[1]=Enthalpy(R$,P=P[1],x=x1); T1=T_sat(R$,P=P_sat_boiler)
//Point 2:
P[2]= P_sat_condenser
n_turbine = (h[1]-h[2])/(h[1]-h2s)   "To find h[2]"
T2 = Temperature(R$,P=P[2], h = h[2])
//Point 2s
s1 = Entropy(R$,P=P[1],x=x1); h2s = Enthalpy(R$,P=P[2],s=s1)
//Point 3:
P[3] = P_sat_condenser
T3= T_sat(R$,P= P_sat_condenser); h[3]=Enthalpy(R$,T=T3,x=x3)
//Point 4:
P[4] = P_sat_boiler; T4 = T3
n_pump = (h4s-h[3])/(h[4]-h[3])   "To find h[4]"
```

```
//Point 4':
h[5] = Enthalpy(R$,P=P[5],x=0); P[5] = P[4]
//Point 4s
s3 = Entropy(R$,P=P[3],x=x3); h4s = Enthalpy(R$,P=P[4],s=s3)
//Energy Balances
W_Turbine = mdot * (h[1] - h[2])
Q_Condenser = mdot * (h[2] - h[3])
W_Pump = mdot * (h[4] - h[3])
Q_Boiler = mdot * (h[1] - h[5])
Q_Preheat = mdot * (h[5] - h[4])
//Efficiency Analysis
n_cycle = (W_Turbine - W_Pump)/(Q_Boiler+Q_Preheat)    "the 1st Law Efficiency"
```

The following results are obtained:

(a) a sketch of the cycle on a *P–h* diagram, showing the ideal (isentropic case): see Figure E.14.1

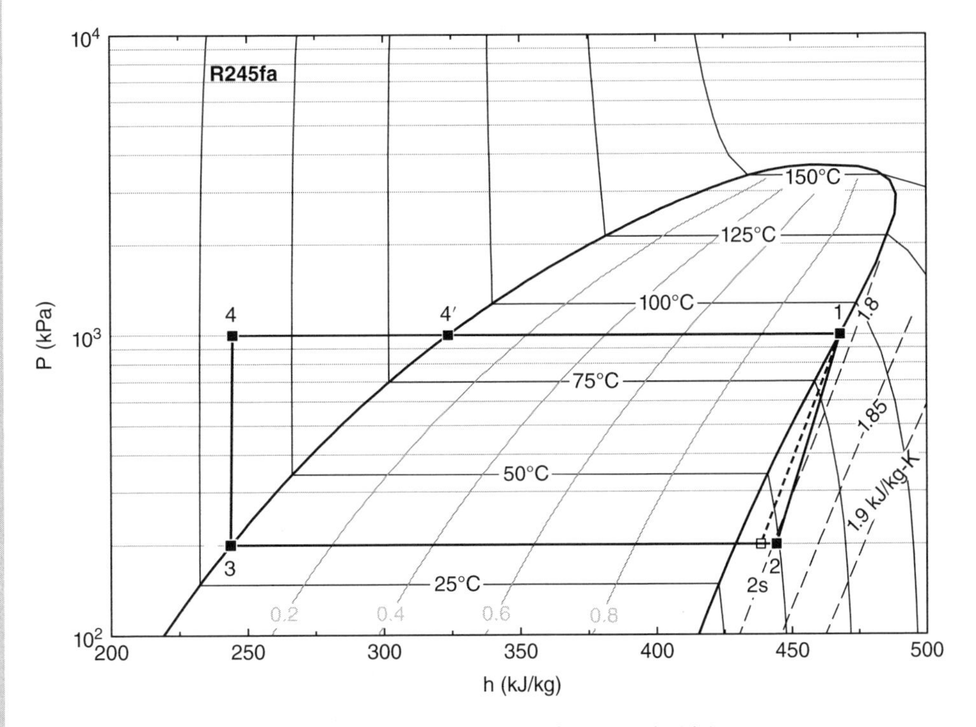

Figure E.14.1 *P–h* diagram for Example 14.1

(b) heat rejected at the condenser: $Q_{Condenser}$=8483 [kW]
(c) heat required to preheat the working fluid: $Q_{Preheat}$=3349 [kW]
(d) heat required to boil the working fluid: Q_{Boiler}=6100 [kW]
(e) working fluid temperature leaving the turbine: $T2$=49.02 [C]

(f) condensing temperature of the working fluid: $T3$=33.45 [C]
(g) boiling temperature of the working fluid: $T1$=89.61 [C]
(h) required mass flow rate of the working fluid: $mdot$=42.31 [kg/s]
(i) thermal efficiency of the cycle: n_{cycle}=0.1022 [-]

***Discussion*:** This example demonstrates that we can make considerable progress by knowing the condensing and boiling pressures, and by making some engineering approximations. Note that knowledge of the working fluid mass flow rate is not necessary to analyze the cycle. When the working fluid mass flow rate is known or assumed, calculation of all heat transfer rates is possible. Note also the relatively high amount of heating needed to raise the temperature of the liquid working fluid to saturation (process 4–4′). Of the total heat added to the working fluid (process 4–1), approximately 35% is attributed to preheating.

14.3.4 Heat Exchanger Analysis

Example 14.1 illustrates an engineering model of an ORC where the operating pressures, turbine power output, and isentropic efficiencies of the turbine and pump were inputs. This simple model results in very useful outcomes, such as the thermodynamic state points, heat transfer rates (when the working fluid mass flow rate is known or assumed), and the cycle efficiency. However, just as was discussed with heat pumps, the heat extraction effectiveness from the source fluid and heat rejection effectiveness to the environment dictate the operating pressures of the working fluid cycle. Thus, we are often interested in the temperatures and flow rates of the heat transfer fluids and the effectiveness of the heat exchange to/from the source/sink.

In Section 12.3.4 the use of the *heat exchanger effectiveness* (ε) concept in heat exchanger analysis was discussed, and in Section 13.3.4 the concept of the approach temperature was used to describe heat exchanger performance. Here, we will use both.

In our ORC analysis, the simple approach temperature concept is used to determine the boiling and condensing temperatures of the working fluid. Thus, the boiling and condensing temperatures of the working fluid can be determined with the use of Equations (13.8) and (13.9) respectively.

In our ORC analysis, the heat exchanger effectiveness is used to determine an important heat exchanger parameter not yet discussed: the heat exchanger UA, where U is the overall heat transfer coefficient ($W{\cdot}m^{-2}{\cdot}K^{-1}$), and A is the heat exchanger area (m^2). The heat exchanger UA is important in heat exchanger design, as it expresses the size (and cost) of the heat exchanger. With knowledge of the heat exchanger effectiveness, UA is determined from Equation (12.13), where C represents the minimum heat capacity rate of the two fluids in the heat exchanger.

Further discussion of the heat balance and heat exchanger effectiveness equations is warranted here for ORC preheaters and boilers. Table 14.1 summarizes the appropriate temperature differences in the heat balance and heat exchanger effectiveness equations. $T_{geo,supply}$ refers to the geothermal fluid entering the boiler, $T_{geo,mid}$ refers to the geothermal fluid exiting the preheater, and $T_{geo,return}$ refers to the geothermal fluid returning to the injection well (see Figure 14.3).

Table 14.1 Summary of Appropriate Temperature Differences in Heat Balance and Heat Exchanger Effectiveness Equations for Geothermal Boiler Analysis

Equation	Preheating	Boiling
Heat Balance on Geothermal Fluid: $q = \dot{m}c_p(T_i - T_o)$	$(T_i - T_o) = (T_{geo,mid} - T_{geo,return})$	$(T_i - T_o) = (T_{geo,supply} - T_{geo,mid})$
Heat Exchanger Effectiveness: $q = \varepsilon C(T_{h,i} - T_{c,i})$	$(T_{h,i} - T_{c,i}) = (T_{geo,mid} - T_4)$	$(T_{h,i} - T_{c,i}) = (T_{geo,supply} - T_{4'})$

Thus, we have the following energy balance equations for the boiler and preheater:

$$\dot{q}_{boiler} = \dot{m}_{geo}c_{p,geo}\left(T_{geo,supply} - T_{geo,mid}\right) \tag{14.9}$$

and

$$\dot{q}_{preheat} = \dot{m}_{geo}c_{p,geo}\left(T_{geo,mid} - T_{geo,return}\right) \tag{14.10}$$

Similarly, we have the following heat exchanger effectiveness equations for the boiler and preheater:

$$\dot{q}_{boiler} = \varepsilon_{boiler}C_{geo}\left(T_{geo,supply} - T_{4'}\right) \tag{14.11}$$

and

$$\dot{q}_{preheat} = \varepsilon_{preheat}C_{min}\left(T_{geo,mid} - T_4\right) \tag{14.12}$$

where C_{min} in Equation (14.12) refers to the minimum heat capacity rate of the geothermal fluid and the working fluid.

To complete our heat exchanger analysis, it is useful to examine yet another thermodynamic diagram: the temperature–heat transfer or T–q diagram for the preheater–boiler combination (Figure 14.5). This was briefly described in Figure 12.7 in relation to the evaporator and condenser of heat pumps. In this case, for the geothermal boiler, the abscissa represents the total amount of heat that is transferred from the geothermal fluid to the working fluid. It can be expressed either in percent or in heat transfer units.

The preheater provides sensible heat to raise the working fluid to its boiling point, state 4′. Boiling occurs from 4′–1 along an isotherm for a pure working fluid. The place in the heat exchanger where the brine and the working fluid experience the minimum temperature difference is typically called the pinch-point, and the temperature difference between $T_{geo,mid}$ and $T_{4'}$ is the pinch point temperature difference. Smaller pinch-point temperature differences correspond to a more effective heat exchange.

A similar analysis could be conducted for the condenser of ORC geothermal plants. However, as the relative amount of desuperheating relative to condensing is small, we will restrict our analysis of condensers to one heat exchanger effectiveness equation.

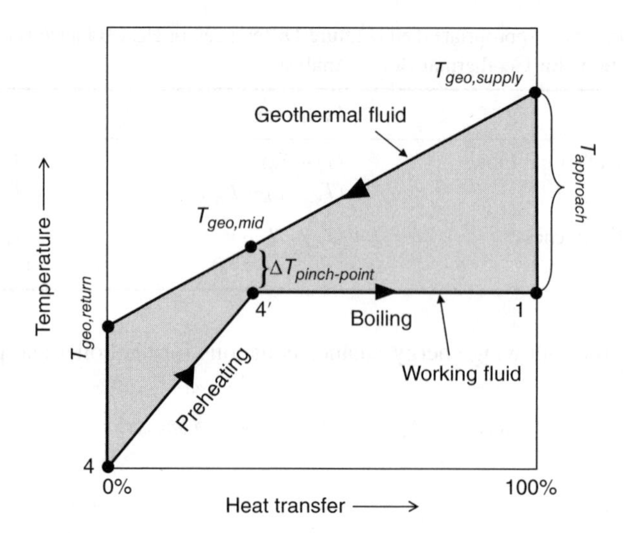

Figure 14.5 Schematic of temperature change vs. percent heat transfer (or *T–q* diagram) in the boiler of a geothermal ORC power plant

14.3.5 Parasitic Loads

Parasitic loads may account for up to a third of the electrical power generation of small geothermal binary power plants. Thus, reduction of these loads is important.

Parasitic loads of significance to binary geothermal power plants include: (i) well pump(s), (ii) the cooling tower pump (if wet tower), and (iii) cooling tower fans or air-cooled condenser fans. Other parasitic loads include lighting and other power requirements needed to operate the plant.

Pumps and fans have two efficiencies to consider: (i) the mechanical work efficiency (what is technically considered to be a pump or fan), and (ii) the motor efficiency. Theoretical mechanical work delivered by a pump or fan is a function of the mass flow rate of the fluid and the pressure (or head) that must be overcome. For pumps, the head includes elevation head, plus friction losses through pipes, valves, etc. For fans, the pressure includes friction losses. Finally, the net work done by the plant includes the net cycle work minus all parasitic loads.

14.4 Organic Rankine Cycle Performance Modeling

We can now consolidate the foregoing to construct a fairly complete mathematical model of an ORC geothermal power plant. The objective is to create a model that is a useful representation of a real ORC. Let's illustrate by constructing a computer simulation model in EES.

Again, we note that there are numerous combinations of possible input and output data for an ORC simulation model, yielding many degrees of freedom. The choice of input and output data depends on what is known, and what is desired to be known. The inputs and outputs are chosen here to obtain as much relationship to real practice as possible, and thus we'll construct a steady-state ORC geothermal power plant model with the data as organized in Table 14.2.

Table 14.2 Summary of Input and Output Data of Interest for Example ORC Geothermal Power Plant Simulation Model (Water-Cooled Plant)

Fluid Cycle	Input Data	Output Data
Geothermal (Boiler)	• temperature of the geothermal fluid • pressure of the geothermal fluid • allowable temperature drop • boiler approach temperature	• required mass flow rate of geothermal fluid • heat transfer rates of boiling and preheating • *T*–*q* diagram • boiler and preheater *UA* values
Working Fluid (ORC)	• working fluid • turbine power output • isentropic efficiency of the turbine and feed pump	• required mass flow rate of working fluid • thermodynamic state of all points in the cycle (and *P*–*h* diagram)
Condenser	• thermal properties of the cooling fluid • allowable temperature rise • condenser approach temperature	• required mass flow rate of cooling fluid • heat rejection rate • condenser *UA* value
Environment	• humidity • atmospheric pressure • dead-state temperature • cooling tower approach temperature	• wet-bulb temperature
System	• parasitic loads	• Carnot efficiency • first-law efficiency • second-law efficiency

Example 14.2 Simulation of a Water-Cooled Geothermal ORC Power Plant with Boiler Preheater

A geothermal binary power plant is being designed with R245fa as the working fluid. The geothermal resource temperature is 95 °C, and an existing well is capable of supplying fluid to the ORC plant at 120 kPa. The plant is to be water cooled. The boiling and condensing approach temperatures are 20 and 10 °C respectively. The allowable temperature drop of the geothermal fluid is 20 °C, and the cooling tower is to be designed for a temperature difference of 7 °C for the cooling water entering and leaving the condenser. The dead-state temperature of the environment is taken as the dry-bulb temperature of 279 K. The relative humidity is 50% and the atmospheric pressure is 100 kPa.

The power plant will be designed with a separate preheater, where the geothermal fluid will enter the boiler first, and then the geothermal fluid exiting the boiler will preheat the working fluid prior to boiling. An isentropic efficiency of 85 and 75% for the turbine and pump, respectively, may be assumed. For a design turbine power output of 1000 kW, and ignoring parasitic loads and stray heat losses, determine/show the following:

(a) a sketch of the cycle on a *P–h* diagram and show the ideal (isentropic case),
(b) a sketch of the heat transfer in the preheater/boiler on a *T–q* diagram,
(c) heat transfer rates in the condenser, preheater, and boiler,
(d) required mass flow rates of the geothermal fluid, working fluid, and cooling fluid,
(e) heat exchanger *UA* values of the preheater, boiler, and condenser, and
(f) Carnot, first-law, and second-law (or utilization) efficiencies of the plant.

Solution

The EES input data are as follows.

```
//GIVEN:
//Cycle Input Data
W_turbine = 1000 [kW]                    "the turbine desired output"
n_turbine = 0.8; n_pump = 0.75           "the turbine and pump isentropic efficiency"
R$ = 'R245fa'                            "string describing the working fluid"
//Boiler Input Data
Tgeo_supply = 95 [C]                     "the geothermal fluid temperature"
DT_geofluid = 20 [C]                     "the design temperature change of the geo-
                                         thermal fluid"
BoilerApproach = 20 [C]                  "the boiler approach temperature"
//Condenser Input Data
CondApproach = 10 [C]                    "the approach between the cooling water and
                                         refrigerant condensing temperature"
DT_condenser = 7 [C]                     "the design temperature change of the con-
                                         denser fluid"
//Environment
CTApproach = 18 [C]                      "the approach between the cooling water and
                                         the ambient wet bulb temperature"
RH = 0.5; Patm = 98.1 [kPa]              "the relative humidity and the atmospheric
                                         pressure"
T0 = 279 [K]                             "the dead state temperature of the environment"

//SOLUTION:
//ASSUMPTIONS:
x1=1; x3 = 0
//Calculate some properties, assuming constant specific heats:
cp_geo = Cp(Water,T=Tgeo_supply, x=0)    "specific heat of the geothermal fluid"
cp_clg_fluid = Cp(Water,T=Tcond_in, x=0) "specific heat of the condenser cooling fluid"
cp_working_fluid = Cp(R$,T=T4, x=0)      "specific heat of the liquid refrigerant"
//For cooling tower
Tdb = Converttemp('K', 'C', T0)
Twb = WETBULB(AIRH2O, P=Patm, T =        "using EES psychrometric function"
Tdb, R = RH)
Tcond_in = Twb + CTApproach              "the condenser inlet fluid temperature"
```

```
//Saturation Temperatures and Pressures of the Working Fluid
T_sat_boiler = Tgeo_supply - Boiler        "saturation temperature in boiler"
Approach
T_sat_condenser = Tcond_in + Cond          "saturation temperature in condenser"
Approach
P_sat_boiler = P_sat(R$, T = T_sat_boiler) "saturation pressure in boiler"
P_sat_condenser = P_sat(R$, T = T_sat_     "saturation pressure in condenser"
condenser)

//Point 1:
T1= T_sat_boiler; h[1]=Enthalpy(R$,T=T1,x=x1); P[1] = P_sat_boiler
//Point 2:
P[2]= P_sat_condenser; T2 = Temperature(R$,P=P[2],h=h[2])
n_turbine = (h[1]-h[2])/(h[1]-h2s)         "this equation is used to find h[2]"
//Point 2s
s1 = Entropy(R$,P=P[1],x=x1); h2s = Enthalpy(R$,P=P[2],s=s1)
//Point 3:
P[3] = P[2]; T3= T_sat_condenser; h[3]=Enthalpy(R$,T=T3,x=x3)
//Point 4:
P[4] = P[1]; T4 = T3
n_pump = (h4s-h[3])/(h[4]-h[3])            "this equation is used to find h[4]"
//Point 4 prime:
h[5] = Enthalpy(R$,P=P_sat_boiler,x=0); P[5] = P_sat_boiler
//Point 4s
s3 = Entropy(R$,P=P[3],x=x3); h4s = Enthalpy(R$,P=P[4],s=s3)
//Energy Balances
W_Turbine = mdot * (h[1] - h[2])
Q_Condenser = mdot * (h[2] - h[3])
W_Pump = mdot * (h[4] - h[3])
Q_Boiler = mdot * (h[1] - h[5])
Q_Preheat = mdot * (h[5] - h[4])
//Geothermal and Cooling Fluid Energy Balance Calculations
Tgeo_out = Tgeo_supply - DT_geofluid
Q_Condenser = mdot_ct * cp_clg_fluid *DT_condenser
Q_Boiler = (mdot_geo * cp_geo * (Tgeo_-    "boiler energy balance"
supply - Tgeo_mid))
Q_Preheat= (mdot_geo * cp_geo *(Tgeo_-     "pre-heat energy balance"
mid - Tgeo_out))
gpm_geo = mdot_geo * 60 [s/min] * convert  "the geothermal flow rate in USgpm"
('kg','lbm')/8.34 [lbm/gal]
gpm_ct = mdot_ct * 60 [s/min] * convert    "the cooling water flow rate in USgpm"
('kg','lbm')/8.34 [lbm/gal]

//Efficiency Analysis
n_cycle = (W_Turbine - W_Pump)/            "the 1st Law Efficiency"
(Q_Boiler+Q_Preheat)
n_max = 1 - T0/converttemp('C', 'K',       "the Carnot Efficiency"
Tgeo_supply)
n_u = n_cycle/n_max                        "the 2nd Law Efficiency"
```

```
//HX Analysis
C_dot_1=mdot_geo*cp_geo; C_dot_2=m-    "heat capacity rate of geofluid and working
dot*cp_working_fluid                   fluid"
C_dot_3 = mdot_ct *cp_clg_fluid        "heat capacity rate of cooling fluid"
Q_Boiler = (epsilon_evap*C_dot_1*      "boiler heat exchanger effectiveness relation"
(Tgeo_supply-T_sat_boiler))
Q_Preheat = (epsilon_preheat * min     "pre-heater heat exchanger effectiveness
(C_dot_1, C_dot_2) * (Tgeo_mid - T4))  relation"
Q_Condenser = epsilon_cond * C_dot_3 * "condenser heat exchanger effectiveness
(T2 - Tcond_in)                        relation"
NTU_evap = -ln(1- epsilon_evap)        "NTU relationship for the boiler"
NTU_cond = -ln(1- epsilon_cond)        "NTU relationship for the condenser"
NTU_Preheat = HX('counterflow', epsi-  "NTU relationship for the pre-heater"
lon_preheat, C_dot_1, C_dot_2, 'NTU')
UA_evap = NTU_evap * C_dot_1           "UA of the boiler"
UA_preheat = NTU_Preheat * min         "UA of the preheater"
(C_dot_1, C_dot_2)
UA_cond = NTU_cond* C_dot_3            "UA of the condenser"
T_pinch = Tgeo_mid - T_sat_boiler      "pinch point temperature difference"
```

The following results are obtained:

(a) a sketch of the cycle on a *P–h* diagram, showing the ideal (isentropic case): see Figure E.14.2a

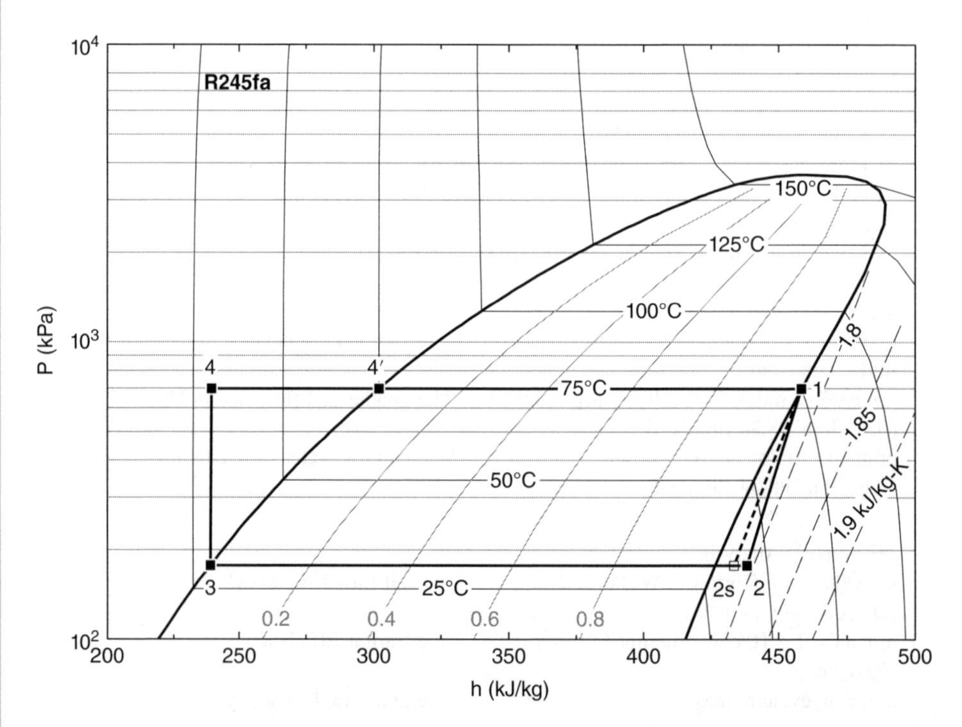

Figure E.14.2a *P–h* diagram for the cycle in Example 14.2

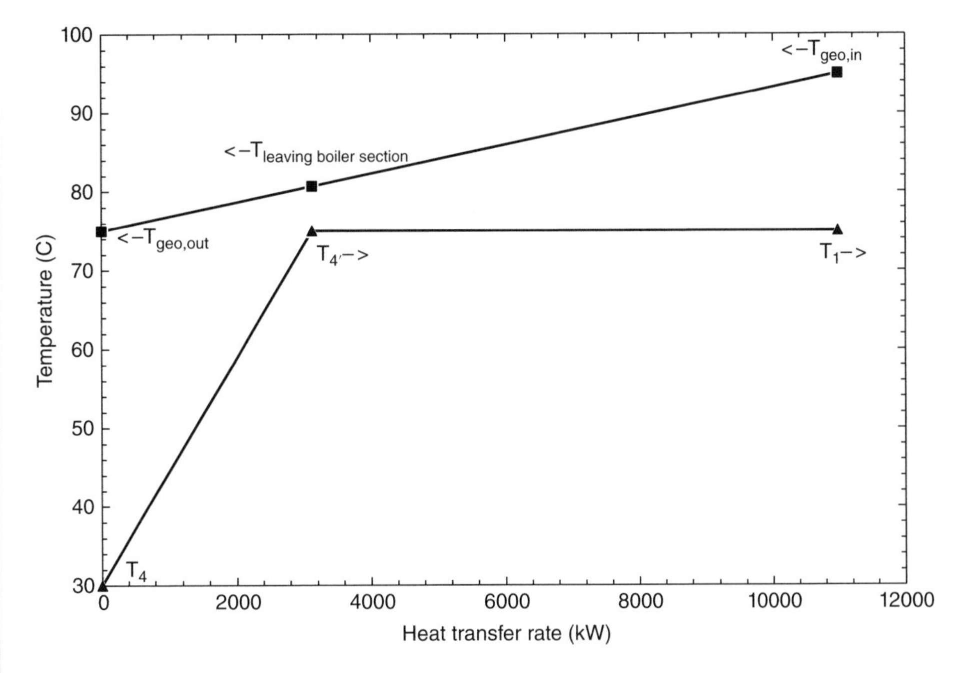

Figure E.14.2b *T–q* diagram for the cycle in Example 14.2

(b) a sketch of the heat transfer in the preheater/boiler on a *T–q* diagram: see Figure E.14.2b
(c) the heat transfer rates in the condenser, preheater, and boiler:

$$Q_{Condenser} = 10002\ [kW],\ Q_{Preheat} = 3123\ [kW],\ \text{and}\ Q_{Boiler} = 7853\ [kW]$$

(d) the required mass flow rates of the geothermal fluid, working fluid, and cooling fluid:

$$mdot_{geo} = 130.3\ [kg/s],\ mdot = 50.18\ [kg/s],\ \text{and}\ mdot_{ct} = 341.6\ [kg/s]$$

(e) the heat exchanger *UA* values of the preheater, boiler, and condenser:

$$UA_{preheat} = 185.3\ [kW/K],\ UA_{evap} = 689.8\ [kW/K],\ \text{and}\ UA_{cond} = 535.9\ [kW/K]$$

(f) the Carnot, first-law, and second-law (or utilization) efficiencies of the plant:

$$n_{max} = 0.2422\ [-],\ n_{cycle} = 0.08872\ [\ dim],\ \text{and}\ n_u = 0.3664\ [\ dim]$$

Discussion: Note the first-law efficiency of the plant, of the order of 9%. This apparently low thermal efficiency is due to the relatively low resource temperature when compared with a fossil-fueled Rankine cycle power plant. The second-law (or utilization) efficiency is of the order of 36%. Note also the low pinch-point temperature difference of 5.7 °C owing to the relatively large *UA* values of the preheater and boiler. Combined, these are about 1.6 times larger than the *UA* of the condenser.

14.5 Chapter Summary

This chapter presented an overview of the Organic Rankine cycle by first comparing it with the Carnot heat engine cycle. Ideal and non-ideal Rankine heat engine cycles were examined in some detail, and cycles were analyzed on pressure–enthalpy diagrams. The mechanics of operation of a geothermal binary Rankine cycle power plant were discussed in some detail. Thermodynamic irreversibilities were considered, specifically a detailed heat exchanger analysis, allowing us to develop detailed computer simulation models of geothermal binary power plants.

Discussion Questions and Exercise Problems

14.1 Using the supplied EES file, repeat Example 14.1, but with the working fluids (i) R134a, (ii) ammonia, (iii) propane, and (iv) isobutane, and compare the results. To compare, use the same condensing and evaporating temperatures of each working fluid and answer/show the following:

(a) a sketch of the cycle on a *P–h* diagram and show the ideal (isentropic case),
(b) heat rejected at the condenser,
(c) heat required to preheat the working fluid,
(d) heat required to boil the working fluid,
(e) working fluid temperature leaving the turbine,
(f) condensing pressure of the working fluid,
(g) boiling pressure of the working fluid,
(h) required mass flow rate of the working fluid, and
(i) thermal efficiency of the cycle.

14.2 The following data have been collected on a geothermal binary power plant using R245fa as the working fluid.

Geothermal Fluid:	T_{in} = 91.5 °C; T_{out} = 68.9 °C; $\dot{m}$ = 28.6 kg·s^{-1}
Cooling Fluid:	T_{in} = 18.1 °C; T_{out} = 23.9 °C; $\dot{m}$ = 86.8 kg·s^{-1}
Turbine Power Output:	214 kW
Environment Temperature:	2.5 °C

Using the supplied EES file for Example 14.1, modify the heat exchanger approach temperatures to 'calibrate' the model to the measured data. Use the calibrated model to determine/show the following:

(a) a sketch of the cycle on a *P–h* diagram and show the ideal (isentropic case),
(b) a sketch of the heat transfer in the preheater/boiler on a *T–q* diagram,
(c) calibrated approach temperatures for the boiler and condenser,
(d) modeled heat transfer rates in the condenser, preheater, and boiler,
(e) modeled mass flow rates of the geothermal fluid, working fluid, and cooling fluid,
(f) heat exchanger *UA* values of the preheater, boiler, and condenser, and
(g) Carnot, first-law, and second-law (or utilization) efficiencies of the plant.

Hint: Owing to an uninsulated boiler, this plant has heat losses that cannot be ignored. Thus, you must modify your energy balance equations accordingly to account for these heat losses.

14.3 Edit the EES file supplied for Example 14.1 to examine further the heat transfer at the condenser. State any assumptions you make. Use your model to determine/show:

(a) the heat transfer rate associated with desuperheating,
(b) the heat transfer rate associated with condensing,
(c) the effectiveness (ε) and UA value of the desuperheating stage,
(d) the effectiveness (ε) and UA value of the condensing stage, and
(e) a sketch of the heat transfer in the condenser on a T–q diagram, showing the desuperheating and condensing processes.

14.4 Edit the EES file supplied for Example 14.1 to model an air-cooled binary geothermal power plant.

Part IV

Energy Distribution

In Part IV, we discuss the distribution of thermal energy to end uses. **Chapter 15** discusses 'inside the building' considerations for forced-air and hydronic heating and cooling systems in buildings. **Chapter 16** concludes with a discussion of economic and environmental impacts of geothermal energy systems, focusing on energy economics and carbon emissions.

Geothermal Heat Pump and Heat Engine Systems: Theory and Practice, First Edition. Andrew D. Chiasson.

Companion website: www.wiley.com/go/chiasson/geoHPSTP

Part IV

Energy Distribution

In Part IV, we discuss the distribution of [illegible] energy [illegible]. Chapter 15 [illegible]. Chapter 16 [illegible].

[illegible]

15

Inside the Building

15.1 Overview

The focus of this chapter is conveying thermal energy to end uses in buildings, with a specific focus on aspects related to geothermal heat pumps. Traditional HVAC topics are not discussed in detail; readers are referred to definitive sources on HVAC topics (e.g., duct design), such as ASHRAE. A series of periodicals on hydronic heating and cooling is published by Caleffi Idronics, and is freely available.

Much of this chapter is a continuation of Chapter 10, where the focus was on the art and science behind the design of the heat exchange fluid transfer piping outside the building, within the GHX itself, in addition to mechanical room components for proper fluid flow management. This present chapter extends that knowledge to complete the picture of hydronic system design for the building services.

Learning objectives and goals:

1. Key principles in laying out distribution piping inside the building.
2. Special considerations for water-to-water heat pump applications.
3. Energy-efficiency measures for ventilation air and geothermal heat pumps.

15.2 Heat Pump Piping Configurations

15.2.1 Single-Zone Systems

Figure 15.1 shows a basic piping layout for a geothermal heat pump in a single-zone, closed-loop system. The expansion tank and air and dirt elimination devices have been discussed in Chapter 10.

Geothermal Heat Pump and Heat Engine Systems: Theory and Practice, First Edition. Andrew D. Chiasson.

Companion website: www.wiley.com/go/chiasson/geoHPSTP

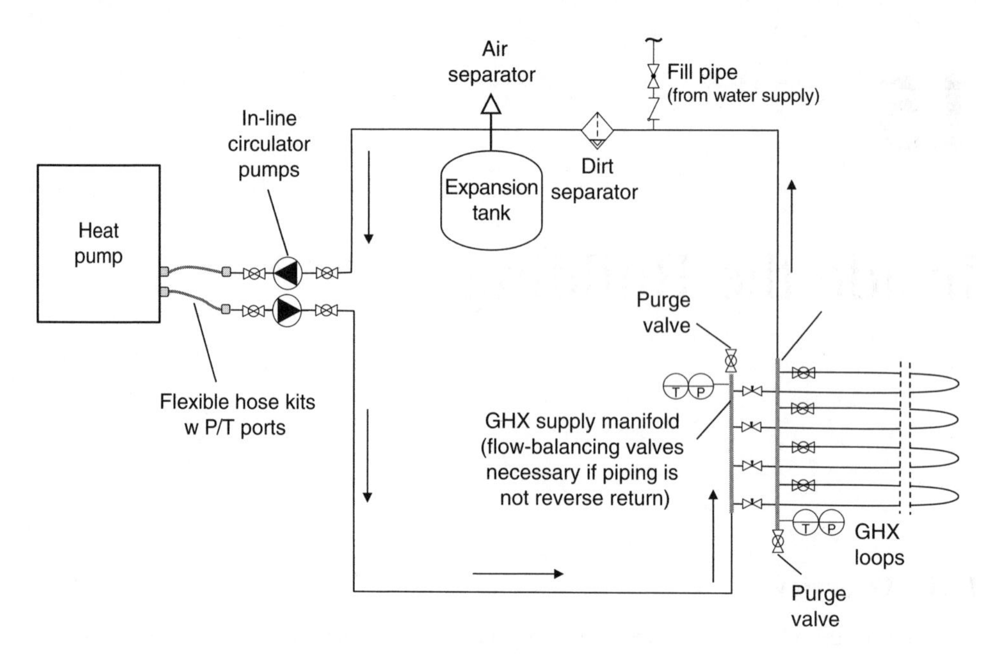

Figure 15.1 Basic piping configuration for a single-zone GHP system

Figure 15.1 depicts two flow circulators in a 'push-pull' arrangement. This is one possible design option where one circulator is deemed impractical to overcome the head losses of the piping system. Circulators in a single-zone application can be variable speed, or be simple on/off control based on when the heat pump is in use. Isolation valves are shown on both sides of the pump so that it can be easily isolated for maintenance.

The heat pump is typically connected to the piping system with flexible hose couplings, as shown in Figure 15.1. These flexible hoses are available as 'kits' that can be specified with various accessories such as pressure/temperature (P/T) ports, y-strainers, and flow control valves. Flow control valves are a prudent option to limit the flow rate to the heat pump design flow rate. These valves can be specified as factory preset or adjustable.

15.2.2 Distributed Geothermal Heat Pump Systems

A distributed geothermal heat pump system is meant here as a system with individual or unitary heat pump units *distributed* throughout the building. The main design choices in these systems include where to place the heat pumps (usually in or near the zone they serve), and the configuration and design of the central building piping loop.

15.2.2.1 Two-Pipe Configuration

A two-pipe system is so named because the piping system contains one main fluid supply pipe and one main fluid return pipe. Figure 15.2 shows a basic two-pipe layout for a distributed

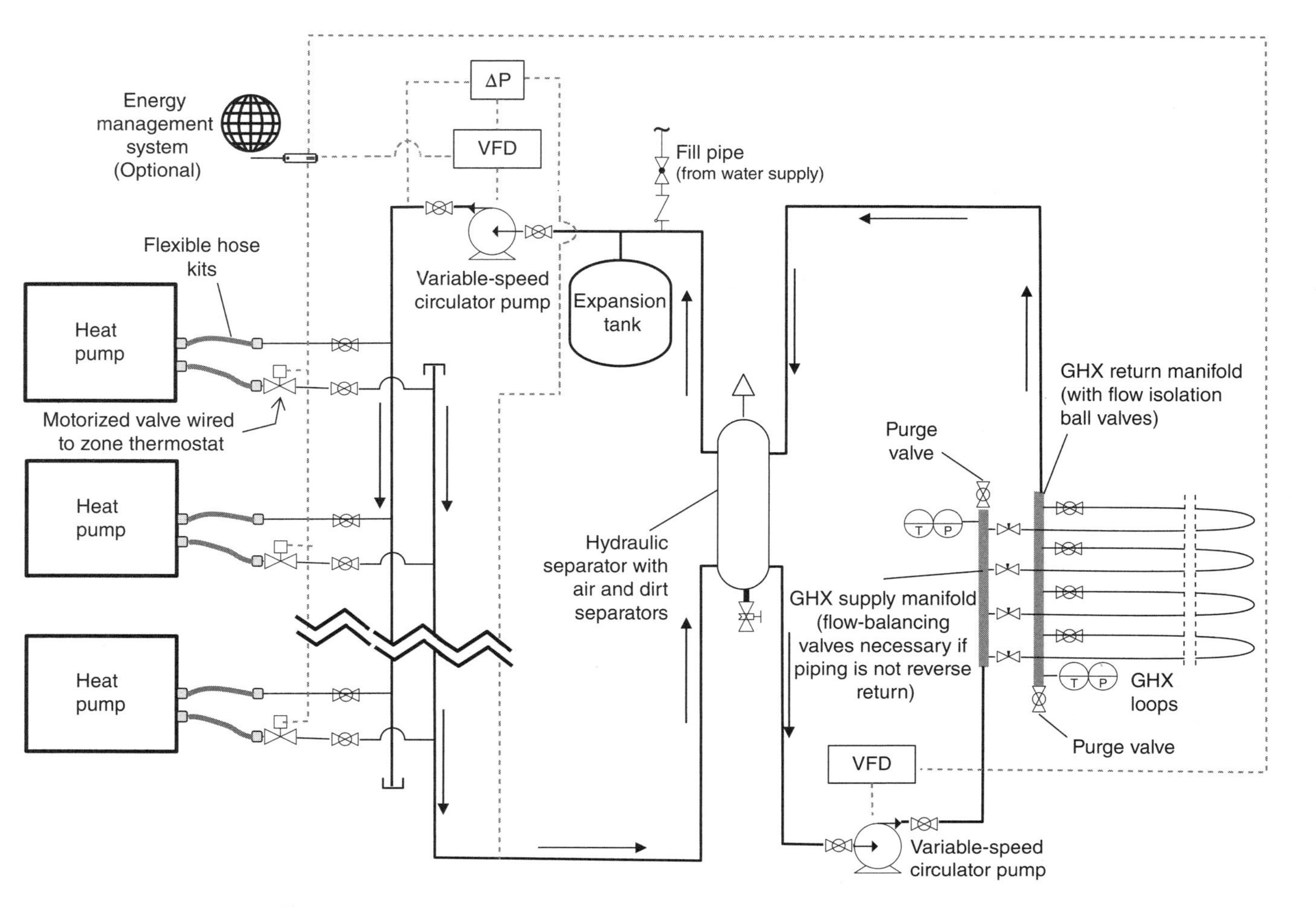

Figure 15.2 Basic piping configuration for a two-pipe, distributed GHP system

geothermal heat pump system in a multizone, closed-loop system. The expansion tank and air and dirt elimination devices have been discussed in Chapter 10.

Figure 15.2 depicts a central building-loop pump. An alternative design option could employ local circulators on each heat pump, as shown in Figure 15.1, so long as head losses through the remainder of the system can be overcome. If not, a small central circulating pump could be added in series with the local circulators.

The central circulating pump shown in Figure 15.2 is controlled to maintain a set differential pressure across the heat pump supply and return headers. This is accomplished with a VFD and a differential pressure transducer (labelled as ΔP), which senses pressure changes in the pipe system as motorized valves on heat pumps open and close. The motorized valves are connected to thermostats in each zone, and are energized to open when the zone calls for heating or cooling. The greatest flow rate through the main supply and return pipes is required when all motorized valves are open. The GHX circulating pump may be controlled in the same way, or by the GHX fluid temperature.

15.2.2.2 One-Pipe Configuration

The one-pipe system, as the name implies, employs a single distribution pipe that acts as the supply and return. These types of system have been historically used in heating systems, particularly in those using steam heat, where a single pipe serves as both the steam supply and condensate return line.

Mescher (2009) describes a one-pipe central distribution loop as a simple pipe system for use with distributed geothermal heat pumps. The design is touted to save first costs and operating costs relative to the more conventional two-pipe designs. In the type of piping design described by Mescher (2009), series-connected heat pumps with a local circulator draw fluid from the one-pipe loop when their local thermostat calls for either heating or cooling. The purpose of the local circulators is to overcome head losses to and from the main distribution loop only. The design described by Mescher (2009) includes no variable-speed drives on the main circulating pumps; two (or more) main pumps are staged and controlled by loop temperature, and circulate the fluid through the building and GHX loops.

In the one-pipe design for a distributed geothermal heat pump, extra attention must be given to the system design to ensure that heat pumps receive the proper fluid temperatures during peak loads. With heat pumps connected in series, each heat pump does not receive fluid at the same temperature as with the two-pipe parallel design; each successive heat pump will receive colder (in heating mode) or hotter (in cooling mode) fluid. To avoid undesirable heat pump entering fluid temperatures, some design considerations for one-pipe systems include: (i) employing them in buildings with large load diversity where it is rare for all heat pumps to peak at the same time, and (ii) adding additional thermal storage, such as larger-diameter pipe or tanks.

15.2.3 Central Plant Geothermal Heat Pump Systems

Central plant applications are essentially opposite in concept to distributed heat pump applications: all heat pump equipment is in a central location. As such, the heat pump equipment is usually of the water-to-water type, and hot and/or chilled water is distributed to fan coils or radiant panels at the end use. One possible design concept was discussed in Chapter 9.6.2, and shown in Figure 9.5.

15.3 Hydronic Heating and Cooling Systems

Thus far, we've been mainly dealing with piping considerations on the 'geothermal side' of the heat pump. In hydronic heating cooling systems, there are similar piping and fluid flow considerations on the 'load side' of the heat pump. In addition to central plant concepts discussed above, there are a number of other possible geothermal hydronic heating and cooling applications, such as:

- radiant floor heating/cooling,
- providing hot/chilled water to fan coil units,
- providing domestic hot water,
- snow-melting,
- swimming pool heating,
- ventilation air preconditioning.

As we know, the COP of any heat pump is directly related to the temperature difference between the source and sink. Thus, in hydronic applications, distribution systems that operate at water temperatures close to the geothermal source temperature are most efficient. This is typically important with heating systems; low-temperature heating systems must employ large heat transfer surfaces relative to high-temperature heating systems (e.g., combustion-based furnaces or boilers). Thus, geothermal hydronic heating applications typically include radiant floor or radiant wall panels at the end use. Water-to-water heat pumps using refrigerant R410A are best suited to heating applications using supply water temperatures less than about 120 °F or 50 °C.

One main design feature associated with hydronic heating and cooling systems is the use of a buffer tank on the load side of the heat pump, particularly when serving multiple zones (see Figure 9.5). The main purpose of the buffer tank is to prevent the heat pump from short-cycling. Thus, the system is controlled such that the heat pump maintains the temperature of the tank within upper and lower design limits. The tank set-points can be varied according to outdoor air temperature (known as 'outdoor reset'). The buffer tank also provides hydraulic separation between the circulator on the load side of the heat pump and the variable-speed circulator in the distribution system. The size of the buffer tank depends on the acceptable minimum run time of the heat pump.

A case study of a multizone community center located in Oregon, USA, heated and cooled by a radiant floor system coupled to a water-to-water geothermal heat pump, is described by Chiasson (2005b).

15.4 Forced-Air Heating and Cooling Systems

Forced-air heating and cooling systems are those that use air as the heat transfer medium to deliver thermal energy to end uses. The majority of GHP systems in North America are forced-air systems, given the need for cooling and dehumidification. These systems distribute air throughout buildings via fans and ductwork.

The principles of fluid flow discussed in Chapter 10 regarding liquid flow in pipes also applies to air flow in ducts. Duct design is another art as well as science, and there is not much that is unique to GHP systems relative to conventional systems. Thus, readers are referred elsewhere for a detailed treatment of duct design.

To summarize duct design procedures briefly, they are very similar to the procedures discussed in Chapter 10 for pipe system design. Air flow requirements and duct distances are typically known and laid out first on a building floor plan drawing, and based on that information the engineer must then select the duct diameter, select the fittings, and determine a duct configuration that results in sufficient flow to the end uses. The total pressure drop caused by the duct system is then determined. The selection of duct diameter generally involves a trade-off between the first cost of the duct and fan energy costs of the lifetime of the system, both of which are highly dependent on duct diameter. Large-diameter ducts have a higher initial cost, but result in reduced friction losses and fan energy costs.

To generate a starting place for sizing ducts carrying less than 40 000 cfm or 68 000 m^3/h of air, select the duct diameter such that friction losses are less than ~0.10 in-H_2O per 100 ft of duct, or less than ~0.80 Pa/m.

The above design guideline ensures that the air velocity is low enough to avoid excess noise, and provides a reasonable balance between the cost of the pipes/ducts and pump/fan energy costs. Using this as a starting place, subsequent design iterations can identify economically optimum duct diameters. In many cases, the economically optimum duct diameter will be larger than that suggested by the design guideline.

In buildings, duct systems almost always include multiple branches, which result in parallel flow. Several methods exist for designing parallel-flow duct systems, but most are variants of the *equal-friction* and *equal-pressure* methods.

15.5 Ventilation Air and Heat Pumps

15.5.1 Ventilation with Outdoor Air

What's so important about outdoor air? Outdoor air is introduced to occupied spaces for the purpose of controlling indoor air quality (IAQ). Fresh, outdoor air replaces stale indoor air, and thus removes air contaminants, odors, and other undesirable particulates and chemicals in building air.

Prior to the first energy crisis in the early 1970s, not much attention was given to indoor air quality. The answer was to open a window, as energy was cheap. Ironically, when energy use became a concern in the 1970s, tighter-envelope buildings drew more attention to IAQ and 'sick building syndrome'. It was discovered that some of the building materials themselves, in addition to synthetic materials in carpets and furnishings, were off-gassing undesirable chemicals into building air. This gave rise to the development of building codes to regulate the amount of fresh air needed in buildings based on building use.

One such standard regarding IAQ is ASHRAE Standard 62, which dictates the amount of outdoor air required to control IAQ. That and other such standards are used to determine the ventilation air requirements based on occupancy and/or carbon dioxide levels in the building air.

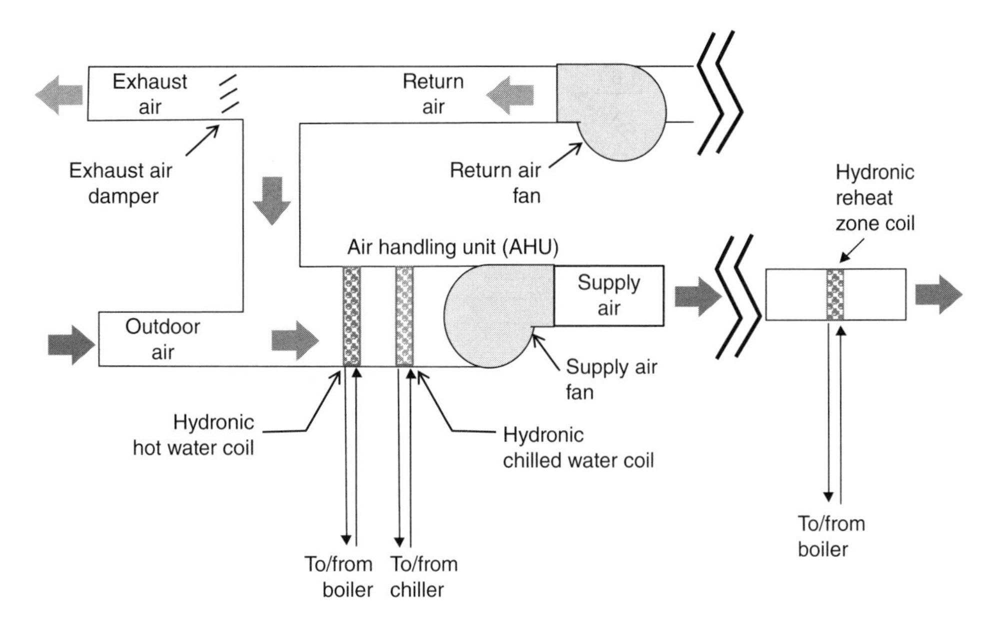

Figure 15.3 Basic configuration of a fan system showing building air exhaust and outdoor air intake

Figure 15.3 is a schematic of a conventional approach to outdoor air management – an HVAC system consisting of a four-pipe configuration (i.e., with separate piping systems for heating and cooling systems). As shown in Figure 15.3, outdoor air is introduced directly to the air-handling unit (AHU), replacing the mass fraction of air exhausted from the building. One energy efficiency measure of this type of system is *demand-controlled ventilation*, which includes controlling the exhaust air damper based on occupancy and/or carbon dioxide levels in the return air stream. However, on a cold winter day or a hot, humid summer day, outdoor air can significantly increase the heating or cooling load on the HVAC equipment.

15.5.2 Outdoor Air and Single-Zone Geothermal Heat Pumps

Introducing outdoor air to any heat pump requires some special considerations regarding the supply air temperature. Recall from Chapter 12 that heat pumps have a relatively fixed capability of changing the air temperature across the refrigerant coil. This fixed capability is, in part, due to the factory fan speed settings and the compressor capacity. Thus, another energy efficiency measure with geothermal heat pumps handling outdoor air is to include a *ventilation air heat recovery* unit (HRU).

Figure 15.4 is a schematic of a geothermal heat pump coupled to a sensible ventilation air heat recovery unit. The HRU shown in Figure 15.4 acts essentially as a cross-flow heat exchanger, recovering sensible heat only. Total energy recovery units (or enthalpy wheels) are also available to remove moisture from humid outdoor air. Figure 15.4 contains numeric labels

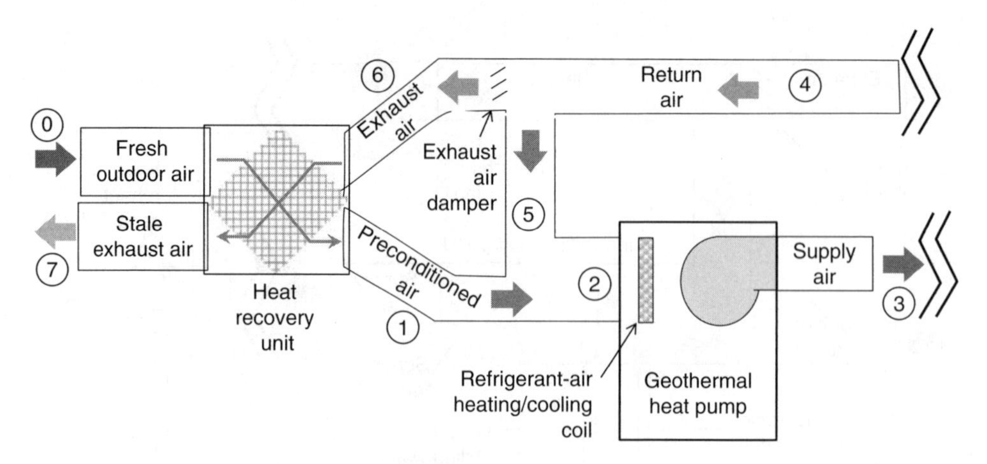

Figure 15.4 Basic configuration of a geothermal heat pump, showing air exhaust and outdoor air intake with a sensible, ventilation air heat recovery unit

for air conditions where the temperature or mass flow rate of the air changes. the following is a description of each point:

- Point 0 = outdoor air,
- Point 1 = preconditioned air after recovering energy from the exhaust air,
- Point 2 = mixed preconditioned air with building return air,
- Point 3 = air supplied to the space,
- Point 4 = air returning from the space, taken as the temperature equal to the thermostat setting,
- Point 5 = the portion of building air returning to the AHU,
- Point 6 = the portion of building air exhausted from the space, entering the HRU, and
- Point 7 = air exiting the HRU from the building.

The benefit of a ventilation air heat recovery unit is best illustrated with the following example.

Example 15.1 Calculating Outdoor Air Loads with and without an HRU

A water-to-air geothermal heat pump is operating in heating mode and serving a meeting room with a thermostat set-point of 22 °C. The calculated space heating load (not including outdoor air loads) is 20 kW. The supply air flow rate to the room is a typical 0.054 $m^3 \cdot s^{-1} \cdot kW^{-1}$.

However, to meet building code requirements, 30% of the building air must be exhausted and replaced with fresh, outdoor air. On a cold winter day when the outdoor air temperature is −10 °C, and assuming a heat exchanger effectiveness of 0.8 for the ventilation heat recovery unit, determine the following *with and without* the HRU:

(a) the additional load on the heat pump due to the outdoor air load, and
(b) the temperature at point 2.

Solution

This problem is readily solved using EES. The EES input code is as follows:

```
"GIVEN:"
"Air Conditions"
T_0 = -10 [C]                               "The outdoor air dry bulb temperature"
T_thermostat = 22 [C]                       "The desired building return air temperature"
Q_dot_air_per_kW = 0.054 [m^3/s-kW]         "Air volumetric flow rate per peak kW heating"
ExhaustFraction = 0.3 [-]                   "Fraction of air exhausted"
q_dot_space = 20 [kW]

"Equipment"
epsilon_HRU = 0.8 [-]                       "HRU HX effectiveness"

"SOLUTION:"
"Assumptions:"
P = 101.1 [kPa]                             "Assume the local atmospheric air pressure"
cp_air = Cp(Air,T = T_thermostat)           "Assume constant cp of air throughout"
rho_air = Density(Air, T = T_thermostat, P = P)  "Assume constant density of air in building"
T_4 = T_thermostat; T_6 = T_4; T_5 = T_4    "By definition of the air conditions"

"Air Mass Flow Calculations"
m_dot_air = Q_dot_air_per_kW * q_dot_space * rho_air   "Mass flow rate of building air"
m_dot_air_exhaust = m_dot_air * ExhaustFraction        "Mass flow rate of exhaust air"
m_dot_air_5 = m_dot_air - m_dot_air_exhaust            "Mass flow rate of air returning to
                                                        heat pump"

"Energy Balance Calculations"

q_outdoor_air = m_dot_air_exhaust * cp_air * (T_4 - T_0) - q_HRU   "Outdoor Air Load"
q_HRU = epsilon_HRU * m_dot_air_exhaust * cp_air * (T_6 - T_0)     "Heat exchanger
                                                                    effectiveness equation.
                                                                    Determines q_HRU"

q_HRU = m_dot_air_exhaust * cp_air * (T_1 - T_0)    "Heat exchanger energy balance on
                                                     outdoor air stream.
                                                     Determines T_1"

m_dot_air * T_2 = m_dot_air_exhaust * T_1 + m_dot_air_5 * T_5      "Determines T_2"
```

The following results are obtained:

	Without HRU	With HRU
Part (a)	Outdoor Air Load = 12.43 kW	Outdoor Air Load = 2.49 kW
Part (b)	T_2 = 12.4 °C	T_2 = 20.1 °C

***Discussion*:** To determine the results with no HRU, the value of ε_{HRU} is simply set to zero.

The assumption of constant specific heat and constant density of air simplifies the problem. Specifically, for the heat exchanger effectiveness equation, the heat capacity rates are then approximately equal on both sides of the heat exchanger.

This example demonstrates the impact of outdoor air loads (under an extreme case) on the heat pump load. The outdoor load in this case is 60% of the space load! Adding an HRU reduces the outdoor air load to only 12.5% of the space load. It would not be realistic simply to consider a larger heat pump with that kind of outdoor air load (with no HRU) because there would be a significant mismatch between the factory fan speed relative to the air flow requirements dictated by the space load. Further, without an HRU, the entering air temperature to the heat pump would be below the manufacturer's recommendations.

For a heat pump in cooling mode, outdoor air loads need to include latent loads, and thus the energy balance would be in terms of enthalpy difference.

15.5.3 Dedicated Outdoor Air Systems

Dedicated outdoor air systems are ducted air-handling systems for the sole purpose of introducing outdoor air to spaces; ventilation air is not part of conditioned air from the heating/cooling system. Thus, heating/cooling equipment serving zones in a multizone building can be sized for the space loads only. Dedicated outdoor air systems can therefore be controlled separately from the heating/cooling system. Figure 15.5 shows a conceptual design for a dedicated outdoor air system for use with a hydronic geothermal heat pump maintaining the supply temperature to the spaces.

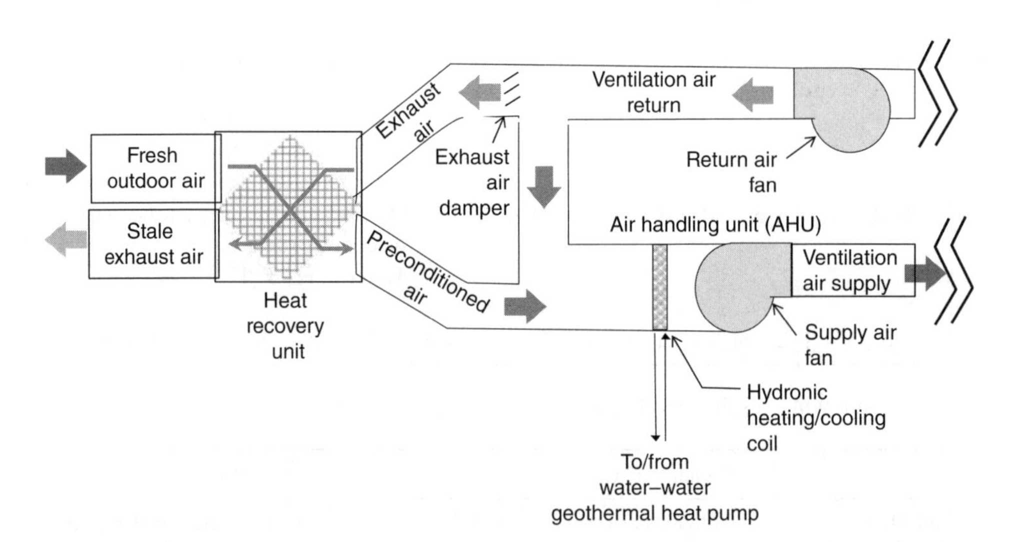

Figure 15.5 Basic configuration of a water-to-water geothermal heat pump in a dedicated outdoor air-handling system

15.6 Chapter Summary

This chapter has provided an overview of 'inside the building' considerations for geothermal heat pump systems. Key principles in laying out distribution piping inside the building were described. Special considerations for water-to-water heat pump applications were discussed, along with a brief overview of duct design. Finally, energy efficiency measures for ventilation air and geothermal heat pumps were examined.

Discussion Questions and Exercise Problems

15.1 Choose to answer this question in either SI or IP units.

Figure P.15.1 shows a one-pipe system in a small office building with four heat pumps conditioning offices located on each side of the building. Consider the system in cooling mode, where the design flow rate is 3 gpm/ton, and a flow control valve (FCV) is installed on each heat pump to control the respective flow rate to each heat pump unit. Circulating pumps are not shown. The GHX has been designed to supply a maximum temperature of 90 °F, or 30 °C, at the design condition. Assuming a Δ of 5 °F, or 2.5 °C, across each heat pump at design conditions, determine the following:

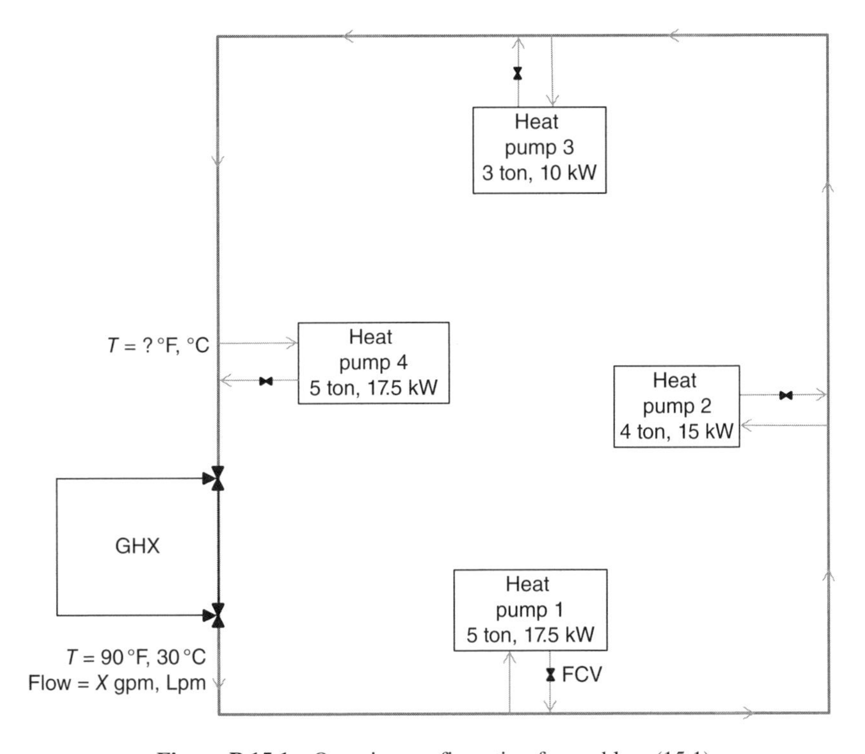

Figure P.15.1 One-pipe configuration for problem (15.1)

(a) The total design flow rate (in gpm, Lpm) through the single pipe, assuming pure water.
(b) A recommended minimum nominal pipe diameter, based on material presented in Chapter 10.
(c) The water temperature entering heat pump 4.
(d) If the COP for heat pump 4 is 4.0 at 90 °F (or 30° C) entering water and 3.4 at 100 °F (or 37 °C) entering water, what is the COP for heat pump 4 at the design condition of this problem?

15.2 Redo Example 15.1, but for a hot summer day with a total energy recovery unit (i.e., a sensible + latent heat recovery unit). The room conditions are 23 °C dry bulb and 50% relative humidity. The outdoor air conditions are 33 °C dry bulb and 90% relative humidity.

16

Energy Economics and Environmental Impact

16.1 Overview

Investments in energy-efficient measures and systems, particularly renewable energy systems, require a relatively large up-front investment in equipment that results in future savings or future income. In order to evaluate these investments, it is necessary to understand how the value of money changes over time. These evaluations are typically studied in a field that has been traditionally called Engineering Economics or, more recently, Energy Economics. Engineers need to be well versed in methods to compare alternatives, or the best ideas may never be realized.

Geothermal energy systems, as well as other renewable energy or energy-efficient systems, may be implemented for one of two main purposes: (i) income generation, such as a power plant or district energy system where some organization acts as a utility and sells energy for profit, or (ii) energy cost savings, such as a geothermal heat pump system installed by a building owner. Depending on the purpose, different economic analyses are appropriate, but both have several common elements and approaches. In brief, costs are incurred at all elements of a project as outlined in this book: resource characterization, energy harnessing, energy conversion, and energy distribution. Renewable energy projects have numerous 'soft costs' that are not necessarily in conventional energy projects. These typically include additional professional fees associated with project feasibility studies, design, and permitting.

In today's highly competitive global economy, energy economics is more important than ever, but there are many other important factors that are difficult to translate into money. Designers are faced with increasing concerns of climate change attributed to carbon and greenhouse gas emissions associated with fossil-fuel energy consumption. Water shortages and water rights are also of increasing concern, and energy engineers are continually under pressure

Geothermal Heat Pump and Heat Engine Systems: Theory and Practice, First Edition. Andrew D. Chiasson.

Companion website: www.wiley.com/go/chiasson/geoHPSTP

to optimize water use in energy systems. Thus, renewable and clean energy projects have other desirable features beyond economics.

Learning objectives and goals:

1. Key principles in evaluating investment alternatives.
2. Basic methods of determining CO_2 emissions from energy systems.

16.2 Simple Payback Period and Rate of Return

The simplest measure of economic feasibility is the simple payback period (SPP), which is the time period required for an investment to create a positive cash flow:

$$\text{SPP} = \frac{\text{incremental cost}}{\text{savings per year}} \tag{16.1}$$

Incremental cost is used here to imply a differential investment between two options. 'Do nothing' is always an option, so in that case the incremental cost would be the full capital cost of an option of doing something. The rate of return (ROR) is the reciprocal of the simple payback and represents the annual return on the investment, expressed in % per year:

$$\text{ROR} = \frac{\text{savings per year}}{\text{incremental cost}} \tag{16.2}$$

Example 16.1 Calculating Simple Payback and Rate of Return

A variable-frequency drive for a pump will cost $US 5000 to install and will save $US 2000 per year in electricity-associated pumping energy costs. Calculate the simple payback period and rate of return for this investment.

Solution

$$\text{SPP} = \frac{\text{incremental cost}}{\text{savings per year}} = \frac{\$\text{US } 5000}{\$\text{US } 2000/\text{year}} = 2.5 \text{ years}$$

$$\text{ROR} = \frac{\text{savings per year}}{\text{incremental cost}} = \frac{\$\text{US } 2000/\text{year}}{\$\text{US } 5000} = 40\% \text{ per year}$$

The simple payback period is exactly that – simple. It is a quick measure of the value of an investment with little calculation effort. However, it ignores the time value of money and periodic cash flows, and thus is not a measure of how profitable one project is compared with another.

16.3 Time Value of Money

The *time value of money* concept is one where money available at the present time is worth more than the same amount in the future owing to its potential to earn interest or to be invested in something that results in future profits. This concept is also useful to compare periodic cash flow to a common time. The most common way to do this is to convert all cash flows into their 'present value', and then compare the present values of alternatives to evaluate investment options. Cash flows involving future amounts of money can be converted to their present values using three methods: (i) the present value of future amounts, (ii) the present value of a series of annuities, and (iii) the present value of an escalating series of annuities.

16.3.1 Present Value of a Future Amount

A common situation is where we invest money (principal or P) in a bank account that pays interest (i) over n compounding periods. The future value (F) of the principal is given by the fundamental equation of exponential growth:

$$F = P(1+i)^{n} \tag{16.3}$$

Equation (16.3) can be rearranged to determine the present value of a future amount:

$$P = F(1+i)^{-n} \tag{16.4}$$

The factor $(1 + i)^{-n}$ is sometimes called the present worth factor, or PWF(i, n).

Example 16.2 Calculating the Present Value of a Future Amount
You want to have \$US 10 000 saved in 10 years. If you find an investment that yields an interest rate of 7% per year, what amount would you need to invest today?

Solution

$$P = F(1+i)^{-n} = \$\text{US}\ 10\,000 \times (1+0.07)^{-10} = \text{US}\ 5083$$

16.3.2 Present Value of a Series of Annuities

An annuity is a regular payment or income that occurs at the end of a fixed period. More specifically, consider investing an annuity of amount A during each of n compounding periods with an interest rate i. The present value of such a series is:

$$P = A \times \left(\frac{1-(1+i)^{-n}}{i}\right) \tag{16.5}$$

The factor $\frac{1-(1+i)^{-n}}{i}$ is sometimes called the series present worth factor, SPWF(i, n). The reciprocal of the series present worth factor is sometimes called the capital recovery factor, CRF(i, n).

Example 16.3 Calculating the Present Value of an Annuity
A standard-efficiency furnace costs $US 5000 and consumes $US 1000 per year in fuel over a 10 year lifetime. A high-efficiency furnace costs $US 7500 and consumes $US 575 per year in fuel over a 10 year lifetime. If interest rates are 10% per year, which is the better investment?

Solution
One solution method could be to convert each series of cash flows to a present-value amount and choose the one with the highest present value. Alternatively, we could look at incremental cost of the high-efficiency furnace and annual savings and choose that option if the present value is positive.

$$P_{incremental} = \Delta P + \Delta A \times \left(\frac{1-(1+i)^{-n}}{i}\right)$$

$$= (\$US\,5000 - \$US\,7500) + (\$US\,1000 - \$US\,575) \times \left(\frac{1-(1+0.10)^{-10}}{0.10}\right)$$

$$= -\$US\,2500 + \$US\,425 \times \left(\frac{1-1.10^{-10}}{0.10}\right)$$

$$= \$US\,111$$

∴ choose the high-efficiency furnace because the present value is positive.

16.3.3 Present Value of an Escalating Series of Annuities

A recurring annuity, A, may sometimes increase over time at some escalation rate e. For example, fuel prices rise with time. The present value of an escalating series is given by

$$P = A \times \left(\frac{1}{i-e}\right)\left(1 - \left(\frac{1+e}{1+i}\right)^n\right), i \neq e \tag{16.6a}$$

$$P = A \times \frac{n}{1+e}, i = e \tag{16.6b}$$

Example 16.4 Calculating the Present Value of an Escalating Series Annuity
Re-evaluate the options in Example 16.3 if the annual fuel escalation rate is 2%.

Solution

$$P_{incremental} = -\$US\,2500 + \$US\,425 \times \left(\frac{1}{0.10-0.02}\right)\left(1 - \left(\frac{1+0.02}{1+0.10}\right)^{10}\right)$$

$$= \$US\,316$$

16.3.4 The Discount Rate and Inflation

Thus far, we have referred to the rate of growth i as the rate of interest. However, engineering economic analyses refer to this growth rate as the *discount rate*, which is the expected blended rate of return from alternative investments. An alternative investment could be interest from a bank, stock market appreciation, or expected profits from one's own company. It is sometimes viewed as an organization's average cost of capital. High discount rates reflect the notion that a large profit can be made from an alternative investment; thus, money today is very valuable and future money is less valuable. High discount rates have the effect of discounting future sums of money or 'discounting the future'. Hence the name 'discount' rate.

Many recurring costs undergo inflation (or deflation) at a fixed percentage each period. Thus, an expense of 1 unit of currency, when inflated at a rate i per time period, will be $1 + i$ at the end of one time period, $(1 + i)^2$ at the end of an additional time period, and so on.

16.4 Cost Considerations for Geothermal Energy Systems

16.4.1 Economic Indicators of Merit

The most comprehensive way to make investment decisions is to consider the cash flows of a system over its life, including planning, design, installation, operation, and decommissioning. Taxes, depreciation, and inflation may also be included. Because the costs and revenues during these phases occur at different times, the time value of money equations described above can be used.

As mentioned at the beginning of this chapter, geothermal energy systems may be installed by an organization for generation of income, or be installed by an entity for energy savings. Thus, there is no universal method for evaluating the economic life cycle of energy projects. The subsections that follow describe a few. The ultimate choice depends on the preference of the person doing the analysis, and more than one indicator is typically examined.

16.4.1.1 Life-Cycle Cost

The life-cycle cost (LCC) of a project is the sum of all the costs associated with the project over its lifetime, in today's currency, and thus considers the time value of money. The basic premise of life-cycle cost is that anticipated future costs are brought back to present cost using a discount rate. This method can include only major cost items or as many details as may be significant, and designers are looking to minimize the LCC of a project.

16.4.1.2 Life-Cycle Savings

Life-cycle savings (LCS) (or net present worth) is defined as the difference between the life-cycle costs of a conventional system and the life-cycle cost of the alternative energy system. A simple version of this concept was presented in Example 16.3. Designers are looking to maximize the LCS of a project.

16.4.1.3 Return on Investment

In general, renewable energy and energy conservation technologies have high first (or capital) costs and low operating (or fuel) costs. Thus, high discount rates (which value the present and 'discount' the future) work against these technologies. Therefore, it may be argued that a discount rate of zero should be used in non-renewable resource decisions; a zero discount rate implies that the future is equally as important as the present. This is where the concept of *return on investment* (ROI) is useful.

The ROI is an economic parameter that represents the discount rate that results in a present value of zero. The power of using this indicator is that it is not necessary to establish the discount rate; an organization interested in a project can compare the internal rate of return with its required rate of return (often, the cost of capital). Another advantage of using the ROI indicator to evaluate a project is that the outcome does not depend on a discount rate that is specific to a given organization; the ROI obtained is specific to the project and applies to all investors in the project.

Example 16.5 Calculating the Return on Investment
Calculate the return on investment for the standard-efficiency furnace of Example 16.3.

Solution
Set $P_{incremental} = 0$ and solve for i. Recall,

$$P_{incremental} = \Delta P + \Delta A \times \left(\frac{1-(1+i)^{-n}}{i}\right)$$

$$0 = (\$US\,5000 - \$US\,7500) + (\$US\,1000 - \$US\,575) \times \left(\frac{1-(1+i)^{-10}}{i}\right)$$

$i = 11\%$ by iteration.

16.4.1.4 Levelized Cost of Energy

The levelized cost of electricity (LCOE) is often cited as a convenient measure to summarize the overall competiveness of different energy-generating technologies. Thus, this metric is appropriate in comparing geothermal power plants or district energy systems with other technologies.

Regarding electricity generation, the levelized cost method represents the cost per kWh or MWh of building and operating a power plant over an assumed financial life and duty cycle. Key inputs to the calculation include capital costs, government or other financial incentives, fuel costs, fixed and variable operating and maintenance (O&M) costs, financing costs, and an assumed utilization rate for each technology.

16.4.2 Costs of Geothermal Energy Systems

Costs are incurred at all elements of a geothermal project, as outlined in this book: resource characterization, energy harnessing, energy conversion, and energy distribution.

Soft costs may be incurred at each step of a project, including professional fees for project management, feasibility studies, environmental impact studies, archaeological studies, cost estimates, permitting, financing, legals, and engineering. Hard costs include materials and labor in the geological, civil, mechanical, structural, electrical, and landscape trades.

Operating costs associated with geothermal energy systems are recurring costs that include cost of energy for operation of main equipment, auxiliary energy costs, additional real estate taxes imposed on the basis of additional assessed value of a building or facility, cost of capital on any funds borrowed to purchase the equipment, labor costs for equipment maintenance, and replacement cost of failed or broken items.

There may be income tax implications in the purchase of renewable or alternative energy equipment. In some countries, such as the United States, interest paid on a loan for its purchase, and extra property tax on an increased assessment due to the equipment infrastructure, may be deductible from income for tax purposes. Equipment purchased by businesses has other tax implications; income-producing resources and equipment may be depreciated, which results in reduced taxable income. If the equipment is for income purposes, there may be investment tax credits or generation tax credits available. Further, the equipment may have salvage or resale value which may result in a capital gains tax.

Thus, annual costs of geothermal energy systems may be summarized as:

$$\begin{aligned}\text{Annual cost} &= \text{energy cost} + \text{parasitic energy cost} + \text{financing or loan payment} \\ &\quad + \text{maintenance cost} + \text{insurance cost} + \text{property taxes} - \text{income tax savings}\end{aligned} \tag{16.7}$$

Actual costs of geothermal installations are scattered and highly variable. Readers are referred to publications summarized in Table 16.1.

16.5 Uncertainty in Economic Analyses

Many assumptions, and therefore uncertainties, are involved in life-cycle economic analyses. One obvious uncertainty is the projection of volatile, future energy costs and escalation rates. A less obvious uncertainty may seem to be capital costs of equipment and installation, but these can also be volatile, depending on contractor availability and desire to undertake a particular project. Thus, it is very important to determine the effects of uncertainties on the calculated economic indicator.

One simple way to quantify the uncertainty of an economic analysis is to conduct a sensitivity analysis on the most uncertain estimates, and show the impact on the economic indicator visually on a scatterplot as shown in Figure 16.1. The procedure involves varying any number of economic parameters of interest (e.g., drilling cost, energy price escalation rate) by, say, ±25%, recalculating the desired economic indicator, and developing a plot of the economic indicator as a function of the percent change in the economic parameter. All curves must pass through the point of 0% change in parameter, because that represents the base case. As implied in Figure 16.1, uncertainties in the parameters with the steepest slope have the greatest sensitivity on the economic indicator. A decision-maker reviewing Figure 16.1 could readily see that the economic indicator is most sensitive to parameter A, and relatively insensitive to parameter C.

Table 16.1 Summary of Sources Documenting Economics of Geothermal Energy Systems

Topic	Reference	Web Link
Economics of GHP systems	Battocletti, E.C. and Glassley, W.E. (2013) *Measuring the Costs and Benefits of Nationwide Geothermal Heat Pump Deployment*. Final Report Prepared for the US Department of Energy	http://cgec.ucdavis.edu/files/2014/08/08-28-2014-DE-EE0002741_Final%20Report_28%20Feb%202013.pdf
Installation costs of commercial GSHP systems	Kavanaugh, S., Green, M., and Mescher, K. (2012) Long term commercial GSHP performance. Part 4: Installation costs. *ASHRAE Journal*, October, pp. 26–36.	http://www.geokiss.com/tech-notes/LongTermGSHPsPt4.pdf
Geothermal heat pump case studies (seven case studies)	*Geo-Heat Center Quarterly Bulletin*, vol. 26, no. 3, September 2005.	http://www.oit.edu/docs/default-source/geoheat-center-documents/quarterly-bulletin/vol-26/26-3/26-3-bull-all.pdf?sfvrsn=4
Direct-use geothermal case studies (seven case studies)	*Geo-Heat Center Quarterly Bulletin*, vol. 25, no. 5, March 2004.	http://www.oit.edu/docs/default-source/geoheat-center-documents/quarterly-bulletin/vol-25/25-1/25-1-bull-all.pdf?sfvrsn=4
Direct-use geothermal case studies (seven case studies)	*Geo-Heat Center Quarterly Bulletin*, vol. 24, no. 2, June 2003.	http://www.oit.edu/docs/default-source/geoheat-center-documents/quarterly-bulletin/vol-24/24-2/24-2-bull-all.pdf?sfvrsn=4

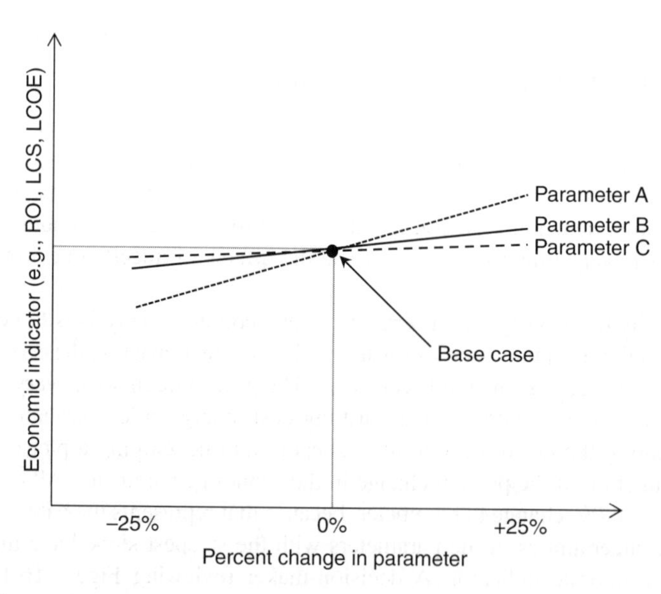

Figure 16.1 Scatterplot showing the sensitivity of three parameters (A, B, and C) on a life-cycle economic indicator

16.6 Environmental Impact

The link between atmospheric greenhouse gases and climate change is well established. Carbon dioxide (CO_2) is the primary greenhouse gas, and to a lesser extent, methane (CH_4) and nitrous oxide (N_2O). While the Earth contains a natural carbon cycle, human activity is altering this natural cycle at a much more rapid rate than normal. Humans influence the ability of natural sinks (e.g., forests and oceans) to remove CO_2 from the atmosphere. The main human activity that emits CO_2 into the atmosphere is the combustion of fossil fuels (coal, natural gas, and oil) for energy.

16.6.1 Fossil Fuel Combustion and CO_2 Emissions

Fossil fuels are combusted to generate electricity in power plants, and to heat buildings. Geothermal energy systems have strong potential to offset current atmospheric carbon emissions relative to those sectors.

In combustion reactions, rapid oxidation of combustible elements of the fuel results in energy release as combustion products are formed. The three major combustible elements common in fossil fuels are carbon, hydrogen, and sulfur. During complete combustion of a hydrocarbon fuel, carbon present in the fuel is converted to carbon dioxide, hydrogen is converted to water, sulfur is converted to sulfur dioxide, and all other combustible elements are oxidized. When these conditions are not achieved, combustion is considered 'incomplete', and primary elements remain in the fuel.

To determine the CO_2 emissions resulting from various energy uses, the mass of CO_2 in the combusted fuel must be known, in addition to the energy content of the fuel, the efficiency of the conversion process, and any other losses in the conversion. Table 16.2 lists greenhouse gas emission factors for common fossil fuels as a function of energy content.

Annual CO_2 emissions may be estimated from energy processes, as illustrated in the following examples.

Table 16.2 CO_2 Emission Factors for Some Common Fossil Fuels

	CO_2 Emission Factor	
Fossil Fuel	(tonne/GJ)	(tonne/MWh)
Coal (average)*	0.095	0.34
Natural Gas	0.055	0.20
Diesel (#2)	0.075	0.27
Fuel Oil (#6)	0.077	0.28

* Note that there are different types of coal: bituminous, anthracite, lignite.

Example 16.6 Calculating CO_2 Emission from a Power Plant
Calculate the annual CO_2 emissions (in tonnes) from a 1000 MW coal-fired power plant with an annual capacity factor of 0.75. Assume the power plant is 35% efficient, and transmission and distribution line losses are 5%.

Solution

The annual electrical energy produced by the plant is given by multiplying the hours per year by the capacity factor by the power plant rating:

$$E_{annual} = 8760\,\text{h} \times 0.75 \times 1000\,\text{MW} = 6.57\text{E}6\,\text{MWh}$$

To calculate the thermal energy input from coal, divide the electrical energy output of the plant by the conversion efficiency, adjusted for transmission and distribution line losses:

$$q_{annual} = \frac{6.57\text{E}6\,\text{MWh}}{0.35 \times (1-0.05)} = 1.98\text{E}7\,\text{MWh}$$

Using the CO_2 emission factor of 0.34 tonne/MWh for coal from Table 16.2:

$$\therefore \text{Annual } CO_2 \text{ emissions} = 1.98\text{E}7\,\text{MWh} \times 0.34 \frac{\text{tonne}}{\text{MWh}} = 6.7 \text{ million tonnes}$$

Example 16.7 Comparing Annual CO_2 Emissions Between a Geothermal Heat Pump and a Gas Furnace

A building with a peak heating load of 20 kW has 1500 equivalent full load heating hours. Compare the annual CO_2 emissions for this building if using a natural gas furnace at 90% seasonal efficiency vs. a geothermal heat pump with a seasonal heating COP of 5. Assume CO_2 emissions attributed to the fans in both systems are offsetting, and that the pumping energy for the geothermal heat pump is 5% of the heat pump energy. Also, assume that electrical energy is provided by the coal plant in Example 16.6.

Solution

The annual heating load for the building is given by multiplying the equivalent full load hours per year by the peak hour heating load:

$$\text{Annual heating load} = 20\,\text{kW} \times 1500\,\text{h} = 30\,\text{MWh}$$

For the Geothermal Heat Pump:

Fuel for the geothermal heat pump and circulating pump is electricity, determined from the heating season COP:

$$E_{geothermal} = \frac{\text{annual heating load}}{\text{seasonal COP}} \times 1.05\,(\text{for pump energy}) = \frac{30\,\text{MWh}}{5} \times 1.05 = 6.3\,\text{MWh}$$

To calculate the thermal energy input from coal at the power plant, divide the electrical energy output of the plant by the conversion efficiency, adjusted for transmission and distribution line losses:

$$q_{geothermal} = \frac{6.3\,\text{MWh}}{0.35 \times (1-0.05)} = 18.95\,\text{MWh}$$

Using the CO_2 emission factor of 0.34 tonne/MWh for coal from Table 16.2:

$$\therefore \text{Annual } CO_2 \text{ emissions} = 18.95\,\text{MWh} \times 0.34\frac{\text{tonne}}{\text{MWh}} = 6.44\,\text{tonnes}$$

For the Gas Furnace:
Fuel for the furnace is natural gas adjusted for seasonal efficiency:

$$q_{furnace} = \frac{\text{annual heating load}}{\eta_{furnace}} = \frac{30\,\text{MWh}}{0.9} = 33.3\,\text{MWh}$$

Using the CO_2 emission factor of 0.20 tonne/MWh for natural gas from Table 16.2:

$$\therefore \text{Annual } CO_2 \text{ emissions} = 33.3\,\text{MWh} \times 0.20\frac{\text{tonne}}{\text{MWh}} = 6.66\,\text{tonnes}$$

Discussion: This example demonstrates the potential of geothermal heat pumps to offset CO_2 emissions relative to conventional systems. Under a somewhat worst-case scenario of 100% electricity produced by a coal-fired plant, CO_2 emissions of a geothermal heat pump are comparable to a natural-gas furnace. A more realistic case of electricity generation would be from a mix of fuels at higher conversion efficiency (e.g., gas turbine plants are of the order of 45% efficient), thus lowering the CO_2 emissions relative to 100% electricity generated by coal-fired plants. As buildings move toward net-zero energy, electricity generated on site from renewable sources, coupled to a geothermal heat pump, will significantly reduce CO_2 emissions from the buildings sector.

16.6.2 Water Consumption

Whenever electrical power is generated at a power plant, or when chilled water is used in a cooling system in a building, the rejection of heat to the environment is an unavoidable result of the second law of thermodynamics. The customary method of discharging waste heat to the environment in chilled water systems and vapor power plants is with evaporative cooling towers. Water is consumed in these systems as a result of evaporative loss of water and 'drift' loss from the tower. Water is also consumed through cooling tower 'blowdown', which is the removal of water concentrated in dissolved constituents that would otherwise result in a build-up of solids in the cooling tower. Make-up water in cooling towers owing to evaporation, drift, and blowdown amounts to about 1–2% of the cooling tower circulation flow.

Geothermal energy systems use little to no water during operation. Regarding cooling systems, geothermal heat pumps obviously do not use wet cooling towers. Regarding the power sector, many geothermal power plants are remote from water sources or water distribution systems, and therefore employ air-cooled condensers.

16.7 Chapter Summary

This chapter has provided an overview of economic and environmental considerations for geothermal energy systems. Key principles in life-cycle economics were discussed, with emphasis on life-cycle cost, life-cycle savings, return on investment, and levelized cost of energy as good economic indicators of project feasibility. Estimation of CO_2 emissions from power plants and heating systems was discussed. Finally, the chapter concluded with a brief overview of water consumption in cooling towers.

Appendix A

Software Used in this Book

A.1 The GHX Tool Box

The GHX Tool Box is a suite of design tools developed by the author to facilitate the study of geothermal heat pump systems. The tools are developed for use in Excel, and include several Visual Basic macros and subroutines. These design tools are meant to be used in conjunction with topics in this book to allow users of the book to apply fundamental principles and solve real-world problems. The Tool Box includes spreadsheets to solve problems related to:

- Eskilson's line source,
- borehole thermal resistance (single U-tube, double U-tube, concentric),
- horizontal trench thermal resistance (two-pipe, four-pipe, six-pipe),
- pressure drop in pipes,
- groundwater heat exchange design,
- vertical ground heat exchanger design (with simple load input),
- vertical ground heat exchanger design and simulation with hybrid options (with detailed load input),
- horizontal ground heat exchanger design,
- Earth tube simulation, and
- surface water heat exchanger simulation.

A.2 Engineering Equation Solver (EES)

Several examples in this book are solved using EES, particularly those requiring iterative solution and use of thermodynamic property data.

Geothermal Heat Pump and Heat Engine Systems: Theory and Practice, First Edition. Andrew D. Chiasson.

Companion website: www.wiley.com/go/chiasson/geoHPSTP

EES (pronounced 'ease') is available through http://www.fchart.com/ees/. It is also packaged and available with many other textbooks. According to the developer, *EES is a general equation-solving program that can numerically solve thousands of coupled non-linear algebraic and differential equations. The program can also be used to solve differential and integral equations, do optimization, provide uncertainty analyses, perform linear and non-linear regression, convert units, check unit consistency, and generate publication-quality plots. A major feature of EES is the high accuracy thermodynamic and transport property database that is provided for hundreds of substances in a manner that allows it to be used with the equation solving capability.*

Several video tutorials are available on the EES website.

A.3 Installing and Using the Excel Solver for Optimization Problems

What is the Excel Solver?

The Excel Solver is an analysis tool that allows users to solve optimization problems.

In mathematics, computer science, or management science, *mathematical optimization* (alternatively, optimization or mathematical programming) is the selection of a best element (with regard to some criteria) from some set of available alternatives. In the simplest case, an optimization problem consists of maximizing or minimizing a real function by systematically choosing input values from within an allowed set and computing the value of the function. The generalization of optimization theory and techniques to other formulations comprises a large area of applied mathematics. More generally, optimization includes finding 'best available' values of some objective function given a defined domain, including a variety of different types of objective function and different types of domain.

Installing the Excel Solver

The Solver is an Add-in program that was installed at the time of Microsoft Office Installation. However, to use it, you must load it.

To load the Solver Add-in:

Click the Microsoft Office Button, and then click Excel Options.

Click **Add-Ins**, and then in the **Manage** box, select **Excel Add-ins**.

Click **Go**.

In the **Add-Ins** available box, select the **Solver Add-in** check box, and then click **OK**.

If you get prompted that the Solver Add-in is not currently installed on your computer, click **Yes** to install it.

After you load the Solver Add-in, the **Solver** command is available in the **Analysis** group on the **Data** tab.

Using the Excel Solver

An optimization model has three parts: the target cell, the changing cells, and the constraints. For the inverse modeling problem described in Section 3.4.2, the Solver dialog box is shown in Figure A.1 below.

In Figure A.1, cell N3 is specified as the target cell, and that is the cell where the sum of squared errors is located. Notice that the **Min** button is selected to specify that the target cell value is to be minimized.

The **Changing Cells** are specified as k and Resistance. Note here the use of meaningful Excel cell labeling. Any cell can be given a convenient name, which is especially helpful when programming numerous equations. Figure A.2 shows how to change cell names.

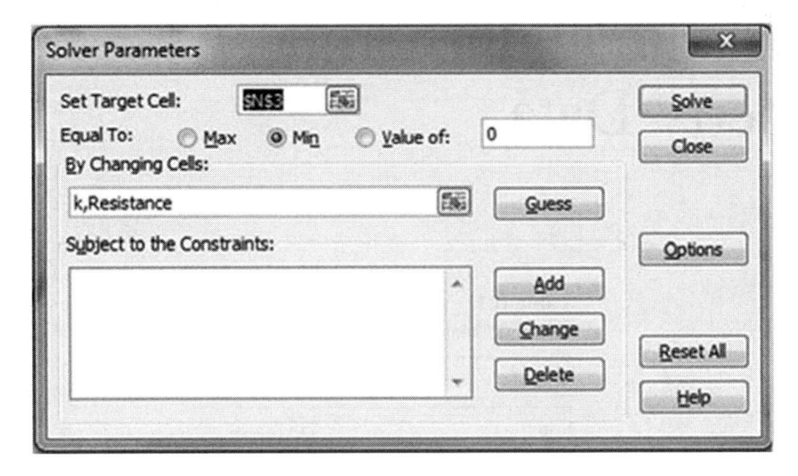

Figure A.1 Solver dialog box for a thermal response test analysis

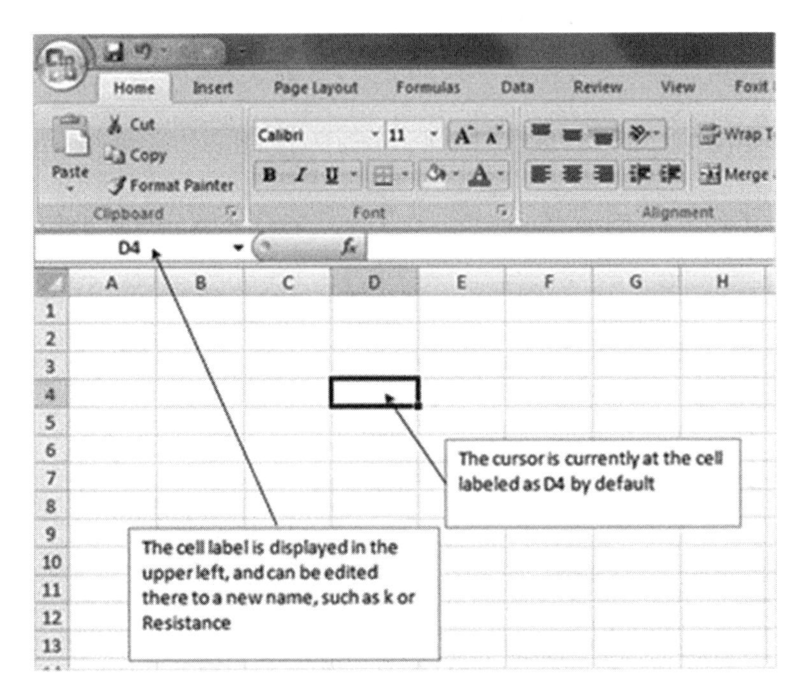

Figure A.2 How to change a cell label in Excel

Appendix B

Hydraulic and Thermal Property Data

Table B.1 Values of Porosity for Geologic Materials

Material	Porosity (%)
Soil	55
Clay	50
Sand	25
Gravel	20
Limestone	20
Sandstone	11
Granite	0.1
Basalt	11

Table B.2 Thermal Conductivity Data for Various Current and Past BHE Materials

	Thermal Conductivity (k)	
Material	Btu/h·ft·°F	$W{\cdot}m^{-1}{\cdot}K^{-1}$
HDPE	0.23	0.40
PEX	0.23	0.40
Polybutylene	0.13	0.22
PVC	0.08	0.14
Black Iron	40	69

Geothermal Heat Pump and Heat Engine Systems: Theory and Practice, First Edition. Andrew D. Chiasson.

Companion website: www.wiley.com/go/chiasson/geoHPSTP

Table B.3 Thermal Conductivity Data for Various BHE Grouting Materials

Material	Thermal Conductivity (k)	
	Btu/h·ft·°F	$W \cdot m^{-1} \cdot K^{-1}$
Bentonite (20–30% solids)	0.42–0.43	0.73–0.75
20% Bentonite/80% silica sand	0.85–0.95	1.47–1.64
15% Bentonite/85% silica sand	0.58–0.64	1.00–1.10
10% Bentonite/90% silica sand	1.20–1.40	2.08–2.42
30% Concrete/70% silica sand	1.20–1.40	2.08–2.42
Graphite–Bentonite mixture	1.20–1.60	2.10–2.77

Table B.4 General Hydraulic and Thermal Property Values of Common Soils and Rocks

Material	Hydraulic Conductivity ($m \cdot day^{-1}$)	Thermal Conductivity**		Volumetric Heat Capacity	
		($W \cdot m^{-1} \cdot K^{-1}$)	(Btu/h·ft·°F)	($MJ \cdot m^{-3} \cdot K^{-1}$)	($Btu/ft^3 \cdot °F$)
Soils					
Clay	10^{-7}–10^{-3}	0.5–1.7	0.3–1.0	1.3	20
Silt	10^{-3}–10^{0}	0.9–1.7	0.5–1.0	1.3	20
Sand	10^{-1}–10^{3}	0.7–2.5	0.4–1.5	1.2	18
Gravel	10^{2}–10^{4}	0.7–1.7	0.4–1.0	1.2	18
Rocks					
Carbonate Rocks	10^{-4}–10^{4*}	2.4–3.8	1.4–2.2	2.0	32
Sandstone	10^{-5}–10^{0*}	2.1–3.5	1.2–2.0	1.9	29
Shale	10^{-8}–10^{-4*}	1.4–2.1	0.8–1.2	2.3	35
Igneous and Metamorphic	10^{-8}–10^{1*}	2.3–3.7	1.3–2.1	2.5	38

* Lower values in the range represent unfractured rocks. Upper values represent fractured or cavernous rocks.

** Lower values of thermal conductivity represent materials with lower water content. Note that thermal conductivity is affected by actual mineral content and material density.

Appendix C

Solar Utilizability Method

The concept of the solar utilizability method is described in detail by Duffie and Beckman (2013) and is based on the premise that a critical solar radiation level can be defined below which no useful energy can be collected. Thus, *utilizability* is essentially a statistic that is a fraction of the total radiation that is received at an intensity higher than a critical level. The average radiation for the period can then be multiplied by this fraction to find the total utilizable energy. The method enables calculation of the monthly quantity of energy delivered, given monthly values of incident solar radiation, ambient temperature, and heating load. Mathematically, the method correlates monthly average daily utilizability ($\bar{\varphi}$) to the monthly average clearness index ($\bar{K}_T$) and two variables: a so-called geometric factor ($\bar{R}/R_n$) and a dimensionless critical radiation level ($\bar{X}_c$).

The critical irradiance level (G_c) which must be exceeded in order for solar energy collection to occur, is given by

$$G_c = \frac{F_R U_L(\bar{T}_i - \bar{T}_a)}{F_R(\bar{\tau}\bar{\alpha})_n} \tag{C.1}$$

where G_c is in units of W·m^{-2}, F_R is the heat removal factor, U_L is the overall heat transfer coefficient for the collector (W·m^{-2}·K^{-1}), T_i is the monthly average inlet fluid temperature to the collector (°C), $\bar{T}_a$ is the monthly average ambient air temperature (°C), and $\bar{\tau}\bar{\alpha}$ is the monthly transmittance–absorptance product. The parameters $F_R U_L$ and $F_R(\bar{\tau}\bar{\alpha})_n$ are readily determined from collector efficiency tests, where the subscript n denotes normal incidence.

Solar collectors are tested, rated, and certified by the Solar Rating and Certification Corporation (SRCC) (Cocoa, Florida) according to industry-accepted standards. The general procedure for testing solar collectors is to operate the collector under steady-state conditions of solar

Geothermal Heat Pump and Heat Engine Systems: Theory and Practice, First Edition. Andrew D. Chiasson.

Companion website: www.wiley.com/go/chiasson/geoHPSTP

radiation at normal incidence, wind speed, ambient temperature, and inlet fluid temperature. The result of a solar collector test is described in terms of collector efficiency, defined as useful energy gained by the circulating fluid divided by the measured solar radiation. Collector efficiency is then plotted versus the ratio of $(T_i - T_a)/I_t$, where T_i (°C) is the collector inlet fluid temperature, T_a (°C) is the ambient air temperature, and I_t is the incident solar radiation ($W \cdot m^{-2}$). Such a plot results is a straight line having a slope equal to $-F_R U_L$ and a y-intercept equal to $F_R(\bar{\tau}\bar{\alpha})_n$. The effect of the heat removal factor (F_R) is to reduce the calculated useful energy gain from what it would be if the whole collector were at the inlet fluid temperature to what it actually is with a fluid that increases in temperature as it flows through the collector.

The monthly useful energy collected (q_u) is given by

$$q_u = A_c F_R (\bar{\tau}\bar{\alpha}) \bar{H}_T \bar{\varphi} \tag{C.2}$$

where q_u is the monthly useful energy collected (J), A_c is the collector area (m^2), $\bar{H}_T$ is the monthly average daily irradiance in the plane of the collector ($J \cdot m^{-2} \cdot day^{-1}$), and $\bar{\varphi}$ is the monthly average daily utilizability.

There are a number of intermediate steps necessary in the calculation of $\bar{\varphi}$. First, the monthly average daily radiation on horizontal ($\bar{H}$) must be known, which is available from weather databases. Next, the monthly average clearness index ($\bar{K}_T$) is calculated from

$$\bar{K}_T = \frac{\bar{H}}{\bar{H}_o} \tag{C.3}$$

where $\bar{H}_o$ is the monthly average extraterrestrial daily radiation, given by

$$\bar{H}_o = \frac{86400 G_{SC}}{\pi}\left(1 + 0.033\cos\left(2\pi\frac{n}{365}\right)\right)(\cos\psi\cos\delta\sin\omega_s + \omega_s\sin\psi\sin\delta) \tag{C.4}$$

where G_{SC} is the solar constant, equal to 1367 $W \cdot m^{-2}$, n is the day of the year, ψ is the latitude of the location, δ is the declination angle, and ω_s is the sunset angle. The declination angle (δ), in degrees, is calculated by

$$\delta = 23.45\sin\left(2\pi\frac{284 + n}{365}\right) \tag{C.5}$$

and the sunset angle (ω_s) is calculated by

$$\cos\omega_s = -\tan\psi\tan\delta \tag{C.6}$$

Next, the monthly average daily diffuse solar radiation ($\bar{H}_d$) is calculated from the global radiation and the clearness index:

$$\bar{H}_d = \bar{H}\left(1.391 - 3.560\bar{K}_T + 4.189\bar{K}_T^2 - 2.137\bar{K}_T^3\right) \text{ for } \omega_s < 81.4° \tag{C.7}$$

$$\bar{H}_d = \bar{H}\left(1.311 - 3.022\bar{K}_T + 3.427\bar{K}_T^2 - 1.821\bar{K}_T^3\right) \text{ for } \omega_s \geq 81.4° \tag{C.8}$$

The monthly average daily beam radiation ($\bar{H}_b$) is computed simply from

$$\bar{H}_b = \bar{H} - \bar{H}_d \tag{C.9}$$

Now, the monthly average radiation in the plane of the collector ($\bar{H}_T$) can be computed from

$$\bar{H}_T = \bar{H}_b\bar{R}_b + \bar{H}_d\left(\frac{1+\cos\beta}{2}\right) + \bar{H}\rho_g\left(\frac{1-\cos\beta}{2}\right) \tag{C.10}$$

where ($\bar{R}_b$) is the ratio of the monthly average beam radiation on the tilted surface to that on a horizontal surface, β is the slope of the collector, and ρ_g is the ground reflectivity, which varies from 0.2 at monthly average air temperatures above 0 °C to 0.7 at monthly average air temperatures below −5 °C, and varies linearly between (Duffie and Beckman, 2013). The first term on the right-hand side of the above equation represents the contribution by beam radiation, the second term represents the contribution by diffuse radiation, and the third term represents the contribution by ground-reflected radiation. $\bar{R}_b$ is defined as

$$\bar{R}_b = \frac{\cos(\psi-\beta)\cos\delta\sin\omega_s' + \pi/180\omega_s'\sin(\psi-\beta)\sin\delta}{\cos\psi\cos\delta\sin\omega_s + \pi/180\omega_s\sin\psi\sin\delta} \tag{C.11}$$

where ω_s' is the minimum of the arccosine of the expression above and $\cos^{-1}(-\tan(\psi-\beta)\tan\delta)$, and all angles are in degrees.

The numerator of the geometric factor ($\bar{R}/R_n$) is defined as the ratio of the monthly average daily radiation on the tilted surface to that on a horizontal surface and is calculated as

$$\bar{R} = \frac{\bar{H}_T}{\bar{H}} \tag{C.12}$$

The denominator (R_n) of the geometric factor is the ratio of the hour centered at noon of radiation on the tilted surface to that on a horizontal surface for an average day of the month and is computed as

$$R_n = \left(1 - \frac{r_{d,n}H_d}{r_{t,n}H}\right)R_{b,n} + \left(\frac{r_{d,n}H_d}{r_{t,n}H}\right)\left(\frac{1+\cos\beta}{2}\right) + \rho_g\left(\frac{1-\cos\beta}{2}\right) \tag{C.13}$$

where $r_{t,n}$ is the ratio of hourly total to daily total radiation for the hour centered around solar noon, and $r_{d,n}$ is the ratio of hourly diffuse to daily diffuse radiation for the hour centered around solar noon. $r_{t,n}$ is computed from the Collares–Pereira and Rable equation (Duffie and Beckman, 2013) as

$$r_{t,n} = \frac{\pi}{24}(a+b)\frac{1-\cos\omega_s}{\sin\omega_s - \omega_s\cos\omega_s} \tag{C.14}$$

where $a = 0.409 + 0.5016\sin(\omega_s - \pi/3)$ and $b = 0.6609 - 0.4767\sin(\omega_s - \pi/3)$. $r_{d,n}$ is computed from the Liu and Jordan equation (Duffie and Beckman, 2013) as

$$r_{d,n} = \frac{\pi}{24} \frac{1 - \cos\omega_s}{\sin\omega_s - \omega_s \cos\omega_s} \tag{C.15}$$

Now, the critical radiation level $(\bar{X}_C)$ can be computed from:

$$\bar{X}_c = \frac{G_c}{r_{t,n} R_n \bar{H}} \tag{C.16}$$

Finally, the monthly average daily utilizability $(\bar{\varphi})$ can be computed from:

$$\bar{\varphi} = \exp\left[\left(a + b\left(\frac{\bar{R}}{R_n}\right)^{-1}\right)\left(\bar{X}_c + c\bar{X}_c^2\right)\right] \tag{C.17}$$

where $a = 2.943 - 9.271\bar{K}_T + 4.031\bar{K}_T^2$, $b = -4.345 + 8.853\bar{K}_T - 3.602\bar{K}_T^2$, and $c = -0.170 - 0.306\bar{K}_T + 2.936\bar{K}_T^2$.

Nomenclature

a	dynamic dispersivity (m [ft])
A	area (m^2 [ft^2])
	value of an annuity (unit of currency)
ACH	air changes per hour
b	$b = Pe\bar{r}/2$
B	borehole spacing (m [ft])
	argument in leaky well function
c	specific heat (J · kg^{-1}·°C^{-1} [Btu/lbm · °F])
C	heat capacity rate (W·°C^{-1} [Btu/h · °F]
	chemical concentration in water (mg/L)
	well loss coefficient [T^2·L^{-6}·L]
cfm	cubic feet per minute
D	hydrodynamic dispersion coefficient (m^2·s^{-1} [ft^2/s])
$D*$	effective thermal diffusion coefficient (m^2·s^{-1} [ft^2/s])
De	Dean number
F	Future value (unit of currency)
Fo	Fourier number ($\alpha t/r^2$) (—)
g	g-function response factors after Claesson and Eskilson (1987)
	gravitational acceleration (m · s^{-2} [ft/s^2])
gpm	gallons per minute
h	dimensionless groundwater flow rate (—)
	convection coefficient (W·m^{-2}·K^{-1} [Btu/h·ft^2·°F])
	enthalpy (J ·kg^{-1} [Btu/lbm])
	hydraulic head (m [ft])
H	borehole depth (m [ft])

Geothermal Heat Pump and Heat Engine Systems: Theory and Practice, First Edition. Andrew D. Chiasson.

Companion website: www.wiley.com/go/chiasson/geoHPSTP

i	hydraulic gradient (—)
k	thermal conductivity (W·m^{-1}·K^{-1} [Btu/h·ft·°F])
K	hydraulic conductivity (m·s^{-1} [ft/s])
K_0	modified Bessel function of the second kind of order 0
Lpm	liters per minute
m	mass (kg [lbm])
$\dot{m}$	mass flow rate (kg·s^{-1} [lbm/h])
N	number of measurements
NB	number of boreholes
NTU	number of transfer units(—)
Nu	Nusselt number (—)
P	pressure (kPa [psi])
	present value (unit of currency)
Pe	Peclet number ($U_{eff}\, r_b/\alpha$) (—)
Pr	Prandtl number (—)
P_w	groundwater flow correction term after Claesson and Eskilson (1987)
$\dot{q}'$	ground thermal load per unit length of vertical bore (W·m^{-1} [Btu/h/ft])
$\dot{q}$	heat transfer rate (W [Btu/h])
	specific discharge or Darcy velocity or Darcy flux (m^3·m^{-2}·s^{-1} [ft^3/ft^2/s])
$\dot{Q}$	volumetric flow rate (m^3·s^{-1} [ft^3/h])
$\dot{Q}'$	volumetric groundwater injection/extraction rate per unit aquifer thickness (m^3·s^{-1}·m^{-1} [ft^3/h/ft])
r	radial distance or radius (m [ft])
$\bar{r}$	dimensionless radius (r/r_b)
R	thermal retardation coefficient (—)
R'_b	borehole effective thermal resistance per unit length of bore (K·m·W^{-1} [h·ft·°F/Btu])
Ra	Rayleigh number (—)
Re	Reynolds number (—)
s	entropy (J·K^{-1} [Btu/°R])
	drawdown (of groundwater level) in wells (m [ft])
S	aquifer storativity or storage coefficient (—)
S_c	well specific capacity [L^3·t^{-1}·L^{-1}]
SG	specific gravity (—)
t	time (s)
T	temperature (°C or K [°F or °R])
ton	refrigeration ton (12 000 Btu/h or 3.5 kW)
U	Darcy groundwater velocity or Darcy flux (m·s^{-1} [ft/s])
	overall heat transfer coefficient (W·m^{-2}·K^{-1} [Btu/h·ft^2·°F])
U_{eff}	effective groundwater velocity (m·s^{-1} [ft/s])
v	average linear groundwater velocity (Ki/n) (m·s^{-1} [ft/s])
	specific volume (m^3·kg^{-1} [ft^3/lbm])
V	volume (m^3 [ft^3])
$\dot{V}$	fluid velocity (m·s^{-1} [ft/s])
W	work (J)
$\dot{W}$	power (W)

$W(u)$	well function for argument u (equivalent to the exponential integral)
$W(u, B)$	leaky well function (after Hantush, 1956) for arguments u and B
x	quality of a two-phase, liquid–vapor mixture
x, y	distance from origin in Cartesian coordinates

Greek Letters

α	thermal diffusivity ($m^2 \cdot s^{-1}$ [ft^2/h])
	hydraulic diffusivity ($m^2 \cdot s^{-1}$ [ft^2/h])
β	aquifer loss coefficient [$T \cdot L^{-3} \cdot L$]
γ	Euler's constant (0.57721566)
ε	heat exchanger effectiveness (—)
	emissivity coefficient (—)
η	first law thermal efficiency (—)
θ	angle between the flow direction and the direction of the horizontal line joining the borehole axis to the point where ground temperature is computed
μ	dynamic viscosity (Pa·s or $kg \cdot s^{-1} \cdot m^{-1}$ [lbm/ft·h])
ν	kinematic viscosity ($m^2 \cdot s^{-1}$ [ft^2/h])
ρ	density ($kg \cdot m^{-3}$ [lbm/ft^3])
	reflectivity coefficient (—)
σ	Stefan–Boltzmann constant = 5.678×10^{-8} ($W \cdot m^{-2} \cdot K^{-4}$ [$Btu/h \cdot ft^2 \cdot {}^\circ R^4$])
τ	dimensionless time in groundwater g-function (—)
ϕ	porosity (—)

Subscripts

avg	average
b	borehole
c, C	cold or cooling
D	dimensionless
E	exergy
eff	effective
f	average fluid
g	undisturbed ground
gw	groundwater
H, h	hot or heating
HW	hot water
i, in	inlet
l	liquid phase
L	longitudunal
o, out	outlet
R	refrigerant
s	solid phase
T	transverse
x,y	coordinate indices

Acronyms

AFUE	annual fuel utilization efficiency
AHRI	Air-Conditioning, Heating, and Refrigeration Institute
AHU	air-handling unit
ANSI	American National Standards Institute
ASHRAE	formerly the American Society of Heating, Refrigerating, and Air-Conditioning Engineers
BHE	borehole heat exchanger
CHP	combined heat and power
COP	coefficient of performance (of a heat pump)
CSA	Canadian Standards Association
DE	district energy
DHE	downhole exchanger
DST	duct ground storage
DSWC	direct surface water cooling
ECM	electronically commutated motor
EER	energy efficiency ratio (of a heat pump) [Btu/W]
FHX	foundation heat exchanger
GCHP	ground-coupled heat pump
GHP	geothermal heat pump
GHX	ground heat exchanger
GWHP	groundwater heat pump
GWP	global warming potential
HDPE	high-density polyethylene
HRU	heat recovery unit
HSPF	heating seasonal performance
HSWHP	hybrid surface water heat pump
HVAC	heating, ventilating, and air-conditioning
IAM	incidence angle modifier
IAQ	indoor air quality
IGSHPA	International Ground Source Heat Pump Association
ISA	international standard atmosphere
ISO	International Standards Organization
LCC	life-cycle cost
LCOE	levelized cost of energy or levelized cost of electricity
LCS	life-cycle savings
LSI	Langelier saturation index
MCWB	mean coincident wet bulb
ORC	organic Rankine cycle
PWF	present worth factor
ROI	return on investment
ROR	rate of return (currency/time)
RSI	Ryznar stability index
SCW	standing column well
SDR	standard dimension ratio
SEER	seasonal energy efficiency ratio (of a heat pump) [Btu/W]

SHR	sensible heat ratio
SPP	simple payback period (years or months)
SWHP	surface water heat pump
SWHX	surface water heat exchanger
TDH	total dynamic head (m [ft])
TDS	total dissolved solids (mg/L)
VFD	variable-frequency drive
VSD	variable-speed drive

References

Adler, F.T., Ingersoll, A.C., Ingersoll, L.R., and Plass, H.J. (1951) Theory of earth heat exchangers for the heat pump, *ASHVE Transactions*, vol. 57, pp. 167–188.

Allan, M.L. and Kavanaugh, S.P. (1999) Thermal conductivity of cementitious grouts and impact on heat exchanger length design for ground source heat pumps, *HVAC&R Research*, vol. 5, no. 2, April. pp. 85–96.

ASHRAE (2009, 2013) *ASHRAE Handbook – Fundamentals*, Atlanta, GA: ASHRAE.

ASHRAE (2011, 2015) *ASHRAE Handbook – HVAC Applications*, Atlanta, GA: ASHRAE.

ASHRAE (2013a) *District Heating Guide*, Atlanta, GA: ASHRAE.

ASHRAE (2013b) *District Cooling Guide*, Atlanta, GA: ASHRAE.

Austin, W., Yavuzturk, C., and Spitler, J.D. (2000) Development of an in-situ system for measuring ground thermal properties, *ASHRAE Transactions*, vol. 106, no. 1, pp. 365–379.

Bear, J. (1972) *Dynamics of Fluids in Porous Media*, New York, NJ: Dover Publications, Inc.

Beckman, W.A., Klein, S.A., and Duffie, J.A. (1977) *Solar Heating Design by the f-chart Method*, New York, NY: Wiley Interscience.

Bennet, J., Claesson, J., and Hellström, G. (1987) *Multipole Method to Compute the Conductive Heat Flows to and Between Pipes in a Composite Cylinder*, Lund, Sweden: Department of Building Technology and Mathematical Physics, University of Lund.

Bergman, T.l., Lavine, A.S., Incropera, F.P., and DeWitt, D.P. (2011) *Introduction to Heat Transfer*, 6th edition, Hoboken, NJ: John Wiley & Sons, Inc.

Bernier, M.A., (2001) Ground-coupled heat pump system simulation. *ASHRAE Transactions*, vol. 107, no. 1, pp. 605–616.

Bertani, R. (2015) Geothermal Power Generation in the World 2010–2014 Update Report, *Proceedings World Geothermal Congress 2015*, Melbourne, Australia.

Bierschenk, W.H. (1964) Determining well efficiency by multiple step-drawdown tests, International Association of Scientific Hydrology, Publication 64, pp. 493–505.

Bloomquist, R.G. (2005) Geothermal heat pump case studies of the West: Inn of the Seventh Mountain, Bend, OR, *Geo-Heat Center Quarterly Bulletin*, vol. 26, no. 3, pp. 15–20.

Bloomquist, R.G. (2006) Economic factors to consider when assessing direct-use geothermal development viability, *Geothermal Resources Council Transactions*, vol. 30, pp. 179–184.

Burch, J., Christensen, C., Salasovich, J., and Thornton, J. (2004) Simulation of an unglazed collector system for domestic hot water and space heating and cooling, *Solar Energy*, vol. 77, pp. 399–406.

Geothermal Heat Pump and Heat Engine Systems: Theory and Practice, First Edition. Andrew D. Chiasson.

Companion website: www.wiley.com/go/chiasson/geoHPSTP

Carslaw, H.S. and Jaeger, J.C. (1947) *Heat Conduction in Solids*, Oxford, UK: Claremore Press.

Chapman, W.P. (1952) Design of snow melting systems, *Heating and Ventilating*, April, p. 95; November, p. 88.

Chiasson, A.D. (2005a) Swimming pool heating with geothermal heat pump systems, *Geo-Heat Center Quarterly Bulletin*, vol. 26, no. 1, pp. 13–17.

Chiasson, A.D. (2005b) Chiloquin Community Center, Chiloquin, Oregon (case study of a hydronic, radiant floor geothermal heat pump system), *Geo-Heat Center Quarterly Bulletin*, vol. 26, no. 3, pp. 12–14.

Chiasson, A.D. (2010) Modeling horizontal ground heat exchangers in geothermal heat pump systems, *Proceedings of the 2010 COMSOL Multiphysics Conference*, Boston, MA.

Chiasson, A.D. (2015) Waste heat rejection methods in geothermal power generation (Chapter 15), in DiPippo R. (ed.), *Geothermal Power Generation: Developments and Innovation*, Cambridge, UK: Woodhead Publishing, Ltd.

Chiasson, A.D. and Yavuzturk, C. (2005) Modeling the viability of underground coal fires as a heat source for electrical power generation, *Proceedings of the Thirtieth Workshop on Geothermal Reservoir Engineering*, Stanford University, Stanford, CA.

Chiasson, A.D. and Yavuzturk, C. (2009a) A design tool for hybrid geothermal heat pump systems in heating-dominated buildings, *ASHRAE Transactions*, vol. 115, no. 2, pp. 60–73.

Chiasson, A.D. and Yavuzturk, C. (2009b) A design tool for hybrid geothermal heat pump systems in cooling-dominated buildings, *ASHRAE Transactions*, vol. 115, no. 2, pp. 74–87.

Chiasson, A.D. and O'Connell, A. (2011) New analytical solution for sizing vertical borehole ground heat exchangers in environments with significant groundwater flow: parameter estimation from thermal response test data, *HVAC&R Research* vol. 17, no. 6, pp. 1000–1011.

Chiasson, A.D., Rees, S.J., and Spitler, J.D. (2000) A preliminary assessment of the effects of groundwater flow on closed-loop ground-source heat pump systems, *ASHRAE Transactions*, vol. 106, no. 1, pp. 380–393.

Chiasson, A. D., Culver, G. G., Favata, D., and Keiffer, S. (2005) Design, installation, and monitoring of a new down-hole heat exchanger, *Geothermal Resources Council Transactions*, vol. 29, pp. 51–56.

Claesson, J. and Eskilson, P. (1987) Conductive heat extraction by a deep borehole. Analytical studies, in Eskilson, P. (ed.), *Thermal Analysis of Heat Extraction Boreholes*, Lund, Sweden: Department of Mathematical Physics, University of Lund.

Claesson, J. and Hellström, G. (2000) Analytical studies of the influence of regional groundwater flow on the performance of borehole heat exchangers, *Proceedings of the 8th international Conference on Thermal Energy Storage, Terrastock 2000*, Stuttgart, Germany, pp. 195–200.

Cooper, H.H. and Jacob, C.E. (1946) A generalized graphical method for evaluating formation constants and summarizing well field history, *Transactions of the American Geophysical Union*, vol. 27, pp. 526–534.

Cristoph, R.E. (2012) Solid sorption cycles: a short history, *International Journal of Refrigeration*, vol. 35, no. 3, pp. 490–493.

Cullin, J and Spitler, J.D. (2010) Comparison of simulation-based design procedures for hybrid ground source heat pump systems, *Proceedings of the 8th International Conference on System Simulation in Buildings 2010*, Liege, Belgium.

Cullin, J.R., Xing, L., Lee, E., Spitler, J.D., and Fisher, D.E. (2012) Feasibility of foundation heat exchangers for ground source heat pump systems in the United States, *ASHRAE Transactions*, vol. 118, no. 1, pp. 1039–1048.

Cullin, J.R., Spitler, J.D., and Gehlin, S.E.A. (2014) Suitability of foundation heat exchangers for ground source heat pump systems in European climates, *REHVA Journal*, January, pp. 36–40.

Culver, G. (2005) A brief history of DHE materials, *Geo-Heat Centre Quarterly Bulletin*, vol. 26, no. 1, pp. 25–28.

Dawson, K.J. and Istok, J.D. (1991) *Aquifer Testing*, Boca Raton, FL: CRC Press.

Deng, Z., Rees, S.J., and Spitler, J.D. (2005) A model for annual simulation of standing column well ground heat exchangers, *HVAC&R Research*, vol. 11, no. 4, pp. 637–655.

Dexheimer, R.D. (1985) *Water Source Heat Pump Handbook*, Dublin, OH: National Water Well Association, 241 pp.

DiPippo, R. (2012) *Geothermal Power Plants: Principles, Applications, Case Studies and Environmental Impact*, 3rd edition, Oxford, UK: Elsevier.

Driscoll, F.G. (2008) *Groundwater and Wells*, 3rd edition, Johnson Screens.

Duffie, J.A. and Beckman, W.A. (2013) *Solar Engineering of Thermal Processes*, 4th edition, Hoboken, NJ: John Wiley & Sons, Inc.

Ellis, P. (1998) Materials selection guidelines, in Lund J.W., Lineau P.J., and Lunis, B.C. (eds), *Geothermal Direct Use Engineering and Design Guidebook*, Klamath Falls, OR: Geo-Heat Centre, Oregon Tech, pp. 191–209.

Epstein, C.M. and Sowers, L.S. (2006) The continued warming of the Stockton geothermal well field, *Proceedings of Ecostock 2006, the 10th International Conference on Thermal Energy Storage*, The Richard Stockton College of New Jersey, Galloway, NJ.

Eskilson, P. (1987) *Thermal Analysis of Heat Extraction Boreholes*, Doctoral Thesis, Department of Mathematical Physics, University of Lund, Lund, Sweden.

Fetter, C.W. (1999) *Contaminant Hydrogeology*, 2nd edition, Long Grove, IL: Waveland Press, Inc.

Fournier, R.O. (1981) Application of water geochemistry to geothermal exploration and reservoir engineering, in Ryback, L. and Muffler, L.J.P. (eds), *Geothermal Systems – Principles and Case Histories*, New York, NJ: Wiley, pp. 109–143.

Freeze, R.A. and Cherry, J.A. (1979) *Groundwater*, Englewood Cliffs, NJ: Prentice-Hall Inc.

GEA (2005) *Factors Affecting Costs of Geothermal Power Development*, Washington, DC: Geothermal Energy Association.

Gehlin, S. (2002) *Thermal Response Test, Method Development and Evaluation*, Doctoral Thesis, Department of Environmental Engineering, Division of Water Resources Engineering, Lulea University of Technology, Lulea, Sweden.

Gnielinski, V. (1976) New equations for heat and mass transfer in turbulent pipe and channel flow, *International Chemical Engineering*, vol. 16, no. 2, pp. 359–368.

Hackel, S. (2008) *Development of Design Guidelines for Hybrid Ground-Coupled Heat Pump Systems. ASHRAE Technical Research Project 1384*, Atlanta, GA: American Society of Heating, Refrigerating, and Air-Conditioning Engineers (ASHRAE).

Hackel, S., Nellis, G., and Klein, S. (2009) Optimization of cooling-dominated hybrid ground-coupled heat pump systems, *ASHRAE Transactions*, vol. 115, no. 1, pp. 565–580.

Hamilton, D. P. and Schladow, S. (1997) Prediction of water quality in lakes and reservoirs. Part I – Model description, *Ecological Modelling*, vol. 96, pp. 91–110.

Hansen, G. M. (2011) *Experimental Testing and Analysis of Spiral-Helical Surface Water Heat Exchangers*, M.S. Thesis, Oklahoma State University, Stillwater, OK.

Hantush, M.S. (1956) Analysis of data from pumping tests in leaky aquifers, *Transactions of the American Geophysical Union*, vol. 37, pp. 702–714.

Hantush, M.S. and Jacob, C.E. (1956) Nonsteady radial flow in an infinite leaky aquifer, *Transactions of the American Geophysical Union*, vol. 36, pp. 95–100.

Heath, R. (1983) *Basic Ground-Water Hydrology*, United States Geological Survey (USGS) Water-Supply Paper 2220, 86 pp.

Hellström, G. (1991) *Ground Heat Storage. Thermal Analyses of Duct Storage Systems*, Lund, Sweden: Department of Mathematical Physics, University of Lund.

Herold, K.E., Radermacher, R., and Klein, S. (1996) *Absorption Chillers and Heat Pumps*, Boca Raton, FL: CRC Press.

Hinde, D., Zha, S., and Lan, L. (2009) Carbon dioxide in North American supermarkets, *ASHRAE Journal*, vol. 51, pp. 18–26.

Hull, J.R., Liu, K.V., Sha, W.T., Kamal, J., and Nielsen, C.E. (1984) Dependence of ground heat losses upon solar pond size and perimeter insulation – calculated and experimental results, *Solar Energy*, vol. 33, no. 1, pp. 25–33.

IGSHPA (1991) *Grouting for Vertical Geothermal Heat Pump Systems*, Stillwater, OK: International Ground Source Heat Pump Association, Oklahoma State University.

IGSHPA (2009) *Residential and Light Commercial Design and Installation Manual*, Stillwater, OK: International Ground Source Heat Pump Association, Oklahoma State University.

Ingersoll, L.R. and Plass, H.J. (1948) Theory of the ground heat pipe heat source for the heat pump, *Transactions of the American Society of Heating and Ventilating Engineers*, vol. 54, pp. 119–122.

Ingersoll, L.R., Zobel, O.J., and Ingersoll, A.C. (1954) *Heat Conduction with Engineering and Geological Applications*, 2nd edition, New York, NJ: McGraw-Hill.

Kasuda, T. and Achenbach, P.R. (1965) Earth temperature and thermal diffusivity at selected stations of the United States, *ASHRAE Transactions*, vol. 71, no. 1, pp. 61–75.

Kavanaugh, S.P. (1998) A design method for hybrid ground source heat pumps, *ASHRAE Transactions*, vol. 104, no. 2, pp. 691–698.

Kavanaugh, S.P. (2006) *HVAC Simplified*, Atlanta, GA: ASHRAE.

Kavanaugh, S.P. and Rafferty, K. (1997) *Ground Source Heat Pumps – Design of Geothermal Systems for Commercial and Institutional Building*, Atlanta, GA: ASHRAE.

Kelvin, Sir W. T. (1882) *Mathematical and Physical Papers, Vol. 1*, Cambridge, UK: Cambridge University Press.

Kishore, V.V.N. and Joshi, V. (1984) A practical collector efficiency equation for non-convecting solar ponds, *Solar Energy*, vol. 33, no. 5, pp. 391–395.

Kühn, A. (2013) Cycle basics of thermally driven heat pumps, in Kühn, A. (ed.), *Thermally Driven Heat Pumps for Heating and Cooling*, Berlin: International Energy Agency, pp. 5–15.

Liu, X. and Hellström, G (2006) Enhancements of an integrated simulation tool for ground-source heat pump system design and energy analysis, *Proceedings of Ecostock 2006, the 10th International Conference on Thermal Energy Storage*, The Richard Stockton College of New Jersey, Galloway, NJ, pp. 331–338.

Lonrenz, E. (2005) Geothermal heat pump case studies of the West: South Cariboo Recreation Centre, 100 Mile House, British Columbia, Canada, *Geo-Heat Center Quarterly Bulletin*, vol. 26, no. 3, pp. 15–20.

Lonrenz, E. (2011) Gibsons, BC district geothermal energy system, *Geo-Heat Center Quarterly Bulletin*, vol. 30, no. 1, pp. 4–8.

Lund, J.W. (2000) Pavement snow melting, *Geo-Heat Center Quarterly Bulletin*, vol. 21, no.2, pp. 12–19.

Lund, J.W. (2007) Characteristics, development and utilization of geothermal resources, *Geo-Heat Center Quarterly Bulletin*, vol. 28, no. 2, pp. 1–9.

Lund, J.W. (2010) Development of direct-use projects, *Geo-Heat Center Quarterly Bulletin*, vol. 29, no. 2, pp. 1–7.

Lund, J.W. and Boyd, T. (1999) Small geothermal power project examples, *Geo-Heat Center Quarterly Bulletin*, vol. 20, no. 2, pp. 9–26.

Lund, J.W. and Chiasson, A. (2007) Examples of combined heat and power plants using geothermal energy, *Proceedings European Geothermal Congress 2007*, Unterhaching, Germany.

Mescher, K. (2009) Simplified GCHP system: one-pipe geothermal design, *ASHRAE Journal*, vol. 51, no. 10, October, pp. 29–40.

Mitchell, M.S. and Spitler, J.D. (2013) Open-loop direct surface water cooling and surface water heat pump systems – a review, *HVAC&R Research*, vol. 19, no. 2, pp. 125–140.

Molineaux, B., Lachal, B., and Guisan, O. (1994) Thermal analysis of five outdoor swimming pools heated by unglazed solar collectors, *Solar Energy*, vol. 53, no. 1, pp. 21–26.

Nelder, J.A. and Mead, R. (1965) A simplex method for function minimization, *Computer Journal*, vol. 7, no. 1, pp. 308–313.

Nydahl, J., Pell, K., Lee, R., and Sackos, J. (1984) *Evaluation of an Earth Heated Bridge Deck*, USDOT Contact No. DTFH61-80-C-00053, University of Wyoming, Laramie, WY.

Piper, A.M. (1944) A graphic procedure in the geochemical interpretation of water analysis, *Transactions of the Geophysical Union*, vol. 25, no. 6, pp. 914–923.

Rafferty, K. (1998a) Greenhouses, in Lineau, P. (ed.), *Geothermal Direct Use Engineering and Design Guidebook*, 3rd edition, Klamath Falls, OR: Geo-Heat Center, Oregon Tech, pp. 307–326.

Rafferty, K. (1998b) Piping, in Lund, J.W., Lineau, P.J., and Lunis B.C. (eds), *Geothermal Direct Use Engineering and Design Guidebook*, Klamath Falls, OR: Geo-Heat Center, Oregon Tech, pp. 241–259.

Rafferty, K. (2000) Scaling in geothermal heat pump systems, *Geo-Heat Center Quarterly Bulletin*, vol. 21, no. 1, pp. 11–15.

Rafferty, K. (2009) Groundwater issues: commercial open loop heat pump systems, *ASHRAE Journal*, vol. 51, no. 3, March, pp. 52–62.

Remund, C.P. and Lund, J.T. (1993) Thermal enhancement of bentonite grouts for vertical GSHP boreholes, *Heat Pump and Refrigeration Systems, Vol. 29*, New York, NJ: American Society of Mechanical Engineers.

RetScreen® International (2004) *Clean Energy Project Analysis*, 3rd edition, RETScreen® Engineering & Cases Textbook, Natural Resources Canada.

Ryan, G. (1981) Equipment used in direct heat projects, *Geothermal Resources Council Transactions*, vol. 5, pp. 233–241.

Rybach, L. and Sanner, B. (2000) Ground-source heat pump systems – the European experience, *Geo-Heat Center Quarterly Bulletin*, vol. 21, no. 1, pp. 16–26.

Sachs, H.M. (2002) *Geology and Drilling Methods for Ground-Source Heat Pump Installations: An Introduction for Engineers*, Atlanta, GA: ASHRAE.

Salimpour, M.R. (2009) Heat transfer coefficients of shell and coiled tube heat exchangers, *Experimental Thermal and Fluid Science*, vol. 33, no. 2, pp. 203–207.

Saloranta, T. M. and Andersen, T. (2007) MyLake – a multi-year lake simulation model code suitable for uncertainty and sensitivity analysis simulations, *Ecological Modelling*, vol. 207, no. 1, pp. 45–60.

Schillereff, H.S., Johnston, K.S., Quibell, J.D., and Kiernan, A. (2008) A method for geo-exchange suitability assessment, *ASHRAE Transactions*, vol. 114, no. 2, pp. 293–315.

Shonder, J. and Spitler, J.D. (2009) Foundation heat exchangers: reducing the first cost of ground source heat pumps, *IEA Heat Pump Centre Newsletter*, vol. 27, pp. 22–23.

Signorelli, S. (2004) *Geoscientific Investigations for the Use of Shallow Low-Enthalpy Systems*, Doctoral Dissertation, Swiss Federal Institute of Technology, Zurich, Switzerland.

Sowers, L.S., York, K.P., and Stiles, L. (2006) Impact of thermal buildup on groundwater chemistry and aquifer microbes, *Proceedings of Ecostock 2006, the 10th International Conference on Thermal Energy Storage*, The Richard Stockton College of New Jersey, Galloway, NJ.

Spitler, J.D. (2005) Ground-source heat pump system research – past, present, and future, Editorial, *International Journal of HVAC&R Research*, vol. 11, no. 2, pp. 165–167.

Spitler, J.D. (2014) *Load Calculations Manual*, Atlanta, GA: ASHRAE.

Spitler, J.D., Rees, S.J., and Xia, X. (2002a) Transient analysis of snow-melting performance. *ASHRAE Transactions*, vol. 108, no. 2, pp. 406–425.

Spitler, J.D., Rees, S.J., Deng, Z, Chiasson, A.D., Orio, C., and Johnson, C. (2002b) *R&D Studies of Standing Column Well Design, ASHRAE 1119-RP*, Atlanta, GA: ASHRAE, 121 pp.

Spitler, J., Xing, L., Cullin, J.R., Fisher, D., Shonder, J., and Im, P. (2010) Residential ground source heat pump systems utilizing foundation heat exchangers, *Proceedings of Clima 2010*, Antalya, Turkey.

Spitler, J.D., Cullin, J.R., Conjeevaram, K., Ramesh, M., and Selvakumar, M. (2012) *Improved Design Tools for Surface Water and Standing Column Well Heat Pump Systems*, Final Report prepared under United States Department of Energy Contract DE-EE0002961, http://www.osti.gov/scitech/biblio/1111113/

Srivastava, R. and Guzan-Guzman, A. (1998) Practical approximations of the well function, *Ground Water*, vol. 36, no. 5, pp. 844–848.

Stoecker, W.F. and Jones, J.W. (1982) *Refrigeration and Air Conditioning*, New York, NJ: McGraw-Hill.

The University of Alabama, (2000) Grout thermal conductivity – bigger in not always better. *Outside the Loop Newsletter*, The University of Alabama, Summer 2000, vol 3, no. 2, Tuscaloosa, AL.

Theis, C.V. (1935) The relation between the lowering of the piezometric surface and the rate and duration of discharge of a well using groundwater storage, *Transactions of the American Geophysical Union*, vol. 2, pp. 519–524.

UNEP (2011) *2010 Report of the Refrigeration, Air-Conditioning and Heat Pumps Technical Options Committee*, United Nations Environment Programme, Montreal Protocol on Substances that Deplete the Ozone Layer, February 2011.

White, D.E. (1973) Characteristics of geothermal resources, Chapter 4, in Kruger, P. and Otte, C. (eds), *Geothermal Energy: Resources, Production, Stimulation*, Stanford, CA: Stanford University Press.

Wickersham, G. (1977) Review of C.E. Jacob's doublet well, *Ground Water*, vol. 15, no. 5, pp. 344–347.

Xing, L., Cullin, J.R., Spitler, J.D., Im, P., and Fisher, D.E. (2012) Foundation heat exchangers for residential ground source heat pump systems – numerical modeling and experimental validation, *HVAC&R Research*, vol. 17, no. 6, pp. 1059–1072.

Xu, X. (2007) *Simulation and Optimal Control of Hybrid Ground Source Heat Pump Systems*, Ph.D. Thesis, Oklahoma State University, Stillwater, OK.

Xu, M. and Eckstein, Y. (1995) Use of weighted least-squares method in evaluation of the relationship between dispersivity and field scale, *Ground Water*, vol. 33, no. 6, pp. 905–908.

Yavuzturk, C. and Chiasson, A. (2002) Performance analysis of U-tube, concentric tube, and standing column well earth heat exchangers using a system simulation approach, *ASHRAE Transactions*, vol. 108, no. 1, pp. 925–938.

Yavuzturk, C. and Spitler, J.D. (1999) A short time step response factor model for vertical ground loop heat exchangers. *ASHRAE Transactions*, vol. 105, no. 2, pp. 475–485.

Index

Geothermal Heat Pump and Heat Engine Systems: Theory and Practice, First Edition. Andrew D. Chiasson.

Companion website: www.wiley.com/go/chiasson/geoHPSTP